Advances in the Understanding
of Crystal Growth Mechanisms

Advances in the Understanding of Crystal Growth Mechanisms

Edited by

T. Nishinaga • K. Nishioka • J. Harada

A. Sasaki • H. Takei

Department of Electronic Engineering
The Faculty of Engineering
The University of Tokyo
Tokyo, Japan

1997

ELSEVIER

Amsterdam • Lausanne • New York • Oxford • Shannon • Tokyo

ELSEVIER SCIENCE B.V.
Sara Burgerhartstraat 25
P.O. Box 211, 1000 AE Amsterdam, The Netherlands

ISBN: 0 444 82504 5

PREFACE

It is our great pleasure to publish the results of the research project entitled "Crystal Growth Mechanisms on an Atomic Scale" which was carried out for 3 years from 1991 with financial support from The Ministry of Education, Science, Sports and Culture. A total of 72 researchers were involved in the project and the total budget was nearly 8 hundred million yen.

Until now, especially in Japan, the technological aspects of crystal growth have been emphasized and only works useful for industry have been thought important. However, the science aspects should also be considered. Otherwise, in future we will be unable to develop the technology of crystal growth. In this project, our main aim was to understand crystal growth and we concentrated on finding growth mechanisms on an atomic scale.

This book consists of 5 parts:
1) Crystal Growth Theory and Simulations,
2) Growth Kinetics,
3) Observations of Growth Surfaces and Interfaces,
4) Mechanisms of Heteroepitaxy, and
5) Crystal Growth and Mechanisms in Complex Systems,

These were edited by Professor K. Nishioka, myself, Professor J. Harada, Professor A. Sasaki and Professor H. Takei respectively. I hope this work contributes to the understanding of crystal growth.

I would like to thank the members of the steering committee of this project, Professor I. Sunagawa (Prof. Em. Tohoku University), Professor M. Fujimoto (Wakayama University), Professor J. Chikawa (Himeji University), Dr. S. Takasu (SEMI Japan) and Dr. H. Watanabe (NEC) for their continuous supports and encouragement. I also thank The Ministry of Education, Science, Sports and Culture for the financial support necessary to execute this project. The publication of this book is financed by a Grant-in-Aid for Publication of Scientific Research Results also from The Ministry of Education, Science, Sports and Culture.

Tatau Nishinaga
Editor-in-Chief
Professor, The University
of Tokyo
February, 1997

Contents

PART III. Observations of Growth Surface and Interface 245

PART I

Crystal Growth Theory and Simulations

Advances in the Understanding of Crystal Growth Mechanisms
T. Nishinaga, K. Nishioka, J. Harada, A. Sasaki and H. Takei (Editors)
© 1997 Elsevier Science B.V. All rights reserved.

Thermodynamic vs. kinetic critical nucleus and the reversible work of nucleus formation

K. Nishioka

Department of Optical Science and Technology, University of Tokushima, 2-1
Minamijosanjima, Tokushima 770, Japan

Reconsideration on the concept of critical nucleus for single component systems leads to the result that the size n_K^h of a kinetic critical nucleus for which the probabilities of its decay and growth balance is not equal to the size n^{h*} of the thermodynamic one for which the reversible work of nucleus formation takes the maximum value. n_K^h is in general smaller than n^{h*} when n^h is treated as a continuous variable. There exist two values for n_K^h, the larger is kinetically unstable but the smaller is stable. The difference between n^{h*} and the larger n_K^h increases but the difference between the two values of n_K^h decreases with the degree of supersaturation or supercooling, and in the critical state two values of n_K^h coincide and it diminishes to $8/27$ of n^{h*} for three dimensional homogeneous nucleation and to $1/4$ of n^{h*} for two dimensional nucleation on a substrate. Beyond this critical state n_K^h does not exist and for a nucleus with any size the probabilty of growth is higher than that of decay.

The Gibbs formula for the reversible work of forming a thermodynamic critical nucleus in multicomponent systems is summarized and the commonly used formula is presented as an approximation to the Gibbs. By taking a binary system as an example, driving force for forming a thermodynamic critical nucleus is clarified. Formula for the reversible work of forming a noncritical nucleus is then derived and it is found that a term due to the difference in the chemical potentials between a nucleus and a parent phase has been missing in the conventional formula. This term can be approximately taken into account by using the value of the interfacial tension for the thermodynamic critical nucleus even for a noncritical nucleus. When determining the critical nucleus from the extremum condition of the reversible work of nucleus formation, the term containing the differential of the interfacial tension does not arise due to the Gibbs-Duhem relation derived for a system of a noncritical nucleus and a parent phase.

1. THERMODYNAMIC VS. KINETIC CRITICAL NUCLEUS

1.1. Definition of thermodynamic and kinetic critical nuclei

Let us take homogeneous nucleation in a three dimensional system for general consideration and extend the results to nucleation on substrates later. Kinetic process at early stage of the first order phase transformation may usually be treated in terms of exchange of monomers between a parent phase and mutually independent clusters of a nucleating phase called nuclei. In the thermodynamic treatment of nucleation, nuclei are specified by the number of molecules n_i^h contained in the corresponding hypothetical clusters [1,2,3], where the subscript i denotes ith component. A critical nucleus is then defined as a one for which the reversible work of nucleus formation takes an extremum value. We consider single component systems in the present section. Then the reversible work takes a maximum value for a critical nucleus, which we call the thermodynamic critical nucleus and denote as n^{h*}. On the other hand, in describing nucleation process it is useful to define a critical nucleus as the size for which the probabilities of decay and growth balance, which we call the kinetic critical nucleus and denote as n_K^h. n^{h*} and n_K^h have been presumed to coincide, but it is recently shown that they do not [4,5].

Suppose a system of a parent phase and nuclei in metastable equilibrium. For simplicity we treat n^h as a continuous variable. Metastable equilibrium number density $c_o(n^h)$ of nuclei with size n^h is given by [6,7,8]

$$c_o(n^h) = \Phi_{LP}c(1)exp[-W^{rev}(n^h)/kT], \quad n^h > 1, \tag{1}$$

where $c(1)$ denotes the monomer density, Φ_{LP} the Lothe-Pound factor [6-8], k the Boltzmann constant and T temperature. $W^{rev}(n^h)$ is given by [2]

$$W^{rev}(n^h) \simeq -n^h \Delta\mu + \gamma^* A, \tag{2}$$

where $\Delta\mu$ represents

$$\Delta\mu = \mu^\alpha - \mu^\beta(T, p^\alpha), \tag{3}$$

and the superscripts α and β denote a parent phase and a nucleating phase, respectively, μ^α and $\mu^\beta(T, p^\alpha)$ the chemical potential of a molecule in a parent phase and that in the bulk β phase under (T, p^α). In (2) n^h denotes the number of molecules contained within the volume enclosed by the surface of tension in a bulk β phase, γ^* the interfacial tension for a thermodynamic critical nucleus and A an area of the surface of tension. If we can neglect size dependence of interfacial tension, then γ^* may be approximated by the value for the planar interface. We employ this approximation here and denote the value as γ. Since n^h dependence of Φ_{LP} is negligible [6-8], $W^{rev}(n^h)$ takes a maximum value at n^{h*}.

The size n_K^h of a kinetic critical nucleus satisfies

$$K^+(n_K^h) = K^-(n_K^h), \tag{4}$$

where $K^+(n^h)$ denotes the attachment rate of monomers to a nucleus with size n^h and $K^-(n^h)$ the detachment rate from a nucleus. It is assumed that growth or decay of a nucleus results from attachment or detachment of monomers and that collision among nuclei or fission of a nucleus may be neglected.

1.2. Relation between n^{h*} and n_K^h

In a metastable equilibrium state, the following relation holds for any n^h due to the principle of detailed balance:

$$c_o(n^h - \delta n^h)K^+(n^h - \delta n^h) = c_o(n^h)K^-(n^h), \tag{5}$$

where δn^h physically represents a monomer. We see from (4) and (5) that a kinetic critical nucleus is determined by the extremum condition of $c_o(n^h)K^+(n^h)$ or equivalently of $W^{rev}(n^h) - kT\ln[K^+(n^h)]$, which may be called the kinetic potential [5]. Employing (1) with an approximation that n^h dependence of Φ_{LP} is negligible and noting that $K^+(n^h)$ is proportional to $(n^h)^{2/3}$, the size of a kinetic critical nucleus is determined by the following equation [5]:

$$X^2 - X^3 = B, \tag{6}$$

where X represents $X = (n^h/n^{h*})^{1/3}$. A parameter B is defined as

$$B = 2kT/(3n^{h*}\triangle\mu) = 9kT(\triangle\mu)^2/[4(\gamma A_o)^3], \tag{7}$$

where A_o denotes $(36\pi v^2)^{1/3}$, v the molecular volume of the bulk of a nucleating phase, and the following equation for n^{h*} has been employed:

$$n^{h*} = [2\gamma A_o/(3\triangle\mu)]^3. \tag{8}$$

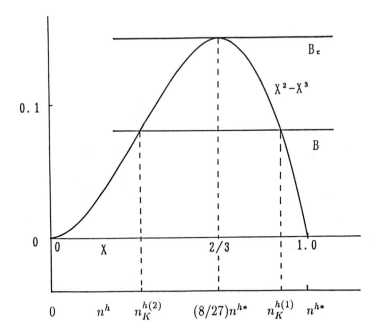

Figure 1. Relation between kinetic critical nucleus n_K^h and thermodynamic one n^{h*} for three dimensional homogeneous nucleation. $B = 9kT(\triangle\mu)^2/[4(\gamma A_0)^3]$

6

Let us study the solution of (6) graphically in Figure 1 [5]. We see that there exist two solutions in general, $n_K^{h(1)}$ and $n_K^{h(2)}$. $K^+(n^h) > K^-(n^h)$ holds for $n^h > n_K^{h(1)}$ or $n^h < n_K^{h(2)}$, whereas $K^+(n^h) < K^-(n^h)$ for $n_K^{h(2)} < n^h < n_K^{h(1)}$. As $\triangle\mu$ diminishes, $n_K^{h(1)}$ approaches n^{h*} while $n_K^{h(2)}$ approaches zero. The difference between $n_K^{h(1)}$ and n^{h*} increases with $\triangle\mu$ and $n_K^{h(1)}$ approaches 8/27 of n^{h*} at the critical value of $\triangle\mu$ given by

$$\triangle\mu_c = (4/9)[(\gamma A_o)^3/(3kT)]^{1/2}. \tag{9}$$

Two values of n_K^h approach as $\triangle\mu$ increases and they coincide with each other at $\triangle\mu_c$. With $\triangle\mu$ beyond this value, where solution for n_K^h does not exist, $K^+(n^h) > K^-(n^h)$ holds for any n^h, i.e., nuclei with any size tend to grow and we call it the runaway instability [5].

Let us consider next nucleation of monatomic substance on a substrate. We assume that adatoms on a substrate are in equilibrium with the parent phase and addition or subtraction of atoms to or from a cluster is dominated by the process via adatoms, i.e., direct interchange of atoms between a cluster and the parent phase may be neglected [9]. We further assume that the shape of a cluster is either spherical cap or circular disc with monatomic height, the latter will be the case when the binding between the substance and the substrate is sufficiently strong such as in homoepitaxy. Since consideration of a spherical cap leads to the results similar to the case of three dimensional homogenous nucleation [5], we discuss only the case of a circular disc in the following.

For a nucleus with the shape of circular disc on a substrate, $W^{rev}(n^h)$ is given by

$$W^{rev}(n^h) = -n^h \triangle\mu + \sigma L_o(n^h)^{1/2}, \tag{10}$$

where σ denotes step free energy, L_o represents $2(\pi a)^{1/2}$, and a the area per atom. The equation which determines n_K^h becomes

$$Y - Y^2 = D, \tag{11}$$

where Y represents $(n^h/n^{h*})^{1/2}$ and D does

$$D = kT/(2n^{h*}\triangle\mu) = 2kT\triangle\mu/(L_o\sigma)^2, \tag{12}$$

and the following expression has been employed:

$$n^{h*} = [L_o\sigma/(2\triangle\mu)]^2. \tag{13}$$

Similarly to the previous cases, there exist two solutions for n_K^h as

$$n_K^{h(1)} = (n^{h*}/4)[1 + (1 - 4D)^{1/2}]^2, \tag{14}$$

$$n_K^{h(2)} = (n^{h*}/4)[1 - (1 - 4D)^{1/2}]^2. \tag{15}$$

Note again that $n_K^{h(1)}$ approaches n^{h*} as $\triangle\mu$ diminishes but the difference between $n_K^{h(1)}$ and n^{h*} increases with $\triangle\mu$ and it approaches $1/4$ of n^{h*} at the critical value of $\triangle\mu$ given by

$$\triangle\mu_c = (L_o\sigma)^2/(8kT). \tag{16}$$

Again, for $\triangle\mu$ beyond this value the runaway instability occurs [5].

1.3. Discussion

The results obtained above are based on the principle of detailed balance (5), which is valid for a system in equilibrium. However, since we assume that interaction among nuclei is negligible, $K^+(n^h)$ and $K^-(n^h)$ are determined only by temperature, size of a nucleus and the state of a parent phase and do not depend on the actual concentration of nuclei in a system. Hence, the results are applicable to nonequilibrium nucleation processes.

Numerically the difference between n^{h*} and $n_K^{h(1)}$ is small and $n_K^{h(2)}$ is almost unity in the usual cases. For example, in a typical case of water nucleation from the vapor, $n^{h*} - n_K^{h(1)} \simeq 0.01n^{h*}$ [4]. As seen in Figure 1, the difference between n^{h*} and $n_K^{h(1)}$ becomes significant for large values of $\triangle\mu$, which results in small n^{h*} as seen in (8). It turns out that n^{h*} itself tends to approach unity before $n^{h*} - n_K^{h(1)}$ becomes significant fraction of n^{h*}. Thus, the practical importance of $n_K^{h(1)}$ and $n_K^{h(2)}$ as distinct from n^{h*} seems to be limited to rather special cases.

Let us consider some examples where the difference between n^{h*} and $n_K^{h(1)}$ becomes appreciable. In the case of water nucleation from the vapor at $T = 560K$, $n^{h*} = 22$, $n_K^{h(1)} = 13$ and $n_K^{h(2)} = 2$ at $p = 1.36p_e$, where p_e denotes the vapor pressure in the bulk equilibrium state, and the runaway instability sets in at $p = 1.45p_e$ with $n^{h*}=12$ and $n_K^h = 4$. However, considering that $T = 560K$ is rather close to the critical temperature of water, $647.3K$, the cluster model itself might become inappropriate. Consider next an example of homoepitaxial growth of Si from the vapor [5]. Since precise data are not available in this case, the following numerical results should be regarded as rough estimates. The areas per atom for (100) and (111) surfaces are $(3.84 \times 10^{-8})^2 cm^2$ and $(3.57 \times 10^{-8})^2 cm^2$, and the step free energies are estimated to be $\sigma_{100} = 10^{-6} erg/cm$ and $\sigma_{111} = 5.7 \times 10^{-6} erg/cm$, respectively [11]. The data on (111) surface lead to the conclusion that appreciable difference between $n_K^{h(1)}$ and n^{h*} is not observed at the temperatures up to the melting temperature of Si [5]. For (100) surface, however, at $T = 586K$ for example, the runaway instability sets in at the supersaturation of $\triangle\mu = 0.125kT$, where $n^{h*} = 16$ and $n_K^h = 4$. For $\triangle\mu = 0.12kT$ at $T = 586K$, $n^{h*} = 17$, $n_K^{h(1)} = 6$ and $n_K^{h(2)} = 3$. Thus, there seems to be a possibility to observe the kinetically stable nucleus $n_K^{h(2)}$ and the runaway instability on (100) surface of Si. Studies of the implication of the difference between n^{h*} and $n_K^{h(1)}$ in describing the kinetic process of nucleation and the extension of the results to multicomponent systems are in progress.

The Gibbs-Thomson equation offers the relation between n^{h*} and the supersaturation or supercooling in a parent phase. This is an equilibrium condition, but it has been employed to take into account the curvature effect in the studies of kinetic processes of crystal growth such as the BCF theory of spiral growth, morphological instability and the Ostwald ripening. However, the required role in these applications is to provide the condition of kinetic balance between attachment and detachment of molecules to and from a curved interface between crystal and a parent phase, hence it must be replaced by the relation between $n_K^{h(1)}$ and the state of a parent phase. For three dimensional systems, the kinetically extended Gibbs-Thomson equation is obtained as [5]

$$\triangle\mu = 2v\gamma/R_K - vkT/(2\pi R_K^3), \tag{17}$$

where R_K denotes the radius of the kinetic critical nucleus defined by $n_K^{h(1)}$. For growth of condensed phases from the vapor, (17) becomes

$$p = p_e \exp[2v\gamma/R_K - vkT/(2\pi R_K^3)]. \tag{18}$$

Note that the term $-vkT/(2\pi R_K^3)$ in (17) and (18) does not exist in the thermodynamic Gibbs-Thomson equation.

2. REVERSIBLE WORK OF FORMING A THERMODYNAMIC CRITICAL NUCLEUS

2.1. Introduction

Since formulation of the nucleation rate in terms of the kinetic critical nucleus remains to be done and the difference between thermodynamic and kinetic critical nuclei is negligible for usual cases, we limit the following consideration to the cases where the steady state nucleation rate is given by the conventional formula which is expressed in terms of the thermodynamic critical nucleus. We neglect from now the difference between kinetic and thermodynamic critical nuclei and call them critical nucleus. We further limit ourselves to three dimensional homogeneous nucleation. For a single component system the steady state nucleation rate J_s is given by [7,9,12]

$$J_s = ZK^+(n^{h*})c_o^*, \tag{19}$$

where Z represents the Zerdovich factor and c_o^* the matastable equilibrium number density of critical nucleus. J_s for a binary system requires more elaborate considerations [13,14,15], but c_o^* remains to be the most important factor. We consider the formula which gives c_o^* in terms of experimentally measurable quantities. c_o^* may be expressed as [6-8]

$$c_o^* = \Phi_{LP}c(1) \exp[-W^*/kT], \tag{20}$$

where $c(1)$ represents total number density of molecules in a parent phase and W^* the reversible work to form a critical nucleus. The formula giving W^* in terms of the interfacial tension γ is given by Gibbs [1], but the formula generally used is not the one due to Gibbs. When the conventional formula is employed in multicomponent nucleation, there have been ambiguity and confusion in evaluating the so-called bulk term. Thus, we summarize first the Gibbs formula, then the conventional formula is derived as an approximation to the Gibbs so that the meaning of the bulk term may be clearly exposed.

2.2. The Gibbs formula

Suppose a supersaturated phase with c components, which we denote as α, under temperature T and pressure p^α and consider the reversible work W^* to form a critical nucleus of another phase β consisting of the same c components within it. We assume isothermal nucleation. Critical nuclei are often extremely small in size so that the homogeneous properties of the bulk β phase are not attained even at the center. Nevertheless, the interfacial thermodynamics remains valid and the following results are applicable to those cases [1,16-18]. The Gibbs formula to obtain W^* is given as follows[1,17,18]:

(a) For a given state $(T, p^\alpha, \{x_i^\alpha\})$ of a parent phase, where x_i^α denotes mole fraction of the ith component, obtain pressure p^β and mole fraction $\{x_i^\beta\}$ for the bulk β phase so that both phases possess the same chemical potentials, i.e., solve the following equations

$$\mu_i^\beta(T, p^\beta, \{x_j^\beta\}) = \mu_i^\alpha(T, p^\alpha, \{x_j^\alpha\}), \quad i = 1, 2, ..., c. \tag{21}$$

(b) Assuming that the value of interfacial tension γ is known, obtain the radius R of the critical nucleus from the Laplace equation

$$p^\beta - p^\alpha = 2\gamma/R, \tag{22}$$

where R denotes the radius of the surface of tension.

(c) Obtain W^* from the following equation

$$W^* = -V^\beta(p^\beta - p^\alpha) + \gamma A, \tag{23}$$

where V^β and A are defined as $V^\beta = 4\pi R^3/3$ and $A = 4\pi R^2$, respectively. Using (22) in (23), W^* may be rewritten as

$$W^* = 4\pi R^2 \gamma/3, \tag{24}$$

or, alternatively, as

$$W^* = 16\pi\gamma^3/3(p^\beta - p^\alpha)^2. \tag{25}$$

The first and the second terms in (23) are called bulk and interface terms, respectively. Their thermodynamic meanings may be understood by employing a thought process devised by Gibbs [1,17,18]. The interface term γA represents the reversible work to form an interface from the two bulk phases which possess the same values of temperature and chemical potentials. The value of γ depends in general on R as well as compositions of the two phases. Note that the composition determined by (21) depends only on bulk properties of parent and nucleating phases and is independent of the value of γ. Although we usually call it the composition of a critical nulceus, we must emphasize that it does not represent the actual composition in a critical nucleus. When a critical nucleus is so large that the properties of a bulk β phase are attained at the center, the composition and the pressure there are given by the values obtained from (21). However, when the size of a critical nucleus is small, the values determined by (21) may not be attained even at its center. The difference in composition is incorporated in the value of γ [1,17-19].

2.3. Commonly used formula as an approximation to the Gibbs

Let us derive the approximate expression for the first term in (23). Integrating the Gibbs-Duhem relation for a bulk β phase

$$s^\beta dT - v^\beta dp + \sum x_i^\beta d\mu_i^\beta = 0 \tag{26}$$

from p^α to p^β under T and $\{x_i^\beta\}$ kept invariant, we obtain

$$\sum x_i^\beta [\mu_i^\beta(T, p^\beta, \{x_j^\beta\}) - \mu_i^\beta(T, p^\alpha, \{x_j^\beta\})] = v^\beta(p^\beta - p^\alpha), \tag{27}$$

where s^β and v^β denote the mean molecular entropy and volume of a bulk β phase, and the pressure dependence of v^β is neglected in the integration. Identifying p^β and $\{x_i^\beta\}$ in (27) with the solutions of (21) and employing(21) in (27), the bulk term in (23) may be rewritten as

$$-V^\beta(p^\beta - p^\alpha) = -n^h \sum x_i^\beta \, \triangle \mu_i, \tag{28}$$

where n^h represents V^β/v^β, and $\triangle\mu_i$ does

$$\triangle\mu_i = \mu_i^\alpha(T, p^\alpha, \{x_j^\alpha\}) - \mu_i^\beta(T, p^\alpha, \{x_j^\beta\}). \tag{29}$$

The RHS of (28) is the commonly used expression for the bulk term.

Employing (22) and (28) to express R in terms of v^β, γ and $\sum x_i^\beta \triangle \mu_i$ and substituting the result in (24), we obtain

$$W^* = \frac{16\pi\gamma^3(v^\beta)^2}{3(\sum x_i^\beta \triangle \mu_i)^2}. \tag{30}$$

(30) is the commonly used formula for W^*. Note that μ_i^β in (29) are the chemical potentials of a bulk β phase with the composition determined by (21) and under the presssure of a parent phase α. This is so even when the critical nucleus is so small that the bulk properties of a β phase are not attained even at the center [17,18]. Thus, the Gibbs procedure (a), (b) and (c) to get W^* may be approximated by the following:

(d) Obtain $\{x_i^\beta\}$ from (21).

(e) Obtain $\{\triangle\mu_i\}$ from (29).

(f) Obtain W^* from (30).

Note that (27) is based on the incompressibility approximation for a bulk β phase, hence (30) is not valid for bubble nucleation [20]. Whereas, the Gibbs procedure is generally valid including bubble nucleation.

2.4. Discussion on the bulk term in the commonly used formula

Let us discuss in some detail the bulk term in the commonly used formula by taking a binary system with species A and B as an example [21]. Consider nucleation of a crystalline phase β from a liquid solution α. We employ the mean molecular Gibbs free energy g vs composition diagram as shown in Figure 2 to illustrate the points of discussion. Solid curves schematically indicate g^α and g^β for the two bulk phases under a pressure p^α as functions of the mole fraction x_B of species B. When the two bulk phases are in equilibrium with the flat interface between them, chemical potentials $\mu_{eq,A}$ and $\mu_{eq,B}$ of each species and the equilibrium compositions $x^\alpha_{eq,B}$ and $x^\beta_{eq,B}$ are found by constructing a common tangent. Suppose that α phase becomes supersaturated to a composition $x^\alpha_B > x^\alpha_{eq,B}$, then chemical potentials μ^α_A and μ^α_B can be found by constructing a tangent line at x^α_B as shown by the dashed line.

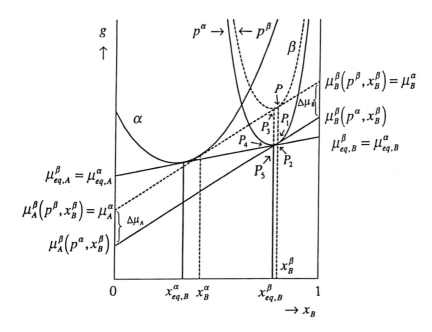

Figure 2. Molecular Gibbs free energies vs mole fraction x_β of a species B for a bulk α and a bulk β phases.

Consider next the bulk term per molecule, which we denote as W^*_{bulk}/n^{h*}, for forming a critical nucleus in the commonly used formula. For a binary system, it follows from (28) that

$$W^*_{bulk}/n^{h*} = -[x^\beta_A \, \triangle \mu_A + x^\beta_B \, \triangle \mu_B], \tag{31}$$

where $\triangle\mu_A$ and $\triangle\mu_B$ are given by (29). x^β_B and p^β are determined by (21), which may be graphically solved as in Figure 2. It follows from (27) that

$$g^\beta(T,p^\beta,x^\beta_B) - g^\beta(T,p^\alpha,x^\beta_B) = v^\beta(T,x^\beta_B)(p^\beta - p^\alpha), \tag{32}$$

where incompressibility of a bulk β phase is assumed. $\triangle\mu_A$ and $\triangle\mu_B$ are obtained as shown in Figure 2, and W^*_{bulk}/n^{h*} can be evaluated from (31). W^* can also be evaluated by substituting the values into (30). If we further assume that the composition dependence of $v^\beta(T,x^\beta_B)$ is negligible, (32) implies that the pressure dependence of g^β is given by simply shifting the curve vertically. The amount of shift required is determined by the condition of common tangent as shown by the broken curve in Figure 2. This step results in the solutions for x^β_B and p^β in this case. Under the assumption of incompressibility for a bulk β phase and composition independence of v^β, the solid and the broken tangent lines at x^β_B are mutually parallel, so that we have

$$\triangle\mu_A = \triangle\mu_B. \tag{33}$$

In this case the composition x^β_B for a critical nucleus may be found from the condition that the tangent line of $g^\beta(p^\alpha)$ at x^β_B is parallel to the tangent line of $g^\alpha(p^\alpha)$ at x^α_B [22].

$-W^*_{bulk}/n^{h*}$ may be called the driving force per molecule for forming a critical nucleus. It follows from (31) that

$$-W^*_{bulk}/n^{h*} = g^\beta(T,p^\beta,x^\beta_B) - g^\beta(T,p^\alpha,x^\beta_B), \tag{34}$$

which is given by the distance PP_1 in Figure 2. Note that

$$\triangle\mu_A \neq \triangle\mu_{eq,A}, \qquad \triangle\mu_B \neq \triangle\mu_{eq,B}, \tag{35}$$

as may be seen clearly in Figure 2, where $\triangle\mu_{eq,A}$ and $\triangle\mu_{eq,B}$ are defined by

$$\triangle\mu_{eq,A} = \mu^\alpha_A(T,p^\alpha,x^\alpha_B) - \mu^\alpha_{eq,A}(T,p^\alpha,x^\alpha_{eq,B}), \tag{36}$$

$$\triangle\mu_{eq,B} = \mu^\alpha_B(T,p^\alpha,x^\alpha_B) - \mu^\alpha_{eq,B}(T,p^\alpha,x^\alpha_{eq,B}). \tag{37}$$

If we incorrectly use $\triangle\mu_{eq,A}$ and $\triangle\mu_{eq,B}$ in (31) instead of $\triangle\mu_A$ and $\triangle\mu_B$, the result corresponds to the distance PP_2 in Figure 2. Thus, although $\triangle\mu_{eq,A}$ and $\triangle\mu_{eq,B}$ themselves are entirely different from $\triangle\mu_A$ and $\triangle\mu_B$, respectively, the value resulting from their use in (31) is not so much different from (34). Nevertheless, the difference between PP_1 and PP_2 may be significant in general. However, when

β phase is a crystal with sharp stoichiometry, then P and P_3 in Figure 2 practically coincide and similarly do the points P_1, P_2, P_4 and P_5. In this case, x_B^β practically coincides with $x_{eq,B}^\beta$ also, hence we can use $\triangle\mu_{eq,A}$ and $\triangle\mu_{eq,B}$ in (30), (31) and (34) with $x_{eq,B}^\beta$ in place of x_B^β. When β phase is practically a pure B crystal with negligible solubility of a species A, we have

$$x_A^\beta \triangle \mu_A + x_B^\beta \triangle \mu_B \simeq \triangle\mu_{eq,B} \simeq kTln(x_B^\alpha/x_{eq,B}^\alpha), \tag{38}$$

which may be used in (30), (31) and (34). Hence, the driving force per molecule for forming a critical nucleus is given in terms of the supersaturation in this case. Similarly, for nucleation of a pure crystal with a species 1 from a multicomponent liquid, we get [21]

$$\sum x_i^\beta \triangle \mu_i \simeq \triangle\mu_{eq,1}. \tag{39}$$

3. REVERSIBLE WORK OF NONCRITICAL NUCLEUS FORMATION AND THE CONDITION OF A CRITICAL NUCLEUS

In the previous section we considered thermodynamic formulae for the reversible work of critical nucleus formation W^*. However, W^* is often derived from the extremum condition of the reversible work for noncritical nucleus formation W^{rev}. In addition, formula for W^{rev} is needed to relate detachment rate of monomers from a nucleus to the attachment rate when we study transient kinetics of nucleation [23]. In this section we consider the formula for W^{rev} and the condition of critical nucleus [2].

Consider a nucleus of a condensed phase β, which consists of $\{n_i^\beta\}$ molecules of each species and is in general noncritical, in a multicomponent vapor phase α. Since difference in the number densities of molecules in a condensed and the vapor phases are 3 to 4 orders of magnitude, $\{n_i^\beta\}$ may be considered as well-defined despite the transition region at the interface. Suppose that temperature T is uniform throughout the system and chemical potentials are uniform within a nucleus, which are denoted as $\{\mu_i^\beta\}$, or in the vapor, denoted as $\{\mu_i^\alpha\}$, separately. In other words, a system is assumed to be in equilibrium under an additional constraint [24] which maintains the number of molecules in a nucleus, and we employ additional independent variables $\{n_i^\beta\}$ compared with the case of a critical nucleus. A parent phase α is assumed to be macroscopic.

Extending the procedure [1,16-18] of formulating interfacial thermodynamics of a critical nucleus to a noncritical one, we obtain the following fundamental equation [2]:

$$dE = TdS + \sum \mu_i^\alpha dN_i^\alpha + \sum \mu_i^\beta dn_i^\beta - p^\alpha dV^\alpha - p^\beta dV^\beta + \gamma dA, \tag{40}$$

where E, S denote internal energy and entropy of a system, $\{N_i^\alpha\}$ number of molecules in a parent phase, γ surface tension, A area of the surface of tension, p^α pressure of a parent vapor, p^β the pressure of a bulk condensed phase which possesses the temperature and the chemical potentials of a nucleus, V^α and V^β the volumes outside and inside the surface of tension, respectively. Employing the standard procedure of thermodynamics, we obtain the Laplace equation (22) and the following Gibbs-Duhem relation for the present system [2]:

$$SdT + \sum N_i^\alpha d\mu_i^\alpha + \sum n_i^\beta d\mu_i^\beta - V^\alpha dp^\alpha - V^\beta dp^\beta + Ad\gamma = 0. \tag{41}$$

The relation (41) is rigorously obeyed for an arbitrary infinitesimal change in a state of the system. The reversible work W^{rev} to form a nucleus with $\{n_i^\beta\}$ molecules from a parent phase under T and $\{\mu_i^\alpha\}$ kept invariant is given by [2]:

$$W^{rev} = \sum n_i^\beta(\mu_i^\beta - \mu_i^\alpha) - V^\beta(p^\beta - p^\alpha) + \gamma A. \tag{42}$$

Note that W^{rev} differs from W^* by the first term in (42). This is due to the difference in chemical potentials between a nucleus and a parent phase for a noncritical nucleus. For a critical nucleus the first term in (42) vanishes because $\mu_i^\alpha = \mu_i^\beta$ holds for all i in this case.

A critical nucleus is defined by the extremum condition of W^{rev} under $(T, \{\mu_i^\alpha\}, V)$ constant, where V represents $V^\alpha + V^\beta$. Taking differential of (42) and employing (41) in the result, we get [2]

$$\sum(\mu_i^\beta - \mu_i^\alpha)dn_i^\beta - (p^\beta - p^\alpha)dV^\beta + \gamma dA = 0. \tag{43}$$

Thus, in effect the intensive variables p^β, $\{\mu_i^\beta\}$ and γ may be treated as constant in taking the extremum condition of (42). They do change with $\{n_i^\beta\}$, but the contribution of their differentials to dW^{rev} mutually cancel exactly due to the Gibbs-Duhem relation (41). Due to the relation $dV^\beta = 4\pi R^2 dR$ and $dA = 8\pi RdR$ together with the Laplace equation (22), the second and the third terms in (43) cancel each other. Hence, we obtain the following relation as the condition for a critical nucleus:

$$\mu_i^\beta = \mu_i^\alpha, \quad i = 1, ..., c. \tag{44}$$

The condition for a critical nucleus thus obtained is equivalent to (21) due to Gibbs. Employing (44) in (42), we obtain (23). Thus, we get a critical nucleus and W^* from the extremum condition of W^{rev}.

Employing the Gibbs-Duhem relation (26) and assuming incompressibility for the bulk condensed phase, we obtain the following expression for W^{rev} from (42) [2]:

$$W^{rev} = \sum n_i^{h\beta}[\mu_i^\beta(T, p^\alpha, \{x_j^\beta\}) - \mu_i^\alpha] + \gamma A + \sum(n_i^\beta - n_i^{h\beta})[\mu_i^\beta - \mu_i^\alpha], \tag{45}$$

where $\{x_i^\beta\}$ denote the composition of a bulk condensed phase having the temperature and the chemical potentials of a nucleus, and $\{n_i^{h\beta}\}$ the number of molecules within a sphere with radius R situated in that bulk condensed phase. The extremum condition of (45) under $(T, \{\mu_i^\alpha\}, V)$ constant results again in the Gibbs formula (21) and (30) [2]. The terms with the differentials of the intensive variables including $d\gamma$ cancel each other by the Gibbs-Duhem relation (41). Hence, in taking the extremum condition of (45), the intensive variables may effectively be treated as constants.

The commonly used formula for W^{rev} may be identified with (45) without the last term, i.e.,

$$W^{rev} = \sum n_i^{h\beta} \left[\mu_i^\beta (T, p^\alpha, \{x_j^\beta\}) - \mu_i^\alpha \right] + \gamma A. \tag{46}$$

Note that a nucleus is specified by $(T, \{n_i^{h\beta}\})$ and $\{x_i^\beta\}$ is determined by $\{n_i^{h\beta}\}$ in (46), though this has not been clearly stated. Since $\{n_i^\beta - n_i^{h\beta}\}$ in the last term of (45) represents the surface excess molecules hence is proportional to A, the last term in (45) is not negligible unless a nucleus is sufficiently close to a critical nucleus. Thus, the commonly used formula (46) is erroneous. Nevertheless, let us see the result of taking the extremum condition of (46). Taking the differential of (46) and employing (26), we obtain [2]

$$dW^{rev} = \sum [\mu_i^\beta (T, p^\beta, \{x_j^\beta\}) - \mu_i^\alpha] dn_i^{h\beta} - (p^\beta - p^\alpha) dV^\beta + \gamma dA + A d\gamma = 0, \tag{47}$$

where we again used the incompressibility approximation for a bulk condensed phase. If we treat as $d\gamma = 0$ in (47), use of the Laplace equation (22) results in (21) and (30), hence the Gibbs formula follows from (47). However, this must be considered as fortuitous, because γ does change in general with $\{n_i^\beta\}$ under $(T, \{\mu_i^\alpha\}, V)$ constant.

When we study transient kinetics of nucleation, W^{rev} must be evaluated to relate the detachment rate of monomers from a nucleus to the attachment rate [23]. However, neither (42) nor (45) can be used for this purpose, because these formulae require knowledge of $\{\mu_i^\beta\}$ for a given nucleus specified by $(T, \{n_i^\beta\})$. Thus, we need to derive a formula for W^{rev} which is useful in practice [3]. Noting that p^β in (41) means the value for a bulk β phase at $(T, \{\mu_i^\beta\})$, dp^β obeys also the following equation derived from (26):

$$-V^\beta dp^\beta + \sum n_i^{h\beta} d\mu_i^\beta = 0. \tag{48}$$

Supposing that the state of a parent α phase is invariant, subtraction of (48) from (41) results in

$$\sum (n_i^\beta - n_i^{h\beta}) d\mu_i^\beta + A d\gamma = 0. \tag{49}$$

It follows from (49) that

$$n_i^\beta - n_i^{h\beta} = -A(\partial \gamma / \partial \mu_i^\beta), \quad i = 1, ..., c. \tag{50}$$

16

Substituting (50) into the last term in (45) and noting that $\{\mu_i^\alpha\}$ coincide with the chemical potentials in a critical nucleus, we obatin

$$\sum(\partial\gamma/\partial\mu_i^\beta)[\mu_i^\alpha - \mu_i^\beta] \simeq \gamma^* - \gamma, \tag{51}$$

where γ^* denotes the surface tension for the critical nucleus corresponding to a given state of a parent phase. Employing (51) in (45), we get the following approximate formula for W^{rev}:

$$W^{rev} \simeq \sum n_i^{h\beta}[\mu_i^\beta(T, p^\alpha, \{x_j^\beta\}) - \mu_i^\alpha] + \gamma^* A. \tag{52}$$

In other words, W^{rev} for a noncritical nucleus is given approximately by the commonly used formula (46) with the surface term in which γ^* for a critical nucleus is used insted of γ for the value of surface tension. Note in (52) that a nucleus is not specified by $\{n_i^\beta\}$, numbers of molecules contained in a nucleus, but by $\{n_i^{h\beta}\}$, numbers of molecules contained in a sphere with the radius of the surface of tension and located in a bulk β phase. This does not cause inconvenience as long as studying the formation rate of supercritical nuclei is concerned, because a critical nucleus is also specified in terms of $\{n_i^{h\beta}\}$.

ACKNOWLEDGMENTS

I wish to thank Dr.I.L.Maksimov of Nizhny Novgorod University for the valuable collaboration which has brought the results presented in Section 1. Colleagues and students of Tokushima University are also acknowledged for the collaborations and discussions. I owe much to Mr.M.Okada in preparing the manuscript. The present work is supported by a Grant-in-Aid for Scientific Research on Priority Areas "Crystal Growth Mechanism in Atomic Scale", No.03243101 and No.04227108.

REFERENCES

1. J. W. Gibbs, The Scientific Papers of J. W. Gibbs, Vol.I, Dover, New York, 1961.
2. K.Nishioka and I.Kusaka, J.Chem.Phys. 96 (1992) 5370.
3. K.Nishioka and A.Mori, J.Chem.Phys. 97 (1992) 6687.
4. K.Nishioka, Phys.Rev.E 52 (1995) 3263.
5. K.Nishioka and I.L.Maksimov, J.Cryst.Growth 163 (1996) 1.
6. J.Lothe and G.M.Pound, J.Chem.Phys. 36 (1962) 2080.
7. W.J.Dunning, in Nucleation, ed. by A.C.Zettlemoyer, Marcel Dekker, New York, 1969, pp.1-67; J.Lothe and G.M.Pound, ibid, pp.109-149.
8. H.Reiss, in Nucleation Phenomena, ed. by A.C.Zettlemoyer, Elsevier, Amsterdam, 1977, pp.1-66; R.Kikuchi, ibid, pp.67-102; K.Nishioka and G.M.Pound, ibid, pp.205-278.
9. J.P.Hirth and G.M.Pound, Condensation and Evaporation, Progress in Materials Science, Vol.11, Pergamon Press, 1963.
10. B.K.Chakraverty, in Crystal Growth:An Introduction, ed. by P.Hartman, North-Holland, 1973, pp.50-104.

11. A.A.Chernov, in Handbook of Crystal Growth, Vol.3b, ed. by D.T.J.Hurle, North-Holland, 1994, pp.457-490.
12. J.Feder, K.C.Russell, J.Lothe and G.M.Pound, Adv.in phys. 15 (1966) 111.
13. H.Reiss, J.Chem.Phys. 18 (1950) 840.
14. G.Shi and J.H.Seinfeld, J.Chem.Phys. 93 (1990) 9033.
15. A.L.Greer, P.V.Evans, R.G.Hamerton, D.K.Shangguan and K.F.Kelton, J. Cryst.Growth, 99 (1990) 38.
16. K.Nishioka, Phys.Rev.A 36 (1987) 4845.
17. K.Nishioka, Met.Trans.A 23 (1992) 1869.
18. K.Nishioka, Physica Scripta T44 (1992) 23.
19. K.Nishioka, H.Tomino, I.Kusaka and T.Takai, Phys.Rev.A 39 (1989) 772.
20. J.P.Hirth, G.M.Pound and G.R.St.Pierre, Metall.Trans. 1 (1970) 939.
21. Jin-Song Li, K.Nishioka and E.R.C.Holcomb, J.Cryst.Growth, in press.
22. C.V.Thompson and F.Spaepen, Acta Metall. 31 (1983) 2021.
23. For example, K.Nishioka and K.Fujita, J.Chem.Phys. 100 (1994) 532.
24. H.Reiss, Method of Thermodynamics, Blaisdell, New York, 1965.

Advances in the Understanding of Crystal Growth Mechanisms
T. Nishinaga, K. Nishioka, J. Harada, A. Sasaki and H. Takei (Editors)
© 1997 Elsevier Science B.V. All rights reserved.

The terrace-step-kink approach and the capillary-wave approach to fluctuation properties of vicinal surfaces

T. Yamamoto [a], N. Akutsu [b] and Y. Akutsu [c]

[a]Department of Physics, Faculty of Engineering, Gunma University, Kiryu, Gunma 376, Japan

[b]Faculty of Engineering, Osaka Electro-Communication University, Neyagawa, Osaka 572, Japan

[c]Department of Physics, Graduate School of Science, Osaka University, Toyonaka, Osaka 560, Japan

For studies of the vicinal surface of a crystal below the roughening temperature, the standard approach has been the one based on the terrace-step-kink model. On the other hand, it has also been known that the vicinal surface is rough and well described by an anisotropic capillary-wave model. We review the both modeling for the vicinal surface and give an explicit mapping between them. As an interpolating modeling between the two, we introduce harmonically-interacting step model which is used to discuss the fluctuation properties of a single step on the vicinal surface.

1. INTRODUCTION

Crystal surface with small tilt angle relative to a facet (a flat surface, mostly with orientation corresponding to the crystal axis) is called vicinal surface. The vicinal surface is well described by the terrace-step-kink model[1]. In the TSK model, the surface is composed of terraces parallel to the facet and steps connecting them. Owning to the large energy cost of multi-height steps and overhangs, steps are prohibited from crossing or overlapping. In the length-scale much larger than the atomic scale a, the TSK model can be regarded as an ensemble of non-crossing strings corresponding to steps[2,3]. Each string has anisotropic properties characterizing the surface anisotropy. Meandering of the non-crossing strings causes the logarithmic divergence of the height-height correlation function which is a characteristic property of a rough surface.

Let the x-y plane be chosen to be parallel to the terraces. We consider the vicinal surface composed of n steps with projected area $M(x\text{-direction})\times L(y\text{-direction})$. By θ, we denote the angle between the y axis and the mean running direction of the steps. The step density along the x-direction is given by $\rho = n/M$. The surface gradient $\vec{p}$ relates to the step density as $\vec{p} = (p_x, p_y) = (-d\rho, -d\rho\tan\theta)$ where d is the unit step height. By $f_s(\tan\theta)$ we denote the step free energy per projected length along the y-direction.

For a single step passing through the two points (x_1, y_1) and (x_2, y_2), we associate the Boltzmann weight $W(x_2 - x_1, y_2 - y_1)$ which is expressed as

$$W(x_2 - x_1, y_2 - y_1) = \exp[-\beta f_s(\theta)(y_2 - y_1)] \quad (\tan\theta = \frac{x_2 - x_1}{y_2 - y_1}), \tag{1.1}$$

where $\beta = (k_B T)^{-1}$ (k_B:the Boltzmann constant, T: the temperature).

The non-crossing (non-penetration) nature of the steps is neatly taken into account by regarding the configurations of steps as world lines of fermions. We can then express the transfer matrix $\hat{A}_{FF}$ of the TSK model[2,3] as

$$\hat{A}_{FF} = \exp[-\hat{H}_{FF}\ell], \tag{1.2}$$

$$\hat{H}_{FF} = \sum_k \epsilon(k) a_k^\dagger a_k, \tag{1.3}$$

$$\epsilon(k) = -\frac{1}{\ell} \ln \Lambda(k; \beta\mu), \tag{1.4}$$

where $a_k^\dagger$ and a_k are the creation and annihilation operators of one-dimensional spinless fermions with wave number $k = 2\pi m/M$ ($m = 0, \pm 1, \pm 2, \cdots$). The quantity Λ is defined by

$$\Lambda(k; \beta\mu) = \int \frac{du}{a} W(u, \ell) e^{\beta\mu u - iku}. \tag{1.5}$$

where we have introduced discretization length unit ℓ which is small but sufficiently larger than the atomic scale a. In what follows, we are implicitly considering the $\ell \to 0$ limit in all the expressions involving ℓ. The "external field" μ causing the tilt of the step runing direction relates to the tilt angle θ via the equation

$$\frac{\partial g(\mu, \rho)}{\partial \mu} = \rho \tan\theta, \tag{1.6}$$

with

$$g(\mu, \rho) = -k_B T \frac{1}{LM} \ln[\sum_{\text{fin.,in.}} < n, \text{fin.}|\hat{A}_{FF}^{L/\ell}|n, \text{in.} >] \tag{1.7}$$

where $|n, \text{in.}>$ and $|n, \text{fin.}>$ are n-fermion states and the summation is taken over all possible states in the n-fermion space. The surface free energy per projected area is given by

$$f(\vec{p}) = g(\mu, \rho) - \mu\rho \tan\theta. \tag{1.8}$$

The capillary wave (CW) model[4–6] is another model describing the rough surface. The CW model which was originally introduced to describe liquid-gas and liquid-liquid interface can also been used to discuss fluctuations of the rough crystal surface (solid-gas interface). This can be justified for the non-vicinal ($\vec{p} = 0$) surface above T_R, because the anisotropy of the surface is not so strong allowing us to regard it as an isotropic liquid surface.

Below the roughening temperature, the crystal anisotropy becomes relevant to surface properties. As for the vicinal surface, regarded as an assembly of unidirectional strings, properties along and across the string-running direction are quite different; the vicinal surface is an extremely anisotropic rough surface. Then, at first sight, applicability of CW modeling seems to be questionable. However, the CW picture of the vicinal surface perfectly explains various fluctuation properties (e.g., the height-height correlation function[7,8] and the single-step fluctuation width[9–11]), as does the TSK model with the free-fermion approach. The TSK model and the "anisotropic" CW model of the vicinal surface have the same fluctuation properties including their anisotropy. It is then natural to expect a direct mapping between the two models.

The Hamiltonian of the CW model correctly expressing the fluctuation properties is given by

$$H_{\mathrm{CW}} = \frac{1}{2} \int dx dy \sum_{i,j=x,y} f_{i,j} \frac{\partial h}{\partial p_i} \frac{\partial h}{\partial p_j}, \tag{1.9}$$

where $h = h(x,y)$ is the surface height difference from the reference plane, $z = p_x x + p_y y$. The constants $\{f_{i,j}\}$ are the components of the "surface stiffness" tensor and are required to be chosen as

$$f_{i,j} = \frac{\partial^2 f(\vec{p})}{\partial p_i \partial p_j}. \tag{1.10}$$

The first purpose of the present article is to review the derivation of the CW Hamiltonian (1,9) with the correct surface stiffness (1.10) from the free-fermion representation (1.2)[12].

It is also interesting to derive the TSK picture from the CW model. Let us choose the y-direction be be parallel to the mean running direction of the steps. The CW Hamiltonian (1.9) can be rewritten as

$$H'_{\mathrm{CW}} = \frac{1}{2} \int dx dy [f_{x,x} (\frac{\partial h}{\partial p_x})^2 + f_{y,y} (\frac{\partial h}{\partial p_y})^2]. \tag{1.11}$$

We can express an instantaneous shape of the $(j+1)$-th step ($j = 0, 1, \cdots, n-1$) as $x = u_j(y)$. We rewrite (1.11) in terms of the step shapes and obtain a new "coarse-grained" TSK model called harmonically-interacting step(HIS) model[13]:

$$H_{\mathrm{HIS}} = \sum_{j=0}^{n-1} \int dy [\frac{1}{2} c_x (x_{j+1}(y) - x_j(y))^2 + \frac{1}{2} c_y (\frac{dx_j(y)}{dy})^2], \tag{1.12}$$

where $x_j(y)$ is the displacement of the $(j+1)$-th step from the reference position $x = j/\rho$; $x_j(y) = u_j(y) - j/\rho$. For the CW model and the HIS model to be equivalent to each other, the constants c_x and c_y should satisfy

$$c_x = \rho^3 d^2 f_{x,x}, \; c_y = \rho d^2 f_{y,y}. \tag{1.13}$$

The HIS model is valid in the length scale much larger than the average step distance $1/\rho$. In this scale, the step-step interactions including the non-crossing nature of the steps are taken into account by the effective harmonic potential between the adjacent steps. The second purpose of the article is to derive the HIS Hamiltonian (1.12) from the CW Hamiltonian (1.11).

The simplicity of HIS Hamiltonian enables us to calculate various quantities expressing surface fluctuation properties. For instance, the single-step fluctuation width is easily obtained. Using the Ginzburg-Laudau-Langevin type equation for the HIS Hamiltonian, we can also discuss the dynamical properties of surface fluctuations. The last purpose of the article is to analyze the long-time behavior of the mean-square displacement of steps.

2. CAPILLARY WAVE REPRESENTATION OF LOW-ENERGY FLUCTUATIONS OF THE FREE-FERMION TSK MODEL

Low-"energy" (or long wavelength) fluctuations described by the free-fermion transfer matrix (1.2) is expressed by the one-dimensional harmonic bose fluid, from which we derive the capillary wave model (1.9) with (1.10).

2.1. Harmonic fluid description of fluctuations

In terms of the "ground state" eigenvalue E_1 of $\hat{H}_{FF}$, which is given by

$$E_1 = \sum_{|k| \leq k_F} \epsilon(k) \quad (k_F = \pi \rho), \tag{2.1}$$

$g(\mu, \rho)$ is simply expressed as

$$g(\mu, \rho) = k_B T \frac{E_1}{M}. \tag{2.2}$$

The corresponding eigenstate $|n, 1 >$ is given by

$$|n, 1 >= \prod_{|k| \leq k_F} a_k^\dagger |vac >, \tag{2.3}$$

where $|vac >$ is the vacuum state for operators $\{a_k\}_k$ satisfying $a_k|vac >= 0$ for all k.

To analyze the excitation spectrum, we pay attention to the "dispersion" $\epsilon(k)$ around $k = \pm k_F$ and linearize it as

$$\epsilon(k) \simeq \epsilon(k_F) + v_1(k - k_F) \quad \text{for } k > 0, \tag{2.4}$$

$$\epsilon(k) \simeq \epsilon(-k_F) - v_{-1}(k + k_F) \quad \text{for } k < 0, \tag{2.5}$$

where $v_1 = \epsilon'(k_F)$ and $v_{-1} = -\epsilon'(-k_F)$.

Introducing the new fermion operators, the right-moving fermion $c_{1,k}$ and the left-moving fermion $c_{-1,k}$, which are related to the original fermion operators as $a_k = c_{1,k}$ for $k > 0$ and $a_k = c_{-1,k}$ for $k < 0$, to describe the low-energy excitations[14,15], we rewrite $\hat{H}_{FF}$ as

$$\hat{H}_{FF} \cong E_1 + \hat{H}'_{FF}, \tag{2.6}$$

$$\hat{H}'_{FF} = \sum_{k \neq 0} \sum_{\sigma = \pm 1} v_\sigma(\sigma k - k_F) : c_{\sigma,k}^\dagger c_{\sigma,k} : . \tag{2.7}$$

In the above : $\cdots$: means the fermion normal-ordering with respect to the ground state given by

$$|FG> = \prod_{k_1 \leq k_F} \prod_{k_2 \geq -k_F} c^\dagger_{1,k_1} c^\dagger_{-1,k_2} |vac'>, \tag{2.8}$$

which corresponds to $|n,1>$ for the original operators a_k. The new "vacuum" state $|vac'>$ is defined by $c_{\sigma,k}|vac'>= 0$ (for all k,σ). Although the operators $c_{1,k}$ with $k < 0$ and $c_{-1,k}$ with $k > 0$ have no corresponding original operators, inclusion of these modes in $\hat{H}'_{FF}$ do not affect the low-energy excitation spectrum.

The "Hamiltonian" operator $\hat{H}'_{FF}$ is similar to that of the Tomonaga-Luttinger (TL) model[15–18]. According to the treatment for the TL model, we introduce the operator,

$$\hat{\rho}_{\sigma,p} = \sum_k c^\dagger_{\sigma,k+p} c_{\sigma,k}, \tag{2.9}$$

which satisfies the commutation relation:

$$[\hat{\rho}_{\sigma,-p}, \hat{\rho}_{\sigma',p'}] = \sigma \frac{Mp}{2\pi} \delta_{p,p'} \delta_{\sigma,\sigma'}. \tag{2.10}$$

From the operator $\hat{\rho}_{\sigma,p}$, we can construct the boson operator as follows:

$$b_q = \sqrt{\frac{2\pi}{M|q|}} \sum_\sigma \Theta(\sigma q) \hat{\rho}_{\sigma,-q}, \tag{2.11}$$

where $\Theta(x)$ is the step function. From (2.10), it is easy to confirm

$$[b_q, b^\dagger_{q'}] = \delta_{q,q'}, \quad [b_q, b_{q'}] = [b^\dagger_q, b^\dagger_{q'}] = 0. \tag{2.12}$$

From the commutation relation

$$[\hat{\rho}_{\sigma,p}, \hat{H}'_{FF}] = -v_\sigma \sigma p \hat{\rho}_{\sigma,p}, \tag{2.13}$$

we have, for $k > 0$,

$$[b_{\sigma k}, \hat{H}'_{FF}] = v_\sigma k b_{\sigma k}. \tag{2.14}$$

Therefore the excitation spectrum of $\hat{H}'_{FF}$ is equivalent to that of the harmonic bose-fluid Hamiltonian:

$$\hat{H}_B = \sum_{k>0} k(v_1 b^\dagger_k b_k + v_{-1} b^\dagger_{-k} b_{-k}). \tag{2.15}$$

2.2. Derivation of the capillary wave Hamiltonian from the harmonic bose-fluid

In the transfer matrix language, the step density operator at the point x is defined by

$$\hat{n}(x) = \frac{1}{M} \sum_{k,k'} e^{i(k-k')x} a^\dagger_{k'} a_k. \tag{2.16}$$

The fluctuation of the step density from the mean step density is expressed by the operator

$$\delta\hat{n}(x) = \hat{n}(x) - \frac{1}{M}\int dx\hat{n}(x) = \frac{1}{M}\sum_{k\neq k'}e^{i(k-k')x}a^{\dagger}_{k'}a_k. \tag{2.17}$$

In terms of the new fermion density operator $\hat{\rho}_{\sigma,p}$, $\delta\hat{n}(x)$ is approximated as

$$\delta\hat{n}(x) \cong \frac{1}{M}\sum_{q\neq 0}e^{-iqx}\sum_{\sigma=\pm 1}\hat{\rho}_{\sigma,q}. \tag{2.18}$$

The operator $\hat{h}(x)$ representing the displacement of the surface height from the reference surface is related to the step-density fluctuation operator $\delta\hat{n}$ as

$$\frac{d\hat{h}(x)}{dx} = -d\delta\hat{n}(x). \tag{2.19}$$

Thus we have

$$\hat{h}(x) = -\frac{d}{M}\sum_{q\neq 0}e^{iqx}\frac{1}{iq}\sum_{\sigma=\pm 1}\hat{\rho}_{\sigma,-q} = -\frac{d}{\sqrt{M}}\sum_{q\neq 0}\sqrt{\frac{|q|}{2\pi}}\frac{1}{\pi}e^{iqx}(b^{\dagger}_{-q}+b_q). \tag{2.20}$$

Introducing its canonical conjugate $\hat{\pi}(x)$ as

$$\hat{\pi}(x) = -\frac{1}{d}\frac{1}{\sqrt{M}}\sum_{q\neq 0}\sqrt{\frac{\pi}{2|q|}}qe^{iqx}(b^{\dagger}_{-q}-b_q), \tag{2.21}$$

we can rewrite the harmonic bose-fluid Hamiltonian $\hat{H}_{\rm B}$ as

$$\hat{H}_{\rm B} = \hat{H}'_{\rm B} - <{\rm BG}|\hat{H}'_{\rm B}|{\rm BG}>, \tag{2.22}$$

$$\hat{H}'_{\rm B} = \frac{1}{2}\int dx[d^2\frac{v_1+v_{-1}}{2\pi}\hat{\pi}^2(x) - (v_1-v_{-1})(\frac{d\hat{h}(x)}{dx}\hat{\pi}(x)+\hat{\pi}(x)\frac{d\hat{h}(x)}{dx})$$

$$+\frac{\pi}{2d^2}(v_1+v_{-1})(\frac{d\hat{h}(x)}{dx})^2], \tag{2.23}$$

where $|{\rm BG}>$ is the ground state of $\hat{H}_{\rm B}$ and satisfies $b_k|{\rm BG}>= 0$ for all k.

In terms of the eigenstate $|\{h(x,y)\}_x >$ of the height operator $\hat{h}(x)$ satisfying $\hat{h}(x')|\{h(x,y)\}_x >= h(x',y)|\{h(x,y)\}_x >$, the Boltzmann weight for a surface height configuration $\{h(x,y)\}$ is given by

$$P(\{h(x,y)\}_{(x,y)}) = \prod_{y=0}^{L-\ell} < \{h(x,y+\ell)\}_x|\hat{A}_{\rm FF}|\{h(x,y)\}_x >$$

$$\propto \prod_{y=0}^{L-\ell} < \{h(x,y+\ell)\}_x|\exp[-\hat{H}_{\rm B}\ell]|\{h(x,y)\}_x >$$

$$\propto \exp\{-\beta\frac{1}{2}\int dxdy[A(\frac{\partial h(x,y)}{\partial x})^2 + B\frac{\partial h(x,y)}{\partial x}\frac{\partial h(x,y)}{\partial y} + C(\frac{\partial h(x,y)}{\partial y})^2]\}, \tag{2.24}$$

where A, B and C are the constants given in terms of v_1 and v_{-1}. From (1.4), (1.5), (1.6), (1.8) and (2.2), we have

$$A = \frac{\partial^2 f(\vec{p})}{\partial p_x^2}, \ B = \frac{\partial^2 f(\vec{p})}{\partial p_x \partial p_y}, \ C = \frac{\partial^2 f(\vec{p})}{\partial p_y^2}. \tag{2.25}$$

It means that the Boltzmann weight $P(\{h(x,y)\}_{(x,y)})$ is given by

$$P(\{h(x,y)\}_{(x,y)}) = \exp(-\beta H_{\text{CW}}). \tag{2.25}$$

Thus, the long wavelength fluctuation of the TSK model is correctly described by the CW Hamiltonian.

3. HARMONICALLY-INTERACTING STEP MODEL

Let the y-direction be chosen to be parallel to the mean running direction of the steps. The CW Hamiltonian describing the vicinal surface is given by (1.11). Through the transfer matrix representation[19,20], we show that the CW Hamiltonian (1.11) is successfully derived from the HIS model (1.12).

3.1. Transfer matrix representation

We impose the periodic boundary condition in the x- and y-directions. In the TSK picture, the periodic boundary condition in the x-direction shows the periodic boundary condition in the "j-space";

$$x_{j+n}(y) = x_j(y). \tag{3.1}$$

The partition function of the HIS model is given by

$$Z_{\text{HIS}} = \int \prod_{j=0}^{n-1} \prod_y dx_j(y) e^{-\beta H_{\text{HIS}}}. \tag{3.2}$$

Let us introduce the operator $\hat{x}_j$ whose eigenvalue is the $j+1$-th step displacement $x_j(y)$, and the operator $\hat{P}_j$ conjugate to x_j. These operators satisfy the following commutation relations:

$$[\hat{x}_j, \hat{x}_{j'}] = [\hat{P}_j, \hat{P}_{j'}] = 0, \ [\hat{x}_j, \hat{P}_{j'}] = i\delta_{j,j'}. \tag{3.3}$$

In terms of these operators, the transfer matrix $\hat{A}_{\text{HIS}}$ of the HIS model is given by

$$\hat{A}_{\text{HIS}} = e^{-\ell \hat{H}_{\text{HIS}}},$$
$$\hat{H}_{\text{HIS}} = \sum_j [\frac{1}{2}\beta c_x(\hat{x}_{j+1} - \hat{x}_j)^2 + \frac{1}{2}\frac{1}{\beta c_y}\hat{P}_j^2]. \tag{3.4}$$

In terms of the transfer matrix, the partition function is given by

$$Z_{\text{HIS}} = (\frac{2\pi\ell}{\beta c_y})^{\frac{nL}{2\ell}} \text{Tr}[\hat{A}_{\text{HIS}}]^{L/\ell}. \tag{3.5}$$

We expand $\hat{x}_j$ and $\hat{P}_j$ in terms of the "Fourier components" $\{\hat{\xi}_Q\}_Q$ and $\{\hat{\Pi}_Q\}_Q$ as

$$\hat{x}_j = \frac{1}{\sqrt{n}} \sum_Q e^{iQj} \hat{\xi}_Q,$$

$$\hat{P}_j = \frac{1}{\sqrt{n}} \sum_Q e^{iQj} \hat{\Pi}_Q, \tag{3.6}$$

where $Q = 2\pi j/n$ $(j = 0, \pm 1, \pm 2, \cdots, n/2)$. From the relations (3.3), we obtain

$$[\hat{\xi}_Q, \hat{\xi}_{Q'}] = [\hat{\Pi}_Q, \hat{\Pi}_{Q'}] = 0, \quad [\hat{\xi}_Q, \hat{\Pi}_{Q'}] = i\delta_{Q,-Q'}. \tag{3.7}$$

In terms of the above Fourier components, the "Hamiltonian" operator $\hat{H}_{\text{HIS}}$ is expressed by

$$\hat{H}_{\text{HIS}} = \sum_Q [\frac{1}{2\beta c_y} \hat{\Pi}_Q \hat{\Pi}_{-Q} + \frac{1}{2}\beta c_x |1 - e^{iQ}|^2 \hat{\xi}_Q \hat{\xi}_{-Q}]. \tag{3.8}$$

3.2. Equivalence between the harmonically-interacting step model and the capillary wave model

In terms of $\{\hat{x}_j\}$, the step density operator $\hat{n}(x)$ is expressed by

$$\hat{n}(x) = \sum_{j=0}^{n-1} \delta(x - (\hat{x}_j + \frac{j}{\rho})) = \frac{1}{M} \sum_q e^{iqx} \sum_{j=0}^{n-1} e^{-iqj/\rho} e^{-iq\hat{x}_j}, \tag{3.9}$$

where $q = 2\pi\nu/M$ $(\nu = 0, \pm 1, \pm 2, \cdots)$. The step density fluctuation operator is given by

$$\delta\hat{n}(x) = \hat{n}(x) - \frac{1}{M} \int dx \hat{n}(x) = \sum_q e^{iqx} \hat{n}_q, \tag{3.10}$$

with

$$\hat{n}_q = \frac{1}{M} \sum_{j=0}^{n-1} e^{-iqj/\rho} e^{-iq\hat{x}_j}. \tag{3.11}$$

Using the relation (2.19), we can express the surface height operator $\hat{h}(x)$ as

$$\hat{h}(x) = \frac{1}{\sqrt{M}} \sum_q \hat{h}_q e^{iqx},$$

$$\hat{h}_q = -d\sqrt{M}\hat{n}_q = -d\frac{1}{\sqrt{M}iq} \sum_{j=0}^{n-1} e^{-iqj/\rho} e^{-iq\hat{x}_j}. \tag{3.12}$$

For small $|q|$, $\hat{h}_q$ can be approximated by[9,21]

$$\hat{h}_q \simeq -d\frac{1}{\sqrt{M}iq}(-iq) \sum_{j=0}^{n-1} \hat{x}_j e^{-iqj/\rho} = d\sqrt{\rho}\hat{\xi}_{q/\rho}. \tag{3.13}$$

Thus,

$$\hat{\xi}_Q = \frac{1}{d\sqrt{\rho}} \hat{h}_{\rho Q}. \tag{3.13'}$$

Let us define a new operator $\hat{\pi}_{\rho Q}$ by

$$\hat{\pi}_{\rho Q} = \frac{1}{d\sqrt{\rho}} \hat{\Pi}_Q. \tag{3.14}$$

For $|k|, |k'| \leq k_F$, $\hat{\pi}_k$ and $\hat{h}_k$ satisfy the following commutation relations:

$$[\hat{h}_k, \hat{h}_{k'}] = [\hat{\pi}_k, \hat{\pi}_{k'}] = 0, \quad [\hat{h}_k, \hat{\pi}_{k'}] = i\delta_{k,-k'}. \tag{3.15}$$

In terms of $\hat{h}_k$ and $\hat{\pi}_k$, $\hat{H}_{\text{HIS}}$ is written as

$$\hat{H}_{\text{HIS}} = \sum_{|k| \leq k_F} \left[\frac{d^2\rho}{2\beta c_y} \hat{\pi}_k \hat{\pi}_{-k} + \frac{1}{2} \frac{\beta c_x}{d^2\rho} |1 - e^{ik/\rho}|^2 \hat{h}_k \hat{h}_{-k} \right]. \tag{3.16}$$

We neglect the fluctuations with the wavelength shorter than the average step spacing $1/\rho$. We can make an approximation, $|1 - e^{ik/\rho}|^2 = 2 - 2\cos(k/\rho) \simeq (k/\rho)^2$, and remove the restriction $|k| \leq k_F$ for the summation in (3.16). Thus, we obtain

$$\hat{H}_{\text{HIS}} \simeq \sum_k \left[\frac{d^2\rho}{2\beta c_y} \hat{\pi}_k \hat{\pi}_{-k} + \frac{\beta c_x}{2\rho^3} \frac{k^2}{d^2} \hat{h}_k \hat{h}_{-k} \right]. \tag{3.17}$$

Restricting ourselves to long wavelength properties, we can write the operator $\hat{\pi}(x)$ conjugate to $\hat{h}$ as

$$\hat{\pi}(x) = \frac{1}{\sqrt{M}} \sum_k e^{ikx} \hat{\pi}_k. \tag{3.18}$$

In terms of $\hat{\pi}(x)$ and $\hat{h}(x)$, the Hamiltonian operator $\hat{H}_{\text{HIS}}$ is rewritten by

$$\hat{H}_{\text{HIS}} = \int dx \left[\frac{\rho d^2}{2\beta c_y} \hat{\pi}^2(x) + \frac{\beta c_x}{2\rho^3} \frac{1}{d^2} \left(\frac{d\hat{h}(x)}{dx} \right)^2 \right]. \tag{3.19}$$

The Boltzmann weight for the surface height configuration $\{h(x,y)\}_{(x,y)}$ is then expressed as

$$P(\{h(x,y)\}_{(x,y)}) = \prod_{y=0}^{L-\ell} < \{h(x, y+\ell)\}_x | \hat{A}_{\text{HIS}} | \{h(x,y)\}_x >$$

$$= \prod_{y=0}^{L-\ell} < \{h(x, y+\ell)\}_x | e^{-\ell \hat{H}_{\text{HIS}}} | \{h(x,y)\}_x >$$

$$\propto \exp\left\{ -\beta \frac{1}{2} \int dx dy \left[\frac{c_x}{\rho^3 d^2} \left(\frac{\partial h(x,y)}{\partial x} \right)^2 + \frac{c_y}{\rho d^2} \left(\frac{\partial h(x,y)}{\partial y} \right)^2 \right] \right\}. \tag{3.20}$$

Hence, choosing the coefficients c_x and c_y as

$$c_x = \rho^3 d^2 f_{x,x}, \quad c_y = \rho d^2 f_{y,y},$$ (3.21)

we have

$$P(\{h(x,y)\}_{(x,y)}) \propto e^{-\beta H'_{\text{CW}}}.$$ (3.22)

This shows the equivalence between the HIS model and the CW model in the length-scale larger than the average step spacing $1/\rho$.

4. DYNAMICAL PROPERTIES OF THE STEP FLUCTUATIONS

The "dynamical" properties of the vicinal surface can be expressed by a time-dependent Ginzburg-Landau-Langevin (GLL) equation based on the HIS Hamiltonian. Let us consider the step displacement as a function of the time t; $x_j = x_j(y;t)$. The GLL equation is given by

$$\frac{\partial x_j(y;t)}{\partial t} = -\Gamma \frac{\delta H_{\text{HIS}}}{\delta x_j(y)} + \eta_j(y;t),$$ (4.1)

where Γ is the kinetic coefficient. The Gaussian white noses $\eta_j(y;t)$ satisfy

$$< \eta_j(y;t) >_{\text{R}} = 0,$$
$$< \eta_j(y;t)\eta_{j'}(y';t') >_{\text{R}} = 2\Gamma k_{\text{B}} T \delta_{j,j'} \delta(y-y')\delta(t-t').$$ (4.2)

The average $< \cdots >_{\text{R}}$ denotes the average over the random process. From (1.12), we have

$$\frac{\partial x_j}{\partial t} = -\Gamma[c_x(2x_j - x_{j-1} - x_{j+1}) - c_x \frac{\partial^2 x_j}{\partial y^2}] + \eta_j.$$ (4.3)

In terms of the Fourier components $\{\xi_{Q,k,\omega}\}$ and $\{\tilde{\eta}_{Q,k,\omega}\}$, which are respectively defined by

$$\xi_{Q,k,\omega} = \frac{1}{\sqrt{n}} \sum_j \frac{1}{\sqrt{L}} \int dy \frac{1}{\sqrt{2\pi}} \int dt e^{-iQj} e^{-iky} e^{i\omega t} x_j(y;t),$$

$$\tilde{\eta}_{Q,k,\omega} = \frac{1}{\sqrt{n}} \sum_j \frac{1}{\sqrt{L}} \int dy \frac{1}{\sqrt{2\pi}} \int dt e^{-iQj} e^{-iky} e^{i\omega t} \eta_j(y;t),$$ (4.5)

the solution of the eq.(4.3) is given by

$$\xi_{Q,k,\omega} = \frac{\tilde{\eta}_{Q,k,\omega}}{i\omega + \Gamma[c_x(2 - 2\cos Q) + c_y k^2]}.$$ (4.6)

Using the solution, we can calculate various dynamical properties of the step systems.

Let us consider the mean-square displacement of the step:

$$D^2(t) = \lim_{t_0 \to \infty} < (x_j(y;t+t_0) - x_j(y;t_0))^2 >_{\text{R}}$$

$$= \lim_{t_0 \to \infty} \frac{1}{nL} \sum_j \int dy < (x_j(y;t+t_0) - x_j(y;t_0))^2 >_{\text{R}}.$$ (4.7)

From the solution (4.6), we have

$$D^2(t) = \frac{1}{nL} \frac{1}{2\pi} \sum_Q \sum_k \int d\omega \frac{2\Gamma k_B T(2 - 2\cos\omega t)}{|i\omega + \Gamma[c_x(2 - 2\cos Q) + c_y k^2]|^2}, \tag{4.9}$$

where we have used the relation,

$$< \tilde{\eta}_{Q,k,\omega} \tilde{\eta}_{Q',k',\omega'} >_R = 2\Gamma k_B T \delta_{Q,-Q'} \delta_{k,-k'} \delta(\omega + \omega'), \tag{4.10}$$

which is derived by rewriting (4.2) in terms of the Fourier components $\{\tilde{\eta}_{Q,k,\omega}\}$.

In the limit $t \to \infty$, $D^2(t)$ behaves as

$$D^2(t) \to \frac{1}{\beta\pi} \frac{1}{\sqrt{c_x c_y}} \ln t = \frac{1}{\rho^2} \frac{1}{\beta\pi d^2} \frac{1}{\sqrt{f_{x,x} f_{y,y}}} \ln t. \tag{4.11}$$

An argument for the universal Gaussian curvature jump at the facet edge[2] shows that $1/(\beta\pi d^2 \sqrt{f_{x,x} f_{y,y}})$ takes the universal value $1/\pi^2$ in the low step-density limit $\rho \to 0$. Therefore, in this limit, we have

$$D^2(t) \to \frac{1}{\rho^2} \frac{1}{\pi^2} \ln t. \tag{4.12}$$

The above asymptotic form does not contain system-specific quantities such as the temperature and the surface stiffness. Hence, the asymptotic behavior is universal.

We should remark here that a similar calculation based on the HIS model gives the *static* (or equal time) behavior[9–11,22]

$$W^2(y) \equiv < [x_j(y,t) - x_j(0,t)]^2 >$$
$$\sim \frac{1}{\rho^2} \frac{1}{\pi^2} \ln y, \tag{4.13}$$

which has originally been derived by different methods[9,22]. Thus the HIS model allows us to discuss the properties, both static and and dynamical, of the vicinal surface in a unified and consistent manner.

5. SUMMARY

We have established the equivalence between the TSK approach and the CW approach to the vicinal surface in the long wavelength scale.

In the length scale much larger than the atomic length scale, the TSK model expressing the vicinal surface is regarded as a non-crossing step system. In the transfer matrix language, the non-crossing step system is described by a one-dimensional free-fermion system. Long wavelength (low-energy) excitations, which correspond to long wavelength fluctuations in the original TSK model, is expressed by a one-dimensional harmonic bose-fluid. From the bose-fluid representation, the correct CW picture is derived.

We have derived the TSK picture from the CW model expressing the vicinal surface. The newly derived TSK model, which is called the harmonically interacting step(HIS) model, validly describes the vicinal surface fluctuations with the length scale much larger than the average step spacing. Since the Hamiltonian of the HIS model is quite simple, we can easily calculate various quantities and can discuss the dynamical properties on the basis of the Ginzuburg-Landau-Langevin equation. We have paid attention to the mean-square step displacement. In the low step-density limit, it shows the universal long-time behavior.

REFERENCES

1. E.E Gruber and W.W. Mullins, J. Phys. Chem. Solids **28**, 875 (1967).
2. Y. Akutsu, N. Akutsu and T. Yamamoto, Phys. Rev. Lett. **61**, 424 (1988).
   T. Yamamoto, Y. Akutsu, and N. Akutsu, J. Phys. Soc. Jpn. **57**, 453 (1988).
3. T. Yamamoto, N. Akutsu and Y. Akutsu, J. Phys. Soc. Jpn. **59**, 3831 (1990).
4. F.P. Buff, R.A. Lovett and F.H. Stillinger, Phys. Rev. Lett. **15**, 621 (1989).
5. J.D. Weeks, J. Chem. Phys. **67**, 3106 (1977).
6. N. Akutsu and Y. Akustu, J. Phys. Soc. Jpn. **56**, 1443 (1987).
7. W.F. Saam, Phys. Rev. Lett. **62**, 2636 (1989).
8. Y. Akutsu, N. Akutsu and T. Yamamoto, Phys. Rev. Lett. **62**, 2637 (1989).
9. T. Yamamoto, Y. Akutsu and N. Akutsu, J. Phys. Soc. Jpn. **63**, 915 (1994).
10. Y. Akustu, N. Akutsu and T. Yamamoto, J. Phys. Soc. Jpn. **63**, 2032 (1994).
11. K. Sudoh, T. Yoshinobu, H. Iwasaki, N. Akutsu, Y. Akutsu and T. Yamamoto, J. Phys. Soc. Jpn. **65**, 988 (1996).
12. T. Yamamoto, J. Phys. Soc. Jpn. **64**, 1945 (1995).
13. T. Yamamoto, submitted to J. Phys. Soc. Jpn.
14. J. Solyom, Adv. Phys. **28**, 201 (1979).
15. F.D.M. Haldane, J. Phys. C **14**, 2585 (1981).
16. S. Tomonaga, Prog. Theor. Phys. **5**, 544 (1950).
17. J.M. Luttinger, J. Math. Phys. **4**, 1154 (1964).
18. D.C. Mattis and E.H. Lieb, J. Math. Phys. **6**, 304 (1965).
19. D.J. Scalapino, M. Sears and R.A. Ferrell, Phys. Rev. **B6**, 3409 (1972).
20. B. Kogut, Rev. Mod. Phys. **51**, 659 (1979).
21. S. Alexander and P. Pincus, Phys. Rev. **B18**, 2011 (1978).
22. N.C. Bartelt, T.L. Einstein and E.D. Williams, Surf. Sci. **276**, 308 (1992).

Advances in the Understanding of Crystal Growth Mechanisms
T. Nishinaga, K. Nishioka, J. Harada, A. Sasaki and H. Takei (Editors)

Fluctuation and morphological instability of steps in a surface diffusion field

M. Uwaha

Department of Physics, Nagoya University,
Furo-cho, Chikusa-ku, Nagoya 464-01, Japan

We study, theoretically and by Monte Carlo simulation, nonequilibrium behavior of steps on the crystal surface growing from vapor. Asymmetry in step kinetics (Schwoebel effect) is very important in determining the morphlogy and the stability of steps. Fluctuation of a step is reduced when the step is receding in sublimation, and enhanced when advancing in growth. There is critical supersaturation above which the step becomes morphologically unstable and shows spatiotemporal chaos. In a vicinal face, in spite of the increase in step fluctuation, the terrace width are more stable in growth. In sublimation, however, steps become unstable and form pairs. Evolution of a step train is discussed with repulsive step interaction taken into account.

1. INTRODUCTION

Invention of the scanning tunneling microscope and the atomic force microscope as well as the development of electron microscopy enables us to observe the motion of atomic steps in real time and at the atomic scale. Helped by the technological demand, the study of silicon surface under vapor growth condition such as the molecular beam epitaxy has made a remarkable advancement. Now we can quantitatively compare atomic theory of crystal surface to real experiment. Also we can discuss physics at the mesoscopic level by measuring statistical fluctuation of steps and kinks [1]. As for equilibrium properties, there are already many experiments which can be compared to the theory of statistical mechanics. Distribution of kinks on a step was measured and the energy of a kink was determined [2]. From the measurement of the width of step fluctuation the step stiffness and the strength of the elastic interaction between steps was estimated [3,4]. Such equilibrium quantities are independent of crystal growth mechanism and determined solely by equilibrium physical quantities. Under nonequilibrium conditions such as in crystal growth, on the contrary, the morphology and fluctuation of steps are determined not only by equilibrium quantities but also by growth mechanisms: dynamics of crystal growth plays an essential role.

In the crystal growth from the vapor, atoms adsorbed on the crystal surface migrate to steps and crystalize at kink sites. Comprehensive theory of vapor growth started with the famous work of Burton, Cabrera and Frank (BCF [5]), which

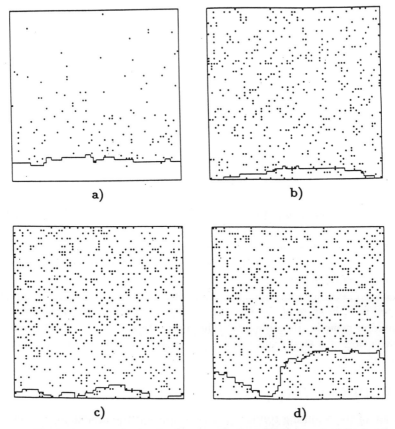

Figure 1. Configuration of adsorbed atoms and an SOS step in Monte Carlo simulation. (a) Sublimation in vacuum with the impingement rate $f = 0$. (b) At equilibrium with $f = f_{eq} = 4.65 \times 10^{-4}$. (c) Growth in supersaturated vapor with $f = 6.5 \times 10^{-4}$. (d) Unstable growth with $f = 7.4 \times 10^{-4}$. (From Ref.9)

gives the standard framework in this field. Recent developement, however, makes more and more clear that several factors, which have been ignored in the BCF theory, such as asymmetry in the step kinetics (so called Schwoebel effect [6,7]) and mechanical interaction between steps [8] bring about many interesting physical phenomena.

In the present article I describe some of the recent development in the study of fluctuation and morphology of steps, in which the asymmetry in step kinetics and the step interaction are crucially important.

2. FLUCTUATION OF A MOVING STEP

Figure 1 shows snapshots of the Monte Carlo simulation of a step growing in a diffusion field [9]. Adsorbed atoms and a step on the surface are viewed from above the surface. In this simulation the upper part is one step lower than the lower part and the boundary condition of the system is periodic except the height change in the vertical direction. Atoms impinge onto the surface with a frequency f per unit area per uint time. They migrate on the surface with the diffusion coefficient D_s and evaporate with a lifetime τ. When atoms are in contact with the step on the lower terrace they try solidification. Here atoms can solidify only from the lower terrace, and migration over the step is forbidden. This condition is called one-sided and represents the extreme case of the asymmetric kinetics. In real systems atoms solidify from both terraces (or migration over the step is allowed) but in general with different rates. This asymmetry in step kinetics is called the Schwoebel effect or the diffusion bias and causes morphological diversities of stepped surfaces in the crystal growth [1]. As one can see in Figure 1 the amplitude of step fluctuation and the morphology change with the impingement rate f. To understand the behavior of a step we develope a continuum theory which is based on a BCF type model with an asymmetry in step kinetics. The adsorbed atoms (we call adatoms) are described by its density $c(\vec{r}, t)$, which obeys the diffusion equation with evaporation and impingement of atoms:

$$\frac{\partial c}{\partial t} = D_s \nabla^2 c - \frac{c}{\tau} + f. \tag{1}$$

The step is a boundary for the diffusion equation. At the step, atoms solidify with the rate proportional to the difference between the adatom density at the step and that at equilibrium [10]

$$\Omega^{-1} v = j_- + j_+ = K_- \left(c(y_{s-}) - c_{eq} \right) + K_+ \left(c(y_{s+}) - c_{eq} \right), \tag{2}$$

where v is the normal step velocity and Ω the atomic area. The suffixes "+" and "−" indicate the lower and the upper terraces locating at $y > y_s(x)$ and $y < y_s(x)$ ($y_s(x)$ is the step position). Here we have assumed, as in Figure 1, the step position y_s is a single-valued function of x. This step is called the solid-on-solid(SOS) step. The equilibrium density c_{eq} for a curved step with the curvature κ is different from that for a straight step c_{eq}^0,

$$c_{eq} = c_{eq}^0 \left[1 + \frac{\Omega}{k_B T} \left(\tilde{\beta} \kappa + F_s \right) \right], \tag{3}$$

where $\tilde{\beta}$ is the step stiffness which is derived from the anisotropic step free energy $\beta(\theta)$ as $\tilde{\beta} = \beta(\theta) + \beta''(\theta)$ ($\theta = \tan^{-1}(dy_s/dx)$). F_s is the force acting on the step from other steps if they are present. In general K_+ is different from and usually larger than K_-. This asymmetry in step kinetics is crucial for the mophology of a growing step or an array of growing steps. Since the advancement of the step (2) occurs with the diffusion flux of adatoms, the solidification flux is also witten as

$$j_\pm = \mp D_s (\hat{n} \cdot \nabla) c(\vec{r})|_{step}. \tag{4}$$

The step is at rest if the impingement rate is that of the saturated vapor, $f = f_{eq}$, and the adatom density is $c_{eq}^0 = f_{eq}\tau$. For a larger impingement rate a straight step advances with a constant velocy v_0. Relative speed of the step kinetics and the diffusion is characterized by the ratio of their resistances, K^{-1} and x_s/D_s, $\lambda \equiv D_s/x_s K$. Large (small) λ implies the kinetics is slow (fast) compared to the diffusion process.

If the step is perturbed we can calculate the growth rate ω of the purterbation,

$$y(x,t) = v_0 t + \delta y_k e^{\omega_k t} \cos kx, \tag{5}$$

by solving the diffusion equation (1) with the boundary conditions (2) and (4). The result can be expressed as a product of a step mobility μ_k and an effective restoring force $\nu_k(f)$, which is a decreasing function of the impingement rate,

$$\omega_k(f) = -\mu_k \nu_k(f). \tag{6}$$

At the long wavelength the restoring force is expressed in terms of the effective stiffness [11]

$$\tilde{\beta}_{eff}(f) = \tilde{\beta}\left(1 - \frac{x_s}{2\xi(f)}\right) \tag{7}$$

as $\nu_k(f) = \tilde{\beta}_{eff}(f)k^2$, where $x_s (= \sqrt{D_s \tau})$ is the surface diffusion length and ξ is a length determined by the density gap Δc_i at the step [12]

$$\xi(f) \equiv \frac{\Omega c_{eq}^0 \tilde{\beta}}{k_B T \Delta c_i(f)}. \tag{8}$$

The length ξ is inversely proportional to the supersaturation $f - f_{eq}$. In the diffusion-limited one-sided model ($K_- = 0$ and $K_+ = \infty$, fast kinetics with the lower terrace) the density gap is $c_{eq}^0/\Delta c_i = f_{eq}/(f - f_{eq})$ and ξ corresponds to the radius of a critical nucleus. ξ is negative in undersaturation and zero at equilibrium. With supersaturated vapor ($f > f_{eq}$) ξ is positive, and the effective stiffness (7) is smaller than that at equilibrium. Therefore the restoring force to make the step straight becomes weak in growth. As the impingement rate is increased up to a critical value

$$f_c = f_{eq}\left(1 + \frac{2\tilde{\beta}\Omega}{k_B T x_s}\right) \tag{9}$$

(for the one-sided model) when ξ becomes $x_s/2$, the step loses its stiffness and becomes unstable. Recently, Latyshev et al. [13] examined the relaxation process of a step deformed by an edge dislocation on a sublimating Si crystal at 1230-1380K. They analyzed the relaxation rate in terms of an effective stiffness and found that the obtained stiffness is much larger than the equilibrium one at lower temperatures. This result as well as the electron microscope observation of smooth steps in sublimation [14] is consistent to the above theory with a negative $\xi(f)$ (in these experiments $f \approx 0$). It is rather difficult to perform a similar experiment in a growing crystal because observation by an electron microscope under nonvanishing vapor is rather difficult.

The change of the effective stiffness results in a change in the step fluctuation. Let us consider a single isolated step. The step is straight (the step position is $y(x) = const.$) if fluctuation is neglected. The fluctuation of the step position $\delta y(x, t)$ may be described by a Langevin equation. The simplest form for its Fourier component is

$$\delta \dot{y}_k(t) = \mu_k \left(-\nu_k(f)\delta y_k(t) + R_k(t) \right), \tag{10}$$

where the random fluctuating force $R_k(t)$ drives the step in addition to the relaxation term $\omega_k \delta y_k = -\mu_k \nu_k \delta y_k$. Assuming that the random force is independent of f and produces the equilibrium thernal fluctuation, which is given by the equi-partition law $k^2 \tilde{\beta}|\delta y_k|^2/2 = \nu_k(f_{eq})|\delta y_k|^2/2 = k_B T/2$, the amplitude of the fluctuaiton is easily calculatied from (10) as [11]

$$\langle |\delta y_k|^2 \rangle = \frac{k_B T}{\nu_k(f)}. \tag{11}$$

The fluctuation of a step in real space is given by

$$\langle (y(x+x', t) - y(x'))^2 \rangle = \frac{1}{L} \sum_k 2(1 - \cos kx)\langle |\delta y_k|^2 \rangle$$

$$\approx \frac{k_B T}{\tilde{\beta}_{eff}} x \tag{12}$$

for $x \gg x_s$. For a finite size system, for instance a step running from $x = 0$ to $x = L$, the step width

$$w^2 \equiv \langle (\delta y(x))^2 \rangle \approx \frac{k_B T}{6\tilde{\beta}_{eff}} L \tag{13}$$

is used to characterize the fluctuation. Both (12) and (13) are inversely proportional to the stiffness so that they are reduced in sublimation and enhanced in growth. Near the instability the fluctuation increases very rapidly and diverges at the critical point $f = f_c$ according to the linear theory. In reality, however, nonlinear effects become important and short wavelength fluctuations renormalize the long wavelength stiffness to keep it finite even at $f \geq f_c$ [15].

The snapshots of the Monte Carlo simulation show this tendency. Figure 1(a) shows sublimation in vacuum ($f = 0$). Atoms emitted from the step migrate on the surface in the distance of the order of x_s and evaporate. Figure 1(b) is at equilibrium ($f = f_{eq} = 4.65 \times 10^{-4}$) and Figure 1(c) is in stable growth. Systematic data is collected in Figure 2 for two different orientations of a square lattice [12]. (The single-value condition for the step position is removed in the Monte Carlo simulation.) The step stiffness $\tilde{\beta}$ can be calculated from the kink energy of the simulation model to be $k_B T/\tilde{\beta} = 0.367$ for [10] orientation and $k_B T/\tilde{\beta} = 0.823$ for [11] orientation. The step free nergy $\beta(\theta)$ is larger in [11] orientation but the stiffness is smaller because of the contribution of the $\tilde{\beta}''(\theta)$. As a result the critical impingement rate f_c for the instability is smaller in [11] orientation. The theoretical

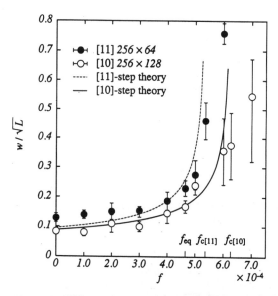

Figure 2. The step width of a step in the [10] and [11] orientations versus the impingement rate f. Simulation results with the indicated system sizes $L \times H$ and theretical results for the infinite system are shown.

curves are for infinite systems and the agreement is not perfect but the dramatic increase of the width is clearly seen in the simulation. The reduction of the width in sublimation ($f < f_{eq}$) is not so large since the simulation is possible only for small systems and the diffusion length is set to a rather small value ($x_s = 16$ for the present simulation). In real systems the diffusion length can be extremely large (on the order of 1000Å for Si(111) surface at 1000K [16]) and the reduction of the step fluctuation is also dramatic [14]. Near and above the instability points the fluctuation is very large and the present linear theory is not satisfactory.

3. MORPHOLOGY OF AN UNSTABLE STEP

Above the critical point ($f > f_c$) the growth rate ω of the fluctuation is positive for wave numbers smaller than a critical one, $0 \leq k \leq k_c$, where

$$k_c \approx \frac{1}{x_s} \sqrt{\frac{4(f - f_c)}{3(f_c - f_{eq})}} \tag{14}$$

and the fluctuation grows exponentially. What kind of shape will the step take beyond the instability? Will it form a periodic structure with the wave number comparable to the most unstable mode (ω_k is maximum at $k_{max} = k_c/\sqrt{2}$) ? After the instability nonlinear effects are impotant and analytic solution of the problem is generally impossible.

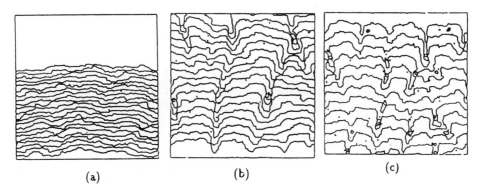

Figure 3. Stroboscopic shots of a growing step. (a) $f = 6 \times 10^{-4} \approx f_c$, (b) $f = 8 \times 10^{-4}$, (c) $f = 10 \times 10^{-4}$.

Figure 3 shows time evolution of an unstable step in the Monte Carlo simulation. The system size is $L \times H = 256 \times 256$. Parameters of the model correspond to $x_s = 16$ ($\tau = 256$), $f_{eq} = 4.65 \times 10^{-4}$ and $f_c = 6.26 \times 10^{-4}$. (a) At $f = f_c$ only the $k = 0$ mode is unstable and the instability does not occur in the finite sytem although the fluctuation is much enhanced. (b) Above the critical point with $f = 8 \times 10^{-4}$, several troughs are seen, and their distances are comparable to the wavelength of the most unstable mode calculated as $\lambda = 118$. (c) At $f = 10 \times 10^{-4}$, the distances between the troughs become smaller and comparable to the the wavelength of the most unstable mode $\lambda = 81$. Characteristic feature of these patterns in (b) and (c) is that the structure is by no means steady: the troughs move left or right and collide to disappear, new troughs appear on large crests, and all happen in a random fashion. The instability may be called a wandering instability. In the [10] direction the step stiffness is large and the pattern has rather flat crests. In the [11] direction the step stiffness is the smallest and the pattern is more pointed than that of Figure 3 [12].

This chaotic behavior is probably not due to the randomness of Monte Carlo algorithm but due to the nonlinearity inherent to the system. Near above the instability, where k_c is small, unstable modes are in the long wavelength and time evolution of the system can be described by a simple continuum nonlinear equation. This evolution equation can be derived from the equation of motion for the continuum step by expanding the deviation from the straight step $\delta y(x, t)$ in terms of the small parameter $\epsilon = (f - f_c)/f_c$. Without the random noise $R_k(t)$, the derived equation [17] in new variables $X \sim \epsilon^{1/2} x$, $Y \sim \epsilon^{-1} \delta y$, $T \sim \epsilon^2 t$ is the Kuramoto-Sivashinsky equation

$$\frac{\partial Y}{\partial T} = -2 \frac{\partial^2 Y}{\partial X^2} - \frac{3}{4} \frac{\partial^4 Y}{\partial X^4} + \left(\frac{\partial Y}{\partial X} \right)^2 . \tag{15}$$

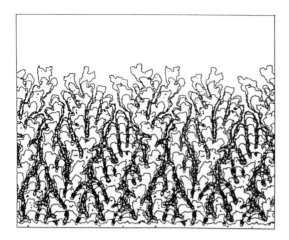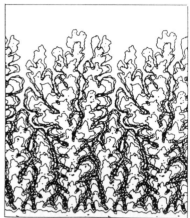

Figure 4. Evolution of unstable [10] (left) and [11] (right) steps with parameters $x_s = 64$, $f = 8 \times 10^{-5}$ ($f_{eq} = 1.16 \times 10^{-5}$). The width of the box shown here is twice the width of simulation system (the pattern is repeated).

The negative signs of the first two terms come from the fact that the effective stiffness is negative and that short wavelength modes are stable: in the linear dispersion $\omega_k \sim k^2 - k^4$. The origin of the third term is the following [18]. A tilted part of the step grows faster than a horizontal part because the velocity parallel to y-direction is

$$v\sqrt{1 + \left(\frac{dy}{dx}\right)^2} \approx v + \frac{v}{2}\left(\frac{dy}{dx}\right)^2, \tag{16}$$

where v is the normal velocity and dy/dx is the slope. The difference $v(dy/dx)^2/2$ is the origin of the nonlinear term, which plays extremly important roles in determining morphology and scaling properties of surfaces [18,1].

The characteristic length of the pattern after the instability is comparable to the linearly most unstable mode. When the impingement rate is increased, the length becomes much smaller than the diffusion length x_s and the pattern looks like a random dendrite or a seaweed (Figure 4). In these patterns arms tend to grow in $\langle 11 \rangle$ directions rather than in $\langle 10 \rangle$ directions, although destabilization by the large shot noise is stronger than stabilization by the crystal anisotropy. Similar patterns are observed in thin film growth experiment of Au on a Ru(0001) substrate [19] and of calcium stearate on an amorphous carbon substrate [20]. Anisotropy effect observed in a tungsten film on a potassium chloride substrate [21] is similar to that we observe in the simulation. Similar patterns are are also observed in the simulation of a two-dimensional crystal at high supersaturation [22] and of an irreversible growth model [23]. Relation to the fractal aggregation [24] is discussed in the article by Saito [25].

4. FLUCTUATION IN A VICINAL FACE

On the surface of real crystals we seldom see isolated steps. Nearby steps are usually within a distance smaller than the surface diffusion length, which can be as large as [16] $1\mu m$. On a vicinal surface nearly straight steps are aligned parallelly with a distance l. If we neglect correlation of fluctuation of these steps the mobility of a step is given by

$$\mu_k \approx \Omega^2 \frac{c_{eq}}{k_B T} \frac{D_s}{x_s} \tanh \frac{l}{x_s}, \tag{17}$$

which is simply $\tanh(l/x_s)$ times the mobility of a free step. The reduction or enhancement term of the stiffness is also multiplied by the same factor [26]:

$$\tilde{\beta}_{eff} = \tilde{\beta} \left(1 - \frac{x_s}{2\xi(f)} \tanh \frac{l}{x_s} \right). \tag{18}$$

Thus apart from the fact that the diffusion field is cut by the step distance l instead of the diffusion length x_s, we expect a similar effect on the fluctuation as that we have found for an isolated step. If we take correlations between neighboring steps, the asymptotic behavior of the step width is not the form of (13) but $w \sim \ln L$ in growing [27] as well as in equilibrium [28,29]. Therefore the above effect, which has been confirmed in the simulation [26,12], is significant in an intermediate distance before step collisions are dominant.

An increase of the impingement rate strengthen the effect of diffusion and thereby the correlation between neighboring steps. This can be seen by looking at the change of the step distance (terrace width) $l(x, t) = y_m(x, t) - y_{m-1}(x, t)$. Its change occurs by the antiphase oscillation of steps

$$y_m(x, t) = ml + v_0 t + (-1)^m \delta y_k e^{\omega_k^{(-)} t} \cos kx. \tag{19}$$

If the step distance is much smaller than the diffusion length, $l \ll x_s$, the growth rate of the mode $\omega_k^{(-)}$ is determined by the effective force

$$\nu_k^{(-)} = \tilde{\beta} \left(k^2 + \frac{2}{\xi l} \right). \tag{20}$$

The term $2\tilde{\beta}/\xi l = (2k_B T/\Omega l)(f - f_{eq})/f_{eq}$ is regarded as the contribution of the diffusion field to the effective inter-step potential. For a growing crystal, $\xi > 0$ and the fluctuation of the terrace width is suppressed, but for a sublimating crystal, $\xi < 0$ and the long wavelength flucuations grow: the vicinal face is unstable to the antiphase fluctuation. Note that $k = 0$ mode is most unstable. This instability is also observed in the Monte Carlo simulation [26,12]. In real crystals there is always a repulsive step interaction such as the elastic interaction due to strain around steps [8]. Otherwise the vicinal face would not be stable even in equilibrium. As a result the instability does not occur immediately, but there appears a critical undersaturation. This type of instability is discussed in the next section.

The effect of surface diffusion on the fluctuation of vicinal face may be summerized as follows. In the growth of a crystal the surface becomes "softer" (weak restoring force) in the parallel direction to the steps and "stiffer" in the perpendicular direction. In sublimation the tendency is the opposite. Softening of the stiffness developes instabilities beyond the critical supersaturation or undersaturation. Here we have assumed $K_+ > K_-$, that is atoms are more easily incorporated to the solid from the lower terrace. If the opposite is the case every effect is simply reversed.

5. BUNCHING OF STEPS IN SUBLIMATION

The instability of the train of steps for the antiphase fluctuation is related to the bunching of steps observed in the vicinal face of various kinds of crystals [30,31]. In this section we consider the instability due to the Schwoebel effect, which is a simplest case though may not be the most realistic example. The instabiliity in the original BCF model has been studied [32] long time ago. The model neither contains the asymmetric kinetics nor the step interaction. The model shows interesting behavior as a result of the symmetry, which is unfortunately unrealistic. Also the step interaction is crucial for the developement of the instability since noninteracting steps cannot avoid unrealistic collision of steps. A long-range elastic interaction between steps has been experimentally detected [3,4,33], and its role is very important in vicinal faces. We study a train of straight interacting steps with the asymmetry in kinetics taken into account.

To understand the role of the asymmetric kinetics, we first assume that the step exchange atoms to the lower terrace and its kinetics is infinitely fast ($K_+ \to \infty$, $K_- = 0$ or $\lambda_{\pm} = D_s/x_s K_{\pm}$) [34]. We suppose the step interaction is elastic and its potential can be written as [8] $A(y_i - y_j)^{-2}$ for steps locating at y_i and y_j. If we choose proper units of time and distance ($\tilde{y} = y/x_s$, $\tilde{t} = t/t_s$, with $t_s = \tau x_s^3 k_B T/(2A\Omega^2 c_{eq}^0)$) the change of the position of ith step ($y_{i+1} > y_i$) is determined by

$$\frac{d\tilde{y}_i}{d\tilde{t}} = \left(\tilde{f} + \sum_j \frac{1}{(\tilde{y}_i - \tilde{y}_j)^3} \right) \tanh{(\tilde{y}_{i+1} - \tilde{y}_i)}, \tag{21}$$

where the dimensionless driving force is defined as $\tilde{f} = \Omega(c_\infty - c_{eq}^0)t_s/\tau$. The first factor is the driving force consisting of supersaturation term and the elastic force term. The other factor reflects the the width of the lower terrace and is roughly proportional to the step distance if $y_{i+1} - y_i \ll x_s$.

If there are only two steps, from (21), their distance l changes as

$$\dot{\tilde{l}} = \left(\tilde{f} + \frac{1}{\tilde{l}^3} \right) - \left(\tilde{f} - \frac{1}{\tilde{l}^3} \right) \tanh{\tilde{l}}. \tag{22}$$

In growth $\tilde{f} > 0$ and the distance always increases, but in sublimation with large negative $\tilde{f}$ the first term wins and the distance has a stable minimum. Thus in growth the diffusion interaction and the elastic interaction are both repulsive and the two steps always replel each other. In sublimation the interference of the two diffusion field effectively produces an attractive force which may win the elastic

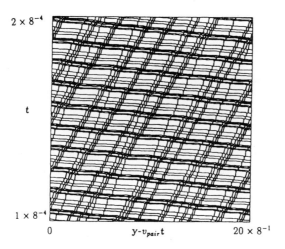

Figure 5. Time evolution of 64 steps in the one-sided model in the moving frame of an isolated pair. $\delta \tilde{f} = -8^4 \times 10^4$.

repulsion. In a vicinal face when the undersaturation is large enough ($|\tilde{f}|$ is large), neighboring steps form a stable bound pair [35]. With increasing undersaturation a hierarchy of bound state of steps may be formed as shown in Figure 5 [36]. The position is in the moving frame of an isolated pair so that steps moving to the left are free steps and those moving slowly to the right are pairs (each discernible line represents a pair of steps). Bound states consisting of two or three pairs are formed.

The one-sided model is, of course, unrealistic. With the general asymmetry of step kinetics the pairing instability turns into a long wavelength instability of step density [37]. The elastic repulsion gives the surface stiffness [38] parallel to the steps $\tilde{\alpha}_{\parallel} \sim A(a/l)$ whereas the step stiffnes gives the surface stiffness perpendicular to the steps $\tilde{\alpha}_{\perp} \sim \tilde{\beta}(l/a)$. The effect of the diffusion field for long wavelength modulation gives a negative contribution in $\tilde{\alpha}_{\parallel}$ similar to the second term of (20). Similar to the wandering instability discussed in **3**, the most unstable wave number k_{max} increases as $|\tilde{f}|$ is increased above the critical value. At first the instability looks similar to the wandering instability in the sense that the linearly most unstable mode grows. But as shown in the numerical simulation (Figure 6) it finally produces a regular array of bunches instead of chaotic motion [39]. This is consistent with the usual observation of periodic structure. By a nonlinear analysis similar to that of **3** it is found [40] that the difference arises from the symmetry of the system. Since the steps are moving to the left the dispersion of the linear mode has a term proportional to k^3 (a term proportional to k can be eliminated by a Galilean transformation) whereas the dispersion for a wandering step does not have odd power terms. Resulting nonlinear equaion has a third derivative term and the form of the Benney equation [41–43]

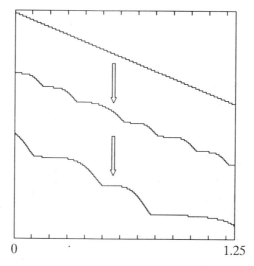

0   ·   1.25

Figure 6. Evolution of the profile of a vicinal face in the bunching instability. The step height is exaggerated.

$$\frac{\partial N}{\partial T} + \frac{\partial^2 N}{\partial Y^2} + \delta\frac{\partial^3 N}{\partial Y^3} + \frac{\partial^4 N}{\partial Y^4} + N\frac{\partial N}{\partial Y} = 0, \tag{23}$$

where N is the change of step density from the initial uniform value and all variables are properly scaled. If the coefficient δ of the third derivative term is large, as is the case near the critical point, the chaotic behavior is suppressed and one may obtain a periodic structure. Therefore origin of the difference between the chaotic behavior of a wandering step and the regular bunching is the difference of the symmetry of the systems.

6. INSTABILITIES INDUCED BY DRIFT

We have studied two kinds of instabilities due to surface diffusion field: wandering of a step and bunching of steps. In both cases the asymmetry in step kinetics produces, in one way or another, difference in the diffusion current of adatoms in front of and in the back of the step. This difference always destabilizes a step or a step train and causes instabilities if it wins the stabilizing force due to the step stiffness or the step repulsion. There is another origin of breaking the symmetry: an external force such as an electric field. It is well known that electric currecnt, which is used for heating the specimen, causes bunching of steps in sublimaing Si(111) surface [16,44]. Theoretical explanation has been attempted [45–48] and is explained by Natori [49]. Here we only summerize the effect of nonlinearity [50,51].

Due to the heating electric current, a constant force F acts on adatoms and produces their drift. The force is very likely to be opposite to the current [45,48] and the recent estimate [52] indicates the effective charge of an adatom is $e_{eff} \approx -0.1e$

at $T = 1400K$, and therefore $F \approx -0.1eE$, where E is the applied electric field. By neglecting the degree of freedom parallel to the steps the change of the adatom density is then given by

$$\frac{\partial c(y, t)}{\partial t} = D_s \frac{\partial^2 c(y, t)}{\partial y^2} - \frac{D_s}{k_B T} F \frac{\partial c(y, t)}{\partial y} - \frac{c(y, t)}{\tau}. \tag{24}$$

the second term on the right is the divergence of the drift current with the mobility of adaotms $D_s/k_B T$. This term breaks the symmetry of the forward and the backward direction even if the asymmetry in step kinetics is absent. Instability of step density occurs if the force F is stronger than a critical value as in the bunching studied in **5**. Nonlinear analysis [50,51] also shows that the instability near the critical point is described by the Benney equation. The third derivative this time is not necessarily large but can change the sign depending on physical parameters [50] although most likely to be the same sign as that of **3** [51]. Recent study [53] shows that the electric current may also causes a wandering instability of an isolated step and the chaotic pattern will appear.

ACKNOWLEGDEMENTS

This article is based on the cooperative work with Y. Saito and M. Sato, to whom I am very much thankful. I am aided by grants from the Ministry of Education, Science and Culture of Japan, and benefited from the interuniversity cooperative research program of IMR, Tohoku University.

REFERENCES

1. For recent reviews: T. Halpin-Healy and Y.-C. Zhang, Phys. Rep. **254** (1995) 215; A.-L. Barabási and H. E. Stanley, Fractal Concepts in Surface Growth, Cambridge, New York, 1995.
2. B. S. Schwartzentruber, Y.-W. Mo, R. Kariotis, M. G. Lagally and M. B. Webb, Phys. Rev. Lett. **65** (1990) 1913.
3. X. S. Wang, J. L. Goldberg, N. C. Bartelt, T. L. Einstein and E. D. Williams, Phys. Rev. Lett. **65** (1990) 2430.
4. C. Alfonso, J. M. Bermond, J. C. Heyraud and J. J. Métois, Surf. Sci. **262** (1992) 371.
5. W. K. Burton N. Cabrera and F. C. Frank, Philos. Trans. R. Soc. London, Ser. A **243** (1951) 299.
6. R. L. Schwoebel and E. J. Shipsey, J. Appl. Phys. **37** (1966) 3682.
7. G. Ehrlich and F. G. Hudda, J. Chem. Phys. 44 (1966) 1039.
8. V. I. Marchenko and A. Ya. Parshin, Zh. Eksp. Teor. Fiz. **79** (1980) 257 [Sov. Phys.-JETP **52** (1980) 120].
9. M. Uwaha and Y. Saito, Surf. Sci. **283** (1993) 366.
10. G. S. Bales and A. Zangwill, Phys. Rev. B **41** (1990) 5500.
11. M. Uwaha and Y. Saito, Phys. Rev. Lett. **68** (1992) 224.
12. Y. Saito and M. Uwaha, Phys. Rev. B **49** (1994) 10677.

44

13. A. V. Latyshev, H. Minoda, Y. Tanishiro and K. Yagi, Phys. Rev. Lett. **76** (1996) 94-97.
14. C. Alfonso, J. C. Heyraud and J. J. Métois, Surf. Sci. Lett. **291** (1993) L745.
15. A. Karma and C. Misbah, Phys. Rev. Lett. **71** (1993) 3810.
16. A. V. Latyshev, A. L. Aseev, A. B. Krasilnikov and S. I. Stenin, Surf. Sci. **213** (1989) 157; Surf. Sci. **227** (1990) 24.
17. I. Bena, C. Misbah and A. Valance, Phys. Rev. B **47** (1993) 7408.
18. M. Kardar, G. Parisi and Y-C. Zhang, Phys. Rev. Lett. **56** (1986) 889.
19. R. Q. Hwang, J. Schröder, C. Günter and R. J. Behm, Phys. Rev. Lett. **67** (1991) 3279.
20. K. Yase, M. Yamanaka, K. Mimura, K. Inaoka and K. Sato, Appl. Surf. Sci. **75** (1994) 228.
21. M. Arita and I. Nishida, unpublished.
22. Y. Saito and T. Ueta, Phys. Rev. A **40** (1989) 3408.
23. Y. Saito, T. Sakiyama, and M. Uwaha, J. Cryst. Growth **128** (1993) 82.
24. M. Uwaha, Y. Saito and S. Seki, *in* Dynamical Phenonena at Interfaces, Surfaces and Membranes, D. Beysens, N. Boccara and G. Forgacs (eds.), Nova Scientific, Commack, 1993, p.177.
25. Y. Saito, article in this book.
26. M. Uwaha and Y. Saito, J. Cryst Growth **128** (1993) 87.
27. D. E. Wolf, Phys. Rev. Lett. **67** (1991) 1783.
28. N. C. Bartelt, T. L. Einstein and E. D. Williams, Surf. Sci. **276** (1992) 308.
29. T. Yamamoto, Y. Akutsu and N. Akutsu, J. Phys. Soc. Jpn. **63** (1994) 915.
30. T. Kimoto, A. Itoh and H. Matsunami, Appl. Phys. Lett. **66** (1995) 3645.
31. H. W. Ren, X. Q. Shen and T. Nishinaga, J. Cryst. Growth ????.
32. W. W. Mullins and J. P. Hirth, J. Phys. Chem. Solids, **24** (1963) 1391.
33. E. Rolley, E. Chevalier, C. Guthmann, Phys. Rev. Lett. **72** (1994) 872.
34. M. Uwaha, Phys. Rev. B **46** (1992) 4364.
35. M. Uwaha, J. Cryst. Growth **128** (1993) 92.
36. M. Sato and M. Uwaha, unpublished.
37. M. Sato and M. Uwaha, Phys. Rev. B **51** (1995) 11172.
38. P. Nozières, *in* Solids Far from Equilibrium, C. Godrèche (ed.), Cambridge University Press, Cambridge, 1992, p.1.
39. M. Uwaha, Y. Saito and M. Sato, J. Cryst. Growth **146** (1995) 146.
40. M. Sato and M. Uwaha, Europhys. Lett. **32** (1995) 639.
41. D. J. Benney, J. Math. Phys. **45** (1966) 150.
42. T. Kawahara, Phys. Rev. Lett. **51** (1983) 380.
43. T. Kawahara and M. Takaoka, Physica D **39** (1989) 43.
44. Y. Homma, R. J. Mcclelland and H. Hibino, Jpn. J. Appl. Phys. **29** (1990) L2254.
45. S. Stoyanov, Jpn. J. Appl. Phys. **30** (1991) 1.
46. A. Natori, Jpn. J. Appl. Phys. **33** (1994) 3538.
47. B. Houchmandzadeh, C. Misbah and A. Pimpinelli, J. Phys. I France **4** (1994) 1843.

48. C. Misbah, O. Pierre-Louis and A. Pimpinelli, Phys. Rev. B **51** (1995) 17283.
49. A. Natori, article in this book.
50. C. Misbah and O. Pierre-Louis, Phys. Rev. E (1996) R4318.
51. M. Sato and M. Uwaha, J. Phys. Soc. Jpn. **65** (1996) 1515.
52. D. Kandel and E. Kaxiras, Phys. Rev. Lett. **76** (1996) 1114.
53. M. Sato and M. Uwaha, J. Phys. Soc. Jpn. **65** (1996) in press.

Advances in the Understanding of Crystal Growth Mechanisms
T. Nishinaga, K. Nishioka, J. Harada, A. Sasaki and H. Takei (Editors)
47

Pattern formation of a crystal growing in a diffusion field

Y. Saito

Department of Physics, Keio University,
3-14-1 Hiyoshi, Kohoku-ku, Yokohama 223, Japan

Morphology of a crystal growing in a diffusion field is discussed by taking various effects into account.

Without the stabilizing capillary effect, the crystal is unstable to small fluctuation, and forms an irregular aggregate characterized by a fractal dimension D_f. The finite density of diffusing gas particles yields the upper limit of the length scale for the fractal self-similarity, and the relation between the growth rate and the gas density is determined. When the aggregate is relaxing to its equilibrium form during a high-temperature annealing, the lower limit for the fractal scaling varies, and the peripheral length of the aggregate relaxes in powers of time.

With an anisotropic capillarity, the tip-stable dendrite is formed when the noise is weak, whereas tip splits when the noise is strong. The phase boundary between the tip-stable and the tip-unstable dendrite is given in the phase space of the noise strength a and the anisotropy ϵ as $\ln a \sim -\epsilon^{-7/8}$.

When there are anisotropies in surface stiffness and kinetic coefficient, the growth is controlled by surface stiffness for slow growth, and by kinetics for fast growth. The scaling relations of the growth rate in both cases are confirmed by the numerical simulation.

1. INTRODUCTION

Crystal growth is controlled by [1–3]

(1) material transport by diffusion,
(2) surface kinetics of material incorporation,
(3) the removal of latent heat by heat conduction.

These processes are connected sequentially, and the slowest process controls the growth and morphology of the crystal.

At very low temperatures where the surface free energy has singularities to some orientations, the crystal is covered with flat surfaces and the surface kinetics (2) is the slowest to control the growth. On the contrary, at high temperatures near the melting point, thermal fluctuation roughens the surface [4] and the surface kinetics (2) proceeds fast. Then, for a crystal with a macroscopic size, the diffusion processes of material (1) or of heat (3) take time and govern the crystal growth.

Also steps on a crystal surface are rough by the thermal fluctuation, and the surface diffusion of adsorbed atoms (or adatoms for short) on the crystal substrate controls the advancement of steps during vapor growth.

In these diffusion-limited crystal growth, the morphological instability takes place [5], and the surface or the step can take complicated structures. I here report our theoretical and numerical studies on the morphology of the surface or steps during the diffusion-limited growth [6].

With only a destabilizing effect of diffusion, crystals take a fine and irregular structure, called the diffusion-limited aggregation (DLA) [7]. The effect of the finite density of the gas through which the DLA is growing is considered [8] as well as the flow effect on it [9,10]. Thermal healing of the aggregate during high-temperature annealing [11,12] is also studied in section 2. If the surface tension is taken into account, it suppresses the fine structure of the aggregate. According to the microscopic solvability theory of dendritic growth [13,14], the isotropic surface tension cannot suppress tip splitting, but with an anisotropy the dendrite tip is shown to be stabilized. Since every crystal is anisotropic reflecting its microscopic regularity in atomic arrangement, there should always be a regular dendrite. In reality, there are irregular structures as DLA. This morphological variety is induced by the competition between the thermal fluctuation and the surface anisotropy [15, 16]. This interplay is summarized in section 3. As for the origin of the anisotropy, we considered so far only the surface tension, but the surface kinetics can also be anisotropic. The effect of surface kinetics [6,17] is considered in section 4. The last section concludes the paper with comments on the related recent developments.

2. FRACTAL AGGREGATION GROWTH AND ITS THERMAL RE-LAXATION

A crystal grows from solution by the incorporation of solute atoms on the crystal surface. Since the material is consumed by the crystallization, the concentration close to the interface c_i decreases from that far away from the interface c_∞. The concentration difference in solution induces the material diffusion

$$\frac{\partial c}{\partial t} = D\nabla^2 c. \tag{1}$$

Here $c(r, t)$ represents the local solute concentration, and D is the diffusion constant. Since the material flows in to the crystal with a diffusion flux $-D\nabla c$, the crystal grows in the normal direction $\mathbf{n}$ with the velocity

$$v_n = \Omega D(\mathbf{n}\nabla)c \tag{2}$$

with Ω being the atomic volume. If the crystal interface is rough and the exchange of atoms at the interface is extremely fast, the interface concentration c_i takes the local equilibrium value including the Gibbs-Thomson effect as

$$c_i = c_{eq}(1 + \frac{\Omega\tilde{\gamma}}{k_B T}\kappa) \tag{3}$$

where c_{eq} is the bulk equilibrium concentration, $\tilde{\gamma}$ is the surface stiffness related to the surface tension, κ is the local curvature. These three equations (1-3) are fundamental for the diffusion controlled growth. For the melt growth, the heat conduction controls the growth, and the concentration c in Eqs.(1-3) should be replaced by the temperature T. For the vapor growth, c represents the adatom concentration on the substrate.

When the flat interface is advancing steadily with a velocity v, the concentration in the solution varies exponentially with a characteristic length

$$l_D = 2D/v, \tag{4}$$

the diffusion length. The steady growth is shown to be possible when the relative supersaturation $\Delta = \Omega(c_\infty - c_{eq})$ is unity. If a part of the interface protrudes forward by fluctuation, it pushes the equi-concentration surface in the solution, and the concentration gradient ∇c increases locally. This leads to the acceleration of the protrusion and the flat interface becomes unstable.

If the small capillary effect is neglected, the crystal interface corresponds to the equi-concentration surface, and Ivantsov showed that the parabolic crystal can grow steadily [18]. The product of the tip radius ρ of the parabolic dendrite and its steady growth rate v is determined as a function p of the relative supersaturation Δ as

$$v\rho/2D = p(\Delta). \tag{5}$$

This so-called Ivantsov relation determines only the product or the Peclet number $P = v\rho/2D$, but the respective values of ρ and v are not determined. Since the structure with a small ρ grows fast with a large v, the Ivantsov paraboloid is unstable to a finer structure. The stabilizing effect is the surface stiffness in Eq.(3).

In order for Eq.(3) to come in real effect, atoms have to try solidification and melting sufficiently often to achieve thermal equilibrium. At a very low temperature, atoms once solidified never melt again. In this irreversible crystallization, the solidification point is determined stochastically, and the microscopic noise associated to the atomic diffusion is frozen in to the aggregate shape. Since the stabilizing effect of the surface tension does not work, the destabilizing effect of the diffusion field reveals itself in the irregular morphology of the aggregate.

In the low density limit ($c \to 0$) where the growth is slow ($v \to 0$), the concentration rearranges itself fast enough such that it follows the Laplace equation; $\nabla^2 c = 0$. Since the diffusion length diverges ($l_D \to \infty$) in this case, the system looses characteristic length and the aggregate becomes self-similar: Its portion looks similar to the whole by enlargement, as is shown in Figure 2a. Thus formed aggregate is called the DLA [7], and the self-similar structure is called fractal. The number of aggregate atoms N within a radius R satisfies the fractal scaling relation

$$N \propto R^{D_f} \tag{6}$$

with a fractal dimension D_f smaller than the space dimension d [7]. In a real aggregate the fractal scaling relation (6) holds only in a limited range of length

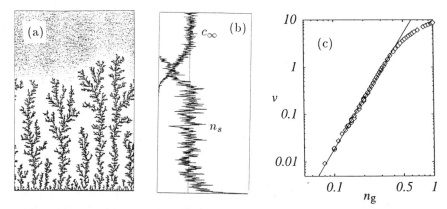

Figure 1. (a) Configuration of an aggregate grown from a linear seed. (b) The density profile. (c) Growth velocity versus gas density.

scales. We study the physical processes to determine the upper and lower cutoff lengths for the fractal scaling (6) [8–12].

According to the scaling relation (6), the density of DLA

$$n_s(R) \sim N/R^d \sim R^{-(d-D_f)} \tag{7}$$

vanishes for a large radius R, reflecting that the DLA grows from a zero-density gas. The real aggregate, however, grows from a gas of finite density c_∞. We simulated the irreversible growth of an aggregate in a finite density gas from a linear seed, as shown in Figure 1a. In this one-dimensional configuration, the steady growth of the aggregate is easily realized. The material conservation requires that the aggregate density n_s cannot be less than c_∞ [8]. Thus the fractal character of the aggregate holds up to an upper cutoff length scale ξ where

$$n_s(\xi) = c_\infty. \tag{8}$$

For scales larger than ξ the aggregate should be compact with the finite density c_∞, as shown in Figure 1b. When the aggregate grows with a finite velocity v, the variation of the gas density is characterized by the diffusion length l_D. Since the gas particles within the range l_D around the aggregate are adsorbed, the gas density there is very low and the situation is similar to the DLA growth. The fractal feature of the aggregate may thus be realized within l_D: ξ is of order l_D. This association determines the relation between the growth rate v and the gas density c_∞ in terms of the fractal dimension D_f as

$$v \sim l_D^{-1} \sim \xi^{-1} \sim c_\infty^{1/(d-D_f)}. \tag{9}$$

The relation (9) is confirmed by the Monte Carlo simulation of the two-dimensional aggregation growth from the lattice gas ($d = 2$ and $D_f = 1.71$), as is shown in Figure 1c [8].

We also studied the effect of flow in the ambient gas phase on the aggregate growth [9,10]. If the flow is toward the aggregate, the aggregate grows with an increased density. If the flow is outward from the aggregate, it ceases growing for a flow velocity larger than the critical value U_c, since the diffusion flow in the crystal surface is less than the drift outflow. The cessation of the growth is shown to be discontinuous with a jump in the growth velocity. The critical drift velocity U_c is shown to depend on the gas density c_∞ in a power law with an exponent determined by the fractal dimension D_f of the DLA. These theoretical predictions are confirmed by the Monte Carlo simulation [10].

Next we studied the lower cutoff length ξ for the fractal scaling (6) [11,12]: It is the width of a single branch of the aggregate. When the aggregate is grown from the adatom gas on a cold substrate, the width of each branch is of atomic size (Figure 2a). With annealing at a high temperature, the aggregated atoms become mobile and the aggregate relaxes to the equilibrium form determined by the surface tension. During this thermal healing, the branch smoothes and thickens as shown in Figure 2b and 2c, and the perimeter length of the aggregate A shrinks so as to decrease the surface free energy. The time dependence of A reflects the relaxation dynamics and the fractal nature of the aggregate structure.

First we consider the structure of aggregate during the thermal healing. For the length scale larger than the branch width ξ, the aggregate looks fractal. Thus by covering the aggregate of a linear size L with a box of linear dimension ξ, the number of boxes depends on the fractal dimension D_f as $N \sim (L/\xi)^{D_f}$. Since the aggregate is compact for a length scale smaller than ξ, the total perimeter length of the aggregate is given by $A \sim \xi^{d-1} N \sim \xi^{d-1-D_f}$ in the space dimension d. Now we consider the time-dependence of the branch width ξ. There are various route of the relaxation of the branch width ξ. Since there is no difference in the bulk chemical potentials, the local driving force is the chemical potential difference $\Delta \mu$ due to

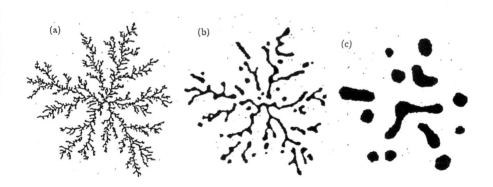

Figure 2. Relaxation of fractal aggregate by annealing at (a) $t = e^8$, (b) $t = e^{14}$ and (c) $t = e^{18}$.

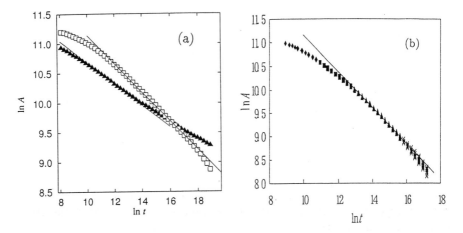

Figure 3. Perimeter length A versus time. (a) Relaxations via surface diffusion and via edge diffusion are observable. (b) Kinetics-limited relaxation.

the curvature κ as is determined by the Gibbs-Thomson effect. If the interface kinetics controls the relaxation, the local growth rate is given as

$$v \sim \Delta\mu \sim \kappa. \tag{10}$$

If the relaxation is controlled by the surface diffusion of atoms detached from the aggregate and reattached to the aggregate at a distant position, the rate is proportional to the diffusion flux as

$$v \sim \nabla c \sim \kappa^2. \tag{11}$$

If the diffusion of atoms along the aggregate edge controls the relaxation, the local velocity is controlled by the gradient of the current $j = \nabla(\Delta\mu)$ along the edge as

$$v \sim \nabla^2(\Delta\mu) \sim \kappa^3. \tag{12}$$

In all these cases, the rate v is proportional to the curvature κ in a power law as $v \sim \kappa^\beta$. From the dimensional consideration, $\kappa \sim \xi^{-1}$ and $v \sim d\xi/dt$, and the time evolution of the branch width is written as

$$\frac{d\xi}{dt} \sim \xi^{-\beta}. \tag{13}$$

By integration, the width of the branch increases as $\xi \sim t^{1/(\beta+1)}$. The time dependence of the aggregate perimeter A is then obtained as $A \sim t^{-\nu}$ with the relaxation exponent $\nu = (D_f + 1 - d)/(\beta + 1)$. In two-dimensional system $(d = 2,\ D_f = 1.71)$, the exponents are $\beta = 1$ and $\nu = 0.36$ for the surface kinetics controlled case, $\beta = 2$ and $\nu = 0.24$ when the two-dimensional surface diffusion controls the relaxation,

$\beta = 3$ and $\nu = 0.18$ when the one-dimensional edge diffusion controls it. These results are confirmed by the Monte Carlo simulation of the two-dimensional lattice gas system, as shown in Figure 3. If the detachment of the atom is suppressed, there is only an edge diffusion and the relaxation exponent is $\nu = 0.18$, as shown in Figure 3a. If the edge diffusion is suppressed with a high edge-diffusion barrier, the surface diffusional relaxation with $\nu = 0.24$ is observed. If the slow transport processes are coupled, we could not observe the kinetics-limited relaxation. Therefore, the simulation with an infinitely fast transport is performed, namely the particle detached from the aggregate tries attachment at an arbitrary position of the aggregate. In Figure 3b, the change of the perimeter length for ten different runs is shown. During $13 < \ln t < 15.5$, the exponent $\nu = 0.38$ is obtained, in consistency with the theoretical prediction $\nu = 0.36$ [12].

3. COMPETITION OF ANISOTROPY AND NOISE IN THE DENDRITIC GROWTH

When the surface stiffness comes into effect at high temperatures, it may suppress the tip instability. For an isotropic surface tension, its effect extends to the range of capillary length $d_0 = \Omega \tilde{\gamma}/k_B T$. From the linear stability analysis [5] of the flat interface against the sinusoidal modulation with a wavelength λ, the short wavelength fluctuation is found to be suppressed by the capillarity effect. But the fluctuation with a wavelength longer than the stability length

$$\lambda_s = 2\pi \sqrt{D d_0 / v} \tag{14}$$

will not be stabilized. The mode with the wavelength

$$\lambda^* = \lambda_s / \sqrt{3} \tag{15}$$

is the most unstable with the maximum amplification rate

$$\omega^* = 4\pi v / 3\lambda^*. \tag{16}$$

Then what kind of crystal shape would be realized?

Because of the instability of a flat interface, its part protrudes and may form a dendrite. As shown in Eq.(5), there is no selection of the tip radius ρ and the tip velocity v unless the surface tension is considered. The surface tension may select the operating point of v and ρ. In the detailed theoretical analysis of the solvability condition [13], the isotropic surface tension cannot stabilize the dendrite tip, and the anisotropy is necessary for the dendritic growth with a stable tip. We assume the anisotropic surface stiffness with four-fold symmetry and the capillary length is written as

$$d = d_0 (1 - \epsilon \cos 4\theta) \tag{17}$$

with the strength of anisotropy ϵ. The product of the growth velocity v and the square of the tip radius ρ is shown to be independent of the relative supersaturation Δ but to depend only on the anisotropy ϵ:

54

$$\sigma \equiv \frac{2Dd_0}{v\rho^2} \sim 2\epsilon^{7/4}. \tag{18}$$

From Eq.(5) and (18), v and ρ are determined separately. The scaling relation $v\rho^2$ =const is confirmed by the numerical simulation [14] and by experiments [19].

According to the solvability theory, the crystal tip is stable against the infinitesimal fluctuation. Even though crystal is always anisotropic due to its microscopic regularity of atomic arrangement, there is a structure where tips are unstable and split as DLA. This morphological variety is caused by the interplay of the anisotropy and the noise with finite strength. We study the time evolution of the dendritic interface growing under a perpetual noise by numerically integrating the interface evolution [15,16].

When the crystal is growing steadily with a constant velocity v, the diffusion field is time-independent in the frame comoving with the growth of the crystal. If the diffusion field relaxes quickly compared to the crystal growth, it is approximated to be time-independent in the moving frame. Using this quasi-stationary approximation, the fundamental set of equations (1-3) are transformed in the integro-differential equation of the crystal interface. It is solved numerically by the boundary integral method to find the normal growth rate of the interface $v_S(r,t)$ [14]. The noise effect by the crystallization is mimicked by the additional random velocity $v_R(t)$ at a particular point r_I close to the tip [17] as

$$v_n(r,t) = v_S(r,t) + v_R(t)\delta(r - r_I). \tag{19}$$

Here the noise is Gaussian and white with the correlation

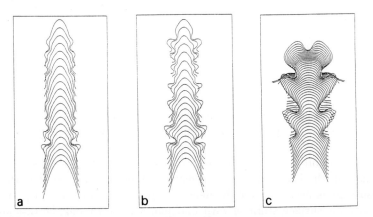

Figure 4. Time evolution of the dendritic profile at different anisotropies ϵ and noise strength Γ: (a) $\epsilon = 0.125$, Γ =0.002, (b) ϵ =0.125, Γ =0.007, (c) ϵ =0.068, Γ =0.002.

$$\langle v_R(t)v_R(t')\rangle = \Gamma^2 \delta(t - t').$$ (20)

We applied the noise near the tip, since the tip stability against the noise is to be studied.

If there is an anisotropy but no noise, the dendrite tip is stable as is confirmed by the previous numerical simulation [14]. Applying the noise, the tip curvature starts to oscillate and the sidebranch amplitude increases as the noise amplitude increases (Figure 4a and 4b). Further increase of the noise amplitude leads to the tip instability and splitting. For a fixed noise amplitude, the tip also splits by decreasing the anisotropy as shown in Figure 4c.

According to the solvability theory, the dendrite is shown to have a smooth needle shape without sidebranches. In the numerical simulation, there is always some numerical noise induced by the interface discretization or by numerical integration, and these noises are expected to produce sidebranches. These noises are, however, uncontrollable. In the present simulation, the noise amplitude is controlled externally, and the noise induced sidebranching is confirmed.

The stability limit of the dendrite tip can also be studied quantitatively [16]. Among fluctuations applied close to the tip, those of unstable modes are amplified, but since the crystal profile is curved parabolically, the fluctuation is convected away from the tip region, as shown in Figure 5. If the initial noise amplitude is small, the fluctuation is convected down far from the tip before it becomes observable. Then the tip is stable. The tip splitting may take place when the amplitude of the most unstable mode becomes the size of the tip radius ρ before it is convected down a distance ρ from the tip. For the crystal growing with the velocity v, it takes time about ρ/v until the fluctuation is convected down from the tip region. The amplitude of the most unstable mode with wavelength λ^* increases with the amplification rate (16) or $\omega^* \sim v/\lambda^*$, and the initial amplitude a increases to $a\exp(\rho/\lambda^*)$ during the time ρ/v. If this amplitude is smaller than the tip

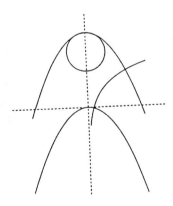

Figure 5. Schematics of the convective motion of a nodal point on the surface of a parabolic dendrite.

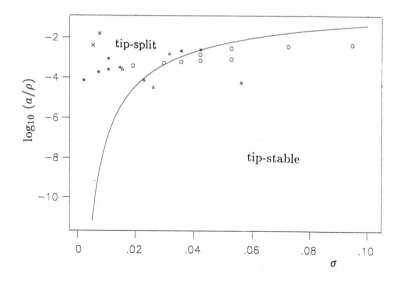

Figure 6. Phase boundary between the tip-stable $(\circ, \triangle)$ and tip-splitting $(*, \times)$ dendrite in the phase space of perturbation amplitude and an anisotropy parameter $\sigma = 2\epsilon^{7/4}$.

radius ρ, the tip is stable. The upper limit of the initial amplitude a for the tip stability is then given as

$$\ln(a/\rho) \sim -\rho/\lambda^* \sim -\sigma^{-1/2} \sim -\epsilon^{-7/8}. \tag{21}$$

The stability limit is shown in Figure 6. The tip is easily split for a small anisotropy.

In our simulation presented in Figure 4, the velocity fluctuation is applied continually. This velocity fluctuation with the strength Γ induces the surface deformation with the amplitude $a \sim \Gamma/\sqrt{\omega^*}\lambda^*$ in a time scale of order $1/\omega^*$. In the parameter space of $\ln(a/\rho)$ and σ in Figure 6, simulations with tip-unstable dendrites are marked by $*$ and $\times$, and those with tip-stable dendrites by $\circ$ and $\triangle$. The boundary between the tip-stable and the tip-unstable dendrites agrees well with the theoretical expectation (21).

4. SURFACE KINETICS AND SURFACE STIFFNESS

So far we assumed that the surface kinetics is fast enough such that the diffusion controls the crystal morphology and the growth mode. Even for a rough interface, however, incorporation of atoms at the surface is not infinitely fast, and the growth needs a finite surface supersaturation. The local equilibrium condition (3) is replaced by the Wilson-Frenkel's condition

$$b\Omega^{-1}v_n = c_i - c_{eq}(1 + \frac{\Omega\tilde{\gamma}}{k_B T}\kappa). \tag{22}$$

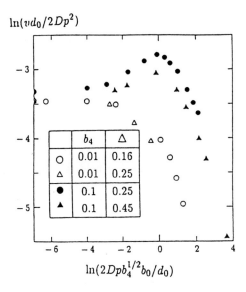

Figure 7. Velocity $vd_0/2Dp^2$ versus the kinetic coefficient $2Dpb_4^{1/2}b_0/d_0$.

Here b is the inverse of the anisotropic kinetic coefficient, and assumed to have the four-fold symmetry as

$$b = b_0(1 - b_4 \cos 4\theta), \tag{23}$$

similar to the capillary length. In the quasi-stationary approximation, the diffusion equation can be transformed to the boundary integral form as before. With the normal growth rate obtained by the boundary element method, the interface is developed in time [17]. For a fixed supersaturation Δ, we varied the kinetic coefficient b_0. With a small b_0, the growth is controlled by the surface stiffness, and the scaling law $v\rho^2 =$const is confirmed. When b_0 increases, the kinetics comes into play, and the growth rate v decreases inversely to b_0, as shown in Figure 7. Simultaneously, the tip radius ρ increases. With a small kinetic anisotropy as $b_4 = 0.01$, the product of v and ρ for large b_0 still follows the Ivantsov relation (5) and is determined by the supersaturation Δ. Instead of the scaling relation (18), the stability parameter $\sigma = 2Dd_0/v\rho^2$ is found to be inversely proportional to b_0, as shown in Figure 7. The solvability analysis [20] derived the scaling relation

$$\sigma = \frac{d_0 b_4^{5/4}}{2Dp(\Delta)b_0} \tag{24}$$

for the kinetics-limited growth. In this case the tip radius ρ and the growth rate v are expected to be determined by the kinetic coefficient as $\rho \sim 2Db_0 b_4^{-5/4}$ and

$v \sim p(\Delta)b_4^{5/4}/b_0$, in qualitative agreement with the simulation result. If both surface kinetics and stiffness are effective, another scaling relation coupled with the stiffness and the kinetic coefficient is derived theoretically [20]. In the present simulation, we could not find the region with that scaling, but it was found in the previous simulation of the unidirectional solidification [21].

5. CONCLUSION

Crystal surface destabilized by the diffusion field is shown to take a variety of morphology under the influence of various stabilizing effect. The problem of the step morphology on the crystal surface during the growth from the vapor phase is also related to the competition between the noise and the anisotropy. The result of our study on the step morphology is summarized in Uwaha's article in this volume [22]. With the diffusion bias or the Schwoebel effect, step is destabilized when it advances too fast. The deformation of the destabilized step can be described by an evolution of the step height $h(x,t)$, called Kuramoto-Sivashinsky equation.

$$\frac{\partial h}{\partial t} = -\frac{\partial^2 h}{\partial x^2} - \frac{\partial^4 h}{\partial x^4} + \left(\frac{\partial h}{\partial x}\right)^2 \tag{25}$$

This equation is known to show the spatio-temporal chaos, and explains the chaotic evolution of step profile obtained by the Monte Carlo simulation [23,24]. We are now studying the system with an anisotropic stiffness, and the step profile is shown to be stabilized into the periodic structure [25].

When the noise is strong or the anisotropy is weak, the dendrite tip is found to be unstable against splitting. The morphology after the instability is irregular but compact due to the finite supersaturation Δ. This structure is called dense-branched morphology (DBM) [26] or compact seaweed (CS) [27] and may correspond to the spherulite. Recently the main building-block of this structure is apparently identified: a double-finger structure or doublon, for short. Doublon consists a dendritic envelope with a central thin groove. Originally, similar structure is obtained in the simulation of a dendrite in a channel confined in reflective walls [28]. Due to the wall constraint the system acquires effective anisotropy, and even the crystal with an isotropic stiffness can grow steadily in the dendritic form in a channel with a finite width. In the simulation using the boundary element integration method, an asymmetric dendrite is discovered to grow steadily closely along one of the wall. Since the reflective wall mathematically means the existence of a mirror image to the asymmetric dendrite in the channel, the double-finger structure is easily imagined to construct a stable growth element for an isotropic or a weakly anisotropic crystal. The doublon is observed in many simulations [29] of weakly anisotropic systems and experiments [30] of spherulites. There is a recent study of dynamical morphology diagram with the considerations of this doublon excitation as well as the regular dendrite [27].

ACKNOWLEDGEMENTS

The works summarized here are mostly supported by Grants-in-Aid for Scientific Research on Priority Areas "Crystal Growth Mechanism in Atomic Scale" No.03243101 and No.04227108, which are provided by the Ministery of Education, Science and Culture in Japan. They are performed in collaboration with many coworkers, whom I sincerely acknowledge: Prof. M. Uwaha, Prof. T. Irisawa, Miss. T. Sakiyama, Mr. K. Shiraishi and Mr. S. Seki. I have benefited from the interuniversity cooperative research program of IMR, Tohoku University, too. Last but not least, many enlightening discussions with and informations from Prof. H. Müller-Krumbhaar, Dr. T. Ihle, Prof. C. Misbah, Dr. E. Brener and Dr. D. Temkin are also greatly appreciated.

REFERENCES

1. A. Ookawa, Crystal Growth (in Japanese), Shokabo, Tokyo, 1977.
2. T. Kuroda, Crystal is Living (in Japanese), Science Publ., Tokyo, 1985.
3. Y. Saito, Statistical Physics of Crystal Growth, World Scientific, Singapore, 1996.
4. W.K.Burton, N.Cabrera and F.C.Frank, Phil. Trans. Roy. Soc., A243 (1951) 299.
5. W.Mullins and R.Sekerka, J. Appl. Phys. 52 (1963) 323.
6. Y.Saito, T.Sakiyama and M.Uwaha, J. Cryst. Growth 128 (1993) 82.
7. T.A.Witten and L.M.Sander, Phys. Rev. Lett. 47 (1981) 1400; Phys. Rev. B 27 (1983) 5686.
8. M.Uwaha and Y.Saito, J. Phys. Soc. Jpn 57 (1988) 3285 and Phys. Rev. A 40 (1989) 4716.
9. M.Uwaha, Y.Saito and S. Seki, *in* Dynamical Phenomena at Interfaces, Surfaces and Membranes, D. Beysens, N.Boccara and G.Forgacs (eds.), Nova Scientific, Jerico, 1993, p.177.
10. Y.Saito, M.Uwaha and S.Seki, *in* Interactive Dynamics of Convection and Solidification, S.H.Davis et al. (eds), Kluwer Academic, Netherland, 1992, p.27.
11. T.Irisawa, M.Uwaha and Y.Saito, Europhys. Lett. 30(1995) 139.
12. T.Irisawa, M.Uwaha and Y.Saito, Fractals 4 (1996) to appear.
13. J.Langer, *in* Chance and Matter, J. Souletie, J. Vannimenus and R.Stora (eds.), Elsevier, Amsterdam, 1987, p.629.
14. Y.Saito, G. Goldbeck-Wood and H. Müller-Krumbhaar, Phys. Rev. Lett. 58 (1987) 1541; Phys. Rev. A 38 (1988) 2148.
15. Y.Saito and K.Shiraishi, *in* Spatio-temporal patterns in nonequilibrium complex systems, P.E.Cladis and P.Palffy-Muhoray (eds.), Addison-Wesley, Reading, 1995, p.157.
16. E.Brener, T.Ihle, H. Müller- Krumbhaar, Y.Saito and K.Shiraishi, Physica 204 (1994) 96.
17. Y.Saito and T.Sakiyama, J. Cryst. Growth 128 (1993) 224.
18. G.P. Ivantsov, Dokl. Akad. Nauk SSSR 58 (1947) 567.

19. S.C.Huang and M.E.Glicksman, Acta Metall. 29(1981) 701 and 717.
20. E.Brener, J. Cryst. Growth 99(1990) 165.
21. A.Classen, C.Misbah, H.Müller-Krumbhaar and Y.Saito, Phys. Rev. A43 (1991) 6920.
22. M. Uwaha, in this volume.
23. M.Uwaha and Y.Saito, Phys. Rev. Lett. 68 (1992) 224; Surf. Science 283 (1993) 366; J. Crystal. Growth 128 (1993) 87.
24. Y.Saito and M.Uwaha, Phys. Rev. B49 (1994) 10677.
25. Y.Saito and M. Uwaha, unpublished.
26. E.Ben-Jacob, G.Deutscher, P.Garik, N.D.Goldenfeld and Y.Lereath, Phys. Rev. Lett. 57 (1986) 1903.
27. E. Brener, H. Müller-Krumbhaar D. Temkin, Europhys. Lett. 17 (1992) 535, and private communications, 1996.
28. E.Brener, H.Müller-Krumbhaar, Y.Saito and D.Temkin, Phys. Rev. E47 (1993) 1451.
29. T. Ihle and H. Müller-Krumbhaar, Phys. Rev. Lett. 70 (1993) 3083 and Phys. Rev. E49 (1994) 2972.
30. P.Oswald, J.Malthête and P.Pelcé, J. Phys. France 50 (1989) 2121.

Advances in the Understanding of Crystal Growth Mechanisms
T. Nishinaga, K. Nishioka, J. Harada, A. Sasaki and H. Takei (Editors)
© 1997 Elsevier Science B.V. All rights reserved.

Step structures of Si(111) vicinal surfaces

A.Natori

University of Electro-Communications, Chofu, Tokyo 182, Japan

The effects of direct current (DC) and surface reconstruction on the step structures of Si (111) vicinal surfaces are investigated. In DC supply above T_c of 7×7 reconstruction, conversion between a regular step structure and a step bunching structure is observed, caused by the inversion of DC direction or by change of temperature. The DC effect is studied by extending Burton-Cabrera-Frank (BCF) theory, to take account of the surface electromigration effect of Si adatoms. The stability condition of a regular step array is obtained and the conversion mechanism to a bunching structure is proposed. Below T_c, Si(111) vicinal surfaces facet into the 7×7 reconstructed (111) surfaces and step bunched surfaces. The faceting mechanism is studied by extending the terrace-step-kink (TSK) model, to consider the energy gain of surface reconstruction. In thermal equilibrium, the local vicinal angle in a step bunching region becomes independent on the global vicinal angle and it increases monotonically as temperature decreases. The faceting dynamics is also studied through the Monte Carlo simulation of rapid cooling.

1. INTRODUCTION

Step structures on crystal surfaces play important roles on surface diffusion, surface adsorption, surface reaction, crystal growth and so on. There exist excellent review works [1,2] on surface morphology of vicinal Si(111) surfaces.

In temperature range above T_c of 7×7 surface reconstruction on a vicinal Si(111), inversion of DC direction induces conversion of the step structures between a regular step structure and a step bunching structure [3,4]. Further, the step structure changes depending on temperature, as seen in Figure 1 [4]. For AC supply, a regular step structure is always observed irrespective of temperature [3]. On the other hand below T_c, Si(111) vicinal surfaces facet into 7×7 (111) surfaces and step bunching regions [5,6]. The temperature dependence of the misorientation angle in a step bunch observed by low-energy electron diffraction (LEED) [5] is shown in Figure 2.

In this chapter, we study the atomic mechanisms of the DC-effect and the surface reconstruction effect in vicinal Si(111) surfaces.

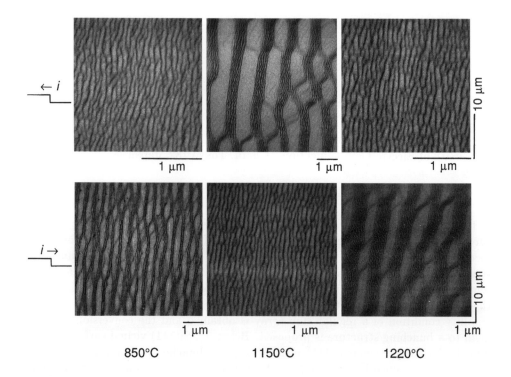

Figure 1. Scanning electron-microscope (SEM) images showing DC direction (step-up direction and step-down direction) and temperature-dependent step-bunching structure [4].

2. DC EFFECT ON STEP STRUCTURE

2.1. Formulation

Burton-Cabrera-Frank (BCF) theory [7] is developed to take account of the following three effects [8,9]; 1) the contribution of the drift flux of adatoms caused by the effective driving force due to current supply [10,11], 2) the anisotropy of capture rates at a step edge [12], and 3) the repulsive interaction between steps [13].

We take a straight step array parallel to the y-axis and the mass transport between steps is carried out by the lateral flux in the x-direction, as shown in Figure 3. On each terrace, the lateral flux J of adatoms can be written as

$$J = -D\frac{\partial \theta}{\partial x} + \frac{DF\theta}{kT}. \tag{1}$$

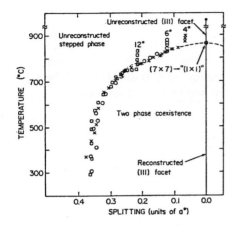

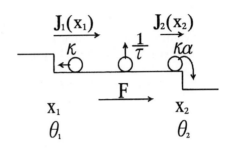

Figure 2. Measured values of the specular beam splittings of LEED spot, which are proportional to surface misorientation, as a function of temperature [5].

Figure 3. Adatom elementary process on a step surface.

Here D is a surface diffusion constant, θ the concentration of adatoms, F the uniform driving force on adatoms, T the temperature and k the Boltzmann constant. In eq.(1), Einstein relation was assumed between the diffusion constant and the mobility of adatoms. The conservation law of adatom concentration and the steady state condition lead to the following diffusion equation on θ [8,9]:

$$D\frac{\partial^2 \theta}{\partial x^2} - \frac{DF}{kT}\frac{\partial \theta}{\partial x} - \frac{\theta}{\tau} = 0, \tag{2}$$

where τ is the evaporation lifetime of an adatom.

As for the boundary condition on $\theta(x)$ on each terrace, we adopted the continuity of a flux [14] at a lower edge x_i and an upper edge x_{i+1} (see Figure 3):

$$J_1(x_i) = \kappa[\theta_i - \theta(x_i)], \tag{3}$$

$$J_2(x_{i+1}) = \kappa\alpha[\theta(x_{i+1}) - \theta_{i+1}]. \tag{4}$$

Here θ_i is the equilibrium adatom concentration at the ith step edge. In this chapter, the positive directions of J_i, F and x are taken to be the step-down direction. If the adatoms attachment-detachment kinetic coefficients, κ and $\kappa\alpha$, are sufficiently large, $\theta(x_i)$ approaches to the equilibrium value θ_i.

With respect to the equilibrium concentration θ_i at the ith step edge, it is affected [15] by the step-step interaction as

$$\theta_i = \theta_0 \exp(-f_i \Omega / kT). \tag{5}$$

Here θ_0 is the equilibrium concentration of adatoms at an edge of an isolated step, f_i the force acting on the ith step per unit length due to step-step interaction and Ω the surface area per atom. The positive direction of f_i is also taken to be the step-down direction. The force f_i on the ith step was calculated from the step-step interaction energy U per unit length as

$$f_i = -\frac{\partial}{\partial x_i}[U(x_{i+1} - x_i) + U(x_i - x_{i-1})]. \tag{6}$$

Concerning to the interaction U on Si(111) surface, the following elastic repulsive interaction [13] was considered.

$$U(x) = \frac{\gamma'}{x^2} \tag{7}$$

On the other hand, the movement of each step position x_i is determined by the lateral fluxes at the step edge from both sides (see Figure 3).

$$\frac{dx_i}{dt} = \Omega[-J_1(x_i, \theta_i, \theta_{i+1}) + J_2(x_i, \theta_{i-1}, \theta_i)] \tag{8}$$

J_i can be calculated analytically as a function of θ_i from eqs.(1),(2),(3) and (4). The coupled differential equations (8) are solved numerically simultaneously with eqs.(5), (6) and (7). In numerical calculation, we took 50 steps in all and assumed the cyclic boundary condition for the step array.

2.2. Stability of a regular step array

We consider the time evolution of small deviations δ_j from a regular step array with a step spacing l.

$$x_j = lj - \Omega[J_1(l) - J_2(l)] + \delta_j \tag{9}$$

Assuming the fluctuation with a wave vector k, δ_j is written as

$$\delta_j \propto \exp(iklj). \tag{10}$$

In the linear approximation with δ_j, eq.(8) is transformed to

$$\begin{aligned}
\frac{1}{\Omega}\frac{d\delta_j}{dt} =\ & \{(\frac{\partial J_1}{\partial l} + \frac{\partial J_2}{\partial l}) \\
& + 2\frac{d^2 V}{dl^2}\frac{\Omega}{kT}[(-\frac{\partial J_1}{\partial \theta_1} + \frac{\partial J_2}{\partial \theta_2}) + (-\frac{\partial J_1}{\partial \theta_2} + \frac{\partial J_2}{\partial \theta_1})\cos(kl)]\}(1 - \cos(kl))\delta_j \\
& + \text{(Imaginary Part)}.
\end{aligned} \tag{11}$$

The stability of a regular step is realized only if the real part of a coefficient of δ_j of the left side in eq.(11) is negative at any values of k.

$$\frac{\partial(J_1 + J_2)}{\partial l} + 2\frac{d^2 V}{dl^2}\frac{\Omega}{kT}(-\frac{\partial J_1}{\partial \theta_1} - \frac{\partial J_1}{\partial \theta_2} + \frac{\partial J_2}{\partial \theta_1} + \frac{\partial J_2}{\partial \theta_2}) < 0 \tag{12}$$

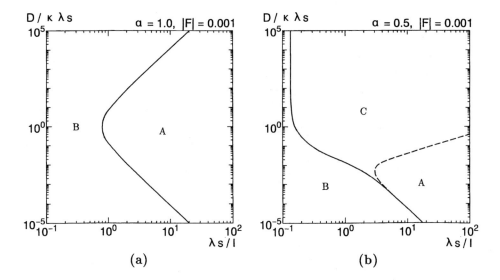

Figure 4. Stability of a regular step for (a) $\alpha = 1$ and (b) $\alpha = 0.5$ [9]. In region A a regular step is unstable for $f > 0$ and stable for $f < 0$, and vice versa in region B. In region C, a regular step becomes unstable irrespective of the direction of f.

The second term of eq.(12) is attributed to the repulsive step-step interaction, and this term is always negative and stabilizes a regular step array. In the case of $\alpha = 1$ (no Schwoebel effect), the first term of eq.(12) becomes an odd function of driving force F and its sign is determined by the sign of F. On the other hand, the magnitude of the first term of eq.(12) depends strongly on both characteristic dimensionless parameters, λ_s/l and $D/\kappa\lambda_s$. Here λ_s is a diffusion length defined as

$$\lambda_s = \sqrt{D\tau}. \tag{13}$$

The stability of a regular step array is calculated in the case of $\gamma' = 0$, in the parameter space spanned by $D/\kappa\lambda_s$ and λ_s/l in Figures 4(a) and (b), for $\alpha = 1$ and $\alpha = 0.5$, respectively. In region A a regular step array is unstable for $f > 0$ and stable for $f < 0$, and vice versa in region B, i.e. the stability is symmetrical concerning to the direction of f in regions A and B. In region C in Figure 4(b), a regular step array becomes unstable irrespective of the direction of f. This is the well-known Schwoebel effect [12].

With respect to the driving force Γ, the dimensionless strength $f = Fl/2kT$ has an order of magnitude of 10^{-3}, assuming $F \sim 20\text{V/cm}$, $l \sim 0.1\mu\text{m}$ and $T \sim 1200\text{K}$.

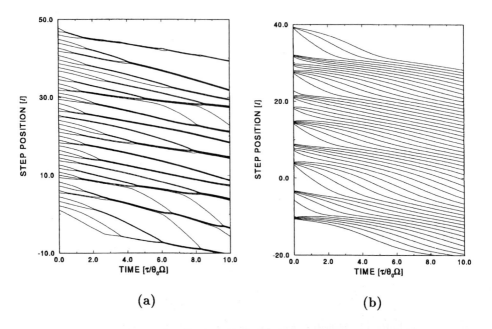

(a) (b)

Figure 5. Time evolution of the step configuration [9], for (a) $f = 0.001$ and (b) $f = -0.001$. Here $\alpha = 1$, $\gamma = 10^{-7}$, $\lambda_s/l = 100$ and $D/\kappa\lambda_s = 10^{-4}$.

The observed step velocity in the temperature range $1090°\mathrm{C} < T < 1200°\mathrm{C}$ is proportional to the average step spacing l [3,16]. This means that λ_s is much larger than l in this temperature range. As for the value of γ', its dimensionless magnitude was estimated [17] as $\gamma = 2\gamma'\Omega/kTl^3 = 10^{-7}$.

2.3. Dynamics of a Step Array

Time evolution of a step array was calculated from eq.(8). As for the initial regular step array, Gaussian terrace width distribution with the standard deviation of $0.2l$ was assumed. The calculated typical result for $\lambda_s \gg l$ in region A in Figure 4(a) ($\alpha = 1$, $\gamma = 10^{-7}$, $\lambda_s/l = 100$ and $D/\kappa\lambda_s = 10^{-4}$) is shown in Figure 5(a) at $f = 0.001$. The subsequent time evolution after inversion of f is shown in Figure 5(b). It is seen clearly that a bunching structure develops in the case of $f > 0$, while a regular step array is recovered in the other case of $f < 0$. In the case of $f = 0$, a regular step array is recovered more slowly due to the repulsive step-step interaction, just as the observed result in AC supply. The time evolution of the case of $l > \lambda_s$ in region B in Figure 4(a), $\lambda_s/l = 0.5$, $D/\kappa\lambda_s = 10$, $f = -1$, is shown in Figure 6. The bunching rate is very slow and it was accelerated by choosing a stronger driving force f. The step bunching is formed for $f < 0$ contrary to the previous case, but the bunching structure is not so stable as seen in Figure 6.

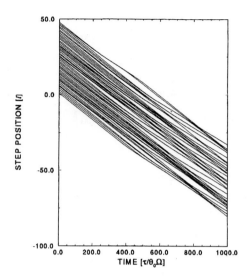

Figure 6. Time evolution of the step configuration [9]. Here $\lambda_s/l = 0.5$, $D/\kappa\lambda_s = 10$, $f = -1$, $\alpha = 1$, $\gamma = 10^{-7}$.

2.4. Mechanism of step bunching

Step bunching is brought about, provided that a wider terrace than an average width has a tendency to extend further.

Step bunching occurs for $f > 0$ in the case of Figure 5, and the terrace width dependence of the flux is shown in Figure 7(a) in the width range including the average width l. For $f > 0$, J_1 at a lower step edge and J_2 at an upper step edge are positive due to the positive drift flux. J_1 is a monotonically increasing function of a terrace width, while J_2 is almost independent on a terrace width around l due to competition of evaporation flux and diffusion flux. The extension velocity of a terrace width is proportional to $J_1 + J_2$, and it becomes a monotonically increasing function of a terrace width in the vicinity of l. Thus a wider tarrace than an average width broaden further. In the case of Figure 6 for $f < 0$, on the other hand, J_1 is a monotonically increasing positive function of a terrace width, while J_2 is negative and almost independent on a terrace width near l due to saturation of the flux to the capture rate at a step edge, as seen in Figure 7(b). This saturation is caused by the negative drift flux. $J_1 + J_2$ becomes a monotonically increasing function of a terrace width in the vicinity of l for $f < 0$, and a wider terrace broaden further.

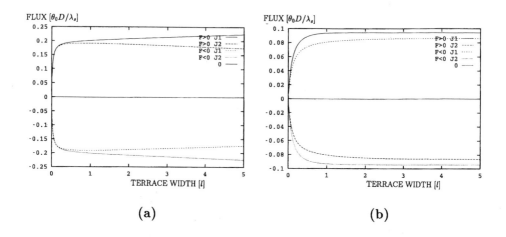

Figure 7. Flux at a step edge as a function of a terrace width in a unit of l, for (a) $f = \pm 0.001$, $\alpha = 1$, $\lambda_s/l = 100$ and $D/\kappa\lambda_s = 10^{-4}$, and (b) $f = \pm 1$, $\alpha = 1$, $\lambda_s/l = 0.5$ and $D/\kappa\lambda_s = 10$ [9].

3. FACETING INDUCED BY SURFACE RECONSTRUCTION

3.1. Model

We adopted a terrace-step-kink (TSK) model, and the vicinal surface is composed of terraces, steps and kinks as shown in Figure 8. We assumed a simple cubic crystal structure with a lattice constant of a. On Si(111) vicinal surfaces, 7×7 reconstruction appears below about 850°C in the first-order phase transition. To describe the surface reconstruction below T_c, the usual TSK model is developed by taking into account of the energy gain of the reconstruction [18]. 7×7 reconstruction units appear below T_c on (111) terrace from the upper edge of each step, and each terrace is nearly completely covered by 7×7 reconstruction units in the temperature range below $T_c - 10°C$ [2]. We consider only this temperature range, and so assume the appearance of maximum possible number of reconstruction units on each terrace. We simplify the 7×7 reconstruction as 7×1 reconstruction, since we address only the 7×7 reconstructed facet size in the direction of a misorientation angle of a vicinal surface.

The Hamiltonian of our model system for Si(111) vicinal surfaces in the temperature region below $T_c - 10°C$ can be written:

$$H = \epsilon \sum_m \sum_y \mid x_{m,y+1} - x_{m,y} \mid - \epsilon_r \sum_m \sum_y \left[\frac{(x_{m+1,y} - x_{m,y})}{7} \right]$$

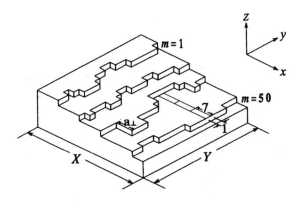

Figure 8. Schematic figure of a vicinal surface in the TSK model. A 7 × 1 reconstruction unit is shown on a (001) terrace.

$$- \epsilon_i \sum_m \sum_y \left[\frac{(x_{m+1,y} - x_{m,y})}{7} - 1 \right] + \lambda \sum_m \sum_y \frac{1}{(x_{m+1,y} - x_{m,y})^2} \tag{14}$$

Here $x_{m,y}$ represents the step position at y along the mth step edge, as seen in Figure 8. All the coordinates are measured in a unit of a lattice constant a. In eq.(14), $[z]$ expresses the Gaussian symbol and it means the maximum integer not exceeding z. The first term represents the usual kink formation energy, the second the energy gain of formation of a 7 × 1 reconstruction unit on each terrace, the third the interaction energy between adjacent 7 × 1 reconstruction units in the x direction on the same terrace (see Figure 8), and the fourth the elastic interaction energy between adjacent steps. In eq.(14), the interaction energy between adjacent 7 × 1 reconstruction units in the y direction was neglected for simplicity.

With respect to the adopted sizes X, Y of a vicinal surface along the x and y directions (see Figure 8), $X = 50la$ and $Y = 100a$ are taken. Here la is the mean step spacing of a vicinal surface and 50 steps are considered in all. In both directions, the periodic boundary conditions are adopted.

$$
\begin{aligned}
x_{m+50,y} &= x_{m,y} + X, \\
x_{m,y+Y} &= x_{m,y}
\end{aligned}
\tag{15}
$$

Finally the assumed values of parameters are discussed. In the high temperature observation by scanning tunneling microscope (STM) of a vicinal surface of Si(111) near T_c [19,20], the atomic structure of 7 × 7 reconstruction is well observed on a

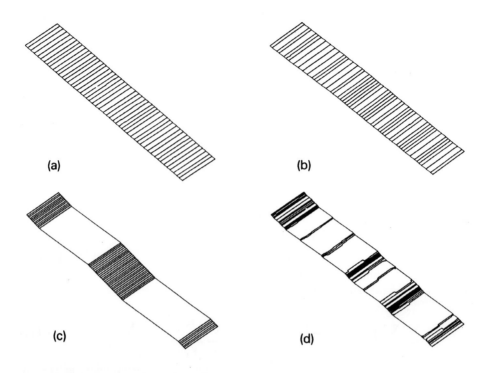

Figure 9. Calculated step configurations at $kT/\epsilon = 0.2$ for $\lambda/\epsilon = 1$ [18]. (a) $\epsilon_r/\epsilon = 0$, $\epsilon_i/\epsilon = 0$; (b) $\epsilon_r/\epsilon = 1$, $\epsilon_i/\epsilon = 0$; (c) $\epsilon_r/\epsilon = 0$, $\epsilon_i/\epsilon = 1$; (d) $\epsilon_r/\epsilon = 1$, $\epsilon_i/\epsilon = 1$.

(111) facet, while no atomic step structure can be seen in a step bunch. Meanwhile at a room temperature, the atomic step structure is cleary seen. This implies the high frequency thermal motion of steps in a step bunch around T_c. It suggests also the large step fluctuation at a temperature around T_c, i.e. $kT_c/\epsilon \sim 1$. The value of ϵ has been approximately estimated as 0.18eV [1]. The critical temperature T_c in our model should have the similar magnitude as ϵ_i, i.e. $kT_c \sim \epsilon_i$. Concerning to the step-step interaction, λ was estimated as 0.07eV [17]. Thus all the normalized quantities kT_c/ϵ, ϵ_i/ϵ and λ/ϵ are thought to be around unity.

3.2. Faceting structure induced by surface reconstruction

The step configurations in the thermal equilibrium were obtained by simulated annealing procedure [21] from a higher temperature than T_c where $\epsilon_r = \epsilon_i = 0$. In Figure 9, the calculated step configurations with $l = 10a$ are presented at a low temperature ($kT = 0.2\epsilon$) below T_c for four parameter sets. A facet formation

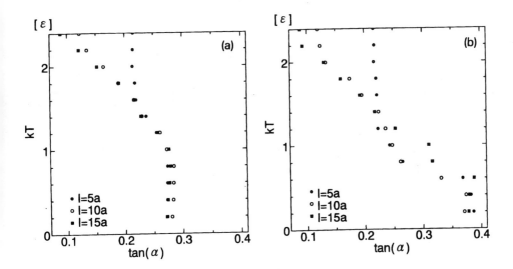

Figure 10. Temperature dependence of misorientation angle in a step bunch, for $\lambda/\epsilon = 1$ [18]. (a) $\epsilon_r/\epsilon = 0$, $\epsilon_i/\epsilon = 1$; (b) $\epsilon_r/\epsilon = 1$, $\epsilon_i/\epsilon = 1$.

accompanied with step bunching can be found, only if ϵ_i is considered. In the faceting cases of Figures 9(c) and 9(d), all the observed widths of (001) facets are quantized by $7a$, i.e. they are constituted of multiples of 7×1 reconstruction units. The temperature dependence of the calculated misorientation angle α of a step bunch are summarized in Figures 10(a) and (b), at $l/a = 5$, 10 and 15, for the previous two faceting cases, respectively. For both cases, the local misorientation angles α are independent on the nominal vicinal angle n which is determined by the relation of $\tan n = 1/l$. α does not depend on l, and increases monotonically as temperature decreases.

The kinetics of transformation, from a regular step structure of 1×1 phase above T_c to a facet structure with a step bunch below T_c, is investigated by quenching simulation. A snapshot of step configuration with $l = 10a$ was presented in Figure 11(a) for $\epsilon_r/\epsilon = 0$ and $\epsilon_i/\epsilon = \lambda/\epsilon = 1$, in quenching to $kT/\epsilon = 1$ below T_c from a temperature higher than T_c. Several (001) facets arise from the critical nuclei constituted of a few 7×1 reconstruction units, and grow toward a constant width determined by both the nucleation density and the temperature. In Figure 11(a) the size and the spacing of (001) facets are rather regular, and the rearrangement of a step, i.e. the movement of a step across a wide terrace between neighbouring step bunches is seen. Time dependence of the widths of relatively wide (001) facets in Figure 11(a) is shown in Figure 11(b). Here 1 Monte Carlo step (MCS) indicates the average loop times required to assign once each step edge of a length a. Monte

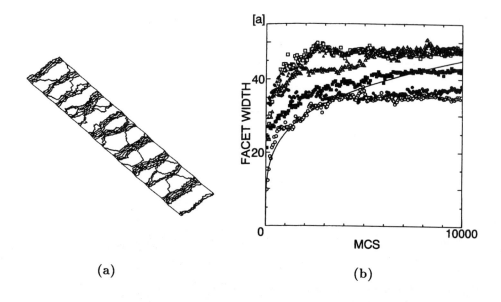

(a)   (b)

Figure 11. (a) The step configuration obtainned by quenching to a temperature of $kT/\epsilon = 1$, for $\epsilon_r/\epsilon = 0$, $\epsilon_i/\epsilon = 1$, $\lambda/\epsilon = 1$, and (b) the time dependence of terrace widths of (001) facets [18].

Carlo step is considered to be approximately proportional to the real time.

3.3. Mechanism of faceting

The faceting with step bunching was reproduced in our simple model, provided the interaction ϵ_i between reconstruction units are considered. In the faceting case, the step fluctuation is suppressed as a 7×1 (001) facet grows with decrease of temperature, and it becomes finally independent on the global vicinal angle. This corresponds to the independence of the misorientation angle α of a step bunch on the global vicinal angle. α increases with decrease of temperature and its temperature dependence is mainly determined by that of the surface free energy of a 1×1 vicinal surface due to entropy effect of step fluctuation. Meanwhile the width of a (001) facet is affected by the global vicinal angle, and it is quantized by $7a$ at low temperatures.

With respect to the faceting kinetics of a 7×1 (001) facet driven by quenching, two stage growth of (001) facets are observed; the first fast stage and subsequent second slow stage. The first stage is advanced by the nucleation and growth of a relatively wide terrace, by reducing neighboring terrace widths. The second stage, on the other hand, is developed by a step rearrangement between neighboring step bunches. The nearly saturated behavior of reconstructed facet widths has been observed on Si (111) vicinal surfaces by low-energy electron- microscope (LEEM) [22]. This behavior is in contrast with the result of Williams et al. [23], in which a facet width increases proportional to $t^{1/4}$ of time t at a temperature below T_c. The $t^{1/4}$ behavior is drawn by a solid line in Figure 11(b).

4. DISCUSSION AND CONCLUSION

First, DC effect on step structure is discussed. A regular step structure is observed on Si(111) above T_c in AC supply, and thus the Schwoebel effect is not so effective. As for the effect of DC on Si adatom, the uniform driving force is expected due to both the electric field force and the wind force of DC [10]. In the parameter region A in Figure 4, we succeeded in explaing conversion between a reguar step structure and a step bunching structure by inversion of DC direction, as suggested by Stoyanov [24]. If a parameter moves into region B by increasing temperature, the step structure exchanges each other for a fixed DC direction [9]. However, the conversion velocity by inversion of DC direction is very slow in region B compared to that in region A [9,25]. As for the step conversion at a higher temperature, electromigration of surface vacancies in the opposite direction to adatoms has been proposed recently [26]. Further, two dimensional step fluctuation effect is studied for step bunching induced by the Schwoebel effect [27] or for an isolated step in DC supply [28]. The calculated step morphology in the former case is very similar to the observed one [3] by reflection electron-microscope (REM) in DC supply. The extension of our one-dimensional treatment to two-dimensional steps is highly desirable.

Second, the surface reconstruction effect is discussed. The calculated results can explain qualitatively the observed feature. In our treatment the applicable temperature range is limited to be below $T_c - 10$ °C, and the self-consistent calculation of both the phase transition and the faceting is desirable. Thermodynamically, faceting can be explained by the difference of surface free energies of coexistent surfaces [1]. Based on the flux caused by the difference of the surface free energy, faceting mechanism through the propagation of nucleation has been proposed for a straight step array [29]. This mechanism can also explain rather regular faceting structure observed in the spinodal decomposition temperature range in the vicinity of T_c [6].

The step structures on a vicinal Si(111) have been clarified only partially, and their comprehensive understanding has not yet been accomplished. Further studies are extensively in progress from both experimental and theoretical sides.

REFERENCES

1. E. D. Williams, R. J. Phaneuf, J. Wei, N. C. Bartelt and T. L. Einstein, Surf. Sci. 294 (1993) 219.
2. K. Yagi, A. Yamanaka, H. Sato, M. Shima, H. Ohse, S. Ozawa and Y. Tanishiro, Prog. Theoret. Phys. Suppl. 106 (1991) 303.

3. A. V. Latyshev, A. L. Aseev, A. B. Krasilnikov and S. I. Stenin, Surf. Sci. 213 (1989) 157.
4. Y. Homma, R. J. Mccleland and H. Hibino, Jpn. J. Appl.Phys. 29 (1990) L2254.
5. R. J. Phaneuf, E. D. Williams and N. C. Bartelt, Phys. Rev.B 38 (1988) 1984.
6. R. J. Phaneuf, N. C. Bartelt, E. D. Williams, W. Swiech and E. Bauer, Phys. Rev. Lett. 71 (1993) 2284.
7. W. K. Burton, N. Cabrera and F. C. Frank, Phil. Trans. Roy. Soc., A243 (1951) 299.
8. A. Natori, H. Fujimura and M. Fukuda, Appl. Surf. Sci. 60/61 (1992) 85.
9. A. Natori, Jpn.J.Appl.Phys. 33 (1994) 3538.
10. H. Ishida, Phys. Rev. B 49 (1994) 14610.
11. H. Yasunaga and A. Natori, Surf. Sci. Rep. 15 (1992) 205.
12. R. L. Schwoebel, J. Appl. Phys. 40 (1969) 614.
13. V. I. Marchenko and A. Ya. Parshin, Sov. Phys. JETP 52 (1980) 129.
14. R. Ghez and S. S. Lyer, IBM J. Res. Develop. 32 (1988) 804.
15. J. M. Blakely and R. L. Schwoebel, Surf. Sci. 26 (1971) 321.
16. C. Alfonso, J. C. Heyraud and J. J. Metois, Surf. Sci. 291 (1993) L745.
17. C. Alfonso, J. M. Bermond, J. C. Heyraud and J. J. Metois, Surf. Sci. 262 (1991) 371.
18. A.Natori, T.Arai and H.Yasunaga, Surf. Sci. 319 (1994) 243.
19. H. Hibino, T. Fukuda, M. Suzuki, Y. Homma, T. Sato, M. Iwatsuki, K. Miki and H. Tokumoto, Phys. Rev. B 47 (1993) 3247.
20. M. Suzuki, H. Hibino, Y. Homma, T. Fukuda, T. Sato, M. Iwatsuki, K. Miki and H. Tokumoto, Jpn. J. Appl. Phys. 32 (1993) 3247.
21. S. Kirkpatric, C. D. Gelatt and M. P. Vecci, Science 220 (1983) 671.
22. R. J. Phaneuf, N. C. Bartelt, E. D. Williams, W. Swiech and E. Bauer, Phys. Rev. Lett. 67(1991) 2986.
23. E. D. Williams, N. J. Phaneuf, N. C. Bartelt, W. Swiech and E. Bauer, Mat. Res. Soc. Symp. Proc. 238 (1992) 219.
24. S. Stoyanov, Jpn. J. Appl. Phys. 30 (1991) 1.
25. S. Stoyanov, H. Nakahara and M. Ichikawa, Jpn. J. Appl. Phys. 33 (1994) 254.
26. C. Misbah, O. Pierre-Louis and A. Pimpinelli, Phys. Rev. B 51 (1995) 17283.
27. D. Kandel and J. D. Weeks, Phys. Rev. Lett. 74 (1995) 3632.
28. M. Sato and M. Uwaha, J. Phys. Soc. Jpn. 65 (1996) 2146.
29. H. C. Jeong and J. D. Weeks, Phys. Rev. Lett. 75 (1995) 4456.

Advances in the Understanding of Crystal Growth Mechanisms
T. Nishinaga, K. Nishioka, J. Harada, A. Sasaki and H. Takei (Editors)
© 1997 Elsevier Science B.V. All rights reserved.   75

Step growth mechanism on (001) surfaces of diamond structure crystals

M. Tsuda * and M. Hata

Laboratory of Physical Chemistry, Faculty of Pharmaceutical Sciences,
Chiba University, 1-33, Yayoi-cho, Inage-ku, Chiba 263, Japan

It is well known that there are two kinds of step structures, S_A and S_B, on the (001) surface of diamond structure crystals [1]. The step growth is important for the preparation of atom-scale flat surfaces by epitaxial growth. In this chapter, elementary processes of the S_A and S_B step growth in homoepitaxy of diamond structure crystals are explained in terms of the behavior of atoms which contribute to the step growth. The behavior of atoms on surfaces was elucidated by quantum theoretical calculations where the Schrödinger equation on model surfaces with adatoms were solved by ab-initio method [9] as well as semi-empirical one [7,8].

1. THE S_B STEP GROWTH MECHANISM [2]

A simple sketch of S_A as well as S_B step is presented in Figure 1.

Now, STM images show that S_B steps of the (2×1) reconstructed Si and diamond(001) surfaces are ragged and a very long dimer row like a finger often runs out forward [3,4], suggesting the independent growth of a dimer row at S_B step edges. It was verified that the S_B step growth mechanism on the (001) surface is very similar in both cases of Si and diamond [5,6] and that the S_A step growth of diamond(001) surface is a composite process initiated by nucleations at the S_A step edge which forms the S_B step type I at both sides of the nucleus. The S_A step growth is a S_B step growth along the S_A step edge starting from the newly produced nucleus [6]. Moreover, it was found that two kinds of different structures, the S_B step type I (Figure 5(g')) and the S_B step type II (Figure 5(c)), appear alternately during the S_B step growth [5,6].

A carbon adatom migrating on the reconstructed (001) surface produces a most stable triangle structure when the adatom comes on the top of reconstructed dimers on terrace. This triangle structure is corresponding to the grand minimum of the potential energy hypersurface for the single adatom migration [6]. Therefore, a migrating adatom arrived at the terminal dimer of the S_B step type I will produce the stable triangle as shown in Figure 2(i). Then, how does this adatom make a growth of the S_B step?

*The authors thank the Computer Center of the Institute for Molecular Science, Okazaki. The computations were also carried out by DRIA System at Faculty of Pharmaceutical Sciences, Chiba University.

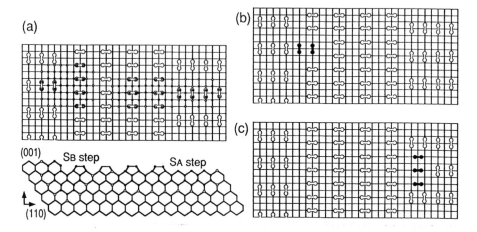

Figure 1. (a) S_A and S_B steps of the (2×1) reconstructed (001) surfaces. (001) surface is normally reconstructed because of the recombination of dangling bonds of the top-most surface atoms. Model molecular systems for the S_A and S_B steps which used for the explanation in this chapter and for the quantum theoretical calculations, are constructed by the replacement of the open circles with hydrogen atoms. The upper figure shows the top view and the lower one is the side view. (b) The independent growth of a dimer row at the S_B step. This figure is corresponding to the structure which appears after one cycle of the S_B step growth mechanism is completed. (c) The real S_A step growth initiates by nucleation. The propagation starting from the nucleus takes place to both sides of the dimer nucleus following the same mechanism as operates in the S_B step growth.

1.1. Single adatom migrations at the S_B step type I on diamond(001) surfaces

The triangle structure, where the adatom is on the top of the terminal dimer at the S_B step type I on diamond(001) surface, is shown in Figure 2(i). The adatom must move to the right side for the growth of the S_B step. Then, there appears a potential energy barrier and the adatom forms the structure of Figure 2(ii) at the saddle point of the lowest energy path of the migration. The adatom goes over the saddle point and arrives at another potential energy minimum. This stable structure is shown in Figure 2(iii). The potential energy change following the single adatom migrations illustrated in Figure 2 is shown in Figure 3(I).

The reason why the adatom is stabilized at the structure of Figure 2(iii) is that the S_B step type I is not a single dimer structure but a diene as shown in Figure 4. It should be noted that the bond length of the dimer at the S_B step type I is longer (1.56Å) than that of the simple dimers on the terrace (1.43Å) [6]. The electron distribution of the second highest molecular orbital at the S_B step type I (Figure 5

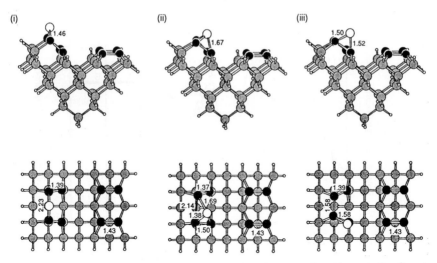

Figure 2. Structure changes following the single adatom migration at the S_B step type I edge of diamond. Numerals are bond lengths in Å. Determined by solving the Schrödinger equation by MNDO method [7,8].

in Ref. [6]) clearly shows the diene structure of this step.

1.2. Adatom cluster migrations at the S_B step type I edge on diamond(001) surfaces

Because of the diene structure illustrated in Figure 4, two stable positions are available for migrating adatoms at the S_B step type I edge as shown in Figure 5(a) where the two adatoms on the two double bonds of the diene structure have already formed a reconstructed dimer. When the adatom cluster (the dimer) moves to the right side for the S_B step growth, there appears a potential energy barrier as shown in Figure 3(II) and then the stable S_B step type II (Figure 5(c)) is formed via the saddle point structure (Figure 5(b)). It should be noted that the potential energy barrier for the cluster migration is very small in this S_B step type II formation process ($\sim$10 kcal/mol) when compared with the potential energy barrier for the single adatom migration on the terrace along the dimer row (30.8 kcal/mol).

1.3. Adatom cluster migration to form a missing dimer on diamond(001) surfaces

From the S_B step type II structure, the cluster (dimer) may migrate to form a missing dimer. The process proceeds in two steps as shown in Figure 3(II). The first step is the bond scission at the left side of the migrating dimer (Figure 5(d)) to form a triple bond cluster(Figure 5(e)). One should remember that the long bond length 3.1Å means the disappearance of bonding characters and the short bond length 1.24Å of the migrating dimer indicates the triple bond character. The

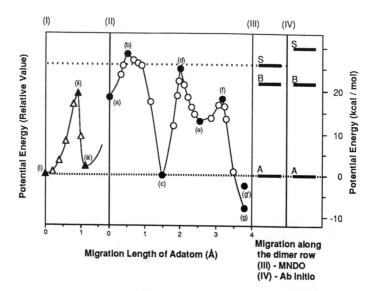

Figure 3. Potential energy changes following adatom migrations. (I) Single adatom migrations at the S_B step type I edge (MNDO); (i), (ii) and (iii) are corresponding to the structures in Figure 2. (II) Cluster (dimer) migrations at the S_B step type I edge (MNDO); (a), (b), (c), (d), (e), (f), (g) and (g') are corresponding to the structures in Figure 5. (III) Single adatom migrations on the terrace of (001) surfaces along the dimer row (MNDO) [12]. (IV) Single adatom migrations on the terrace of (001) surfaces along the dimer row (ab initio [9]) [12]; A is corresponding to the grand minimum structure. S is the saddle point and B is a shallow minimum structure. The relative values obtained using the semiempirical MNDO calculations were scaled by the absolute values obtained using the ab initio calculations at two points: i.e., the A and B which are the minima of the potential energy hypersurface.

second step is the triple bond cluster migration (Figure 5(f)) to produce a dimer nucleus and a missing dimer (Figure 5(g)). The first bond scission process is the rate limiting step for this cluster migration, which requires the approximately same activation energy as requires in the single adatom migration along the dimer row on diamond(001) surface terraces (Figure 3(II) and (III)).

There are two stable structures for the missing dimer with the dimer nucleus produced here (Figure 5(g) and (g')). The structure (g) is more stable than the structure (g'). (g') has the S_B step type I (a diene structure) at both sides of the dimer nucleus whereas the dimer in the structure (g) has a real double bond (1.42Å) which merely binds two most stable triangle structures.

A missing dimer with a dimer nucleus at the S_B step edge appeared in Figure 5 (Figure 5(g) or (g')) is often observed in the pictures of STM images of Si(001)

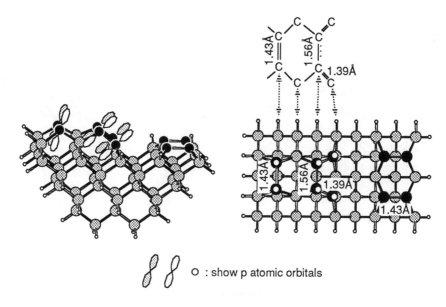

Figure 4. Diene structures at the S_B step type I.

surfaces appeared in literatures (e.g., Refs.[10,11]).

1.4. Dimer formation at the missing position

The S_B step growth is completed when a dimer is produced at the missing position in Figure 5(g'). The research has not been carried out on this problem, but we can imagine that the adatom migration along the dimer rows produces the structure (a) in Figure 5 at the right side of the missing position in the structure (g') in Figure 5, via the process shown in Figure 2. The structure (a) produced at the missing position of the structure (g') may form a dimer to make up the missing position. Thus, one cycle of the S_B step growth mechanism is completed as illustrated in Figure 1(b).

2. THE S_A STEP GROWTH MECHANISM [13,14]

It was found on (001) surfaces of diamond structure crystals that the adatom migration passing through the S_A step easily occurs from the upper terrace to the lower one, but hardly takes place from the lower terrace to the upper one, because the going-up migration of an adatom at the S_A step requires a twice amount of the activation energy than that of the terrace migration [13]. This result means that the migrating adatoms stay for a longer time at the down side of S_A steps than the other places of (001) surfaces. This adatom staying at the down side of S_A steps has to give a higher probability for producing a new dimer especially at the S_A step with another migrating adatom on (001) surfaces. The newly produced

80

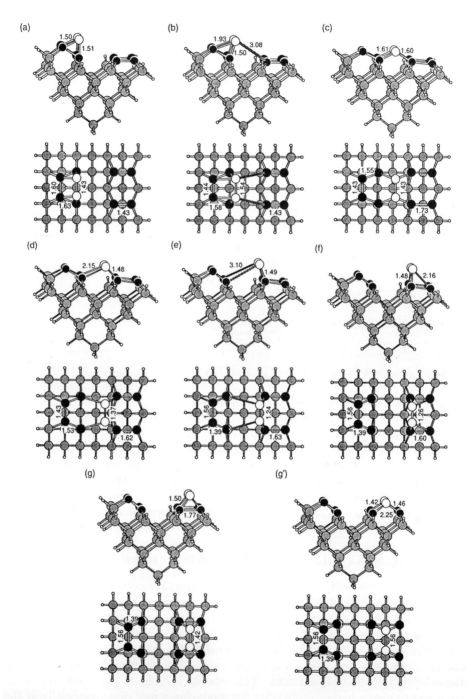

Figure 5. Structure changes following cluster migrations at the S_B step type I. Numerals are bond lengths in Å.

dimer at the S_A step may act as the smallest nucleus [6] for the S_A step growth in a sense, because the S_B step growth along the S_A step follows the smallest nucleus formation as illustrated in Figure 1(c).

2.1. A dimer formation at the S_A step with the second adatom migrating on the lower side terrace

Suppose that the first adatom stands still at the down side of S_A steps and the second adatom migrating on the lower side terrace approaches the first one to produce a dimer as shown in Figure 6(a). The lowest potential energy change was shown in Figure 7(I) along the adatom migration process illustrated in Figure 6. The lowest potential energies of the adatom migrating system were determined by the energy minimization under the conditions that those atoms indicated by black and white circles in Figure 6 are freely movable and other underlayer atoms of the model are fixed in order to hold the characteristics of crystals. The parameter indicating the adatom migrations is the distance between the first adatom and the second one. In Figure 7(I), (a), (b), (c), (d), (e), (f) and (g) are corresponding to the potential energies of the structures having the same notation in Figure 6. The first saddle point appears in the structure (b) where the second adatom occupies the same position that an adatom occupies in the structure S which is the saddle point in the single adatom migration on the terrace (Figure 7(II) and (III)).

A shallow minimum appears at the structure (c) which is corresponding to the shallow minimum structure B (Figure 7(II) and (III)) during the single adatom migration on the terrace. The second saddle point appears again in the structure (d) in Figure 6 which is corresponding to the saddle point structure S in Figure 7(II) and (III) during the single adatom migration.

Passing through the structure (d), the potential energy is rapidly stabilized to form the dimer structure (g). The structure (e) is corresponding to a point of inflexion of the lowest potential energy change for the adatom migration. The isolated dimer structure produced here is often observed in experiments as discussed later. On the other hand, it is already known that the dimer serves as the smallest nucleus for the S_B step growth [6]. For this reason, the S_A step growth is not considered to be a simple event, but a complex one; i.e., the isolated dimer formation is followed by the S_B step growth along the S_A step starting from the newly formed isolated dimer. Therefore, the dimer structure (g) produced at the S_A step is the smallest nucleus for the S_A step growth in a sense.

2.2. A dimer formation at the S_A step with the second adatom migrating on the upper side terrace

Since the going-up migration of an adatom at the S_A step requires a twice amount of the activation energy than that of terrace migrations [13], it was postulated that the first adatom stands still at the down side of S_A steps and the second adatom migrating down from the upper side terrace approaches the first one to form a dimer as illustrated in Figure 8.

Suppose that an adatom migrates on the upper side dimer row next to the S_A step. This adatom goes down the S_A step easily with a similar amount of

82

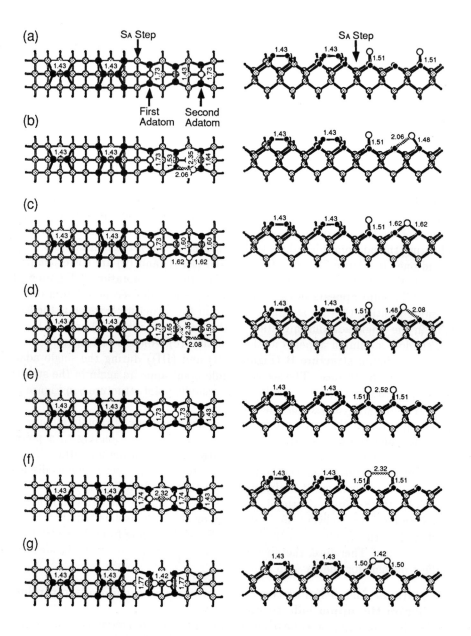

Figure 6. Structure changes following the second carbon adatom migration on the lower side terrace at the S_A step. The migrating second adatom forms a dimer with the first adatom standing still at the down side of the S_A step.

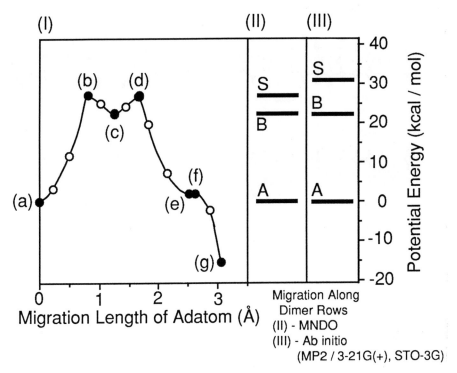

Figure 7. The lowest potential energy changes following the adatom migrations. (I) The potential energy variation calculated by MNDO method following the structure changes (a) → (g) illustrated in Figure 6. The abscissa shows the migrated distance of the second adatom. On the others, see the caption of Figure 3.

activation energy required for the terrace migration of an adatom [13]. Therefore, the second adatom at the structure (a') may migrate down the S_A step as illustrated in Figure 8. The lowest potential energy change along the adatom migration process of Figure 8 was shown in Figure 9(I) where the notations (a'), (b'), (c'), (d'), (e'), (f'), (g'), (h') and (i') are corresponding to the optimized structure shown in Figure 8.

The first saddle point (b') appears with the bond fission of the triangle structure of (a'). Passing through a plateau of the potential energy corresponding to the structure (c') and (d'), a minimum is observed when a new triangle is produced in the structure (e'). The second saddle point appears when the bond fission of this newly produced triangle occurs in the structure (f'), then the dimer row next to the S_A step is reproduced again in the structure (g') at the potential energy minimum. It is interesting that the second adatom is positioned on the top of the first adatom in the structure (g'). Passing through the third saddle point (h'), a new dimer, which operates as a smallest nucleus [6] for the step growth, is produced at the

84

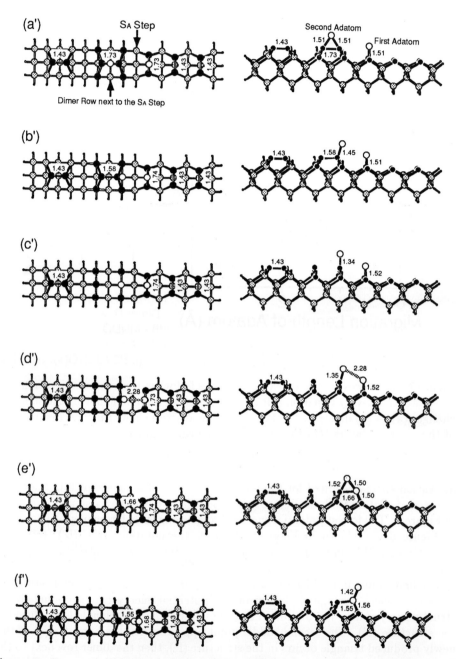

Figure 8. Structure changes following the second carbon adatom migration from the upper side terrace at the S_A step to the first adatom. The migrating second adatom forms a dimer with the first adatom standing at the down side of the S_A step.

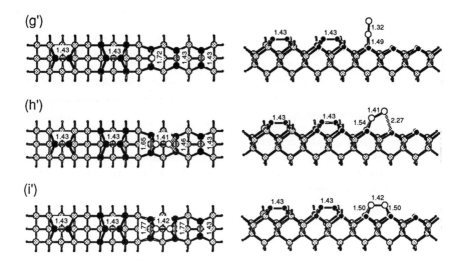

(g')

(h')

(i')

Figure 8. Continued.

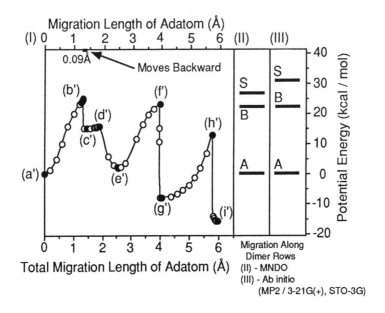

Figure 9. The lowest potential energy change following the adatom migrations. (I) The potential energy variation following the structure changes (a') → (i') illustrated in Figure 8. On the others, see the caption of Figure 3.

structure (i'). The energy required for passing through each of three saddle points, from (a') to (b'), from (e') to (f') and from (g') to (h'), are a similar value. Since this value is not different from the activation energy required for the single adatom migration on terrace (from A to S in Figure 9(II)), the dimer formation process at the S_A step investigated in this section occurs easily. This newly produced dimer is followed by the S_B step growth along the S_A step, and one step propagation of the S_A step is accomplished.

2.3. Conclusion on the S_A step growth mechanism

A mechanism is elucidated that dimers are favorably produced from migrating adatoms at the S_A step on (001) surfaces of diamond structure crystals. The dimers produced at the S_A step operate as the smallest nucleus for the S_B step growth, because both sides of the dimer act as the S_B step [6]. It was concluded that the S_A step growth is not a simple event, but a complex one; i.e., the isolated dimer formation at the S_A step is followed by the S_B step growth starting from the newly formed isolated dimers. The S_B step growth along the S_A step means one step propagation of the S_A step growth. This mechanism repeats itself during the S_A step growth.

A dimer row produced on the terrace has two S_A steps at both sides of the dimer row. Since migrating adatoms produce favorably dimers at the S_A steps, it has to be observed that short dimer rows gather each other apparently to form small monolayer islands on the terrace. Experiments [10] support this theoretical prediction.

3. COMPARISON WITH EXPERIMENTS

It is well known that the most stable binding site of a single Si adatom is different from the position that is required by epitaxial growth [15,16]. On the other hand, two adatoms interacting in order to produce a dimer were proved by ab initio quantum-theoretical calculations [16] to take just the positions required for epitaxial growth [16]. Moreover, the (2×1) reconstructed structures were observed by the scanning tunneling microscope (STM) experiments of diamond(001) surfaces, where the same two types of steps, S_A and S_B, were clearly shown[4] as observed on the epitaxially grown reconstructed Si(001) surfaces. For these reasons, the theoretical results on the finally produced dimer in this paper will be comparable with the STM images on the Si dimer on the reconstructed Si(001) surfaces.

It is often observed in STM photographs that many isolated dimers are formed at S_A steps of Si(001) surfaces produced by MBE (molecular beam epitaxy) [11]. The dimer formation mechanism at the S_A step elucidated in this research explains the origin of the isolated dimers at the S_A step observed in the experiments. The isolated dimers operate as the S_B step for epitaxial growth [6]. The S_B step growth progressing along the S_A step means the S_A step growth. It is known that short dimer rows which were produced by a small amount of deposition of Si atoms (0.07 mono-layer) [10] on terraces of Si(001) surfaces often get together and form a small monolayer island. Provided that the dimer formation on Si(001) surfaces takes place only by an encounter of two Si adatoms migrating on the same dimer row, the gathering of short dimer rows may scarcely occur under this experimental condition [10]. However, an adatom stays for a longer time at the next side of a dimer row than the other places on Si(001) surfaces as elucidated in the present research, since a dimer row on the terrace forms two S_A steps at both sides of the dimer row itself and gives to a migrating adatom the same circumstances that a S_A step does. Therefore, dimers are produced favorably at the next side of dimer rows, and as the result the formations of small monolayer islands have to be observed as was in experiments.

REFERENCES

1. D. J. Chadi, Phys. Rev. Lett. 43 (1979) 43.
2. M. Hata, C. Ohara, S. Oikawa and M. Tsuda, Appl. Surf. Sci. 75 (1994) 21.
3. C. C. Umbach, M. E. Keeffe and J. M. Blakely, J. Vac. Sci. Technol., B9 (1991) 721.
4. T. Tsuno, T. Imai, Y. Nishibayashi, K. Hamada and N. Fujimori, Jpn. Appl. Phys., 30 (1991) 1063.
5. M. Tsuda, S. Oikawa and S. Furukawa, J. Cryst. Growth, 115 (1991) 556.
6. M. Tsuda, S. Oikawa, S. Furukawa, C. Sekine and M. Hata, J. Electrochem. Soc., 139 (1992) 1482.
7. M. J. S. Dewar and W. Thiel, J. Am. Chem. Soc., 99 (1977) 4899.
8. J. J. P. Stewart, MOPAC Ver. 6, QCPE Bull. 9, (1989) 10.
9. M. J. Frisch, G. W. Trucks, M. Head-Gordon, P. M. W. Gill, M. W. Wong, J. B. Foresman, B. G. Johnson, H. B. Schlegel, M. A. Robb, E. S. Replogle, R. Gomperts, J. L. Andres, K. Raghavachari, J. S. Binkley, C. Gonzalez, R. L. Martin, D. J. Fox, D. J. Defrees, J. Baker, J. J. P. Stewart, and J. A. Pople, Gaussian 92, Revision A (Gaussian, Inc., Pittsburgh PA, 1992).
10. Y. W. Mo, J. Kleiner, M. B. Webb and M. G. Lagally, Phys. Rev. Lett., 66 (1991) 1998.
11. R. J. Hamers, V. K. Köbler and J. E. Demuth, J. Vac. Sci. Technol., A8 (1990) 195.
12. M. Tsuda, M. Hata and S. Oikawa, Appl. Surf. Sci., 107 (1996).
13. M. Tsuda, M. Hata, Y. Shin-no and S. Oikawa, Surf. Sci., 357/358 (1996) 844.
14. M. Tsuda, M. Hata and Y. Shin-no, to be published.
15. G. Brocks, P. J. Kelly and R. Car, Phys. Rev. Lett. 66 (1991) 1729.
16. M. Tsuda, M. Hata, E. Araki and S. Oikawa, Appl. Surf. Sci., 107 (1996).

Advances in the Understanding of Crystal Growth Mechanisms
T. Nishinaga, K. Nishioka, J. Harada, A. Sasaki and H. Takei (Editors)

Monte Carlo simulation of MBE growth

T. Irisawa[a]* and Y. Arima[b]

[a]Computer Center, Gakushuin University,
1-5-1 Mejiro, Toshima-ku, Tokyo 171, Japan

[b]Department of Physics, Gakushuin University,
1-5-1 Mejiro, Toshima-ku, Tokyo 171, Japan

The periodic changes in the structure of a surface grown by MBE (molecular beam epitaxy) are investigated theoretically by means of a Monte Carlo simulation. The conditions for oscillations in RHEED (reflection high energy electron diffraction) intensity are interpreted in terms of the lifetime τ_c of an adatom before capture by other adatoms and the mean diffusion length λ_c during τ_c. If J is the incident beam flux and D_s is the surface diffusion coefficient of adatoms, $\tau_c = (JD_s)^{-1/2}$ and $\lambda_c = (D_s/J)^{1/4}$. We obtain a diagram that can be used to predict the growth conditions under which the periodic changes in a growing surface that cause RHEED oscillation occur. The characteristic surface diffusion field is obtained for each growth mode. Two-dimensional nucleation on terraces is investigated at the atomic level by considering the time evolution of the density of clusters. A stoichiometric A-B crystal grown layer-by-layer is obtained under the conditions $\phi_{AA} = \phi_{BB} = \phi_{AB}/r$, $r \geq 1$, $J_A = J_B$, where ϕ_{AA}, ϕ_{BB} and ϕ_{AB} are the bond energies of A-A, B-B and A-B bonds, and J_A and J_B are the fluxes of a A and B atoms, respectively. It is found that the degree of ordering S of A-B crystal is a function of $(2\phi_{AB}/kT - \Delta\mu/kT)$, where $\Delta\mu/kT$ is a parameter representing the intensity of the flux.

1. INTRODUCTION

Since it was reported that RHEED (reflection high-energy electron diffraction) intensities for a GaAs crystal surface growing by MBE (molecular beam epitaxy) oscillate with a period corresponding to the completion of a monolayer [1], this phenomenon has been applied to thin layer growth of superlattices. The MBE method is a crystal growth process in which all the incident atoms are incorporated into the lattice because substrate temperature is low enough that evaporation from the surface can be negrected. Recently, the surface structure has been studied in detail using STM (scanning tunneling microscopy), with which one can see not

*This work was partly supported by Grants-in-Aid for Scientific Research from the Ministry of Education, Science and Culture, Japan, No. 03243101 and No. 04227108, and the "Foundation for the Promotion of Materials Science and Technology" of Japan (MST Foundation).

only the monolayer steps but also some atomic clusters.

We carried out a Monte Carlo study to determine the structural features of a surface during MBE growth, because we cannot apply the conventional vapor growth theory [2]. However, we had to develop a new simulation method because of the slow growth rate under MBE conditions.

We describe the simulation methods used in this work in the next section. In section 3, we determine the characteristic length for MBE growth by considering taking into account the kinematics of the system, and the periodic changes and flatness of a surface growing under MBE conditions [3,4]. The two growth modes (layer by layer and step flow) of a vicinal surface under MBE conditions are distinguishable by whether or not nucleation occurs. In section 4, we discuss the diffusion fields and nucleation on a growing vicinal surface under several growth conditions. Finally in section 5, we discuss aspects of a two-component system under MBE growth which are not seen in a one-component system.

2. SIMULATION MODEL

First, we discuss the simulation models and the algorithm used in this work. There are the two models, and we simulate the growth of the (001) face of a simple cubic crystal. The processes to be considered are adsorption, evaporation and surface diffusion of the atoms.

Model 1 is for a one-component system and anisotropic bond energy (ϕ_x, ϕ_y and ϕ_z) and surface diffusion are taken into account. When we set $\phi = \phi_x = \phi_y = \phi_z$, this model is equivalent to Gilmer and Bennema's model [5]. The result in the next section is obtained with this simple model with $\phi = \phi_x = \phi_y = \phi_z$.

The ratio of evaporation rate K_{ij}^- to adsorption rate K^+ is given by

$$K_{ij}^-/K^+ = \exp[(1-i)\phi_x/kT + (1-j)\phi_y/kT - \Delta\mu/kT] \tag{1}$$

and the ratio of the surface diffusion rate K_{ij}^D to the adsorption rate is given by

$$K_{ij}^D/K^+ = (\lambda_s/a)^2 K_{ij}^- / K^+. \tag{2}$$

Here, $\Delta\mu$ is the difference between the chemical potential of the solid and the vapor, a is the lattice constant, k is the Boltzmann constant, T is the temperature, and i and j ($i, j = 0, 1, 2$) are the numbers of nearest neighbor bonds in the x and y directions, respectively. The surface diffusion length λ_s is given by

$$\lambda_s^2 = D_s \tau_s = a^2 exp[(\phi_z - E_D)/kT]$$
$$= a^2 exp[(1-\varepsilon)\phi_z/kT], \tag{3}$$

where the surface diffusion constant is given by $D_s = a^2\nu \exp[-E_D/kT]$ and the lifetime of adatoms is given by $\tau_s = \nu^{-1} \exp[\phi_z/kT]$. Here, ν is the vibrational frequency and E_D is the activation energy of surface diffusion. E_D is given by $E_D = \varepsilon\phi_z$ ($0 < \varepsilon < 1$).

Model 2 is for an A-B two-component system. The adsorption rates of A and B atoms are determined by the fluxes J_A and J_B and are given by

$$K_X^+ = J_X a^2, \tag{4}$$

where the suffix X is A or B. The evaporation rates of A and B atoms are given by

$$K^-_{Xij} = \nu_X \exp[(-i\phi_{XA} - j\phi_{XB})/kT].\tag{5}$$

Here i and j denote the numbers of A and B atoms bonded to X $(0 < i + j < 6)$, respectively, ϕ_{AA}, ϕ_{BB} and ϕ_{AB} are the bond energies of A-A, B-B and A-B pairs, and ν_A and ν_B are the vibrational frequencies, which are assumed to be equal. Surface diffusion is an important process in MBE growth. We define the surface diffusion rates of A and B atoms as

$$K^D_{Xij} = (\lambda_{sX}/a)^2 K^-_{Xij},\tag{6}$$

where λ_{sA} and λ_{sB} are the surface diffusion lengths of A and B adatoms, respectively.

The simulation procedure is used the waiting time method [6]. That is, the probability that the surface structure changes is given by

$$P_{mn} = N_{mn}(K^+ + K^-_{mn} + K^D_{mn})\tau_w\tag{7}$$

$$\tau_w = 1/\sum_{i,j} N_{ij}(K^+ + K^-_{ij} + K^D_{ij}),$$

where N_{mn} is the number of the lattice site with (m,n) bonds, and τ_w is the waiting time taken for the surface structure to chage. The structure changes in proportion to P_{mn}. We cannot simulate the movement of adatoms using any other method when the substrate temperature is low and the surface diffusion length is large, as they are under MBE growth conditions, because the surface structure hardly changes if the conventional time constant simulation is used ($\tau = 1/N(K^+ + K^-_{00} + K^D_{00})$; N is the total number of lattice sites).

3. PERIODIC CHANGES IN AND FLATNESS OF A SURFACE GROWING UNDER MBE CONDITIONS

In the conventional vapor growth theory [2] used for ordinary growth conditions, evaporation of adatoms from the surface is included as an important process. Thus, a key parameter which determines the growth rate or the surface structure is the average waiting time τ_s for an adatom to evaporate (i.e. the mean lifetime on the surface) or the average diffusion length λ_s of an adatom during that time. On the other hand, under MBE growth conditions no evaporation occurs, and hence, the conventional theory cannot be applied. Under such conditions, the important parameter is the waiting time, τ_c, of an adatom to be captured by another adatom (i.e. the actual mean lifetime of an adatom) or the expected diffusion length λ_c of an adatom during that time.

In a one-component system, the time τ_c and the length λ_c are defined [3,4] by considering the kinematics of the system [7], and neglecting the evaporation of adatoms and the dissociation of two-atom clusters (dimers) on the surface. Since an adatom migrates over an area D_s in one second, it encounters $D_s N_1$ adatoms per second, where N_1 is the adatom density. Therefore, τ_c is given by

$$\tau_c = 1/D_s N_1. \tag{8}$$

Then the decrease in the adatom density due to encounters between two adatoms is $-N_1/\tau_c = -D_s N_1^2$ per unit time.

If the incident beam flux J is so large that the number, i^*, of atoms in the critical two-dimensional nucleus is one, dissociation of the dimers does not occur. Moreover, since we assume that $\lambda_c < \lambda_s$, the decrease in the adatom density, $-N_1/\tau_s$, due to evaporation from the surface can also be ignored. Therefore, the adatom density under steady state conditions is given by

$$N_1 = (J/D_s)^{1/2}. \tag{9}$$

From equations (8) and (9), τ_c is expressed as

$$\tau_c = (J/D_s)^{-1/2}. \tag{10}$$

Stable clusters $(i > 2)$ are created at $t \simeq \tau_c$, and there are as many as $N_s = (N_1/\tau_c)\tau_c = N_1$. Then

$$N_s = (J/D_s)^{1/2}. \tag{11}$$

Therefore, the territorial area $\lambda_c^2 = 1/N_s$ of each stable nucleus is expressed as

$$\lambda_c^2 = (D_s/J)^{1/2}. \tag{12}$$

The mean distance λ_c between neighboring stable nuclei given by eq. (12) is equal to the mean diffusion distance $\lambda_c = (D_s\tau_c)^{1/2}$ of an adatom during its lifetime τ_c. Since $N_s = N_1$, there is only one adatom in territorial area of the stable nucleus, on average, and no further nucleation by encounters of adatoms occurs. In other words, stable nuclei grow by capturing all of the adatoms from the vapor which migrate over the surface, and no further nucleation occurs. Thus growth continues as follows. A dimer (i.e. a stable nucleus) is created in each territory with an area of λ_c^2 on the flat surface. Then, the nuclei grow by in incorporating all atoms from the vapor which are migrating over the surface. When the nuclei grow to the maximum diameter λ_c and coalesce with each other, the growth layer is completed and the surface is again flat. This periodic change in the surface structure is detected experimentally by observing oscillations of the RHEED intensity. However, when the incident flux J is so large that the length λ_c becomes as short as the lattice constant a, the surface is no longer flat because the impinging atoms crystallize without a surface diffusion. Therefore, the necessary condition for layer growth with a flat surface and without evaporation of the adatoms is given by

$$a < \lambda_c < \lambda_s. \tag{13}$$

As shown in Figure 1, a diagram of the growth mode can be obtained using the computer simulations described in the last section. The periodic changes in the

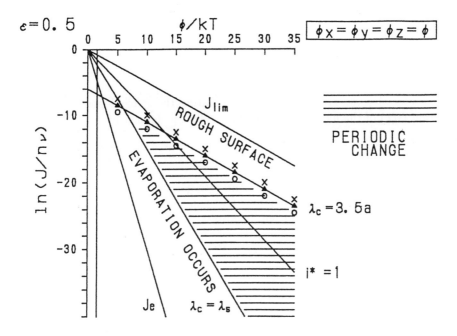

Figure 1. Growth mode corresponding to various growth conditions ($\phi/kT, \ln J/n\nu$) for $\varepsilon = 0.5$, where n is the density of the surface lattice points.

surface structure occurs in the hatched region. Evaporation of only 5 percent of the adatoms is occurs at the lower limit of the hatched region showing that $\lambda_s = \lambda_c$. At the upper limit of the hatched region, λ_c is about 3.5 times the lattice constant a. This simulation result agrees with our theory described above.

4. GROWTH OF VICINAL SURFACE

4.1. Decrease in magnitude of RHEED oscillations

On a vicinal surface, there is a series of parallel steps separated by a distance λ. The advance of these parallel steps contributes to the growth of the vicinal surface. The two-dimensional nucleation on the terraces also contributes to the growth. If the contribution of the former is much greater than that of the latter, the surface grows continuously. Under step flow growth, the surface structure does not change periodically and oscillation of the RHEED intensity is no longer observed.

If the characteristic length λ_c is greater than the distance λ between neighboring steps, almost all of the adatoms are incorporated by the step before being captured by other adatoms. This means that two-dimensional nucleation does not occur on the terraces. Therefore, oscillation of the RHEED intensity is not observed under the condition

$$\lambda < \lambda_c. \tag{14}$$

A higher substrate temperature or a smaller incident flux results in a greater diffusion length λ_c.

We determine the growth mode by calculating the space-dependent correlation function $G(r)$ of the local surface height z_i. This is defined by

$$G(r) = <\Delta z_i \cdot \Delta z_j> \tag{15}$$
$$\Delta z_i = z_i - <z>,$$

where the suffix i denotes the i-th lattice site on the surface, and $r = |r_i - r_j|$ is the distance between two lattice sites i and j in the direction perpendicular to the steps [8]. Δz_i is the deviation from the average height. The base of the local height z is the height of the initial surface with straight parallel steps. Figure 2 shows the correlation function at the point when half a layer has grown. The local maximum of the correlation function at the middle of the terrace indicates the occurrence of two-dimensional nucleation on the terrace. Then, we define the boundary of the two growth modes by the criterion that the local maximum of $G(r)$ is equal to 0. As shown in Figure 3, the critical value of λ_c^* which corresponds to the boundary of the growth mode has a linear dependence on the step distance λ. Therefore, the growth mode of the vicinal surface under MBE conditions is characterized by λ_c, which is in agreement with our theory.

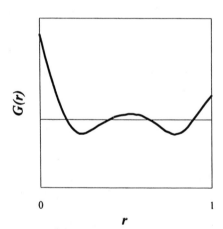
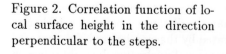

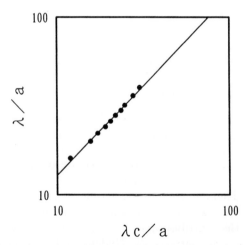

Figure 2. Correlation function of local surface height in the direction perpendicular to the steps.

Figure 3. Plot of the critical values of λ_c corresponding to the boundary between the growth modes.

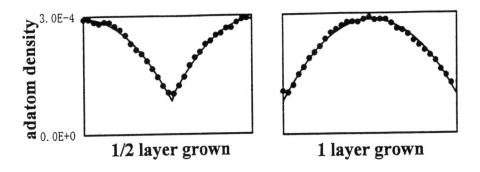

Figure 4. Distribution of adatom density $N(x)$ during step flow growth. The thin lines are obtained by least squares fits with parabolas.

4.2. Surface diffusion and nucleation on terrace

Since no evaporation of adatoms occurs under MBE conditions, the growth rate R depends only on the incident flux J. Thus, it is given by

$$R = \Omega J, \tag{16}$$

where Ω is the atomic volume. If the growth mode is step flow, the advance velocity v of the parallel steps is proportional to the step distance λ as

$$v = R\lambda/a. \tag{17}$$

This is much smaller than the rate of surface migration of an adatom. Hence, the adatom density N_1 satisfies the following steady state equation, neglecting the advancement of steps.

$$D_s d^2 N_1 / dx^2 = J. \tag{18}$$

Then, the distribution $N_1(x)$ of adatom density perpendicular to the steps , i.e. the surface diffusion field, is a parabolic function of x. As shown in Figure 4, we obtain the surface diffusion field for step flow growth using the simulation. Since the adatom density is very small, we determine the surface diffusion "field" by taking an average of over many samples. In Figure 4, we choose the initial step positions as the boundaries of the horizontal axis. Then, the step lies half way along the axis when half of one layer has grown. The distribution $N_1(x)$ is carved with the maximum at the center of the terrace and the minimum at the step. The minimum point moves as the steps advance. The thin lines in Figure 4 show the least squares fits with parabolas. Good agreement is obtained.

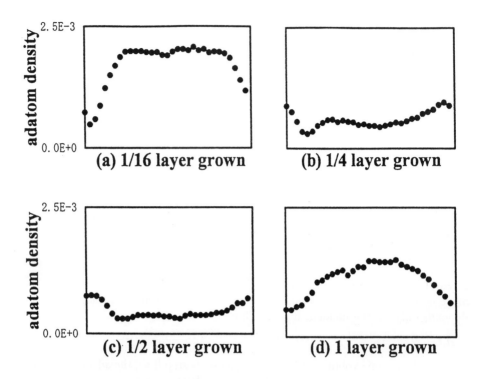

Figure 5. Distribution of adatom density $N(x)$ for nucleation growth. As the growth proceeds, the distributions change from (a) to (d).

It is very difficult to calculate the surface diffusion field when the growth mechanism includes two-dimensional nucleation on the terraces, because the edges of the nuclei are also the sinks for adatoms. However, we can obtain an approximation using simulations as shown in Figure 5. In the early stage of growth (Figure 5a), the maximum of the diffusion field distribution collapses due to two-dimensional nucleation on the terraces. As the growth continues with nucleation and the spread of nuclei (Figures. 5b and 5c), the diffusion field collapses entirely. However, when the growth of a monolayer is complete (Figure 5d), the field again has a parabolic shape on the flat terrace. Thus, the surface grows periodically by repetition of this process. These results support the proposal of Nishinaga et al. [9] that two-dimensional nucleation occurs on the terrace when the supersaturation at the center of the terrace exceeds a certain critical value.

The conventional thermodynamic nucleation theory cannot be applied to the two-dimensional nucleation process under MBE growth conditions in which the critical nuclei consist of only a few atoms. Therefore, we investigate the nucleation process using simulations and analysis of the master equation.

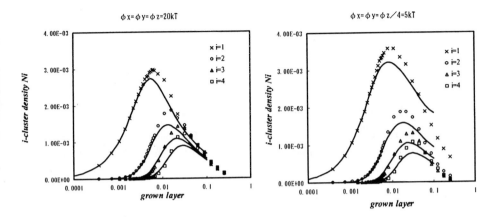

Figure 6. Change in the density $N_i(i = 1, 2, 3, 4)$. Fitting curve is obtained by solving the master equations.

First, we count the number of clusters on the surface generated in the simulation. In order to remove statistical fluctuations, we take the average over 50 samples. The clusters which are incorporated into the steps or merge with other clusters are not counted. The time evolutions of the densities N_i ($i = 1, 2, 3$ and 4) of the clusters are shown in Figure 6. Since the evaporation of adatoms is negligible, we represent the time in results of grown layers. We considered two cases:

case 1: isotropic atomic interaction ($\phi_x = \phi_y = \phi_z = 20kT$),

case 2: weak lateral atomic interaction ($\phi_x = \phi_y = \phi_z/4 = 5kT$).

In both cases, the clusters are formed at a very early stage of the growth at which the adatom density N_1 is a maximum. In case 1, three-atoms clusters have not yet been created when N_1 is a maximum. Thus, it is assumed that a two-atom cluster (i.e. the dimer) is a stable nucleus. On the other hand, three- or four-atom clusters have already been created when N_1 reaches a maximum in case 2. The delay in the formation of the N_1 peak implies that dissociation of atoms from the clusters occurs. Therefore, a critical nucleus consists of three or more atoms in case 2. In addition, it is noted that the density N_4 of four-atom clusters exceeds that of three-atom clusters in the later stages of growth. This is considered to be due to the shape of the clusters.

Secondly, we analyze the nucleation process in the early stage of growth using the master equations. The density N_i of a cluster of size i is given by;

$$dN_i/dt = w_{i-1}^+ N_{i-1} + w_{i+1}^- N_{i+1} - (w_i^+ + w_i^-)N_i, \quad (i = 2, 3, ...) \tag{19}$$

$$dN_1/dt = J + \sum_{i=1}^{\infty}(w_{i+1}^- N_{i+1} - w_i^+ N_i) + w_2^- N_2 - w_1^+ N_1, \tag{20}$$

where w_i^+ and w_i^- denote the rates of attachment and detachment of adatoms to a cluster of size i, respectively. These rate are taken to be

$$w_i^+ = D_s N_1 \sqrt{i},$$

and

$$w_i^- = (2D_s/a^2)\exp(-\phi_x/kT)/4,$$

where the factor $\sqrt{i}$ originates from the length of the edge of a cluster of size i. The numerical solutions of equations (19) and (20) are shown by the solid lines in Figure 6. Good agreement with the simulation results is obtained.

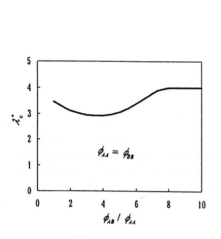

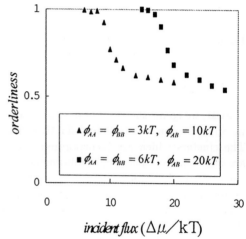

Figure 7. Change in the critical condition λ_c^*.

Figure 8. Plots of orderliness vs. incident flux.

5. A-B TWO-COMPONENT SYSTEM

Many multicomponent crystals are grown by MBE. However, the growth mechanisms are not yet well understood because the growth conditions are controlled by many parameters [10]. In this section, we consider a two-component system namely a stoichiometric A-B crystal for which $\phi_{AA} = \phi_{BB} < \phi_{AB}$, $J_A = J_B$, $\lambda_{sA} = \lambda_{sB}$ and the growth conditions are simulated to those for MBE.

Under these conditions, there are 4 growth mechanisms obtained by simulation. The crystal grows a) with a large degree of ordering and layer by layer growth, b) with a large degree of ordering but not layer by layer growth because of the larger incident flux, c) with a small degree of ordering due to the smaller ratio of the bond energies ϕ_{AA} and ϕ_{AB} and layer by layer growth because of the small flux, and d) with a small degree of ordering and not layer by layer growth. Here, the degree of ordering, S, denotes the fraction of adatoms the six nearest neighbors of which are all different species, normalized by the number of atoms which constitute the layer. As for the one-component system, the characteristic lengths can be defined by

$$\lambda_{Xc} = [D_X/(J_A + J_B)]^{1/4}, \tag{21}$$

where the suffix X is A or B. Since we consider a stoichiometric system, we omit this suffix and simply write λ_c. We simulated the growth with various fluxes J_A and J_B and found the critical value, λ_c^*, periodic changes in the surface structure. This corresponds to the upper limit of the hatched region in Figure 1. Figure 7 shows the critical length λ_c^* versus the bond energy ratio $r = \phi_{AB}/\phi_{AA}$. We can see that the growth mode cannot be understood in terms of λ_c alone in the case of a two-component system, unlike a one-component system, because the plot of λ_c^* is not a straight line. That is, λ_c^* has a minimum value near $r = 4$ at $\phi_{AB}/kT = 20$. For large r, the lifetime of an adatom is shorter when it is the same species as the substrate atom than when it is different. That is, the territory of a stable cluster has a larger area than that given by eq. (21) because it must diffuse over a large distance before encountering other atoms of different species. Therefore, the surface structure becomes multilayered, because nuclei can be created on large stable clusters when r is too large. In contrast, when $r < 4$, the surface structure is stabilized, because the territory of a stable nucleus is not so large and adatoms can move easily in comparison with the case of $r = 1$. Therefore, adatoms are caught at the edge of the stable cluster before they encounter other adatoms.

It is clear that a layer-by-layer grown crystal does not always have a large degree of ordering, considering the case for $r = 1$. The degree of ordering S changes with the ratio of the bond energies ($\phi_{AA}, \phi_{BB} : \phi_{AB}$), temperature, surface diffusion length and incident atom flux.

Figure 8 shows the degree of ordering versus the incident flux J, which is given by $\exp[\Delta\mu/kT - 3\phi_{AB}/kT]$. When the flux is small, since an adatom has enough time to migrate towards a stable position, the degree of ordering is large. On the other hand, when the flux is large, since many adatoms collide with other adatoms and crystallize in unstable positions, the degree of ordering is small. This transition seems to be a second order one. In this case, the degree of ordering S is a function of $(2\phi_{AB}/kT - \Delta\mu/kT)$, because the variable $(2\phi_{AB}/kT - \Delta\mu/kT)$ determines K_{A01}^-/K_A^+, which is important in determining the surface structure under stoichiometric growth conditions. Here, $\Delta\mu$ is not the actual chemical potential difference but a control parameter of the flux. Therefore, we assume that this transition exists in some region of $\phi_{AA}, \phi_{BB} < \phi_{AB}$.

6. CONCLUSIONS

Several growth features are analyzed by Monte Carlo simulation in order to understand MBE growth on an atomic scale. First, the surface diffusion fields are clarified for the step growth mode on a vicinal surface, and the cluster distributions are also clarified. We can observe the cluster distributions using recent experimental techniques. We obtained several results for two-component systems which can be applied to many systems, although we considered only a simple case. However, there are many problems for two-component systems, such as the dependence on partial pressure and the nonsymmetric bonding energy. In order to take account of these, we must wait for future research and development.

REFERENCES

1. J.J.Harris and B.A.Joyce: Surf. Sci. 108 (1981) L90.
2. W.K.Burton, N.Cabrera and F.C.Frank: Phil. Trans. Roy. Soc. A243 (1951) 299.
3. T.Irisawa, Y.Arima and T.Kuroda: J. Crystal Growth 99(1990) 491.
4. T.Irisawa, A.Ichimiya and T.Kuroda: Surf. Sci. 242 (1991) 148.
5. G.H.Gilmer and P.Bennema: J. Appl. Phys. 43 (1972) 1347.
6. A.B.Borz, M.H.Kalos and J.L.Lebowitz: Comput. Phys. 17 (1975) 101.
7. B.Lewis and D.S.Campbell: J. Vacuum Sci. Tecnol. 4 (1967) 209.
8. Y.Arima and T.Irisawa: J.Crystal Growth 115 (1991) 428.
9. T.Nishinaga and K.I.Cho: Jpn. J. Appl. Phys.27 (1988) L12.
10. M.Takata and A.Ookawa: J. Crystal Growth 24/25 (1974) 515, M.Takata and A.Ookawa: J. Crystal Growth 42 (1977) 35.

Advances in the Understanding of Crystal Growth Mechanisms
T. Nishinaga, K. Nishioka, J. Harada, A. Sasaki and H. Takei (Editors)

Crystal growth kinetics on stepped surface by the path probability method

K. Wada and H. Ohmi

Department of Physics, Faculty of Science, Hokkaido University,
Sapporo 060, Japan

The crystal growth kinetics on the stepped surface is studied on the basis of the kinetic equation derived by the path probability method (PPM) of non-equilibrium statistical method. First, the kinetic equation is applied to the crystal growth on a stepped surface by molecular-beam epitaxy. The growth mode changes from the nucleation mode on the terrace to the step propagation mode with increasing temperatures. It agrees qualitatively with the RHEED experiments on the GaAs(001) stepped surface. Second, the current-induced domain conversion phenomena on the Si(001) stepped surface with two types of terraces are discussed. The anisotropy of diffusion rates along with the Schwoebel effect at steps reproduces the domain conversion under heating current.

1. INTRODUCTION

For many semiconductor films grown by MBE (molecular-beam epitaxy), the intensity oscillations of RHEED (reflection high-energy electron diffraction) have been commonly utilized as a technique for monitoring the layer-by-layer crystal growth. The stepped surface as well as the flat surface is often used as a substrate for MBE to fabricate excellent thin films. Neave et al.[1] have first demonstrated experimentally the temperature variation of the intensity oscillations of RHEED corresponding to growth modes on the stepped surface of GaAs(001). On the other hand, in order to obtain a clean and smooth crystal surface on an atomic level, the annealing through the DC heating current is often used after crystal growth such as by MBE. Latyshev et al.[2] observed a current-induced conversion of two types of domains on the Si(001) stepped surface during annealing. It is important to explain these phenomena theoretically based on the microscopic models in order to understand the crystal growth and annealing mechanism on an atomic level.

We apply the path probability method (PPM)[3] to the crystal growth problems on stepped surfaces. The PPM was devised by Kikuchi as a systematic variational method of non-equilibrium statistical mechanics and first applied to a crystal growth problem from vapor by Temkin[4]. The advantage of this method consists in that once the Hamiltonian and the elementary kinetic processes are set up, the kinetic equation of the system can be written down systematically within a chosen approximation of the PPM.

First, we show the derivation of a kinetic equation for crystal growth by MBE and for a current-induced domain conversion on the basis of the SOS (solid-on-solid) model in the site-dependent point approximation of the PPM[5,6], though some modifications are required for the latter case. Second, we apply the kinetic equation to crystal growth by MBE on the stepped surface. Third, the current-induced conversion on the Si(001) stepped surface is treated by the above kinetic equation with some modifications[7].

2. FORMULATION

Consider the (001) face of a simple cubic crystal. Then the surface is defined as a SOS model by regarding the crystal as an assembly of columns on a two-dimensional square lattice, each column being composed of solid atoms. The Hamiltonian of the SOS model is written as

$$H = J \sum_{<\boldsymbol{R}, \boldsymbol{R}'>} \left| h(\boldsymbol{R}) - h(\boldsymbol{R}') \right|, \tag{1}$$

where $h(\boldsymbol{R})$ is the height of a column at the lattice site $\boldsymbol{R}$ on the square lattice (hereafter termed as $\boldsymbol{R}$th column), the summation runs over all the pairs of nearest neighbor columns and $J(> 0)$ denotes the excess energy of an atom-vacancy pair. In the SOS model only the topmost atoms of columns can take part in the crystal growth kinetics. As the basic kinetic processes of atoms we consider (i) the impingement onto the surface, (ii) the diffusion along the surface and (iii) the re-evaporation from the surface. However, in MBE the evaporation process is neglected due to the long residence time of adatoms for typical MBE conditions[8]. On the other hand, the impingement of atoms onto the surface is neglected in the annealing phenomena.

Now we proceed to the formulation of the site-dependent point approximation of the PPM. $F_n(\boldsymbol{R}, t)$ is the state variable which represents the probability of finding a height n in the $\boldsymbol{R}$th column at time t with a normalization condition $\sum_{n=-\infty}^{\infty} F_n(\boldsymbol{R}, t) = 1$. In order to describe the temporal change of the system, in addition to the state variables, we require path variables which denote the joint probability connecting a state at time t and another state at $t + \Delta t$ of the $\boldsymbol{R}$th column. According to an elementary rule of probability, the state variable $F_n(\boldsymbol{R}, t)$ is written as a linear combination of path variables connecting two states at t and $t + \Delta t$:

$$\begin{aligned} F_n(\boldsymbol{R}, t) &= X_{n,n}(\boldsymbol{R}) + X_{n,n+1}(\boldsymbol{R}) + X_{n,n-1}(\boldsymbol{R}) \\ &+ \sum_{i=1}^{4} [X_{Dn,n-1}^i(\boldsymbol{R}) + X_{Dn,n+1}^i(\boldsymbol{R})], \end{aligned} \tag{2}$$

where the first path variable $X_{n,n}(\boldsymbol{R})$ on the right-hand side denotes the joint probability that the $\boldsymbol{R}$th column keeps the same height n both at t and $t + \Delta t$. A convention is taken that the argument $(t, t + \Delta t)$ in the path variable is omitted for simplicity. The second path variable $X_{n,n+1}(R)$ denotes the joint probability that the Rth column takes a height n at t and $n+1$ at $t + \Delta t$ due to the impingement of atoms onto the surface, the third one $X_{n,n-1}(R)$ representing the probability that an atom at a height n at t evaporates at $t + \Delta t$. The path variables with a subscript D are those for diffusion processes. $X^i_{Dn,n-1}(R)$ is the probability that the Rth column takes a height n at time t and $n-1$ at $t+\Delta t$ due to moving of the topmost atom to its nearest neighbor column in the ith direction and $X^i_{Dn,n+1}(R)$ denotes the similar probability that an atom moves into the Rth column with a height n from the ith direction where $i = 1 \sim 4$ stands for four directions $\pm x$ and $\pm y$. In just the same fashion the state variable $F_n(R, t + \Delta t)$ is also written down in terms of path variables defined above:

$$
\begin{aligned}
F_n(R, t + \Delta t) &= X_{n,n}(R) + X_{n-1,n}(R) + X_{n+1,n}(R) \\
&+ \sum_{i=1}^{4}[X^i_{Dn+1,n}(R) + X^i_{Dn-1,n}(R)].
\end{aligned}
\tag{3}
$$

The idea of the PPM is to calculate a transition probability $T(t, t + \Delta t)$ of an ensemble of L equivalent systems along a path in a short time interval from time t to $t + \Delta t$, the transition probability $T(t, t + \Delta t)$ being called the path probability function in the PPM. Then the kinetic equation is derived as the most probable path by maximizing $T(t, t + \Delta t)$ with respect to independent path variables with state variables at t fixed. The path probability function is composed of three factors[3]. The first is the kinetic factor of the ensemble along a path

$$
\begin{aligned}
T_1 &= \prod_{R} \prod_{n} (\theta_I \Delta t)^{LX_{n,n+1}(R)} (\theta_E \Delta t)^{LX_{n,n-1}(R)} \\
&\times (\theta_D \Delta t)^{L \sum_{i=1}^{4}(X^i_{Dn,n-1}(R) + X^i_{Dn,n+1}(R))} (1 - \Theta(R) \Delta t)^{LX_{n,n}(R)},
\end{aligned}
\tag{4}
$$

where θ_I denotes an impingement rate, θ_E an evaporation rate, θ_D a migration rate of atom and $1 - \Theta(R)\Delta t$ is the residual probability of the atom in Δt at R to be determined self-consistently. The second is the activation probability for the path of the ensemble

$$
T_2 = \exp(-L\Delta E/k_B T),
\tag{5}
$$

where ΔE is the activation energy per system along the chosen path traversed by the ensemble from time t to $t + \Delta t$:

$$
\begin{aligned}
\beta \Delta E &= -K \sum_{R} \sum_{n} (X_{n,n-1}(R) + \sum_{i} X^i_{Dn,n-1}(R)) \sum_{j} [1 - 2C_n(R + \rho_j, t)] \\
&- \sum_{R} \sum_{n} \sum_{i} L^i(R) X^i_{Dn,n+1}(R).
\end{aligned}
\tag{6}
$$

Here $K = \beta J$, $\beta = 1/k_B T$, ρ_j is the nearest neighbor vector in the jth direction on the square lattice and $C_n(R, t)(\equiv \sum_{m=n}^{\infty} F_m(R, t))$ is the atomic density in the nth layer at the Rth column. Further, $L^i(R)$ is a parameter corresponding to an effective chemical potential to be determined later. However, since there are a number of equivalent paths corresponding to the temporal change of the ensemble specified by T_1T_2, the third factor is the number of equivalent paths, the logarithm of which being called the path entropy:

$$T_3 = \prod_R \prod_n \frac{[LF_n(\boldsymbol{R})]!}{[LX_{n,n}(\boldsymbol{R})]![LX_{n,n\pm1}(\boldsymbol{R})]! \prod_i [LX^i_{Dn,n\pm1}(\boldsymbol{R})]!}. \tag{7}$$

Now the path probability function $T(= T_1T_2T_3)$ is maximized with respect to independent path variables with state variables at time t fixed to yield, up to the order Δt,

$$X_{n,n+1}(\boldsymbol{R}) = \theta_I \Delta t F_n(\boldsymbol{R}) \tag{8}$$
$$X_{n,n-1}(\boldsymbol{R}) = \theta_E \Delta t \exp[-\beta \Delta \epsilon_n(\boldsymbol{R})] F_n(\boldsymbol{R}) \tag{9}$$

for impingement and evaporation path variables, respectively. Here it is defined that $\beta \Delta \epsilon_n(\boldsymbol{R}) \equiv K \sum_i [2C_n(\boldsymbol{R} + \rho_i) - 1]$.

Similarly we have path variables for diffusion

$$X^i_{Dn,n-1}(\boldsymbol{R}) = \theta_D \Delta t \exp[-\beta \Delta \epsilon_n(\boldsymbol{R})] F_n(\boldsymbol{R}) \tag{10}$$
$$X^i_{Dn,n+1}(\boldsymbol{R}) = \theta_D \Delta t \exp[L^i(\boldsymbol{R})] F_n(\boldsymbol{R}) \tag{11}$$

for $i = 1 \sim 4$. Since in the point approximation a set of path variables $X^i_{Dn,n-1}(\boldsymbol{R})$ is equivalent to another set of path variables $X^{\bar{i}}_{Dn,n+1}(\boldsymbol{R} + \rho_i)$, the number conservation rule of atoms exists between them:

$$\sum_n X^i_{Dn,n-1}(\boldsymbol{R}) = \sum_n X^{\bar{i}}_{Dn,n+1}(\boldsymbol{R} + \rho_i), \tag{12}$$

where $\bar{i}$ is the opposite direction of ith direction. Substituting the obtained forms (10) and (11) into eq. (12), we can determine $\exp[L^i(\boldsymbol{R})]$ as

$$\exp[L^i(\boldsymbol{R})] = \sum_m F_m(\boldsymbol{R} + \rho_i) \exp[-\beta \Delta \epsilon_m(\boldsymbol{R} + \rho_i)]. \tag{13}$$

By making a difference between (2) and (3) with the help of equations (8) $\sim$ (13), we finally obtain in the limit $\Delta t \to 0$ the kinetic equation for the present model:

$$\frac{dF_n(\boldsymbol{R}, t)}{dt} = \theta_I[F_{n-1}(\boldsymbol{R}) - F_n(\boldsymbol{R})]$$
$$+ (\theta_E + 4\theta_D)[F_{n+1}(\boldsymbol{R}) \exp[-\beta \Delta \epsilon_{n+1}(\boldsymbol{R})] - F_n(\boldsymbol{R}) \exp[-\beta \Delta \epsilon_n(\boldsymbol{R})]]$$
$$+ \theta_D[F_{n-1}(\boldsymbol{R}) - F_n(\boldsymbol{R})] \sum_{i=1}^{4} \exp[L_i(\boldsymbol{R})]. \tag{14}$$

We see that the increment of the state variable $F_n(\boldsymbol{R}, t)$ is composed of three kinds of terms, that is, the impingement term with θ_I, the evaporation term with θ_E and the migration term with θ_D, respectively. In the following the diffusion rate θ_D is assumed to take the Arrhenius form

$$\theta_D = \nu \exp(-E_D/k_B T), \tag{15}$$

where E_D is the activation energy of a migrating atom independent of column sites, ν being the vibrational frequency. Owing to the temperature dependence upon the diffusion rate θ_D, the atomic migration is increased as the temperature is increased.

3. RHEED INTENSITY ON THE STEPPED SURFACE

First we apply the above formula to the crystal growth by MBE on a stepped surface where the positive x-direction is taken to be the step down direction. By integrating the kinetic equation (14) numerically, we calculate the time development of the surface roughness $q(t)$ defined by

$$q(t) \;=\; \sum_{i=1}^{4} \sum_{R} \sum_{n,m} |n - m| F_n(R) F_m(R + \rho_i)/8N \qquad (16)$$

where N is the number of columns in the system. The surface roughness $q(t)$ counts the average number of lateral atom-vacancy pairs per bond which reflects the specular RHEED intensity. For the surface roughness $q(t)$ we call $1 - q(t)$ the surface smoothness. It is expected that when the surface becomes smooth, the RHEED intensity becomes strong and when the surface becomes rough, the RHEED intensity becomes weak.

For numerical calculations each terrace is taken to be completely flat on initiation. During growth the average inclination of the stepped surface is maintained by the boundary condition while the homogeneity is assumed in the y-direction. In Figure 1 we show the evolution of the smoothness parameter $1 - q(t)$ for different temperatures $k_B T/J = 0.30, 0.33, 0.35$ and 0.37 with values $\theta_I/\nu = 1.18 \times 10^{-9}$ and $E_D/J = 9.2$. In order to see the recovery process the deposition is off at time $t = 17.0/\theta_I$. We see that as the temperature is raised, the amplitude of oscillation decreases more rapidly and at $k_B T/J = 0.37$ the oscillatory behavior cannot be seen from the beginning. In Figure 2 we show the typical experimental data of the specular RHEED intensity by Neave et al.[1] for GaAs (001) MBE growth on the stepped surface. The present calculation is qualitatively in good agreement with the experimental data.

Next, in order to see the growth mode change directly we show in Figure 3 the temporal change of surface morphology in one period corresponding to a monolayer growth period after the steady state is achieved. At $k_B T/J = 0.30$ where the oscillation pattern is stable, the crystal growth proceeds on each terrace. The height of the terrace is increased by one layer without changing the position of steps after one monolayer deposition time. On the other hand, at $k_B T/J = 0.37$ where the oscillation cannot be seen from the beginning, the crystal growth proceeds by step propagation mode. The height of each terrace is increased by one layer through the movement of steps after one monolayer deposition time. Thus we have confirmed that as the temperature is raised, the growth mode changes from the nucleation mode on each terrace to the step propagation mode.

We can estimate the transition temperature for this growth mode change approximately by calculating the diffusion length of adatom until capture by other

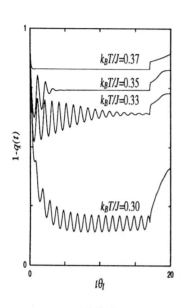

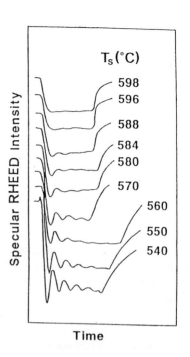

Figure 1. Evolution of the surface smoothness parameter $1 - q(t)$ at temperature $k_B T/J = 0.30$, 0.33, 0.35 and 0.37 with values $\theta_I/\nu = 1.18 \times 10^{-9}$, $E_D/J = 9.2$ and $\theta_E = 0$. Initial terrace length is 10 columns.

Figure 2. RHEED measurement of GaAs(001) MBE on a stepped surface by Neave et al.[1], showing the eventual disappearance of the oscillation as the temperature is raised.

adatoms[6]. For this purpose we rewrite the kinetic equation (14) in the form of number conservation law for the average number of atoms in the Rth column in the x-direction by taking a continuum limit. With the average height of the Rth column defined by $H(R,t) \equiv \sum_n n F_n(R,t)$, the number conservation law for atoms can be written as

$$\frac{dH(R,t)}{dt} = \theta_I - \frac{\partial}{\partial R} J(R,t) \tag{17}$$

with the flux density $J(R,t) = -\frac{\partial}{\partial R}\theta_D \sum_n F_n(R,t)e^{-\beta \Delta \epsilon_n(R,t)}$. Further the flux density can be expressed in terms of concentration gradients as

$$J(R,t) = -\sum_n D(C_n(R,t))\frac{\partial C_n(R,t)}{\partial R}, \tag{18}$$

where $D(C_n(R,t))$ is the concentration-dependent diffusion coefficient given by

$$D(C_n(R,t)) = \theta_D[(1 - 8K F_n(R,t))e^{4K(1-2C_n(R,t))} - e^{4K(1-2C_{n-1}(R,t))}]. \tag{19}$$

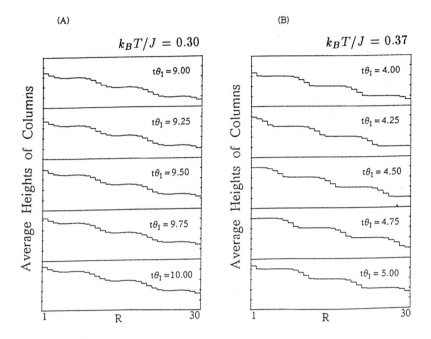

(A) $k_BT/J = 0.30$

(B) $k_BT/J = 0.37$

$t\theta_1 = 9.00$

$t\theta_1 = 9.25$

$t\theta_1 = 9.50$

$t\theta_1 = 9.75$

$t\theta_1 = 10.00$

$t\theta_1 = 4.00$

$t\theta_1 = 4.25$

$t\theta_1 = 4.50$

$t\theta_1 = 4.75$

$t\theta_1 = 5.00$

Figure 3. The change of the surface morphology in one period corresponding to a monolayer growth at (A) $k_BT/J = 0.30$ and at (B) $k_BT/J = 0.37$. Terrace length is 10 columns. In (A) the step positions are unchanged during one monolayer growth. In (B), on the other hand, the crystal growth proceeds through the step propagation.

With the help of eq. (19), we can calculate the diffusion length X_S of an adatom on a terrace with a height $n = 0$ during growth under the initial condition that the chosen terrace is completely flat. Taking into account the Einstein relation for diffusion, the diffusion coefficient is related to the adatom diffusion length X_S by $X_S^2 = 2 \int_0^\tau D(C(t))dt$. Here $C(t)$ is the density of atoms on the terrace at time t and the integral comes from the fact that the diffusion coefficient decreases as the density increases with time. The integral limit τ is the life time of an adatom until capture by other adatoms which is identified as the time when the diffusion coefficient $D(C(t))$ becomes zero. Now the transition temperature T_C above which the oscillation in the roughness cannot be seen from the beginning is estimated as the temperature at which the adatom diffusion length becomes a half of the terrace length. Then the transition temperature for the stepped surface in the present case is obtained as $k_BT/J = 0.372$, which compares well with the result in Figure 1.

4. CURRENT-INDUCED CONVERSION ON THE Si(001) STEPPED SURFACE

Latyshev *et al.* [2] observed a current-induced conversion of two types of domains on the Si(001) stepped surface during annealing. When the DC heating current is supplied to the [110] step-down direction, one type of domains dominates over another type of domains. When the current direction is changed towards the step-up direction, the role of domains is reversed (Figure 4). We apply our kinetic equation (14) with some modifications[7] to the present domain conversion phenomena.

Suppose that a Si crystal is cut with a slight tilt from the exact (001) surface towards the [110] azimuth. Because of the diamond structure of the Si crystal, two types of reconstructed faces are formed on the Si(001) stepped surface according to the direction of dimers rows which rotates by 90° on every next terrace. Following Chadi's notation[9], monatomic steps which are parallel to the dimers rows on the upper terrace are labelled S_A and those perpendicular to the dimers rows on the upper terraces are labelled S_B. The corresponding upper terraces with the 2×1 and 1×2 reconstructed surfaces are labelled type A and type B terraces, respectively.

In our model the diamond-type structure of the Si crystal is simplified into a simple cubic lattice and the SOS model is also adopted to represent the (001) surface of the Si crystal (Figure 5). Since only the topmost atom of each column participates in kinetic processes, we must specify the types of terraces on which the topmost atom of a column exists. The topmost atom with a height $n = 2l + 1$(l : integer) is assumed to be on the A terrace and the topmost atom with a height $n = 2l$ on the B terrace. Although the dimer formation or dissociation process is ignored, we assume that the underlying layer upon which atoms migrate forms a complete dimers rows structure.

Detailed studies[10] show that atoms can migrate much more easily along the direction parallel to the underlying dimers rows than along that perpendicular to the underlying dimers rows. In the present treatment we include the effect of anisotropy of surface migration into the migration rates $\theta_{D\parallel}$ and $\theta_{D\perp}$ of an isolated atom, where $\theta_{D\parallel}$ and $\theta_{D\perp}$ stand for migration rates parallel and perpendicular to the underlying dimers rows, respectively ($\theta_{D\parallel} \gg \theta_{D\perp}$). Denoting the position of an atom at site R and height n by (R, n), the virtual migration rate of an atom at the position $(R, 2l)$ in the x-direction is given by

$$\Lambda_x(R, 2l) = \theta_{D\parallel} \exp[-2Jm(R, 2l))/k_B T] \tag{20}$$

and similarly the rate in the y-direction is given by

$$\Lambda_y(R, 2l) = \theta_{D\perp} \exp[-2Jm(R, 2l)/k_B T], \tag{21}$$

where $2J$ is the binding energy with a lateral nearest neighbor atom and $m(R, n)$ is the number of lateral nearest neighbor atoms of an atom at (R, n). Since the

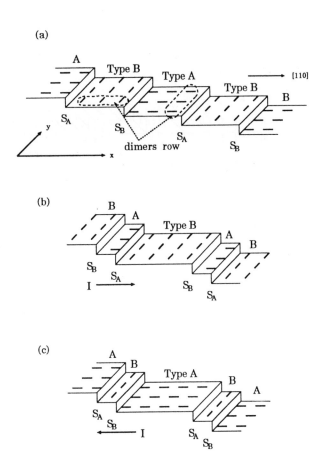

Figure 4. Schematic illustrations of (a) a model of the Si(001) stepped surface and (b), (c) the stepped surface with the DC heating current. In (b) when the DC current is in the step-down direction, the B terraces are dominant. In (c), on the other hand, the A terraces are dominant when the DC current is in the step-up direction. A short segment on the surface denotes a dimer.

direction of dimers rows rotates by 90° on every next layer, the following relations hold for the atom at $(\boldsymbol{R},2l+1)$ in contrast to eq. (20) and (21):

$$\Lambda_x(\boldsymbol{R},2l+1)/\theta_{D\perp} = \Lambda_y(\boldsymbol{R},2l+1)/\theta_{D\parallel} = \exp[-2Jm(\boldsymbol{R},2l+1)/k_BT]. \qquad (22)$$

Further, the evaporation rate $\theta_E = \nu\exp(-\beta E_e)$ is assumed with a bond breaking energy E_e from the substrate. In addition to the anisotropic migration effect and the evaporation we have to take into account the electromigration[11] and the Schwoebel effect[12] in order to reproduce the domain conversion phenomena. Un-

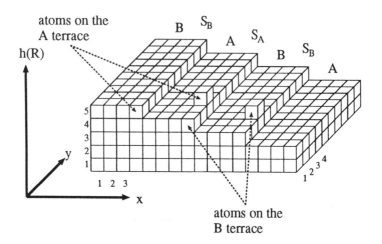

Figure 5. The Si(001) stepped surface on the basis of the SOS model. The atom with odd height is on the A terrace while the atom with even height is on the B terrace.

der the electromigration induced by the DC current in the step-down direction, if an atom migrates to the nearest neighbor site, the migration rate is multiplied by $\exp(\beta F/2)$ in the step-down direction through an effective force $F(>0)$ and by $\exp(-\beta F/2)$ in the step-up direction, respectively. On the other hand, the Schwoebel effect is the effect that the capture rate of an adatom at a step depends upon the direction from which the atom approaches the step edge, i.e., from above or below the step. When an atom on the upper surface jumps down a step, the migration rate is multiplied by $\exp(-\beta V_d^A)$ at the S_A step owing to potential barrier V_d^A and by $\exp(-\beta V_d^B)$ at the S_B step owing to potential barrier V_d^B, respectively. Similarly, when an atom jumps up a step, the migration rate is multiplied by $\exp(-\beta V_u^A)$ at the S_A step and by $\exp(-\beta V_u^B)$ at the S_B step, respectively. Recent studies[13] show that the potential barrier at the S_A step is much higher than that at the S_B step. We assume $V_{u,d}^A \gg V_{u,d}^B$ in the present calculation.

Now, with the above mentioned modifications we apply the kinetic equation (14) to the present annealing model. First, in Figure 6 we show the time evolution of surface morphology. When the DC current is in the step-down direction, the B terraces become dominant over the A terraces while the A terraces become dominant over the B terraces when the current is reversed. We see that the domain conversion by the direction change of the DC heating current is reproduced. However, it should be noted that though the role of A and B terraces is reversed, the complete symmetry of terraces is not recovered due to asymmetry of the Schwoebel effects at steps. Next, we show in Figure 7 the time evolution of the positions of the steps when the DC current is first supplied in the step-up direction and afterwards in the step-down direction. We see that when the current is in the step-up direction, the S_B steps recede in the step-up direction while the S_A steps scarcely move. The situation is reversed when the current is in the step-down direction. These facts are in good agreement with experiments[2,14]. We can explain these phenomena with the combinations of the anisotropy of migration, the evaporation of atoms and the Schwoebel effects[7].

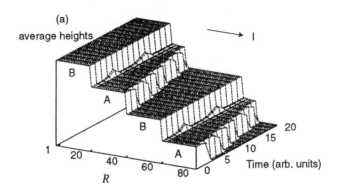

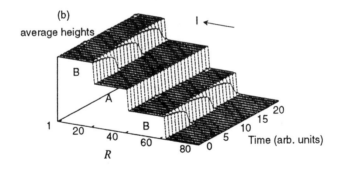

Figure 6. The time evolution of surface morphology $H(R)$ with parameters $E_D/J = 9.20$, $k_B T/J = 0.470$, $\theta_{D\parallel}/\theta_{D\perp} = 1.00 \times 10^{-3}$, $\theta_E/\theta_{D\parallel} = 2.39 \times 10^{-3}$, $\beta V_u^A = 4.50$, $\beta V_u^B = 2.00 \times 10^{-1}$, $\beta V_d^A = 2.00$ and $\beta V_d^B = 1.00 \times 10^{-1}$. (a) The B terraces are expanding when the current direction is in the step-down direction ($\beta F/2 = 1.00$). (b) The A terraces are expanding when the current direction is in the step-up direction ($\beta F/2 = -1.00$).

112

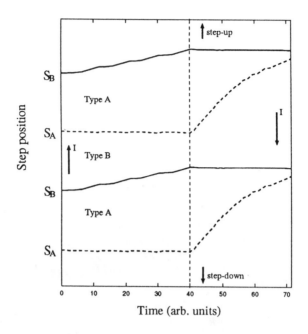

Figure 7. The temporal change of step positions during the domain conversion with same parameters as those used in Figure 6. Before the reversal of current the S_B steps move in the step-up direction while the S_A steps do not move. On the other hand, after reversal of current the S_A steps move in the step-up direction while the S_B steps do not move.

5. CONCLUSIONS

We applied the PPM in non-equilibrium statistical mechanics to the crystal growth kinetics on the stepped surface. The first one is the analysis of the intensity oscillations of RHEED monitoring the crystal growth by MBE. The present calculation showed that the growth mode on the stepped surface changes from the nucleation mode on the terrace to the step propagation mode as the temperature is raised and was in qualitatively good agreement with the RHEED measurements by Neave *et al.* on GaAs(001). The second one is the current-induced domain conversion on the Si(001) stepped surface. We showed that the combinations of anisotropy of migration, evaporation of atoms and the Schwoebel effects bring about domain conversions when the direction of DC heating current is reversed.

Along with the calculations of the kinetic equations derived by the PPM we had also carried out the Monte Carlo simulations in each case based on the same model[7,15]. Since the results of the present calculations are also in qualitatively agreement with those of the Monte Carlo simulation, the PPM is expected to be a promising method to study even the kinetics of heterogeneous crystal growth on the flat and stepped surfaces.

The authors would like to thank Prof. T. Uchida for the collaborative work and useful advice on this work. One of the authors (H.O.) is supported by Research Fellowships of the Japan Society for the Promotion of Science for Young Scientists.

REFERENCES

1. J.H. Neave, P.J. Dobson, B.A. Joyce and J. Zhang, Appl. Phys. Lett. 47 (1985) 100.
2. A.V. Latyshev, A.B. Krasil'nikov and A.L. Aseev, JETP Lett. 48 (1988) 526.
3. R. Kikuchi, Prog. Theor. Phys. (Kyoto) Suppl. 35 (1966) 1.
4. D.E. Temkin, Sov. Phys. Crystallogr. 14 (1969) 344.
5. T. Uchida, Phys. Lett. A 161 (1992) 373.
6. K. Wada, TA. Uchida and TE. Uchida, Phys. Lett. A 162 (1992) 346.
7. H. Ohmi, T. Uchida and K. Wada, Jpn. J. Appl. Phys. 35 (1996) 226.
8. S. Clark and D.D. Vvedensky, J. Appl. Phys. 63 (1988) 2272.
9. D.J. Chadi, Phys. Rev. Lett. 59 (1987) 1691.
10. Y.W. Mo, J. Kleiner, M.B. Webb and M.G. Lagally, Phys. Rev. Lett. 66 (1991) 1998.
11. T. Doi and M. Ichikawa, Jpn. J. Appl. Phys. 34 (1995) 25.
12. R.L. Schwoebel and E.J. Shipsey, J. Appl. Phys. 37 (1966) 3682.
13. C. Roland and G.H. Gilmer, Phys. Rev. B 46 (1992) 13437.
14. H. Kahata and Y. Yagi, Jpn. J. Appl. Phys. 28 (1989) L858.
15. H. Ohmi, T. Uchida and K. Wada, Trans. Mat. Res. Soc. Jpn. 16A (1994) 303 (Advanced Materials '93, III/A: Computation, Glassy Materials).

PART II

Growth Kinetics

Advances in the Understanding of Crystal Growth Mechanisms
T. Nishinaga, K. Nishioka, J. Harada, A. Sasaki and H. Takei (Editors)

Inter-surface diffusion of cation incorporation in MBE of GaAs and InAs

T. Nishinaga and X.Q. Shen*

Department of Electronic Engineering, Graduate School of Engineering, The University of Tokyo, 7-3-1 Hongo, Bunkyo-ku, Tokyo 113, Japan

The surface diffusion of group III atom incorporation in MBE of GaAs and InAs is studied. First is discussed the diffusion length of incorporation on the (001) top surface with the(111)A or (411)A side surfaces of V-grooves. It is shown that the diffusion length takes the same value for both cases and is inversely proportional to the arsenic pressure while it is independent of Ga pressure. The same relationships are also obtained for the diffusion of In in InAs MBE. However, the diffusion length of Ga on (111)B shows an inverse parabolic dependence of the arsenic pressure. It is suggested that on (001) surface two As_4 molecules meet to give active As atoms for the growth. On the other hand, the behavior of As_4 molecule on (111)B surface is still not clear.

The ratio of surface diffusion coefficient on (111)B and that on (001) is calculated. It is found that the ratio takes the value of around 140. With this ratio, the incorporation life times τ_{inc} on (111)B and (001) surface are calculated as functions of arsenic pressure. It is found that the lines of the incorporation life time intersect almost at an arsenic pressure where flow inversion occurs.

1. INTRODUCTION

Recent semiconductor industries require the devices with the structure of nano-meter scale. For instance, modern devices such as quantum well laser, high frequency transistors with modulation doping, hetero-bipolar transistor and etc. use a layer structure with the accuracy of atomic layer thickness. To improve drastically the device performance, the structures with nano-scale dimension has been required. Quantum wire and dot structures have been proposed to realize laser diode of low threshold current and single electron transistor is believed being another promising device, which requires the three dimensional structure of nano-meter scale.

There are many techniques proposed to fabricate such structures. Photolithography followed by chemical and physical etchings is one of those techniques. Although one can control the positions of such structures like wires and dots rather precisely,

---

*Present address: Semiconductor Reasearch Lab, The Institute of Physical and Chemical Research, 2-1 Hirosawa, Wako-shi, Saitama, 351-01, Japan

to get the size accuracy with atomic scale is rather difficult. The lattice defects possibly introduced when the physical etching is employed is another disadvantage of this technique. Ion implantation is also possible to use for the fabrication of nano-structures. However, in this case, the structure is composed of regions of the same materials with different conductivity type. Hence, the kinds of the structures are limited and again we have the problem of damages in the implantation process.

On the other hand, the fabrication of nano-structures by crystal growth can be almost free from the defects and in addition by using naturally appearing facets one can get the structures with atomic scale precision. On the contracy, the positional control of the structure is rather difficult by crystal growth. Hence, the combination of these techniques might be most promising.

As for the technique to employ crystal growth, Petroff et al.[1] proposed to use step edges on vicinal surface as preferential growth site and GaAs wire structures were made in AlAs by MBE. Fukui et al.[2,3] have used this idea and with MOCVD they fabricated quite regular and straight wire structures.

Another growth technology is to use nonplanar substrate. By making use of the different growth rate between the crystallographically different faces, Kapon et al.[4,5] fabricated wire structures in the bottoms of V-grooves by MOCVD. Mesa structure is also employed for the fabrication of wire structure on the top. It is possible to extend this technique to obtain dot structure by using the patterned substrate with pyramids or pyramidal depressions. Another important technique for the microstructure fabrication is the selective epitaxy[6,7]. By opening a small window in a mask, one can grow the crystal only in a limited area.

Most of the above mentioned techniques require the precise control of the growth and therefore deep understanding in the elemental growth processes is extremely important. Among them surface diffusion is one of the key processes which govern the local growth behavior. There have been strong discussions among the scientists about the true diffusion length of incorporation, λ_{inc} in MBE of III-V compounds. In MBE on vicinal surface, it is believed that λ_{inc} is of the order of inter-step distance, namely, of the order of a few tens of nm[8,9]. On the other hand, it has been shown that the diffusion length in MBE on the non-planar substrates takes the value of a few μm[10,11]. The present authors have shown that the diffusion length of incorporation is a strong function of arsenic pressure and a great discrepancy between these two diffusion lengths can be explained by taking into account the ratio of Ga and As surface fluxes which enter the growth step[12,13].

In the present artical, we discuss the surface diffusion of cation incorporation and try to give a physical picture for the elementary growth process in MBE taking GaAs and InAs as examples.

2. EXPERIMENTS

The surface diffusion length of incorporation has been measured experimentally by using microprobe-RHEED/SEM MBE. Hata et al. have measured the diffusion length from the exponential change in local growth velocity as a function of distance from the corner where two low index planes meet and across which the inter-surface diffusion takes place[10,11]. They found the diffusion length varies with the change of arsenic pressure and the growth temperature. To find the functional dependency of the diffusion length upon the arsenic pressure, the present authors repeated their experiments by changing the arsenic pressure systematically by using also similar microprobe-RHEED/SEM MBE system the detail of which has been described in ref.[14].

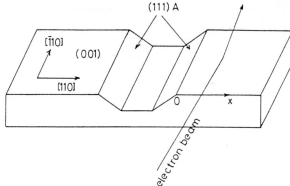

Figure 1. Schematic picture of the non-planar substrate with V-groove and the position of electron beam.

Figure 1 shows schematically the relation of an electron beam to the non-planar substrate when a groove is cut in [$\bar{1}10$] orientation in a wafer with (100) surface. In this case, (111)A face appear on the side wall of the groove. On the other hand, if a groove is cut in [110] direction, (111)B faces appear. Before the growth rate measurements, a thin buffer layer is grown to get a smooth surface. The minimum size of electron beam is $\sim$ 5 nm in diameter.

The electron beam is focused on the sample and by RHEED intensity oscillation technique one can measure the local growth velocity with the spatial accuracy of 100 nm. The growth rate, $R(x)$, on the top surface of non-planar substrate with V-grooves is shown schematically in Figure 2. Here, the origin of x is chosen at the top corner of a V-groove. $R(x)$ is composed of $R_L(x)$ and R_V which are respectively the rates of the growths caused by lateral surface flux and vertical direct flux from Knudsen cell. As shown in Figure 2, positive and negative values of $R_L(x)$ means that the surface diffusion occurs from side to top and top to side respectively.

It is important to note that although the growth rate changes exponentially as a function of position, this does not mean the growth occurs on a vicinal surface. On V-grooved substrate, a buffer layer is at first deposited to get smooth surface with the thickness of 20nm. Then, the RHEED oscillation is recorded as a function of position. Final thickness of the grown layer after several series of experiments is typically 35-40nm, maximum 20% of which is the part of exponentially changing at $x = 0$. Hence, the slope caused by the exponential change is less than 0.5°, which means the surface is nearly flat and the growth is done in 2D nucleation mode. This has been confirmed by RHEED intensity oscillation.

120

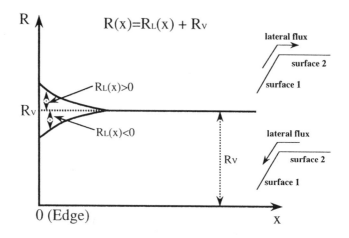

Figure 2. Schematic illustration of the growth rate distribution.

2.1. Diffusion length of Ga on (001) GaAs

Inter-surface diffusion occurs if there is a difference in surface adatom concentration between two adjacent planes. The adatom concentration is inversely proportional to the adatom life time, namely, the average incorporation time, τ_{inc}. Hence, the inter-surface diffusion occurs when there is a difference in τ_{inc} between these faces.

Here, τ_{inc} employed here has no local sencitivity of atomic scale, althouth it is a function of position. On the surface, 2D nuclei first appear, then the sizes increase and finally they coalesce to give a flat surface. In such a way , the surface condition changes periodically during the growth. Hence, in principle τ_{inc} should be also the function of time since the local step density changes periodically when the growth is taken place in 2D-nucleation mode like this experiment. However, we define τ_{inc} as that averaged for a cirtain period of time. If the period is much longer than that of the oscillation, we can get τ_{inc} independent of time. We also take the spatial average of τ_{inc} for the area much longer than several nm. Thus, we obtain τ_{inc} as a function of macroscopic position with an accuracy of the order of 100 nm. The concept of τ_{inc} here is rather close to that of minority carrier life time in semiconductors although the number of recombination center is fixed and not changes temporally.

Figure 3 shows the diffusion length of Ga incorporation, λ_{inc} on (001) surface when the side surface was (111)A or (411)A. Irrespective of the kinds of the side surfaces, λ_{inc} takes the same value and is almost inversely proportional to arsenic pressure. By integrating $R_L(x)$ from $x = 0$ to infinity, the total amount of surface lateral flux can be evaluated. Figure 4 shows thus calculated total flux. As seen in the figure, the flux from (411)A is more than twice of that from (111)A although the Ga flux is kept at the same value. This means τ_{inc} on (411)A is much longer and hence the surface concentration is much higher than those on (111)A.

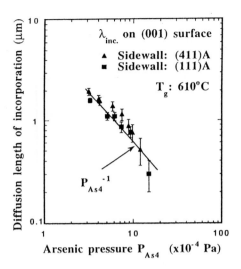

Figure 3. Diffusion length of Ga incorporation v.s. arsenic pressure on (001) surface when the side surface is (111)A or (411)A.

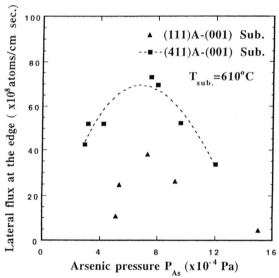

Figure 4. The amount of total lateral flux passing the boundary at $x = 0$ from (111)A or (411)A to (001) surface.

Another interesting behavior of λ_{inc} is its dependency on Ga flux. In Figure 5, it is shown that λ_{inc} on (001) top surface shows no change in the range of the Ga pressure employed in the present experiments for both (111)A and (411)A side surface. Hence, it can be said that Ga does not play any role in determining the λ_{inc}

122

nor τ_{inc}. The reason why both side surfaces gave the same λ_{inc} is clear. Even the kind of side surface is different, incorporation life time on the same (001) surface should be the same since it is not the function of Ga flux.

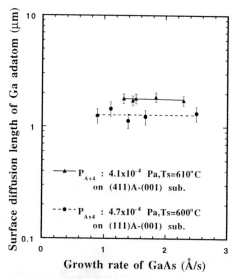

Figure 5. Incorporation diffusion length of Ga on the (001) surface with the side surfaces of (111)A and (411)A as a function of growth rate.

When (111)B was chosen as the side surface, it was found that the flow direction was reversed as the arsenic pressure was changed. This occurs at an arsenic pressure between 3.5 and 6×10^{-4} Pa for the growth temperature of 600 °C as shown in

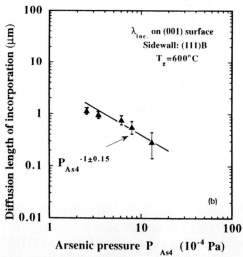

Figure 6. Incorporation diffusion length of Ga on the (001) surface as a function of As pressure with the side surface of (111)B.

Figure 6. In this case, $R_L(x)$ becomes positive and negative respectively for lower and higher arsenic pressures. This happens because τ_{inc} on (111)B and (001) has different arsenic pressure dependencies as we will discuss also in the next section. Although the flow direction is opposite to each other, the diffusion length of Ga incorporation lies on a single line which is inversely proportional to P_{As_4}. This is reasonable since the diffusion length was measured both on (001) surface.

2.2. Surface diffusion of In on (001) InAs

Similar to GaAs MBE on V-grooved substrates with (111)B side surface, and unlike those with (111)A or (411)A side surfaces, InAs MBE with (111)A side surface gives the reversible lateral flow of In when the arsenic pressure is changed. Figure 7 shows the area for each region in the diagram of P_{As_4} v.s. $1/T$. If the arsenic pressure is low, In atoms diffuse from (111)A to (001) while they diffuse in the opposite direction if the arsenic pressure is high.

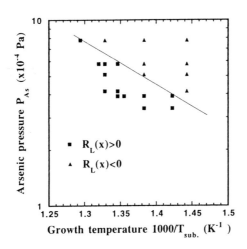

Figure 7. The direction of the lateral flux and growth conditions. In a region below the line, In atoms diffuse from (111)A to (001) and in the above the direction is reversed.

Systematic measurements for obtaining λ_{inc} of In was carried out varying the arsenic pressure and the results are given in Figure 8. It was found that λ_{inc} is again inversely proportional to the arsenic pressure and is independent of In flux intensity[15]. Figure 9 shows the dependence of λ_{inc} on the growth temperature. As the growth temperature is increased the incorporation diffusion length is increased. This can be understood as follows. Since λ_{inc} is given by the square root of the product of D_s and τ_{inc}, the increase of either D_s or τ_{inc} should be the origin for the increase of λ_{inc}. Although D_s increases with the temperature but the increase is not large for such small increase of the temperature. Hence, the increase of τ_{inc} should be the origin. This can happen because the increase of temperature gives

124

the strong evaporation of As_4 which will lead to the decrease of As_4 concentration on the surface. The dependency of λ_{inc} on In flux is shown in Figure 10. Again it is shown that λ_{inc} is independent of In flux like the case of Ga diffusion shown in Figure 5.

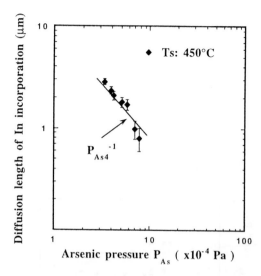

Figure 8. Surface diffusion length of In incorporation v.s. arsenic pressures.

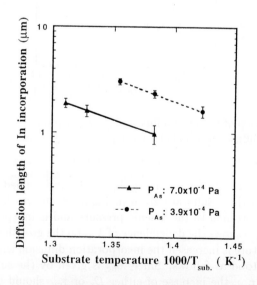

Figure 9. Surface diffusion length of In incorporation as a function of growth temperatures.

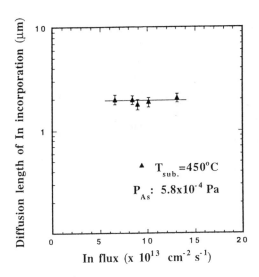

Figure 10. Diffusion length of In incorporation v.s. In flux on (001) InAs surface with (111)A sidewall.

2.3. Surface diffusion of Ga on (111)B GaAs

So far we have discussed about λ_{inc} on (001) surface. In this section, the diffusion length of Ga on (111)B surface is described. Nomura et al. have measured λ_{inc} on (111)B GaAs with (001) side surface[16]. They found that λ_{inc} varies quickly with arsenic pressure and that λ_{inc} takes different activation energy for the diffusion length when the surface reconstruction is changed from $\sqrt{19} \times \sqrt{19}$ to 1×1. The present authors repeated their experiment to obtain λ_{inc} on both of (111)B and (001) in the same MBE machine and with the same method to measure the arsenic pressure to avoid the errors in the calculation given in the next section. λ_{inc} measured at 545 °C on (111)B surface with (001) side surface is shown in Figure 11. Differing from λ_{inc} on (001) shown in Figure 6, λ_{inc} on (111)B is proportional to $P_{As_4}^{-2}$. Unfortunately, RHEED intensity oscillation on (111)B could not be detected at 600°C at which λ_{inc} was measured on (001). Hence, λ_{inc} measured at 545°C was extrapolated to 600°C and this value was employed for the calculation. There was no change in the surface reconstruction ($\sqrt{19} \times \sqrt{19}$) when the temperature was increased from 545 to 600°C. Thus obtained λ_{inc} is shown in Figure 11 by a closed square.

3. DISCUSSION

Most important thing to be mentioned here is the existence of special growth situations in the region near the corner where two low index planes meet. As we have discussed in our previous paper[17], there is a large lateral flux of the group

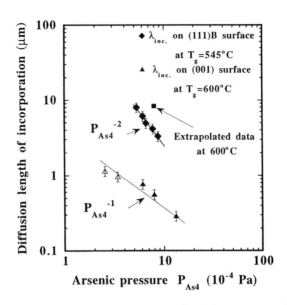

Figure 11. Diffusion length of Ga incorporation v.s.arsenic pressure. Closed rhombuses and triangles show respectively the lengths on (111)B with (001) side surface and (001) with (111)B side surface. In the latter case, open and closed triangles denote respectively that the diffusion flow of Ga occurs from (111)B to (001) and from (001) to (111)B. A closed square shows the extrapolated λ_{inc} to 600 °C.

III atoms passing over the corner. A simple calculation shows the lateral flux of Ga in the region is more than 10 times larger than the surface flux induced by the vertical flux directly supplied from the Knudsen cell. This means that the ratio of group III surface flux to that of group V entering the step is larger than unity. Namely, the group III atoms enter the step much more frequently than the group V atoms.

However, it has been shown that (2×4) surface reconstruction always appears on (001) during the MBE experiments for both of GaAs and InAs. Therefore, the surface should be arsenic rich. Hence, we should understand that a group III rich lateral flux exists on an arsenic rich surface. But, this special situation appears only in the vicinity of the boundary between two different faces. Although a relatively big amount of group III atoms enter from one face to the other, these atoms are consumed for the crystal growth in a semi-infinite area next to this special region. The presence of such a large area makes the surface concentration of group III atoms near the boundary rather low so that the surface can be kept under the arsenic rich condition.

When group III surface flux entering the step is higher than that of group V, once the group III atom enters a kink of the step, there are not enough atoms of group V to fix it at the step edge. In this case, the group III atoms are easily detached and moves on to the next step. Thus the group III atom can migrate far away from the boundary.

The vertical growth rate R should be proportional to the product of the surface concentrations of group III and V atoms and is given by,

$$R = C N_{III} N_V, \tag{1}$$

where N_{III} and N_V denote the adatom concentrations of group III and V elements respectively. On the other hand, the growth rate is expressed by using the life time of group III in the form of

$$R = N_{III}/\tau_{inc}, \tag{2}$$

Figure 3 and the similar relationship for InAs show that λ_{inc} is proportional to $P_{As_4}^{-1}$, which gives the following equation,

$$\lambda_{inc} = D_s \tau_{inc} = C' P_{As_4}^{-1}, \tag{3}$$

With Eqs. (1), (2) and (3), one gets

$$N_{As} = C'' P_{As_4}^2, \tag{4}$$

In the above equations, C, C' and C'' are all constants which depend on temperature. Eq.(4) shows that the reaction which gives active arsenic atoms for the growth is of the second order. This has been shown by Foxon et al.[18] many years ago. According to their model two As_4 molecules give free As atoms for the GaAs growth on the (001) surface. A model for the reaction is schematically illustrated in Figure12. From Eqs.(1) and (2) it is clear that τ_{inc} depends only on N_V and not on N_{III}, which means λ_{inc} is independent of N_{III} and hence of group III flux. This agrees very well with the experiments.

In the case of the growth on (111)B, it is found that λ_{inc} is proportional to $P_{As_4}^{-2}$. Therefore, the reaction should be of the fourth order. This means that four As_4 molecules should meet to give active As atoms for the growth on the (111)B surface if single arsenic atom is the rate determining species. However, if As_2 which is produced by the reaction of single arsenic is the rate determining species on (111)B surface, the reaction should be the fourth order with As_4 although only two As_4 molecules are required to meet to produce single arsenic atoms. Anyway, in this moment, the reactions occurring on (111)B surface are not clear.

By solving one dimensional diffusion equations with the assumption of no diffusion barrier between the adjacent faces, one can get the growth rate as a function of the position on the top surface of the V-groove as[12],

$$R(x) = \frac{d}{N_0} \left[J_{III} + \frac{J_{III} D' \lambda (\tau' \cos\theta - \tau)}{(D'\lambda + D\lambda')\tau} exp\left(-\frac{x}{\lambda}\right) \right], \tag{5}$$

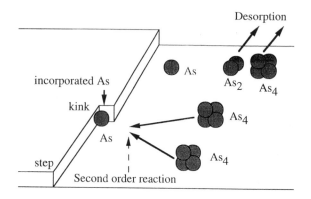

Figure 12. Schematic illustration of the reaction producing active arsenic atoms from As_4 molecules for the growth of GaAs and InAs.

where d, N_0, J_{III}, D, λ, τ and θ are respectively the distance between the successive group III atomic layer in vertical direction, the number of group III atoms per unit area, the vertical flux of the group III atoms, diffusion coefficient, incorporation time and an angle between the top and the side surfaces. D, λ, and τ without and with a prime mean respectively the values on the top and on the side surfaces. The other assumptions needed to obtain Eq.(5) can be gleaned from the original derivation in ref.[12]. The first and the second terms in the right hand side of Eq. (5) stand for R_V and $R_L(x)$ respectively. Hence, if the top and the side surfaces are respectively (001) and (111)B, $R_L^{(001)}(x)$ is given as

$$R_L^{(001)}(x) = \frac{R_V^{(001)}(\lambda_{(111)B}^2 \cos\theta - M\lambda_{(001)}^2)}{(M\lambda_{(001)} + \lambda_{(111)B})\lambda_{(001)}} exp\left(-\frac{x}{\lambda_{(001)}}\right), \tag{6}$$

where each suffix denotes the value for the surface indicated and M is $D_{(111)B}/D_{(001)}$. From Eq.(6), M is given in the form of

$$M = \frac{\lambda_{(111)B}(R_V^{(001)}\lambda_{(111)B}\cos\theta - R_L^{(001)}(0)\lambda_{(001)})}{\lambda_{(001)}^2(R_L^{(001)}(0) + R_V^{(001)})}, \tag{7}$$

Hence, if we measure both of $\lambda_{(111)B}$ and $\lambda_{(001)}$ simultaneously we can calculate M with Eq.(7). By using $\lambda_{(111)B}$ extrapolated to 600 °C as given in Figure 11 and other values obtained experimentally, M is calculated being as large as 140.

Once M is determined, $\tau_{(001)}$ and $\tau_{(111)B}$ are calculated as functions of arsenic pressure if we use appropriate value of $D_{(001)}$ and assume that both $D_{(111)B}$ and $D_{(001)}$ are independent of arsenic pressure. Here, we tentatively have chosen the value of $D_{(001)}$ as $2 \times 10^{-8} \mathrm{cm}^2/\mathrm{sec}$[19]. The calculated results are given in Figure 13. As seen in the figure, the lines of $\tau_{(001)}$ and $\tau_{(111)B}$ intersect at one arsenic

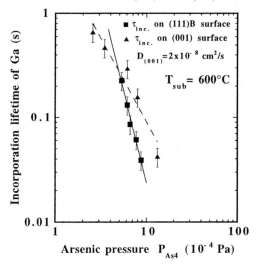

Figure 13. Incorporation time of Ga on (111)B and (001) as a function of arsenic pressure.

pressure. At this point, the flow direction of Ga is reversed if the arsenic pressures on both surfaces are the same.

Eqs. (5)-(7) is derived under the assumption of no diffusion barrier existing between the two faces. Probably, this assumption is not always correct. Nevertheless, at this moment we should satisfy with this assumption since there is no reported data on the height of the potential barrier if any. However, there are no influences of this assumption in determining the diffusion length of Ga incorporation since this has been measured on the surfaces outside of this barrier. In Figure13, the arsenic pressure at which the lines of $\tau_{(001)}$ and $\tau_{(111)B}$ cross is almost the same as the pressure at which the Ga lateral flux changes its direction. This means, Eq.(5) does not contain so big errors.

4. CONCLUSIONS

Surface diffusion of group III atom incorporation in MBE of III-V compounds was studied taking GaAs and InAs as examples. It was shown that for both of GaAs and InAs the surface diffusion length of incorporation, λ_{inc} is a strong function of group V pressure, while it is independent of group III pressure.

When the side surfaces are (111)A or (411)A and the top surface is (001), intersurface diffusion in GaAs MBE occurs always from side to the top. On the other hand, when (111)B side surface was employed the direction of the flow was reversed as the arsenic pressure was increased. The same reversible flow was observed in InAs MBE even when (111)A was used as a side surface. The origin of the flow

130

inversion was explained in terms of the different arsenic pressure dependence of the incorporation life time of group III atoms on the top and the side surfaces.

By measuring λ_{inc} simultaneously on (111)B and (001), it was possible to calculate the ratio of the diffusion coefficient on (111)B to that of (001). The ratio was found being around 140 which means the diffusion on (111)B is much faster than on (001) surface.

ACKNOWLEDGEMENTS

The present authors wish to thank Dr. M. Tanaka of the University of Tokyo for his valuable discussions. This work was supported by Grant-In-Aid for Scientific Research on Priority Areas "Crystal Growth Mechanism in the Atomic Scale" Nos. 03243102 and 04227101 from the Ministry of Education, Science and Culture, Japan.

REFERENCES

1. P.M. Petroff, A.C. Gossard and W. Wiegmann, Appl. Phys. Letters 45(1984)620.
2. T. Fukui and H. Saito, J. Vac. Sci. Technol. B6(1988)1373.
3. T. Fukui and H. Saito, Jpn. J. Appl. Phys. 29(1990)L731.
4. E. Kapon, S. Simhony, R. Bhat and D.M. Hwang, Appl. Phys. Letters 55(1989)2715.
5. E. Kapon, K. Kash, E.M. Clausen, Jr., D.M. Hwang and E.Colas, Appl. Phys. Letters 60(1992)477.
6. S. Ando, T. Honda and N. Kobayashi, Jpn. J. Appl. Phys. 32 (1993)L104.
7. Y. Nagamune, S. Tsukamoto, M. Nishioka and Y. Arakawa, J. Cryst. Growth 126(1993)707.
8. J.H. Neave, P.J. Dobson, B.A. Joyce and Jing Zhang, Appl.Phys. Letters 47(1985)100.
9. T. Nishinaga and K.I. Cho, Jpn. J. Appl. Phys. 27(1988)L12.
10. M. Hata, T. Isu, A. Watanabe and Y. Katayama, Appl. Phys. Letters 56(1990)2542.
11. M. Hata, A. Watanabe and T. Isu, J. Cryst. Growth 111 (1991) 83.
12. X.Q. Shen, D. Kishimoto and T. Nishinaga, Jpn. J. Appl. Phys. 33(1994)11.
13. T. Nishinaga and X.Q. Shen, Applied Surface Science 82/83 (1994)141.
14. T. Suzuki and T. Nishinaga, J. Cryst. Growth 142(1994)49.
15. X.Q. Shen and T. Nishinaga, J. Cryst. Growth 146(1995)374.
16. Y. Nomura, Y. Morishita, S. Goto, Y. Katayama and T. Isu, Appl. Phys. Letters 64(1994)1123.
17. T. Nishinaga and T. Suzuki, J. Cryst. Growth 115(1991)398.
18. C.T.Foxon and B.A.Joyce, Surface Science 50(1975)434.
19. T. Ohno, K. Shiraishi and T. Ito, MRS Symposium Proc. 362 (1994)27.

Advances in the Understanding of Crystal Growth Mechanisms
T. Nishinaga, K. Nishioka, J. Harada, A. Sasaki and H. Takei (Editors)
© 1997 Elsevier Science B.V. All rights reserved.   131

Study of surface chemical reactions in GaAs atomic layer epitaxy by *in situ* monitoring methods

A. Koukitu

Department of Applied Chemistry, Tokyo University of Agriculture and Technology, Koganei, Tokyo 184, Japan

In situ monitoring of the growth process in atomic layer epitaxy (ALE) is essential for understanding the growth mechanism. The present paper reports the investigation of the ALE growth mechanism using two *in situ* monitoring methods, a gravimetric method using a microbalance and an optical method using surface photo-absorption (SPA). The reaction mechanisms that occur on the surface in each step of the gas sequence are discussed in relation to GaAs ALE growth using the halogen transport system. Furthermore, As desorption from the (001) As surface and chemisorption of hydrogen atoms on the (001) Ga surface are discussed.

1. INTRODUCTION

In situ monitoring of the crystal growth process is impotant for the investigation of the growth mechanism of epitaxy. In atomic layer epitaxy (ALE), crystal growth is induced by the alternating supply of reactants on the substrate surface. The chemical reactions that generally occur using conventional epitaxial methods are divided into sub-reactions in ALE. *In situ* monitoring of the layer-by-layer growth is possible for each sub-reaction. As a result of this monitoring, we can investigate the crystal growth process at the atomic level.

Previous studies have investigated the growth process in halogen transport ALE of GaAs using two *in situ* monitoring methods: gravimetric and optical [1-7]. The *in situ* gravimetric monitoring system, which is equipped with a halogen transport ALE growth reactor and a recording microbalance, provides direct information on the growth rate (monolayer/cycle) and the weight corresponding to the chemical species adsorbed on the surface under real ALE growth conditions. The advantage of this method is that it can directly monitor the growth rate without influencing the process such as by enhancing the reaction by an electron or photon probe. On the other hand, the optical method using surface photo-absorption (SPA) with an ALE growth system provides real-time information on the growth reactions during ALE. Using these methods, growth of a mono-molecular layer unit is demonstrated to occur during each cycle of ALE, and the self-limiting mechanism of halogen transport ALE is shown to result from the complete coverage of surface As sites by an adsorbed complex that includes GaCl. In the present paper, our investigation of the GaAs ALE process using two *in situ* monitoring methods is reviewed.

2. EXPERIMENTAL PROCEDURE

ALE of III-V compounds can be performed using either metalorganic vapor-phase epitaxy (MOVPE) or halogen transport VPE. For practical purposes, the latter method is more suitable for investigation of the growth mechanism, because it easily induces monolayer growth over a wide range of growth conditions. Therefore, in the present investigation, we employed the halogen transport method.

Figure 1 shows the gravimetric *in situ* monitoring system, which uses a recording microbalance. The wafer used for the experiment was approximately 2.6cm^2 (001)GaAs, and 50 μm thick, and was suspended from the microbalance by a fused quartz fiber. One ALE cycle consisted of four stages: GaCl supply (step 1), H_2 purge (step 2), AsH_3 supply (step 3) and H_2 purge (step 4). The standard gas flow sequence was as follows: GaCl supply for 8 s, H_2 purge for 8 s, AsH_3 supply for 8 s, and H_2 purge for 8 s. Generally, the time constant for the gas sequence in halogen transport ALE is longer than that in the ALE using organometallic sources, because GaCl remains stable at higher temperatures than organometallic compounds such as trimethylgallium. The typical partial pressures of GaCl and AsH_3 at the substrate position were maintained at 1.0×10^{-4} and 5.0×10^{-4} atm, respectively. The GaCl obtained by reacting metallic Ga and HCl at 780°C was fed intermittently onto the substrate surface using a suction pump. AsH_3 was used as the arsenic source and was introduced intermittently from a separate tube using air-operated run-and-vent valves. AsH_3 was decomposed at 780°C.

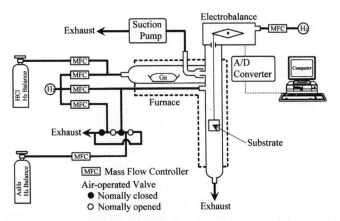

Figure 1. Schematic diagram of *in situ* gravimetric monitoring system in halogen transport ALE.

The most important factor in the gravimetric monitoring of ALE is the maintenance of a constant gas flow rate in sequential runs in order to avoid weight change. All stages should occur at a uniform gas velocity in the substrate zone. In this study, the mass flow controllers, run-and-vent valves, and suction pump were controlled externally by computer. The quartz reactor, which had a 20 mm internal diameter, was heated using a resistance furnace with 5 heating zones. The substrate temperature varied from 400 to 600°C. Before monitoring, ALE growth was performed approximately 50 times in order to grow a buffer layer at 500°C after preheating the GaAs substrate for 10 min above 580°C. The microbalance had a sensitivity of the 0.025μg with a maximum capacity of 0.3 g. The noise level from the balance was minimized by electronic filtering. Using this system, successive measurement of the growth rate is possible in sequential processes.

Figure 2 shows the optical *in situ* monitoring system, which consisted of a surface photo-adsorption (SPA) monitoring system and a halogen transport ALE reactor. The internal diameter of the reaction tube was 20 mm. The feed system for GaCl and AsH$_3$ was very similar to that of the gravimetric monitoring system. The SPA system was similar to that used by Kobayashi et al. [8, 9]. The p-polarized 488 nm laser beam (2 - 6 mW) or 670 nm semiconductor laser beam (1mW) irradiated the (001) GaAs substrate surface at a 70° angle of incidence, which is approximately the Brewster angle of GaAs. Reflected light was detected to a Si p-i-n photodiode (PD), the output of which was amplified by a preamplifier and a lock-in-amplifier and to which reference pulses from the mechanical chopper were supplied.

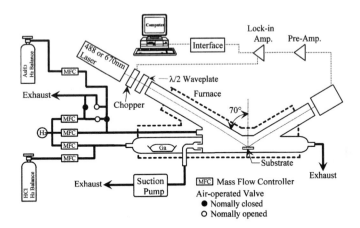

Figure 2. Schematic diagram of *in situ* SPA monitoring system in halogen transport ALE.

3. RESULTS AND DISCUSSION

3.1. IN SITU GRAVIMETRIC MONITORING

In situ gravimetric monitoring was performed under the condition of an atmospheric pressure. Figure 3 shows a typical output trace during halogen transport ALE cycles for the following gas sequence: GaCl for 8 s, H_2 purge for 40 s, AsH_3 for 8 s, and H_2 purge for 8 s. The growth temperature was 450°C. As shown by the ordinate, the output trace corresponds to the weight change of the substrate. The output trace rapidly increased when GaCl was supplied, and subsequently decreased when purged. The increase was due mostly to the flow rate change induced by the addition of GaCl flow. A similar result was observed when AsH_3 was supplied. In these sequential processes, the difference, ΔW_{GaAs}, in the output trace level between two successive cycles during H_2 purge after AsH_3 supply corresponds to the change in substrate weight in one ALE cycle. Thus, we can monitor the growth rate with a monolayer scale in each cycle. In addition to the weight, ΔW_{GaAs}, we can also monitor the weight difference, ΔW_S, during H_2 purge after GaCl supply, as shown in the figure. The value of ΔW_S corresponds to the weight of the chemical species on the As surface during H_2 purge after GaCl supply.

As mentioned above, the gravimetric monitoring method yields direct information regarding real-time growth rate and surface coverage, and therefore is a powerful tool for researching the growth mechanism of ALE.

3.2. IN SITU OPTICAL MONITORING

In situ optical monitoring methods such as reflectance difference spectroscopy (RDS) [10] and SPA [8] have been performed in order to analyze the surface process of ALE growth in real time. Figure 4 shows typical SPA signals for halogen transport ALE using GaCl and AsH_3 sources. The optical light sources were 488 nm and 670 nm. As shown in Figure 4(a), the reflective intensity decreases rapidly from the As surface value when GaCl is supplied, and then slowly increases and reaches a plateau at position (1). When H_2 purging begins, the intensity increases to position (2), and behavior similar to that of the first-order reaction is observed. When AsH_3 is supplied, the intensity rapidly increases to position (3). Figure 4(b) shows the SPA signal after irradiation with 670 nm light. A similar signal trace asin 488 nm light is obtained, but the intensity difference that is normalized by the intensity of As surface, $\Delta R/\Delta R_{As}$, is smaller than that following irradiation with 488 nm light. The value of $\Delta R/\Delta R_{As}$ at position (2) (approximately -1%) is consistent with the SPA signal during H_2 purge after triethylgallium (TEGa) supply, as reported by Kobayashi and Horikoshi [11]. In addition, after irradiation with 670 nm light, $\Delta R/\Delta R_{As}$ at position (5) is the same as the extrapolation from the spectral dependence of the SPA signal [8]. As shown in the figure, the reflection intensity changes with more complicated behavior than that of MBE or MOVPE [8. 9], suggesting that complicated reactions occur on the surface during halogen transport ALE.

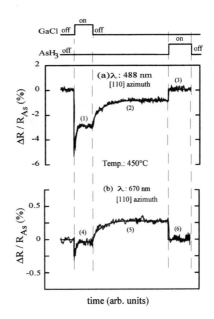

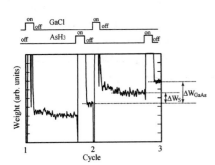

Figure 3. Typical output trace obtained from *in situ* gravimetric monitoring system. Gas sequence: GaCl for 8 s, H_2 purge for 40 s, AsH_3 for 8 s and H_2 purge for 8 s.

Figure 4. Typical SPA signal intensity during halogen transport ALE cycles. Optical light sources used were (a) 488 nm or (b) 670 nm. ΔR is $(R_{Ga} - R_{As})$, R_{Ga} and R_{As} are the reflected light intensities of Ga- and As-stabilized GaAs surfaces.

3.3. SURFACE REACTION DURING GaCl SUPPLY

Based on the results of these two *in situ* studies, we propose that the following reaction mechanisms occur on the surface for each step of the halogen transport ALE sequence. Figure 5 shows the number of grown layers based on the output trace of the gravimetric monitoring method as a function of sequential cycles. The first 4 cycles show the growth rate without AsH_3 supply. No increase in the substrate weight is observed during these cycles, presumably because GaCl molecules cannot be adsorbed on the Ga-terminated surface. After the 4th cycle, AsH_3 was supplied. The number of grown layers increases linearly as the number of cycles increased. On the other hand, in the SPA signal trace of halogen transport ALE shown in Figure 4, the SPA signal rapidly decreased when GaCl was supplied. This process is probably induced by random adsorption of GaCl at the As sites, as shown schematically in Figure 6(a). The SPA signal then slowly increased and reached a plateau. At this stage, the randomly adsorbed GaCl becomes tightly adsorbed species by desorbing excess GaCl. All surface As sites then become covered by this complex (Figure 6(b)), because layer-by-layer growth occurs in ALE, as shown in Figure 5. The complex is stabilized by the partial pressure of GaCl over the substrate surface. Excess GaCl molecules physically adsorbed over the monolayer complex may exist during GaCl supply. However, excess molecules are desorbed from the substrate surface in the initial period of H_2 purge due to the lower binding energy of physical adsorption. According to this mechanism, the adsorbed surface complex completely covers surface As sites and prevents further adsorption of GaCl. Thus, the self-limiting mechanism of halogen transport ALE is thought to be this complete coverage. Since GaCl is stable and the complex is formed with stability under a wide variety of growth conditions, a wide ALE windows exist for halogen transport ALE. On the other hand, during ALE growth with organometallic compounds, the source compounds easily decompose. Thus, the ALE window depends on the lifetime of the organometallic compounds.

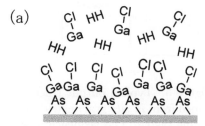

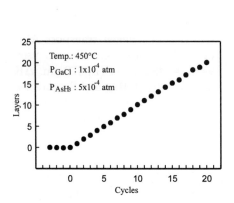

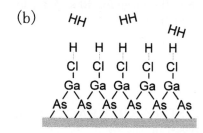

Figure 5. Number of grown layers obtained from output trace of gravimetric monitoring method as function of sequential cycles.

Figure 6. Schematic diagram of proposed mechanism of GaAs (001) surface during GaCl supply.

3.4. SURFACE REACTION DURING H_2 PURGE AFTER GaCl SUPPLY

When H_2 purge begins, the reflective intensity of SPA during H_2 purge slowly changes from the value of the complex surface and eventually plateaus as shown in Figure 4. Corresponding to the increase in SPA signal during H_2 purge, the output signal of the gravimetric monitoring gradually decreases before leveling off, as seen in Figure 3. The gradual decrease in weight is thought to correspond to the desorption of HCl from the surface during H_2 purge, because the surface becomes covered with Ga after H_2 purge, as described below. Figure 7 shows the complex coverage on the As surface as a function of the H_2 purge time after GaCl supply at 450°C. The complex coverage was converted from SPA signal in step 2, assuming that the value of the complex coverage was equal to 1.0 at position (1) in Figure 4 and that the value at position (2) was 0.0. The coverage, as determined using a first-order rate equation, is represented by a solid line for comparison. An integrated rate equation of the form $\theta(t) = \theta_0 e^{-kt}$ was used. The complex formed during GaCl supply reacts at a first-order rate during H_2 purge, because the first-order rate equation fits well with the signal.

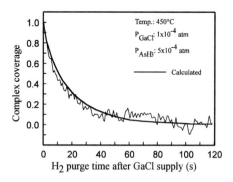

Figure 7. Complex coverage calculated from SPA signal and theoretical curve (solid line) obtained by first-order rate equation as function of H_2 purge time after GaCl supply.

Next, we determined the surface chemical species during H_2 purge after GaCl supply using the *in situ* gravimetric monitoring method. Figure 8 shows weight ratios ($\Delta W_S/\Delta W_{GaAs}$) at 425, 450, 475 and 500°C as a function of the sequential cycles. In these sequential processes, we confirmed that growth of the monomolecular layer occurred in each cycle. The $\Delta W_S/\Delta W_{GaAs}$ ratios are approximately equal to 0.48, and are irrespective of the growth temperature. Compared to the surface chemical species that are capable of binding to the As surface indicated on the right side of the figure, we can conclude that the surface chemical species during H_2 purge after GaCl supply is a single monolayer of Ga atoms[4]. The fact that the surface after H_2 purge is Ga-covered is consistent with the SPA signal value of $\Delta R/\Delta R_{As}$ at position (2) in Figure 4(a), because this value, approximately -1%, agrees with that of the Ga surface in organometallic sources such as triethylgallium[8].

Based on these two *in situ* observations, we can assume the following mechanisms for steps 1 and 2. In step 1, all surface As sites are completely covered by the adsorbed complex shown in Figure 6(b), which is stabilized under the partial pressure of GaCl. With H_2 purge after GaCl supply (step 2), the adsorbed complex

138

is reduced to a single monolayer of Ga atoms, shown in Figure 9(a). The Ga atoms on the surface probably are terminated by hydrogen atoms. However, during typical halogen transport ALE, both complexes and Ga atoms may exist on the surface (Figure 9(b)) after H_2 purge due to the short purge time. Since the purge time required to reach complete Ga surface is longer than 30 s at 450°C, the amount of complex remaining on the surface depends on the temperature and purge time.

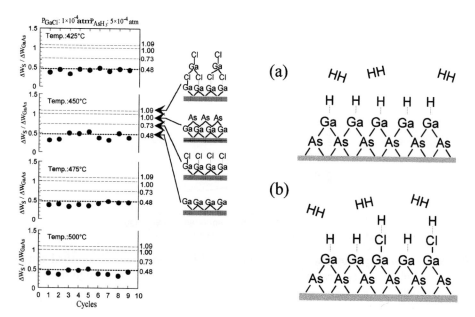

Figure 8. $\Delta W_S/\Delta W_{GaAs}$ ratio at temperature from 475 to 550°C. Partial pressure of GaCl and AsH_3 were 1.0×10^{-4} and 5.0×10^{-4} atm, respectively.

Figure 9. Schematic diagram of proposed mechanism of GaAs (001) surface during H_2 purge after GaCl supply.

3.5. SURFACE REACTION DURING AsH_3 SUPPLY

During this stage, AsH_3 source gas is supplied over the substrate surface where Ga atoms and complexes exist. The growth rates observed by changing the H_2 purge time after GaCl supply were consistently a single monolayer/cycle, and were independent of H_2 purge time at the temperature below 450°C. Therefore, ALE growth is accomplished by both reactions of

$Ga(ad) + 1/4As_4 \rightarrow GaAs$ and

$GaCl(adsorbed\ complex) + 1/4As_4 + 1/2H_2 \rightarrow GaAs + HCl$.

Based on these results, the reaction mechanism that occurs on the surface during AsH_3 supply shown in Figure 10 was proposed. The surface, covered by Ga atoms

(Figure 10(a)) and / or complexes (Figure 10(b)), reacts with As species to form a single monolayer of GaAs on the surface (Figure 10(c)).

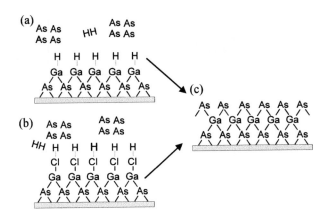

Figure 10. Schematic diagram of proposed mechanism of GaAs (001) surface during AsH₃ supply.

3.6. SURFACE REACTION DURING H$_2$ PURGE AFTER AsH$_3$ SUPPLY

During epitaxial growth of GaAs, various reconstructions of the (001) GaAs surface are known to result from the desorption of arsenic atoms from the surface[12-17]. However, most previous investigations have been indirect observations using an electron, an X-ray, or a photon as a probe. In this section, we report the As desorption from the (001) GaAs surface in ALE using the *in situ* gravimetric monitoring method, which provides direct information regarding the growth rate at the monolayer level.

As shown at section3.5, the growth rates were independent of H$_2$ purge time at the temperature below 450°C. However, the growth rates became dependent on the H$_2$ purge time after GaCl supply at higher temperature. Figure 11 shows growth rate as a function of H$_2$ purge time after AsH$_3$ supply, step 4[5]. The growth temperature was 500°C, and duration of GaCl supply (step 1), H$_2$ purge after GaCl supply (step 2), and AsH$_3$ supply (step 3), were 8 s each. Approximately 10 cycles of the standard gas sequence (8-8-8-8 s) were performed in order to ensure growth of a single monolayer between each measurement. Growth of a single monolayer/cycle occurs in each cycle with H$_2$ purge times of less than 20s. The growth rate per cycle, however, decreased slowly to 0.75 monolayers/cycle, and then remained constant as the H$_2$ purge time after AsH$_3$ supply increased.

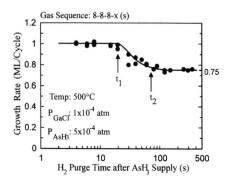

Figure 11. Growth rate as function of H_2 purge time after AsH_3 supply. Durations of GaCl, H_2 purge and AsH_3 were kept constant.

These results indicate that the growth rate depends on the surface structure, and that at least two types of surface structures exist in this ALE process. One is the surface structure that exists during the H_2 purge time , $t < t_1$ in Figure 11, at which time the a single monolayer/cycle growth is attained. This surface is probably a $c(4\times4)$ or an ideal (1×1) structure, in which the excess As atoms on the single-monolayer As plane do not contribute to the formation of GaAs. Yamauchi et al. [18] described a very similar role for an As dimmer on the $c(4 \times 4)$ surface.

With increasing H_2 purge time, the growth rate decreased due to As desorption from the surface, and reached a constant value, 0.75, for long purge times. At t_2 in Figure 11, another stable surface reconstruction was formed. The surface structure during the H_2 purge time between t_2 and 350 s (the longest experimental time) is thought to be a (2×4) reconstruction, because the growth rate of 0.75 monolayers/cycle corresponds to the As coverage structure with a (2×4) surface[19]. A stable growth rate of 0.75 monolayers / cycle can be obtained with a prolonged purge time at 500°C. In the H_2 purge time between t_1 and t_2, the surface may consist of mixed structures of $c(4\times4)$ or ideal (1×1) and (2×4) reconstructions. We previously investigated the surface morphology of ALE-grown layers using atomic force microscopy (AFM)[5]. AFM images of the GaAs surface were obtained at three H_2 purge times after AsH_3 supply: 10 s (1.0 monolayer/cycle), 30 s (0.9 monolayer/cycle) and 90 s (0.75 monolayer/cycle). The film thickness was 300 monolayers for all samples. The AFM images showed that a much smoother surface with less than 0.5 nm of roughness could be obtained with H_2 purge times of 10 and 90 s, whereas a very rough surface with large islands was obtained at a H_2 purge time of 30 s. We proposed a model to explain why a smooth surface could be grown on the (2×4) reconstructed surface [5].

3.7. HYDROGEN CHEMISORPTION ON THE (001) Ga SURFACE

As described above, the GaAs (001) Ga surface is covered with a single monolayer of Ga atoms during H_2 purging after GaCl supply. However, whether the monolayer Ga atoms on the surface are terminated by hydrogen atoms is not clear. In this section, we describe an investigation of hydrogen adsorption on (001) Ga surface. We employed the optical method in order to monitor hydrogen atoms adsorption on the (001) Ga surface[7]. The equipment used is shown in Figure 2. Purified H_2, He, and N_2 were used as carrier gas. *In situ* monitoring was performed by controlling the flow ratio of H_2 to the inert gas, He or N_2, during the period after step 2 in order to vary the partial pressure of H_2 in the carrier gas. The total flow rate was constant.

Figure 12 shows a typical SPA signal observed on the (001) GaAs surface in the hydrogen-inert gas carrier system. The growth temperature was 450°C, and the H_2 partial pressure in the carrier gas was 0.41 atm. As shown in Figure 4, the reflection rapidly decreased from the value for an As surface when GaCl is supplied in step 1, as shown at point A, and then slowly increased and reached a plateau at point B. The surface at point B was completely covered by the adsorbed complex. In step 2, the intensity increased until point C, when the complexes on the surface were decomposed to Ga atoms which desorbed HCl from the surface. At point C, the surface consisted of a single monolayer of Ga atoms, as shown in Figure 8.

In order to investigate whether the Ga atoms on the surface are terminated by hydrogen atoms under the condition of an atmospheric pressure, the H_2 partial pressure in the carrier gas was reduced from 0.41 atm to zero by increasing the partial pressure of the inert gas over the Ga surface at point D in Figure 12. The intensity slowly decreased and then reached a plateau at point E. When the H_2 partial pressure was increased to 0.41 atm again, the reflection increased to the value at point F, which equals the same as the initial value at point D. The level of the SPA signal was kept constant from point D to F without changing the carrier gas. The intensity rapidly increased to point G when AsH_3 was supplied on the surface, forming a single monolayer of GaAs.

The same optical change shown in Figure 12 were observed using either He or N_2 as the inert gas, and the magnitude of the optical intensity at point E was dependent on the H_2 partial pressure. Furthermore, the same optical change was observed between points D and F using H_2, He or N_2 as a carrier gas. These findings may be due to reaction with and termination of Ga atoms on the (001) surface by hydrogen in the carrier gas. Furthermore, desorption of hydrogen from the surface appears to occur as a result of the reverse reaction in the inert carrier gas.

Interestingly, different SPA signals were observed between two samples at room temperature (RT): one with the surface cooled to RT in H_2 and the other cooled in the inert gas. Therefore, we monitored the Ga surfaces, which were frozen in inert gas or H_2, using temperature-programmed desorption (TPD) with SPA. Figure 13 shows the SPA signal traces of three different samples, (a), (b), and (c), in the inert carrier gas during temperature increase. $\Delta T/\Delta t$ was approximately 1.4

142

deg/s. Sample (a) was cooled in inert carrier gas to RT after Ga surface formation at 450°C in H_2, and (b) was cooled in H_2 to RT after Ga surface formation at 450°C in H_2. Sample (c) was cooled to RT in the inert carrier gas, following which the sample was exposed to H_2 for 10 min and the carrier gas was switched to an inert gas just before TPD measurement. All of the SPA traces increase linearly due to the change in substrate reflectivity with increasing temperature.

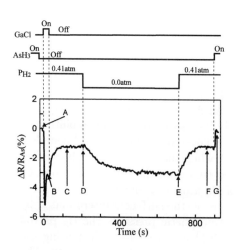

Figure 12. Typical SPA signal trace on (001) GaAs surface at 450°C for halogen transport ALE. 488 nm laser light was irradiated with [110] incidence azimuth. GaCl and AsH_3 partial pressures were 1.0×10^{-4} and 5.0×10^{-4} atm, respectively.

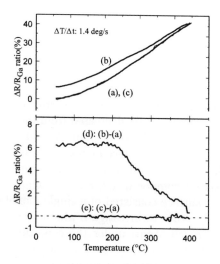

Figure 13. SPA signal traces obtained from the three different surfaces in inert carrier gas by TPD method: (a) cooled to RT in inert gas after Ga surface formation at 450°C in H_2; (b) cooled in H_2 after Ga formation in H_2; (c) exposed to H_2 for 10min at RT after cooling in inert gas, then with subsequent of inert gas; (d) signal difference (b) − (a); and (e) signal difference (c) − (a).

Figure 13 shows two curves, (d) and (e), obtained by subtracting (a) from (b) and from (c), respectively. Trace (d) remains constant up to approximately 200°C, and then decreases almost to zero with increasing temperature. On the other hand, no change in the difference is observed in trace (e). These results indicate the occurrence of hydrogen chemisorption over the Ga surface and hydrogen desorption from the Ga surface under the condition of an atmospheric pressure. As a matter of fact, the surface of sample (b) was chemisorbed by hydrogen, while surfaces (a) and (c) were not chemisorbed by hydrogen. The reaction rate between Ga atoms and hydrogen depends closely on temperature, and is slow enough to cause hydrogen chemisorption for a short period, (10 min, at RT) At present, it is not clear whether the Ga-Ga dimer bonds, or the Ga-H bonds, or both, are the origin of the SPA signal change. Using RDS, Kamiya et al.[20] reported a similar optical change after switching the carrier gas due to the reaction of Ga dimers with N_2 or impurities on the (001) GaAs surface. However, we believe that hydrogen desorption is reflected in the signal change displayed by trace (d), because reproducible results are obtained using either He or N_2 as the inert gas, and the initial temperature (approximately 200°C) at which desorption occurs is in good agreement with the results of the D_2 desorption experiment using QMS spectra in an ultrahigh vacuum (UHV), as reported by Creighton[21].

Next, the influence of H_2 partial pressure on hydrogen chemisorption was investigated. Figure 14 shows the number of Ga atoms adsorbed by hydrogen in a unit area, n_H (left axis), as a function of the H_2 partial pressure. The value of n_H is defined as:

$$n_H = \frac{N_H \left(R - R_{inert} \right)}{R_{hydrogen} - R_{inert}}, \quad (1)$$

where R indicates the SPA signal intensity of the surface in mixed gas, and R_{inert} and $R_{hydrogen}$ indicate the SPA signal intensity of the surfaces in the inert gas and in H_2, respectively. N_H in eq. (1) is the number of Ga atoms terminated with hydrogen in a unit area where $P_{H_2}=1.0$. The temperature was kept constant at 450°C, and the reflection R indicated in eq.(1) was measured from point E in Figure 12, varying the H_2 partial pressure from 0.0 to 1.0 atm. The value of n_H increases with a behavior similar to that of the Langmuir adsorption isotherm.

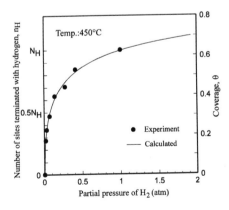

Figure 14. Number of Ga atoms chemisorbed by hydrogen atoms in unit area as function of H_2 partial pressure with relationship between surface coverage and P_{H_2} calculated using Langmuir isotherm with dissociation.

144

Generally, atomic hydrogen is the species presumed to bond to the semiconductor surfaces, because H_2 molecules do not adsorb at RT[22,23]. Therefore, we employed the dissociative model in the following calculation. In the Langmuir adsorption model, the dissociative isotherm[24] can be expressed as

$$\theta = \frac{\sqrt{KP_{H_2}}}{1 + \sqrt{KP_{H_2}}}, \quad (2)$$

where θ is the surface coverage adsorbed by hydrogen through dissociation, and K is an equilibrium constant for the adsorption isotherm. On the other hand, θ denotes n_H/N_T, where N_T is the number of sites of Ga atoms terminated with hydrogen in a unit area at $\theta = 1.0$. Therefore, we can denote the relationships between θ (right-hand axis) and P_{H_2} and between θ and n_H (left-hand axis) from eq (2), as plotted in Figure 14. In this calculation, the value of K is assumed to be $2.79 atm^{-1}$. The experimental value is in close agreement with the calculated result. Therefore, based on Figure 14 two important conclusions can be drawn regarding hydrogen adsorption on the (001) Ga surface under atmospheric pressure. First, hydrogen in the carrier gas reacts with Ga atoms, which have dimer bonds, through dissociative adsorption (Ga-Ga(sur) $+H_2$(g) $\rightarrow$ 2Ga-H(sur)), and the desorption of hydrogen occurs by the reverse reaction in the inert carrier gas. Second, approximately 63% of the Ga atoms on the (001) Ga surface are terminated with atomic hydrogen and approximately 37% of Ga-Ga dimer bonds remain when the H_2 partial pressure is 1.0 atm.

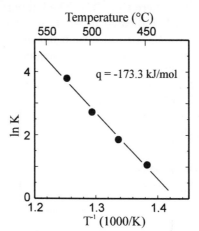

Figure 15. Arrhenius relationship for hydrogen adsorption from 450 to 525°C. K was obtained by fits of data to Langmuir adsorption with dissociation.

In addition, the heat of adsorption, q, can be obtained by varying temperature according to the following equation:

$$K = A \exp\left(\frac{q}{RT}\right), \quad (3)$$

where K is the equilibrium constant for the adsorption, A the frequency factor. Figure 15 shows Arrhenius relationship for the hydrogen adsorption on the surface in the temperature range from 450 to 525°C. The value of K was obtained by the fits of the experimental data to Langumuir adsorption isotherm with dissociation. The heat of adsorption, q, is -173kJ/mol and this value denotes an endothermic. This fact supports that Ga atoms on the (001) surface are chemisorbed with atomic hydrogen because generally energy transfer for the dissociative adsorption is an endthermic reaction.

4. CONCLUSIONS

In situ monitoring techniques are essential for understanding the ALE growth mechanism. The present paper reviews recent studies of the growth mechanism of halogen transport GaAs ALE. The primary results are summarized as follows.

1. GaCl supply
    Surface As sites are completely covered by a complex that includes GaCl. The growth of a mono-molecular layer unit occurs in each cycle of ALE, and the self-limiting mechanism of halogen transport ALE is ascribed to the complete coverage of surface As suites by an adsorbed complex.

2. H_2 purge after GaCl supply
    The complex on the surface is reduced to a single monolayer of Ga atoms. However, the amount of complex remaining on the surface at the end of the purge depends on the temperature and the purging time.

3. AsH_3 supply
    The surface covered by Ga atoms and / or complexes reacts with As species to form a single monolayer of GaAs on the surface.

4. H_2 purge after AsH_3 supply
    In the short purge time required for common ALE growth, the growth rate is unity during each cycle. However, the growth rate decreases to a constant value, 0.75, for long purge times, which corresponds to the As coverage on the (2×4) reconstructed surface.

5. Hydrogen chemisorption
    Under atmospheric pressure conditions, hydrogen molecules dissociatively react with Ga atoms on the (001) surface, and Ga surface is terminated by hydrogen

atoms. The desorption of hydrogen atoms occurs by the reverse reaction in the inert carrier gas. The relationship between the surface coverage of adsorbed hydrogen atoms and the H_2 partial pressure can be clearly elucidated by the Langmuir equation of the dissociative isotherm.

ACKOWLEDGEMENTS

The author would like to extend thanks to Prof. H. Seki at Tokyo University of Agriculture and Technology for his helpful discussion and suggestions. The author would also like to thank Prof. T. Nishinaga at University of Tokyo for his encouragement and assistance in writing this review.

REFERENCES

1. A.Koukitu, H.Ikeda, H.Yasutake and H.Seki: Jpn. J. Appl. Phys. 30 (1991) L1847.
2. A.Koukitu, H.Ikeda, H.Suzuki and H.Seki: Jpn. J. Appl. Phys. 30 (1991) L1712.
3. N.Takahashi, M.Yagi, A.Koukitu and H.Seki: Jpn. J. Appl. Phys. 32 (1993) L1277.
4. A.Koukitu, N.Takahashi, Y.Miura and H.Seki: Jpn. J. Appl. Phys. 33 (1994) L613.
5. A.Koukitu, N.Takahashi, Y.Miura and H.Seki: J. Crystal Growth 146 (1995) 239.
6. A.Koukitu, N.Takahashi and H.Seki: J. Crystal Growth 146 (1995) 467.
7. A.Koukitu, T.Taki, N.Takahashi and H.Seki: Jpn. J. Appl. Phys., 35 (1996) L710.
8. N. Kobayashi and Y.Horikoshi: Jpn. J. Appl. Phys. 28 (1989) L1880.
9. N. Kobayashi, T.Makimoto, Y. Yamauchi and Y. Horikoshi: J. Cryst. Growth 107 (1991) 62.
10. D.E.Aspnes, J.P.Harbison, A.A.Studna and L.T.Florez: Phys. Rev. Lett. 59 (1987) 1687.
11. N. Kobayasi and Y. Horikoshi: Jpn. A. Appl. Phys. 29 (1990) L702.
12. M. A. Mendez, F. J. Palomares, M. T. Cuberes, M. L. Gonzalez and F. Soria, Surf. Sci. 251/252(1991)145.
13. C. Deparis and J. Massies, J. Cryst. Growth 108(1991)157.
14. D. E. Aspnes and A. A. Studna, Phys. Rev. Lett. 54(1985)1956.
15. P. Drathen, W. Ranke and K. Jacobi, Surf. Sci. 77(1978)L162.
16. M. Sauvage-Simkin, R. Pinchaux, J. Massies, P. Calverie, N. Jedrecy, J. Bonnet and I. K. Robinson, Phys. Rev. Lett. 62(1989)563.
17. S. A. Chamber, Surf. Sci. Lett. 248(1991)L274.
18. Y. Yamauchi, K. Uwai and N. Kobayashi, Japan. J. Appl. Phys. 32(1993)3363.
19. P. K. Larsen and D. J. Chadi, Phys. Rev. B37(1988)8282.

20. I.Kamiya, D.E.Aspnes, H.Tanaka, L.T.Florez, J.P.Harbison and R.Bhat: Phys. Rev. Lett. 68 (1992) 627.
21. J.R.Creighton: J. Vac. Sci. Technol. A 8 (1990) 3984.
22. H.Lüth and R.Matz: Phys. Rev. Lett. 46 (1981) 1652.
23. L.H.Dubois and G.P.Schwartz: Phys. Rev. B 26 (1982) 794.
24. P.W.Atkins: Physical Chemistry (Oxford University Press, Oxford, 1990) 4th ed.

Advances in the Understanding of Crystal Growth Mechanisms
T. Nishinaga, K. Nishioka, J. Harada, A. Sasaki and H. Takei (Editors)

Surface Kinetics and Mechanism of Atomic Layer Epitaxy of GaAs Using Trimethylgallium

H. Ohno

Laboratory for Electronic Intelligent Systems
Research Institute of Electrical Communication
Tohoku University, Sendai 980-77, Japan

Time resolved *in situ* Auger Electron Spectroscopy (AES) revealed the desorption time constant of C-related species from clean molecular beam epitaxially grown GaAs surfaces exposed to trimethylgallium (TMGa). The measured time constant was of the order of 100 s in the temperature range where Atomic Layer Epitaxy (ALE) of GaAs takes place. Similar time constant was also observed in the change in the reflection high energy diffraction patterns. Results on other surface orientations and surface reconstructions were also presented. Growth of GaAs by ALE using TMGa and arsine was carried out under various conditions. A model based on the adsorbate-inhibition mechanism with a long time constant as observed by AES measurements was developed and shown to be capable of explaining the ALE growth data. Remaining issues were also discussed.

1. INTRODUCTION

The ultimate monolayer thickness control offered by Atomic Layer Epitaxy (ALE) is expected to play a major role in fabricating quantum nano-structures for the future electronic and optical devices, where monolayer control of layer thickness is vital in obtaining the designed performance of the devices. Since the pioneering work by Nishizawa et al. [1], who demonstrated ALE of GaAs using trimethylgallium (TMGa) and arsine, extensive studies on the nature of the self-limiting process have been conducted by a number of research groups [2–11]. The self-limiting process of ALE results in one-monolayer of growth regardless of the supplied amount of source materials if the amount is above a certain threshold. To understand this self-limiting mechanism, one has to clarify the surface reaction and surface kinetics occurring on the semiconductor surface with the impinging source materials that enable the ALE process. Currently, there are three widely known models for the self-limiting process, which are 1) the Adsorbate-inhibition model [2], 2) the Selective adsorption model [3,4], and 3) the Flux balance model [5]. The latter two are based on the fast desorption time constant (< 1 s) of CH_3 from GaAs exposed to TMGa. The fast desorption time constant of CH_3, resulting from the dissociative adsorption of TMGa on GaAs, has been experimentally observed by

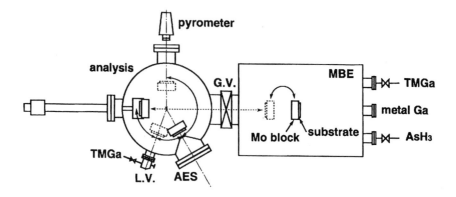

Figure 1. Schematic diagram of the Auger-MBE apparatus used in in situ Auger measurements.

several groups [6–8], and modeling of the ALE growth process using rate equations has been reported [5,9]. On the other hand, the presence of long desorption time constant of the order of 100 s at ALE temperatures [10,11], which is not compatible with the models based on the high desorption rates, has been overlooked and not much attention has been paid to the modeling based on it.

Here we review the experimental results obtained by *in situ* Auger Electron Spectroscopy (AES) and by Reflection High Energy Electron Diffraction (RHEED), the ALE growth results of GaAs using TMGa and arsine, and the results of modeling of the ALE growth processes based on the long time constant. The model is based on the assumption that the adsorbed CH_3 and monomethylgallium (MMGa), which are reaction products of TMGa on GaAs, blocks further reaction/adsorption of TMGa on the surface, i.e. the adsorbate-inhibition model. The model rate equations reproduce the GaAs growth rate dependence on the TMGa supply period as well as the growth rate of the double TMGa pulse experiments with varying purge time in-between the two TMGa pulses. The remaining issues are also discussed.

2. TIME RESOLVED *IN SITU* AUGER ELECTRON SPECTROSCOPY

2.1. Experimental

Figure 1 shows the apparatus used in the present work, which consists of an introduction chamber, an analysis chamber, and an MBE growth chamber. The MBE chamber is equipped with TMGa and AsH_3 as well as Ga and As sources. The grown sample can be transferred to the analysis chamber without exposing it to air.

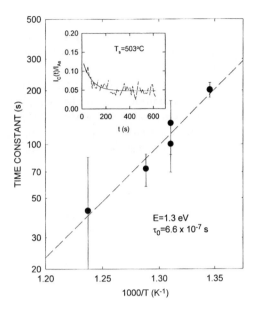

Figure 2. Time constant, τ, of desorption of C-related species from (100) GaAs surfaces determined by in situ Auger measurement as a function of substrate temperature. Broken line represents $\tau = \tau_0 \exp(E/kT)$. Inset shows a typical time variation of C Auger signal intensity after exposure to TMGa. Broken lines are experimental data normalized by As Auger signal intensity. Solid line is a fit using eq.(1)

All the substrates used in experiments were n-type GaAs (001) $\pm$ 0.1°. After growing buffer layer of undoped GaAs for about 200 nm in the MBE chamber, samples were transferred to the analysis chamber (with cryo-panel), where Auger spectrometer was located, and then heated to a pre-determined temperature without As beam. The temperature stability was $\pm$ 3 °C. TMGa was provided directly into the analysis chamber via variable leak valve; 30 s at 3×10^{-6} Torr. Auger spectra recording was started within 10 s after the termination of TMGa. The time constant of the initial decay of the pressure in the analysis chamber after terminating the TMGa supply was about 4 s and the chamber pressure at the start of Auger measurements was at least an order of magnitude lower than that during TMGa supply ($< 3 \times 10^{-7}$ Torr). For AES, 3 keV electron beam with 3 μA absorption current was used and the scan was made continuously between 220 - 320 eV for 3 - 15 s, where C KLL signal (273 eV) is present.

2.2. Time dependence of C-AES signal on GaAs (100) surface

The inset of Figure 2 shows a typical time variation of C signal intensity recorded by the *in situ* Auger measurements. The C peak-to-peak signal intensity $I_C(t)$ was normalized by the As peak-to-peak intensity I_{As} at 1229 eV, which did not show significant time variation in the time scale of interest here. $t = 0$ corresponds to the time when the introduction of TMGa was terminated. One broken line shows the results of a series of measurements on the same electron beam spot (3 mm diameter), and the sample was moved for about 0.5 mm between the broken lines. This was done to minimize the effect of electron beam on the decomposition / desorption of C-related species on the surface of GaAs. The electron beam induced effect is believed to be absent judging from the overall smooth shape of the measured data.

Before the introduction of TMGa, no C was observed by AES. After termination of TMGa, the C signal intensity first decreased and then saturated at a steady state level. The time dependence of the C intensity, $I_C(t)$, can be fitted well by the following form, from which the desorption time constants of C-related species at various substrate temperatures were determined as in Figure 2.

$$I_C(t) = \{I_C(0) - I_C(\infty)\} \exp(t/\tau) + I_C(\infty), \tag{1}$$

The desorption time constant $\tau(T)$ in Figure 2 can then be seen to have temperature dependence of the form $\tau(T) = \tau_0 \exp(E/kT)$, where, determined from fitting, $E = 1.3$ eV and $\tau_0 = 6.6 \times 10^{-7}$ s. The amount of C at $t = 0$ obtained from the curve fit shown in the inset of Figure 2 is about 2 times that of the steady state value. The steady state value is of the order of 1 monolayer (ML) according to the sensitivity factor table supplied by the AES system. The steady state value was always about 1/2 of the initial value in the temperature range investigated here. It should be noted that the sensitivity factor is the major source of large uncertainty and the absolute value should be used with caution.

Figure 3 compares the present results with those reported previously as methyl radical desorption time constants for the GaAs-TMGa system. The reported time constants can be categorized into two; one longer than the time sequence involved in the ALE growth (typically of the order of 5 s) and the other shorter. The time constant obtained in the present study falls into the long time constant group (of the order of 100 s at temperature range where ALE is realized), and is in good agreement with those obtained by Kobayashi and Horikoshi [10] and by Sato and Weyers [11]. Sato and Weyers used quadropole mass analyzer to determine the amount of desorbing CH_3 as a function of time interval between the end of TMGa exposure and the AsH_3 introduction; the adsorbed CH_3 on GaAs desorbed immediately when AsH_3 was introduced. Since the present results are in good agreement with the results by Sato and Weyers, the time constant observed in present experiments is believed to be that characteristic of CH_3 desorption.

Surface reconstruction monitored by RHEED during the exposure experiments showed change in the RHEED pattern with time constant similar to that obtained by the AES measurements. The starting RHEED pattern before TMGa exposure was always (2×4) which changed to (1×1) during exposure and then to vague

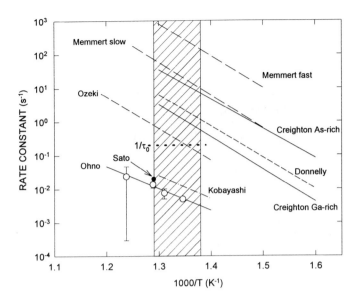

Figure 3. Comparison of reported rate constants $(1/\tau)$ for desorption of CH_3 from (100) GaAs surfaces after exposure to TMGa as a function of temperature. The shaded temperature range is where the ALE growth takes place. τ_0 is the characteristic time scal involved in the ALE growth. Since typical supply/purge periods are 5 s, τ_0 is set to 5 s. Ohno et al. represents the data of present study determined by time resolved AES measuring surface directly. The data that fall in the same region are those by Sato-Weyers (by time resolved mass spectroscopy) and those by Kobayashi-Horikoshi (by surface photo absorption), indicating that the three are measuring the same desorption process of CH_3.

(2×4) like pattern after the exposure. The final steady state pattern which has $6\times$ reconstruction in $[01\bar{1}]$ direction appeared 150 s after exposure when the substrate temperature was at 490 °C.

2.3. Results on other reconstructions and orientations

Time resolved AES measurements with TMGa exposure done on Ga-stabilized (4×6) (100) GaAs surfaces (see Figure 4 (a) for the raw experimental data) showed similar time constant to that of As-stabilized (2×4) surfaces. The time constant obtained by AES measurements was 75 s at 500 °C for C-related desorbing species. The time constant observed in the transition of RHEED pattern was again 150 s at 490 °C.

The same measurements done on Ga-stabilized $(\sqrt{19} \times \sqrt{19})(111)$B GaAs surfaces revealed that the desorption of C-related species can also be described by eq. 1 but with slightly shorter time constant obtained for (100) surfaces. (1×1) (110) and (2×2) (111)A surfaces showed no C-AES signals even right after the termination of

154

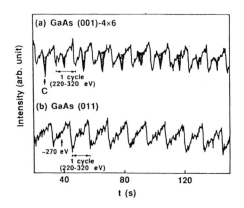

Figure 4. Raw experimental data showing continuous recording of AES signal in the range of 220 - 320 eV. The upper trace (a) is from (100) (4 × 6) surface at 500 °C, where C signal (shaded) is indicated by an arrow. The lower trace (b) is from (110) surface at 456 °C, where no C signal above the noise level is seen. The position of C signal is also indicated by an arrow in (b).

TMGa exposure. Typical raw data were shown in Figure 4 (b), where no C related peaks were observed. Judging from the very small growth rate of (110) GaAs presented in the next section, the absence of C-AES signal can be interpreted as the result of short residence time of TMGa on the surface rather than short decomposition time of TMGa on the surface.

3. ALE OF GaAs

Figure 5 shows the growth rate of GaAs grown by alternating supply of TMGa and AsH_3. The growth sequence is shown in the inset of Figure 5. Growth rates were determined from the thickness of GaAs ALE layers (approximately 15 nm) grown in between AlGaAs marker layers, measured by a high resolution scanning electron microscope. Growth on (100) surfaces showed a plateau of growth rate in the temperature range of 450 - 510 °C, indicating that the self-limiting growth, characteristic of ALE, is taking place at this temperature range. The growth rate plateau was not equal to 1 ML/cycle expected from an ideal ALE growth but rather 0.7 ML/cycle, which will be discussed later. The growth rate on (111)B surface is equal to the growth rate of normal growth (by co-supply of TMGa and AsH_3) showing that all of the supplied TMGa decomposed and contributed to the film growth. The growth rate on (110) surface was low and the surface of the grown films was rough.

Figure 6 is the dependence of growth rate of ALE (100) GaAs on the duration of TMGa supply period at the substrate temperature of 490 °C. After sufficient

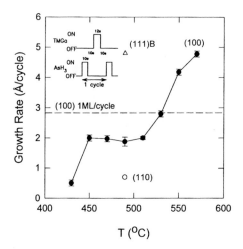

Figure 5. Growth rate as a function of substrate temperature. Growth sequence used for ALE growth is outlined in inset. Flow rates of TMGa and AsH3 were 0.75 and 1.5 sccm, respectively.

supply of TMGa (> 5 s), growth rate saturated and showed no significant increase up to 40 s of TMGa supply, clearly showing the self-limiting nature of the ALE growth. It should be noted that one can obtain ALE-like growth rate plateau that does not depend much on the amount of supplied TMGa when the reaction is kinetically limited. To differentiate the true ALE from those ALE-likes, one needs to do the experiments shown in Figs. 5 and 6 together. In Figure 6, the saturation value is 0.8 ML/cycle, lower than that expected from the ideal ALE.

Figure 7 shows the result of double TMGa pulse experiments done to examine the time scale involved in the self-limiting behavior of GaAs ALE growth. The growth condition is the same as in Figure 6, except for the second pulse sequence shown in inset. If the time constant of the mechanism that self-limits Ga deposition is short compared to the interval between the two pulses, t_W, TMGa in the second pulse decomposes to Ga, resulting in increased growth rate. The fact that growth rate gradually increases with increasing t_W from 10 s to 200 s indicates that the time scale involved in the self-limiting mechanism of ALE growth is of the order of 100 s at $T_s = 490$ °C.

4. MODELING OF ALE PROCESSES

Even in the ordinary GaAs gowth, there is a self-limiting mechanism that prevents As deposition on As. This is because of the weak ineraction between the adsorbed As and the topmost As, which allows fast desorption of As adatoms at elevated temperatures. Here we discuss about the self-limiting mechanism and

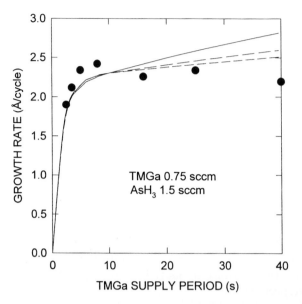

Figure 6. Growth rate of ALE GaAs as a function of TMGa supply period. Closed circles are experimental data. Solid and broken lines are calculated by eqs.(2). Solid line; $n = 0$, long broken line; $n = 1$, short broken line; $n = 2$. $T_s = 490\,^{\circ}\mathrm{C}$.

modeling of the formation of Ga layers.

In the adsorbate-inhibition model, dessociative decomposition of TMGa produces MMGa adsorbed on the Ga sites, which prevents further adsorption/decomposition of the supplied TMGa. The site-selective decomposition model invokes a mechanism that decomposition of TMGa does not occur on Ga because of the short residence time, whereas it decomposes to Ga on As, thereby leaving 1 monolayer (ML) of Ga on GaAs surface after TMGa exposure. The flux-balance model is rather complicated, in which TMGa decomposes to Ga but CH_3 generated during the process reacts with Ga on surface and forms MMGa that desorbs from the surface. The incoming TMGa flux supplies Ga to the surface and the outgoing MMGa flux removes Ga on the surface, and the final coverage determined by the balance of the two fluxes. The key factor that differentiates the models is whether TMGa decomposes to MMGa or to Ga within the time period that is characteristic to the ALE growth (about 5 s).

As summarized in Figure 3, a wide range of time constant has been reported for the desorption time constant of CH_3, which can be divided into two groups; faster than 5 s (short time constant) and slower than 5 s (long time constant). Two models, the site- selective and flux-balance models, are not compatible with the presence of long time constant, since methylgallium and CH_3 are assumed to leave

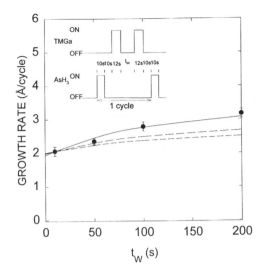

Figure 7. Growth rate of ALE GaAs with two TMGa pulses in one cycle. Horizontal axis shows the interval between the two pulses. Closed circles are experimental data. Solid and broken lines are calculated by eqs.(2). Solid line; $n = 0$, long broken line; $n = 1$, short broken line; $n = 2$. $T_s = 490$ °C.

the surface quickly. On the other hand, the adsorbate-inhibition model can explain the presence of short and long time constant quite naturally; TMGa decomposes quickly to MMGa and releases CH_3 first (short time constant), and then MMGa decomposes to Ga releasing the remaining CH_3 (long time constant).

In order to explain the growth results by the adsorbate-inhibition model, a set of rate equations based on the adsorbate-inhibition model is solved and compared with experimental growth results in the following. The following reactions are assumed to take place during ALE growth. First, the incoming TMGa adsorbs on both As- and Ga-stabilized surfaces [14]. Upon adsorption, TMGa decomposes immediately to MMGa and two CH_3. The time constant of this process is short (< 0.1 s) and corresponds to the short time constant observed in experiments (see Figure 2). Thus produced MMGa prevents further adsorption of other molecules and radicals. MMGa, however, decomposes to $CH_3 +$ Ga and releases CH_3 with a characteristic time constant (τ_1 on As and τ_2 on Ga), which is believed to be what we have observed in our in situ measurements (long time constants). Since this process is longer than the preceding process [7], we assumed that the preceding process takes place immediately. For more detailed study, the process of the dissociative adsorption of TMGa and its time constant may have to be included explicitly in the model. Since the substrate temperatures of ALE growth are too low for Ga to desorb, it remains on surface. When As is supplied subsequently, Ga and MMGa on surface react with As and become GaAs; MMGa releases a CH_3. The release of CH_3 from TMGa exposed (001) GaAs surfaces upon AsH_3

introduction was reported by Sato and Weyers [11]. Since the number of CH_3 left on surface upon TMGa adsorption, n, is not known, three possible cases are considered here; 1) no CH_3 adsorption (all released CH_3 desorbs immediately), 2) one CH_3 adsorption (one CH_3 adsorbs on surface and the other leaves surface), and 3) two CH_3 adsorption (no immediate desorption of CH_3). The last case might be unlikely because it cannot explain the presence of the short time constant. If CH_3 on surface is on Ga it forms MMGa, whereas if it is on As it desorbs with a time constant of τ_3. No desorption of MMGa is considered in the present model. The set of rate equations shown in eqs.(2) represents the processes described above. Here, subscripts indicate the site on which the atom/molecule resides. For example, $[Ga]_{Ga}$ represents adsorption of Ga on Ga. Ga or MMGa adsorption on GaGa (and beyond) is neglected for simplicity: As long as our results do not appreciably exceed 1 ML, this is a good approximation. The arrival rate of TMGa is I (monolayer (ML)/s).

$$\frac{d[MMGa]_{As}}{dt} = I \cdot (1 - [MMGa]_{As} - [Ga]_{As} - [MMGa]_{Ga} - [Ga]_{Ga}$$

$$-[CH_3]_{As}) - \frac{[MMGa]_{As}}{\tau_1} + n \cdot I \cdot [Ga]_{As}$$

$$\frac{d[Ga]_{As}}{dt} = \frac{[MMGa]_{As}}{\tau_1} - (n+1) \cdot I \cdot [Ga]_{As}$$

$$\frac{d[MMGa]_{Ga}}{dt} = I \cdot [Ga]_{As} + n \cdot I \cdot [Ga]_{Ga} - \frac{[MMGa]_{Ga}}{\tau_2}$$

$$\frac{d[Ga]_{Ga}}{dt} = \frac{[MMGa]_{Ga}}{\tau_2} - n \cdot I \cdot [Ga]_{Ga}$$

$$\frac{d[CH_3]_{As}}{dt} = n \cdot I \cdot (1 - [MMGa]_{As} - [Ga]_{As} - [MMGa]_{Ga} - [Ga]_{Ga}$$

$$-[CH_3]_{As}) - \frac{[CH_3]_{As}}{\tau_3} \tag{2}$$

In the calculation, the time constant for decomposition of MMGa on As- and Ga-terminated surfaces (τ_1 and τ_2) are kept equal and 100 s, which corresponds to the experimentally observed time constants at substrate temperature of 490 °C (see Figure 2). The third time constant, τ_3, is assumed to be 0.1 s, which does not affect the result much as long as it is short compared to τ_1 and τ_2. $I = 0.7$ (ML/s) is used for calculation. Solving the set of rate equations with the specified parameters gives the results shown in Figs. 6 and 7 as solid and dashed lines. Growth rate is obtained from the calculated amount of Ga, in the form of MMGa or Ga) on surface after TMGa supply cycle. In Figure 6, saturation of growth rate with increasing the supply period of TMGa is reproduced by calculation, where the calculated results are normalized to 0.23 nm at 10 s. The gradual increase of growth rate after 5 s is due to adsorption of MMGa on Ga; Ga is produced by decomposition of MMGa on As. The rate of increase is reduced with increasing n because adsorbed CH_3 prevents further adsorption of MMGa. The rate decreases with increasing

τ_1 and τ_2. Although, it is not clear whether the experimental results of Figure 6 show this gradual increase due to scatter of data, precise determination of the increase in growth rate with TMGa supply period is believed to lead to accurate determination of kinetic parameters involved in ALE growth. The gradual increase of growth rate observed in the double pulse experiments shown in Figure 7 is well accounted for by the present set of rate equations. The increase in growth rate with increasing t_W takes place, because CH_3 desorption from MMGa on surface, which has prevented further adsorption of MMGa, proceeds during t_W producing Ga-terminated surface on which MMGa can adsorb further. The calculated results in Figure 7 are normalized with the experimental data at $t_W = 10$ s.

5. DISCUSSION

It has been shown in Figs.6 and 7 that the set of rate equations based on the adsorbate-inhibition model involving long time constant can reproduce the basic characteristics of ALE growth. The model, however, still needs refinement in several aspects. One of the major points is the number of CH_3, n, remaining on surface upon dissociative decomposition of TMGa. In Figure 3, $n = 2$ gives better agreement with calculation and experiment, whereas in Figure 7 the calculation with $n = 0$ results in excellent agreement. Refinement in both experiment and model is necessary to resolve this subtle but important discrepancy. Another point of discussion is the role of the steady state C observed by in situ Auger spectroscopy (see inset of Figure 2). This indicates that there still exists appreciable amount of C of unknown form on surface. Effort to remove such C by hydrogen radical beams has resulted in reduced C concentration in ALE GaAs layers, showing that the residual C is the source of C contamination [15]. At the same time, it was found that ALE growth rate increases with hydrogen radical beam applied after TMGa supply period [16]. This anomalous growth rate increase strongly suggests that Ga-containing species, that do not contribute to GaAs growth, exist on the surface after TMGa supply period. Although the detailed nature of this 'second' adsorption layer is currently being studied, the presence of the 'second' adsorption layer if it really exists points out the limitation of the present model. The discrepancy between the present model and experimental results discussed above may be reconciled when the effect of the 'second' adsorption layer is included in the model. It seems quite possible that the real growth environment is somewhat different from and more complicated than the simple single layer adsorbate-inhibition model. It is stressed, however, that the presence of the long time constant can only be accounted for by the adsorbate-inhibition model.

Another point of discussion is the absolute value of growth rate per cycle. The growth rate of ALE of GaAs under MBE environment tends to saturate at about 0.7 to 0.8 ML/cycle [1,13] as shown in Figure 3, whereas that under CVD environment is very close to 1 ML/cycle. The difference in the growth rate can at first glance be explained conveniently by the difference in the surface reconstruction. The MBE environment results in (2×4) reconstruction with As coverage of 0.75 ML [17] and the CVD environment (due to its higher As pressure) leads to c(4×4) reconstruction where the As coverage is more than 1 ML [18]. The excess As in c(4 × 4) probably has small binding energy and readily desorbs after the As cycle during ALE, leading to the 1 ML/cycle growth rate. The CVD based ALE is not totally immune to the surface reconstruction effects. Growth rate of InAs in the CVD based ALE was shown to vary with purge time after As exposure [19]. There, only with very short purging time, was it possible to realize 1 ML/cycle growth: otherwise the growth rate was 0.6 to 0.75 ML/cycle. Similar observation was also made by Koukitu et al. [20] in GaAs ALE based on chloride vapor phase epitaxy. It is also worth mentioning that it has recently been reported that even in the MBE environment one can obtain 1 ML/cycle growth by supplying sufficient AsH_3 [21].

Although the growth rate of 0.75 ML/cycle observed in MBE and CVD environments may be related to the presence of surface reconstruction, there is still a missing process not fully appreciated and understood behind the relation between growth rate and surface reconstruction. Suppose that we have a complete (2x4) reconstruction with As coverage of 0.75 ML. The remaining 0.25 ML is where the missing dimers are found and hence where the underlying Ga is exposed. In order to realize the observed 0.75 ML/cycle growth rate, MMGa or Ga produced by decomposition of TMGa has to cover only the portion where As is present. This suggests that the Ga-exposed portion of the surface is either not accessible for the Ga-containing species or they have a very short residence time constant and/or reevaporation time constant on such sites. Specific atomic configuration of the missing dimer structure is probably not responsible for the observation because those dimers disappear during the TMGa supply cycle. The hidden process behind the 0.75 ML/cycle growth rate suggests, again, that the whole ALE mechanism is more complicated than what all had hoped in the beginning.

6. CONCLUSION

Time resolved in situ Auger electron spectroscopy revealed the presence of long time constant of the order of 100 s in the desorption process of C-related species from GaAs surfaces exposed to TMGa. The RHEED pattern change also showed similar time constant. Growth of GaAs on (100) GaAs was carried out by ALE and shown to be able to be explained by a set of rate equations based on the adsorbate-inhibition model. The model can readily explain the presence of long and short time constants. The remaining discrepancies indicate that the self-limiting process is more complex than that expected from the simple single layer adsorbate-inhibition model.

ACKNOWLEDGEMENTS

The author thanks S. Goto, Y. Nomura, Y. Morishita, and Y. Katayama of Optoelectronics Technology Research Laboratory for fruitful collaboration resulted in the work described in this review article. This work was partly supported by a Grant-in-Aid for Scientific Research on Priority Areas, "Crystal Growth Mechanism in Atomic Scale" (No. 03243202, 04227202, 5211202), from the Ministry of

Education, Science and Culture, Japan.

REFERENCES

1. J. Nishizawa, H. Abe, and T. Kurabayashi, J. Electrochem. Soc., 132, (1985) 1197.
2. J. Nishizawa, T. Kurabayashi, and H. Abe, Surf. Sci., 185, (1987) 249.
3. A. Doi, Y. Aoyagi, and S. Namba, Appl. Phys. Lett., 49, (1986) 785.
4. M. Ozeki, K. Mochizuki, N. Ohtuka, and K. Kodama, Appl. Phys. Lett., 53 (1988) 1509.
5. M.L. Yu, J. Appl. Phys., 73 (1993), 716.
6. U. Memmert and M.L. Yu, Appl. Phys. Lett., 56 (1990) 1883.
7. J. R. Creighton and B.A. Banse, MRS Symp. Proc., 222 (1991) 15.
8. V. M. Donnelly, J. A. McCaulley and R.J. Shul, MRS Symp. Proc., 204 (1991) 15.
9. J. R. Creighton and B.A. Bansenaer, Thin Solid Films, 225 (1993) 17.
10. N. Kobayashi and Y. Horikoshi, Jpn. J. Appl. Phys., 30 (1991) L319.
11. M. Sato and M. Weyers, Jpn. J. Appl. Phys., 30 (1991) L1911.
12. H. Ohno, S. Goto, Y. Nomura, Y. Morishita, A. Watanabe, and Y. Katayama, Appl. Phys. Lett., 62 (1993) 2248.
13. S. Goto, H. Ohno, Y. Nomura, Y. Morishita, A. Watanabe, and Y. Katayama, J. Crystal Growth, 127 (1993) 1005.
14. A. Watanabe, T. Isu, M. Hata, T. Kamijoh, and Y. Katayama, Jpn. J. Appl. Phys., 28 (1989) L1080.
15. S. Goto, Y. Nomura, Y. Morishita, Y. Katayama, and H. Ohno, J. Crystal Growth, 144, (1994) 126.
16. S. Goto, unpublished.
17. M.D. Pashley, K.W. Haberern, W. Friday, J.M. Woodall, and P.D. Kirchner, Phys. Rev. Lett., 60, (1988) 2176.
18. F.J. Lamelas, P.H. Fuoss, P. Imperatori, D.W. Kisker, G.B. Stephenson, and S. Brennan, Appl. Phys. Lett., 60, (1992) 2610.
19. Y. Sakuma, M. Ozeki, and K. Nakajima, J. Crystal Growth, 130 (1993) 147.
20. A. Koukitu, N. Takahashi, Y. Miura, and H. Seki, J. Crystal Growth 146 (1995) 239.
21. J. Nishizawa and T. Kurabayashi, J. Vacuum Science and Technology, B13, (1995) 1024.

Advances in the Understanding of Crystal Growth Mechanisms
T. Nishinaga, K. Nishioka, J. Harada, A. Sasaki and H. Takei (Editors)
© 1997 Elsevier Science B.V. All rights reserved.   163

Atomic ordering in epitaxial alloy semiconductors: from the discoveries to the physical understanding

H. Nakayama, T. Kita and T. Nishino
Faculty of Engineering, Kobe University,
1-1 Rokkodai, Nada, Kobe 657 Japan

The discoveries of atomic long-range ordering (LRO) in III-V alloy semiconductors opened a new research field which encompasses solid state physics of atomically ordered semiconductors, epitaxial growth mechanism and optoelectronics applications. A number of LRO phases have been observed for various kinds of III-V alloy semiconductor systems. The mechanism of LRO formation in epitaxial alloy semiconductors has not been fully understood. However, it can be acceptable to consider that the role of surface epitaxial processes such as surface reconstruction and step structure/motion are key processes. In this review, the historical brief review of the discovery of famatinite type LRO phase in InGaAs alloy grown by liquid phase epitaxy on InP(001) substrate was given, followed by the review of our optical studies of LRO in InGaP and related materials. Final section was devoted to the modeling and the simulation of binary epitaxial growth, which gave us a clue to understand an order-disorder transition in epitaxial alloy semiconductors.

1. INTRODUCTION

Long-range ordering (LRO) phenomena had not been observed in alloy semiconductors until present authors[1,2] and Kuan et al.[3] discovered the LRO phases in $In_{1-x}Ga_xAs$ and $Al_{1-x}Ga_xAs$ in 1983 to 1984. This was quite natural since the equilibrium phase diagram of such system showed us that the low temperature phase is phase separation state instead of the ordering state. Nobody believed the presence of LRO in such alloy semiconductors. In terms of physics of metallic alloy, it is generally considered that the clustering, phase separation in other words, and ordering is an exclusive phenomenon in the sense that either phase separation or ordering can be realized at low temperatures below the critical temperature. The choice is determined uniquely by the sign of the ordering energy as

$$V_{\text{order}} = \phi_{AB} - \frac{1}{2}[\phi_{AA} + \phi_{BB}], \tag{1}$$

where ϕ_{AB}, for example, is the pair bond energy of A and B atoms. Bond energies are defined to be negative in the stable bonds. Then, negative ordering energy means that the A-B bonds are more stable than those of A-A or B-B bonds. Let us consider a lattice of A-B alloy, say body-centered cubic (bcc) lattice. Here we define "sublattices" as follows;the corner sites of bcc lattice are named as α sublattice sites and similarly the body-center sites are denoted by β sublattice sites. The stable structure of ordered phase is generally determined by the sign and the amplitude of these interatomic interaction parameters[4]. Each stable structure has their own sublattices. For example, in the well known Cu-Zn system, Cu and Zn atoms occupy either of the above mentioned α and β sublattices in the LRO state at low temperature below the critical temperature of order-disorder transition. The occupation is not generally complete and some atoms mis-occupy the sublattice. Therefore, the order-disorder phenomenon is a matter of statistical mechanics. Bragg-Williams treatment[5] of order-disorder phenomena of A-B binary alloy system gives us the simplified description of free energy of alloy system with finite lattice sites as,

$$
\begin{aligned}
F = E - TS = \frac{zN}{2}\phi_{AA}x + \frac{zN}{2}\phi_{BB}(1-x) + zNV_{\text{order}}x(1-x) \\
+ zNV_{\text{order}}\frac{S_{\text{LRO}}^2}{4} \\
- \frac{k_{\text{B}}TN}{4}\{(2x + S_{\text{LRO}})\ln(2x + S_{\text{LRO}}) \\
+ [2(1-x) + S_{\text{LRO}}]\ln[2(1-x) + S_{\text{LRO}}] \\
(2x - S_{\text{LRO}})\ln(2x - S_{\text{LRO}}) \\
+ [2(1-x) - S_{\text{LRO}}]\ln[2(1-x) - S_{\text{LRO}}]\},
\end{aligned}
\tag{2}
$$

where N is the total number of atoms (total lattice sites) and z is the coordination number of atom A or B. When the sign of V_{order} is negative LRO state having the finite value of LRO parameter, S_{LRO}, gives the lower energy state. Then order-disorder transition occurs with increasing the temperature, resulting in $S_{\text{LRO}} = 0$ at

$$
T_{\text{critical}} = \frac{-V_{\text{order}}z}{2k_{\text{B}}}.
\tag{3}
$$

On the contrary, positive ordering energy causes the phase separation between A and B elements below the critical temperature. This is due to the third term of Eq.(2), $zNV_{\text{order}}x(1-x)$, which is the positive mixing enthalpy with the energy peak at $x = 0.5$. Therefore, the choice between LRO or phase separation (clustering) is an exclusive phenomenon. Namely, low-temperature phase of III-V alloy semiconductors, at least in the bulk form, is considered to be phase-separation state having the positive mixing enthalpy. In fact, infrared reflectivity study of a series of III-V alloy semiconductors[6] showed the presence of clustering in these alloy states. These epitaxial alloy semiconductors are grown by liquid phase epitaxy and generally grown at temperatures just below the solidification temperature in order to get an appropriate super saturation. Some alloy systems have the very narrow window between the solidification temperature and the critical temperature of phase separation because of the large lattice-constant mismatch between the constituent compounds. Therefore, clustering is possible to occur during the crystal growth near the critical temperature of the phase separation. However, discoveries of LRO phases in such alloy semiconductors showed completely opposite phenomenon, suggesting the strong question about the applicability of the equilibrium phase diagram of these materials. This was the dawn of so-called LRO problem in alloy semiconductor physics. As was mentioned earlier, many experimentalist tried to find LRO phases in various kinds of III-V alloy systems. They discovered a number of LRO phases having various kinds of ordered structures. LRO phases were discovered in InGaP[7], GaAsSb[8], GaAsP[9] as well as InGaAs[1] and AlGaAs[3]. The complete summary of the papers concerned with the discoveries of LRO can be available from the recent comprehensive review of Zunger and Mahajan[10]. Extensive theoretical works have been also achieved by Zunger and coworkers of NREL. Zunger et al. predicted the presence of a kind of LRO in their pioneering paper[11] at the same time when the first reports of experimental discoveries were just submitted by present authors[1] and Kuan et al.[3] on October of 1984.

2. DISCOVERY OF ORDERED PHASE

The discovery of the LRO phase in semiconductor alloys was an unexpected results of the study on the interface-structure of InGaAs/InP(001) heterostructure. One of the authors (H.N.) was performing transmission electron microscopy (TEM) observation on the interfaces of the liquid-phase-epitaxy (LPE) grown InGaAs layers and InP(001) substrates with the several different alloy compositions. These samples were prepared by present author (T.N.) and coworkers[12]. $In_{1-x}Ga_xAs$ samples for TEM study were grown by conventional horizontal LPE on InP(001) substrates at $630°C$. The composition x ranged from 0.463 to 0.498. The sample with $x = 0.478$ satisfied the lattice-matching condition at the growth temperature, $630°C$. TEM observation was made with plane view on the thin foil of epitaxial $In_{1-x}Ga_xAs$ with removing the substrate by using selective etchant like $HCl-H_3PO_4$ solution. The details about the LPE growth conditions were published by Yagi et al.[12]. Misfit dislocations and stacking faults were observed for the InGaAs samples having the large lattice-constant mismatch between the epitaxial layers and the substrates. On the contrary, samples with small lattice mismatch did not show such structural defects. However, in the TEM bright-field (BF) images, there appeared some areas of the samples having small and dark-image islands with the sizes of an order of 10nm. The area of the specimen showed the characteristic diffraction patters which includes a number of superreflection diffraction spots. This suggested the presence of some phases having the different atomic structure from the host zincblende(ZB) alloy structure. This was actually the first observation of LRO phase in alloy semiconductors. It was the night of Dec. 24, 1983.

Figure 1 shows the TEM data of $In_{1-x}Ga_xAs$ with $x = 0.478$. The aggregation of dark islands can be observed. The corresponding electron diffraction pattern includes a series of weak diffraction spots in addition to those of ZB structure. The superreflection spots are indexed as $(\pm\frac{1}{2}, \pm1, 0)$, $(\pm1, 0, 0)$, $(\pm1, \pm\frac{1}{2}, 0)$, $(0, \pm1, 0)$,

Figure 1. TEM micrographs together with the electron diffraction patterns. Dark island-shape areas correspond to the LRO phases. The corresponding electron diffraction shows the presence of a number of superreflection spots.

and $(\pm 1, \pm 1, 0)$. The $(\pm 1, \pm 1, 0)$ spot intensities are very weak. The structural image micrograph is shown in Figure 2. The diffracted beams within $(2, 0, 0)$ diffractions were included in the formation of the structural image. Set of lattice fringes are observed in the islands which correspond to the dark islands in BF image in Figure 1. There were two types of islands having different fringe patterns. A set of the fringes correspond to $(\pm \frac{1}{2}, \pm 1, 0)$ and $(\pm 1, 0, 0)$ diffractions, whose islands is marked by the horizontal arrows in Figure 2. The other set of fringes correspond to $(\pm 1, \pm \frac{1}{2}, 0)$ and $(0, \pm 1, 0)$ diffractions, which is also marked by the vertical arrows in Figure 2. As can be seen in the left-bottom corner of Figure 2, these two kinds of islands are located closely. Twin-like structures composed of these two types of structure was also observed. These observation led to a conclusion that the net electron diffraction patters are made of superposition of three types of sets of diffractions; $(1)(\pm \frac{1}{2}, \pm 1, 0)$, $(\pm 1, 0, 0)$, $(2)(\pm 1, \pm \frac{1}{2}, 0)$, $(0, \pm 1, 0)$, and $(3)(\pm 1, \pm 1, 0)$. (1) and (2) correspond to those mentioned above. The structure images corresponding to $(\pm 1, \pm 1, 0)$ diffractions could not be observed, however the diffraction spots can be weakly observed in electron diffraction patterns.

The results are summarized in Figure 3 together with the structural model of the ordered lattice. ZB lattice is composed of two face-centered cubic (fcc) sublattices

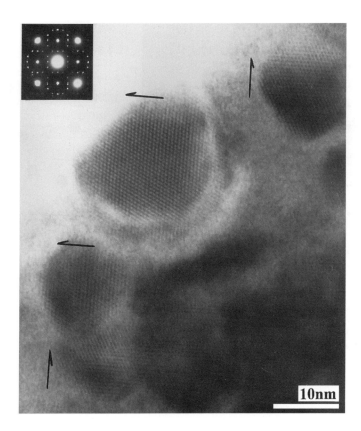

Figure 2. Structural images of LRO phases. Arrows denote the different types of LRO phases. Arrows correspond to the c-axis of the ordered structure, which is described in Figure 3.

of column-III and column-V elements. In the alloy state of $In_{1-x}Ga_xAs$, In and Ga atoms distribute randomly in the column-III fcc sublattice. On the contrary, the ordered states can be formed by the ordered arrangements of In and Ga atoms in the fcc sublattice. Therefore, the presence of the set of the superreflection spots as shown here means the presence of ordered arrangements of In and Ga atoms in the fcc sublattice. We called this as sublattice ordering in the first paper[2]. The proposed LRO structure was a famatinite structure[13]. The possible compositions are $In_{4-n}Ga_nAs_4(n = 1, 3)$. The ordered column III sublattice of the famatinite structure is a fcc-based LRO structure known as $D0_{22}$ structure[4].

The presence of three types of sets of diffraction fringes corresponds to the three possible directions of c-axis of the famatinite structure, namely, $c \parallel [100]_{InP-sub.}$, $c \parallel [010]_{InP-sub.}$ and $c \parallel [001]_{InP-sub.}$. The most controversial points related to

168

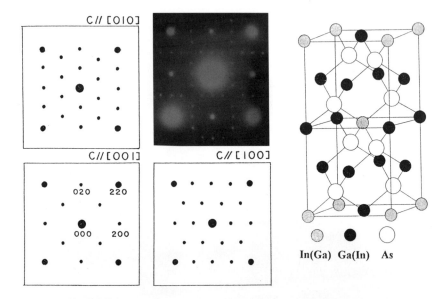

Figure 3. Analysis of the electron diffraction pattern of InGaAs including the LRO phases. The net diffraction patterns are composed of three types of the set of diffractions. These three types of net patterns correspond to the direction of the c-axis of the famatinite type LRO structure of InGaAs. The atomic structure model corresponds to the composition of $In_{4-n}Ga_nAs_4(n = 1,3)$.

the presence of famatinite-type ordering is that the local compositions of the LRO phases are $x = 0.25$ and $x = 0.75$. This means that a kind of phase separation must be accompanied with the LRO formation because the average composition of the alloy state is 0.478 as mentioned above. There are two explanations. Structure image of Figure 2 showed that the ordered island areas are always accompanied with the surrounding boundary area with slightly bright image in the vicinity of the ordered islands. The surrounding areas may correspond to the locally phase-separated regions. Another explanation is that two types of islands marked by the horizontal and vertical arrows correspond to either $x = 0.25$ or $x = 0.75$. This problem is not solved yet. It might be valuable to make a comment about the similarity of the chalcopyrite-type LRO phases observed by Jen et al. in $GaAs_{0.5}Sb_{0.5}$[14]. The net diffraction patterns reported are quite similar to those observed by us. However, the set of diffraction fringes and the corresponding set of the diffraction like $(\pm\frac{1}{2}, \pm1, 0)$ and $(\pm1, 0, 0)$ observed in ordered InGaAs can not be assigned as due to chalcopyrite structure because chalcopyrite structure can be characterized by the $(\pm\frac{1}{2}, \pm1, 0)$ and $(\pm\frac{3}{2}, \pm1, 0)$ diffractions and moreover, $(\pm1, 0, 0)$-type spots do not appear in chalcopyrite structure. Jen et al. assigned their electron diffraction as a superposition of the mixed phases of CuAu-I type LRO and chalcopyrite-type LRO.

Therefore, the similarity in the net electron-diffraction patterns is occasional, which means that atomic structure can not be simply determined by electron diffraction patterns. It should be noted that it is required to observe structural images to assign the structure of the LRO phases. It may lead to mis-assignment to derive the structure model only on the basis of the electron diffraction patterns[15].

3. ELECTRONIC STRUCTURE

3.1. Band structure of LRO phase

Early works of electronic and optical properties of LRO phases in III-V alloy semiconductors have been mainly made for organomatallic vapor-phase epitaxy (OMVPE) grown InGaP alloys[16–18]. LRO is experimentally manifested by the appearance of superlattice diffraction spots, removal of the valence-band degeneracy, and the resultant light-polarization dependence in optical spectra. The characteristic feature of the electronic and optical properties of ordered alloy semiconductors is the order-parameter dependence. The fundamental energy gap and valence-band splitting are functions of the order parameters. Especially LRO parameter is a key parameter in the determination of electronic structures. The electronic structure near the optical-band gap in long-range ordered $Ga_{0.5}In_{0.5}P$ alloys is the subject which has been most extensively studied both experimentally and theoretically.[16–26] Ordered $Ga_{0.5}In_{0.5}P$ has the CuPt type ordered sublattices having a kind of monolayer superlattice along {111} directions. There are, in general, four equivalent {111} directions, however, only two kinds of monolayer superlattices along [-1,1,1] and [1,-1,1] directions can be observed[16]. It was also found that the band gap energy of ordered $Ga_{0.5}In_{0.5}P$ alloys strongly depends on the OMVPE-growth conditions such as growth temperature and gas-flow ratio of column-V to -III sources[16]. Figure 4 plots photoluminescence(PL)-peak

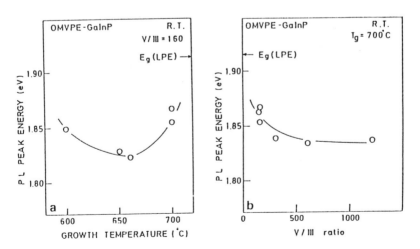

Figure 4. PL-peak energy as functions of growth temperature and V/III-gas-flow ratio at room temperature.

170

energy measured at room temperature as functions of growth temperature and input gas-flow ratio, together with PL-peak energy positions of LPE-grown random $Ga_{0.5}In_{0.5}P$ alloys[17]. These peak energy positions roughly correspond to the E_0-band gaps of these alloys. The PL-peak energies of the OMVPE-grown $Ga_{0.5}In_{0.5}P$ alloys are much lower than those of the LPE-grown random alloys, and vary with growth temperature and gas-flow ratio. The PL-peak energy shows a minimum value around the growth temperature of 650 °C and decreases monotonously with increasing gas-flow ratio. Extensive optical studies such as PL [16,17,19,20,22], electroreflectance (ER)[17,18,21] as well as structural studies such as TEM , x-ray diffraction study and Raman-scattering spectroscopy have been performed to clarify the origin of the band-gap reduction (BGR) relative to the random alloy states. The periodic layered structure of the ordered alloy causes the band fold-

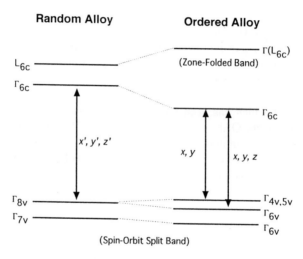

Figure 5. Schematic depiction of band diagrams of ordered and disordered $Ga_{0.5}In_0.5P$.

ing phenomena in the Brillouin zone and a crystal-field splitting of the valence bands.[25,26] Figure 5 draws band diagrams near the band gap of random and ordered alloys. A level repulsion between different symmetry states of the binary constituents folding into equal-symmetry states in the ordered ternary structure causes strong variation in the band structures. The fundamental-band gap of the ordered alloys is usually smaller than the linear average of the band gaps of the binary constituents. A periodic crystal field in the ordered $Ga_{0.5}In_{0.5}P$ also splits the four-fold degenerated $\Gamma 8v$ valence-band maximum into two doubly degenerated bands (at $k = 0$) represented by $\Gamma 4v/\Gamma 5v$ and $\Gamma 6v$. These splitting energies are functions of the spin-orbit splitting in a cubic field and the crystal-field splitting in the absence of spin-orbit interactions. This valence-band splitting (VBS) also reduces the band gap. The magnitude of the symmetry-enforced change of the band

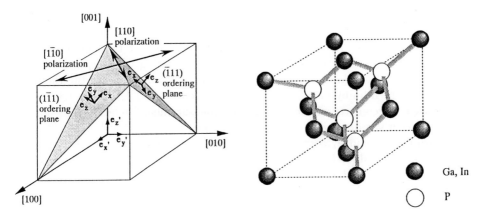

Figure 6. Relations among polarization vectors and ordering direction.

structures strongly depend on order parameter[20,22,24–26]. LRO $Ga_{0.5}In_{0.5}P$, is considered to be composed of a composition-modulated monolayer superlattice with the alternate compositions of $Ga_{0.5+\frac{\eta}{2}}In_{0.5-\frac{\eta}{2}}P$ and $Ga_{0.5-\frac{\eta}{2}}In_{0.5+\frac{\eta}{2}}P$ along the ordering vectors. Here, η in this compositional modulated superlattice is equivalent to LRO parameter, S_{LRO}. Continuous variations of band gaps and valence-band splittings with the growth conditions originate from the variation of LRO parameters with these growth conditions[20].

3.2. Modulation and photoluminescence measurements

The samples used in this experiment were ordered $Ga_{0.5}In_{0.5}P$ alloys grown on exact n^+-GaAs (001) substrates by OMVPE. The growth temperature was in the range of 600-700 °C. The gas-flow ratio of column-V to -III sources was ranged from 160 to 1212. The thickness of the undoped epitaxial layer was 0.7 mm. The alloy composition was determined by measuring precisely the lattice constant perpendicular to the (001) interface plane between $Ga_xIn_{1-x}P$ and GaAs using a double-crystal x-ray diffracto-meter. The epitaxial films were horizontally lattice matched to the GaAs substrate within ±0.1the OMVPE-grown $Ga_{0.5}In_{0.5}P$ alloys. The undoped $Ga_{0.5}In_{0.5}P$ showed n-type conductivity. The residual carrier concentration measured by the capacitance-voltage method was less than $10^{16}cm^{-3}$. For ER measurements, a semitransparent $Au/Ga_{0.5}In_{0.5}P$ Schottky-barrier diode was used for a modulation of the surface-electric field. The modulating voltage was 400 mV at a zero-bias condition. The modulating frequency was 1 kHz. A monochromatic light was polarized by a polarizer with an extinction coefficient of 10^{-4}. The polarized light irradiated the (001) surface of a sample. The reflectance was detected by a silicon photo diode. The incident angle of the light was less than 10°because an anisotropy of the Seraphin coefficients for the p and s waves becomes negligible at this angle. The polarization of the incident-probe light was set between the [1,1,0] and [1,-1,0] direction in the (001) plane. The [1,1,0] and [1,-1,0] directions were determined from etch pit anisotropy characteristics. PL spectra were measured in temperature range from 13 K to 300 K. The excitation light was the 488 nm line of an Ar-ion laser. The Laser-spot area at the sample surface was about 20 mm^2. The spontaneous CuPt-type ordering of $Ga_{0.5}In_{0.5}P$ alloys cause a symmetry breaking. F43m of the ZB structure changes into the rhombohedral symmetry R3m by monolayer periodicity along the [-1,1,1] or [1,-1,1]. So as to make clear a geometry of the polarization measurements, relationships among the polarization vectors and the ordering directions are shown in Figure 6[20]. In this figure, e_x, e_y, and e_z corresponds to unit vectors along the [-1,1,2] ([-1,1,2]), [1,1,0] ([-1,-1,0]), and [1,-1,1] ([1,-1,1]) directions for the [-1,1,1] ([1,-1,1]) ordering, respectively. e'_x, e'_y, and e'_z correspond to those along the [1,0,0], [0,1,0], and [0,0,1] directions, respectively.

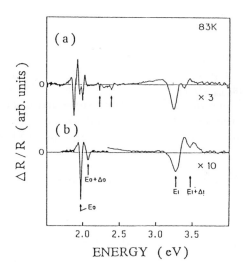

Figure 7. ER spectra of ordered (a) and disordered (b) $Ga_{0.5}In_{0.5}P$.

3.3. Electroreflectance spectra

Figure 7 (a) and (b) show typical ER spectra of ordered and disordered $Ga_{0.5}In_{0.5}P$, respectively. The disordered LPE- and OMVPE-grown $Ga_{0.5}In_{0.5}P$ alloys show similar ER structures at the E_0 fundamental and E_1 higher-interband edges as shown in this figure[17].

The ER spectrum for the ordered alloy shows very complicated ER structures as compared with those of disordered alloy and has additional structures around 2.2 eV and 2.4 eV (indicated by arrows). These additional structures are explained by transitions related to the X6c originated from a relaxation of momentum conservation and the folded Γ(L6c), respectively. Furthermore, small red shifts of the E_1

transitions were observed, because of interactions between the X6c and the other three L6c at the Brillouin-zone edge. Around the band gap, we can see a BGR of about 100 meV and the fine structure caused by the VBS. The selection rules of electronic-dipole transitions in the band structures are shown by x (x'), y (y'), and z (z') in Figure 5 for the ordered and random alloys. ER-polarization spectra mea-

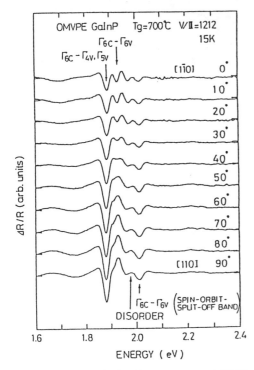

Figure 8. ER-polarization spectra at 15 K.

sured at various polarization angles are shown in Figure 8. The measured sample was grown at 700 °C and V/III ratio of 1212. The angles of 0° and 90° correspond to the [1,-1,0] and [1,1,0]. Each spectrum reveals signals due to three transitions; the $\Gamma 6c - \Gamma 4v, \Gamma 5v$ transition at 1.89 eV, the $\Gamma 6c - \Gamma 6v$ transition at 1.93 eV and the $\Gamma 6c - \Gamma 6v$ (spin-orbit-split-off band) transition at 2.01 eV. The ER signal due to the $\Gamma 6c - \Gamma 6v$ transition at 1.93 eV shows the drastic change of the line shape at the different polarization angles[21]. On the other hand, the signal due to the $\Gamma 6c - \Gamma 4v, \Gamma 5v$ transition only shows a gradual change in the intensity. The VBS energy evaluated from these spectra is 40 ± 2 meV, which is in agreement with the value determined from the temperature dependence of polarized photoluminescence intensities[20]. From the optical selection rule, the $\Gamma 6c - \Gamma 4v, \Gamma 5v$ transition has isotropic components of x and y. Then, the $\Gamma 6c - \Gamma 4v, \Gamma 5v$ signal changes only in the intensity according to the observed component of the allowed

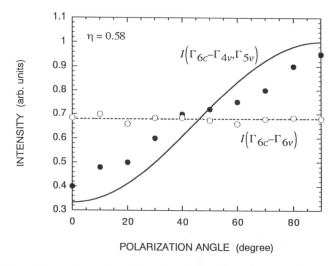

Figure 9. Transition intensity of ordered $Ga_{0.5}In_{0.5}P$ as a function of linear-polarization angle. Here 0°denotes light polarization along [110]. The solid and open circles are the experimental data. The solid and dashed lines plot calculated intensities[26].

polarizations. The $\Gamma 6c - \Gamma 6v$ transition has the z component as well as the x and y. With decreasing the polarization angle, the x and z components gradually increase and the y component decreases. Since the isotropic components don't change the line shape, the observed change in the ER-line shape originates from the z component. This result directly reveals a presence of a dielectric anisotropy in the ordered $Ga_{0.5}In_{0.5}P$ alloys. The signal observed at 1.98 eV is a well established $\Gamma 6c - \Gamma 8v$ transition in residual disordered variants. This signal is very small as compared with the ordered one and is independent on the polarization direction. Each ER spectrum coming from the four signals can be separated by a line-shape analysis. The signal intensities obtained from the analysis are plotted as a function of the polarization angle in Figure 9. The solid and open circles plot the experimental data for the $\Gamma 6c - \Gamma 4v, \Gamma 5v$ and $\Gamma 6c - \Gamma 6v$ ER signals, respectively. The solid and dotted lines are theoretical results given by Wei and Zunger[26]. Here we use order parameter of 0.58. The measured intensities show good agreements with the theoretical trends. ER measurements have been performed for various ordered $Ga_{0.5}In_{0.5}P$ alloys prepared at different growth conditions. Figure 10 summarizes the BGR and VBS as a function of order parameter. The lines shown in this figure depict the prediction of the theory of Wei et al. . This theory can be compared with the our experimental data by fitting the measured band-gap reductions in given samples. The measured results by the ER are plotted by the closed circles and triangles for the VBS and BGR, respectively. The agreement with theory is rather good.

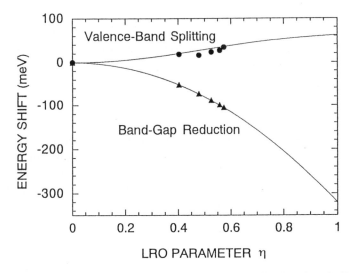

Figure 10. Experimental and calculated BGR and VBS of ordered $Ga_{0.5}In_{0.5}P$. The lines shown in this figure depict the prediction of the theory of Wei et al. The measured results by the ER are plotted by the closed circles and triangles for the VBS and BGR, respectively.

4. STOCHASTIC MODEL OF EPITAY AND ORDERING

4.1. Stochastic theory of epitaxy

Growth system considered here is ZB based $(A_x^{III}B_{1-x}^{III})C^V$ system formed on the (001) substrate plane. The (001) growing plane is a face-centered cubic (fcc) (001) sublattice plane of the ZB lattice. Therefore, in order to describe the order-disorder phenomena in the (A,B) column-III sublattice, this section begins with the treatment of the growth of (A,B) (001) two-dimensional (2D) square lattice. Interatomic interactions which are key factors to the description of ordering phe-nomena in binary growing system are taken into account within the framework of effective Ising Hamiltonian of surface atoms. In the present study, the effective Ising parameters were given a priori as calculating parameters.

The master equation for the site-occupation probability describing the binary growth system is given by[27–29]

$$\frac{\partial p_\mu(\mathbf{x}, \mathbf{t})}{\partial \mathbf{t}} = -\sum_{\mathbf{r}} p_\mu(\mathbf{x}, \mathbf{t})\bar{p}(\mathbf{x} - \mathbf{r}, \mathbf{t})w_\mu(\mathbf{x} \to \mathbf{x} - \mathbf{r})$$
$$+ \sum_{\mathbf{r}} p_\mu(\mathbf{x} - \mathbf{r}, \mathbf{t})\bar{p}(\mathbf{x}, \mathbf{t})\mathbf{w}_\mu(\mathbf{x} - \mathbf{r} \to \mathbf{x})$$
$$+ J_\mu p_{\mathrm{ad},\mu}(\mathbf{x}, \mathbf{t})\bar{p}(\mathbf{x}, \mathbf{t}), \tag{4}$$

where $p_\mu(\mathbf{x}, \mathbf{t})$ is the occupation probability of atom μ=(A or B) at site $\mathbf{x}$ and time t. $\mathbf{r} \in \{(\mathbf{a}, \mathbf{0}), (-\mathbf{a}, \mathbf{0}), (\mathbf{0}, \mathbf{a}), (\mathbf{0}, -\mathbf{a})\}$ is a site displacement vector of jumping atom. J_μ is the molecular beam flux of atom μ. First two terms in the right-hand side show the site-exchange (jumping) terms and the last term comes from the adsorption of impinging atoms. Here

$$\bar{p}(\mathbf{x}, \mathbf{t}) = 1 - \sum_\mu \mathbf{p}_\mu(\mathbf{x}, \mathbf{t}) \tag{5}$$

is the probability of finding a vacant site at$(\mathbf{x}, t)$. In this study, the master equation, Eq.(4), was used so as to take the local atomic configuration into account in the formulation of adsorption and atomic jump probabilities on the basis of square-lattice model.

The effective Ising Hamiltonian of growing surface, which is characterized by the atomic configuration $\xi(\sigma)$ on 2D square lattice, is given by

$$\mathcal{H}[\xi(\sigma)] = -J_x \sum_i \sigma_{i,j}^\xi \sigma_{i+1,j}^\xi - J_y \sum_j \sigma_{i,j}^\xi \sigma_{i,j+1}^\xi \tag{6}$$

Ising variables in Eq.(6) are defined as $\sigma = +1$ for A atom, $\sigma = -1$ for B atom and $\sigma = 0$ for vacant site. The coupling of the effective Ising Hamiltonian and the master equation can be simply formulated by describing adsorption and atomic-jump probabilities in terms of the Ising Hamiltonian[30] . Thus, the adsorption probability of μ atom ($\mu =$(A or B) at the adsorption site (i_0, j_0) is approximately given by the occupation probability of the μ atom at the adsorption site with neglecting the thermal evaporation of atoms. Namely,

$$[p_{\mathrm{ad}}(i_0, j_0)] = \frac{\exp\{-\beta\mathcal{H}[\xi_{\mathrm{loc}}(\sigma), \sigma(i_0, j_0) = \sigma_\mu]\}}{\sum_{\xi_{\mathrm{loc}}(\sigma)} \exp\{-\beta\mathcal{H}[\xi_{\mathrm{loc}}(\sigma), \sigma(i_0, j_0) = \sigma_\mu]\}} \tag{7}$$

We assumed the local equilibrium at the adsorption site (i_0, j_0). $\beta = 1/k_B T$ as usual. The local energy was estimated , with considering the nearest-neighbor interaction, as

$$\mathcal{H}[\xi(\sigma)] \simeq -J_x \sum_{i=i_0-1}^{i_0} \sigma_{i,j_0}^\xi \sigma_{i+1,j_0}^\xi - J_y \sum_{j_0-1}^{j_0} \sigma_{i_0,j}^\xi a_{i_0,j+1}^\xi \tag{8}$$

Atomic-jump probability from (i_0, j_0) site to the nearest-neighbor vacant (i_0', j_0') site can be defined by using the local activation energy as

$$w_\mu[(i_0, j_0) \rightarrow (i_0', j_0')] = v_0 \exp\{-\beta\Delta E_\mu[(i_0, j_0) \rightarrow (i_0', j_0')]\}, \tag{9}$$

where the pre-exponential factor, v_0, was assumed to be unity in the present calculation. The activation energy for atomic jump was defined here by using the diagram shown in Figure 11. For example, in the case of atomic jump from (i_0, j_0) site to $(i_0 + 1, j_0)$ site, the initial state energy $\mathcal{H}(i_0, j_0)$ is given by

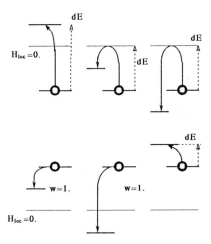

Figure 11. Definition of the activation energies related to various types of diffusion paths. The definition of the local energies concerned with the atomic jump is shown in the text.

$$\mathcal{H}(i_0, J_0) = \mathcal{H}[\xi(\sigma), \sigma(i_0, J_0) = \sigma_\mu, \sigma(i_0\prime, j_0\prime) = 0]$$

$$\simeq -J_x \sum_{i=i_0-1}^{i_0+1} \sigma^\xi_{i,j_0} \sigma^\xi_{i+1,j_0} - J_y \sum_{j=j_0-1}^{j_0} \sigma^\xi_{i_0,j} \sigma^\xi_{i_0,j+1}$$

$$-J_y \sum_{j=j_0-1}^{j_0} \sigma^\xi_{i_0+1,j} \sigma^\xi_{i_0+1,j+1} \tag{10}$$

The key point in the definition of Eq.(9) is that the activation energy for atomic jump between nearest-neighbor bound states corresponds to the activation energy to zero-level energy from the initial bound state instead of the energy difference between initial and final states. Therefore, it corresponds to the binding energy of the initial state itself. The detailed balance is realized among the transitions between local bound states[30].

4.2. Structural evolution during monolayer epitaxy

Monte Carlo (MC) calculation based on the master equation, Eq.(4), has been made for monolayer growth process of binary (A,B) alloy. The algorithm of MC MBE-growth simulation is as follows;

(1) choice of impinging atom, A or B, depending on the molecular-beam-flux ratio, (2) choice of a vacant adsorption site, (3) calculation of local site energy of adsorption site on the basis of Ising Hamiltonian, (4) calculation of adsorption probability, (5) calculation of atomic-jump probabilities from nearest-neighbor occupied sites into the selected vacant site, (6) choice of events among adsorption, atomic

jump and scattering (without adsorption nor atomic jump), (7) determination of new atomic configuration, $\xi(\sigma)$, (8) continue to the next event.

MC calculation has been made for three types of surface with (J_x, J_y)=(F,F), (AF,F) and (AF,AF) interatomic interactions. "F" means the ferromagnetic-type interatomic interaction $(J > 0)$, where AA and BB nearest-neighbor pairs are more preferable than AB pair. "AF" means the antiferromagnetic-type interatomic interaction $(J < 0)$, where AB nearest-neighbor pair is more preferable than AA or BB pairs. As can be easily understood, (F,F) interaction results in a surface phase separation in 2D surface. Figures 12 and 13 show the calculated atomic arrangements of (A,B) alloy with the four different surface coverages. The interaction parameters are $(J_x/k_BT, J_y/k_BT)$=(AF,AF)=(-0.5, -0.5) for chalcopyrite-type surface in Figure 12. Another surface has the Ising parameters of $(J_x/k_BT, J_y/k_BT)$=(AF,F)=(-0.5, 0.5) for CuPt type surface as shown in Figure 13. The super cell size is 50x50=2,500 atomic sites. It is well demonstrated that the adsorbed atoms gradually form the ordered arrangement with repeating adsorption and diffusion processes as can be seen from Figures 12 and 13. In this condition, adsorption and diffusion (atomic jump) processes cooperatively act to form the preferential atomic ordering in the (A,B) monolayer.

Warren-Cowley short-range order (SRO) parameters[31] are also calculated on the growing surface as a function of the surface coverage for the atomic arrangement evolution as shown in Figures 12 and 13. As is shown in Figure 14, the data of Warren-Cowley SRO parameters clearly show the evolution of SRO formation with increasing the surface coverage. The highly ordered atomic arrangement with full coverage ,as shown in Figures 12 and 13, corresponds to the surface of chalcopyrite type (001) LRO surface and CuPt-type LRO surface.

4.3. Order-disorder transition in epitaxial binary alloy

LRO parameters of the atomic arrangements have been also evaluated as a function of the normalized temperature, $k_BT/|J|$. Considering the α and β sublattices for A and B atoms corresponding to the perfect superlattice of LRO phase, the LRO parameters of (A_x, B_{1-x}) system can be estimated by using

$$S = \frac{([A_\alpha] + [B_\beta]) - ([A_\beta + [B_\alpha])}{([A_\alpha] + [A_\beta] - ([B_\alpha + [B_\beta])}$$
$$= \frac{1}{N_t}\{([A_\alpha] + [B_\beta]) - ([A_\beta + [B_\alpha])\}, \tag{11}$$

where $[T_\lambda]$ means the number of T=(A or B) atom in the λ=(α or β) sublattice and N_t is the total number of surface atoms. The LRO parameter is defined here as

$$S_{LRO} = |S| \tag{12}$$

In the case of A_1B_1 stochiometry in the perfect LRO state, we obtain

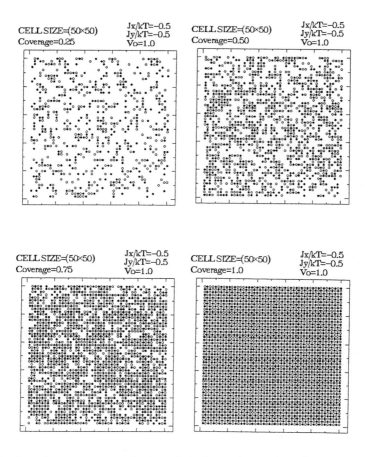

Figure 12. Atomic structure evolution of chalcopyrite-type surface.

$$0 \leq S_{\mathrm{LRO}} \leq 2x \quad \text{for} \quad (0 \leq x \leq 0.5)$$
$$0 \leq S_{\mathrm{LRO}} \leq 2(1-x) \quad \text{for} \quad (0.5 \leq x \leq 1.0) \tag{13}$$

The results of the present calculation of S_{LRO} value for (AF,F) type alloy system as a function of substrate temperature, $k_B T/|J|$, is shown in Figure 15. As can be seen from Figure 15, LRO parameter is quite small at low growth temperatures. This feature is quite strange as compared with usual feature of order-disorder transition, where the LRO parameter is close to 1.0 at low temperature. This point is considered to be essential in the order-disorder phenomena during epitaxial growth. The value of molecular-beam flux is set to be constant in the present algorithm for growth simulation. Therefore, the lack in the chance of surface diffusion at low temperature makes the LRO parameter reduced. This is the case of conventional

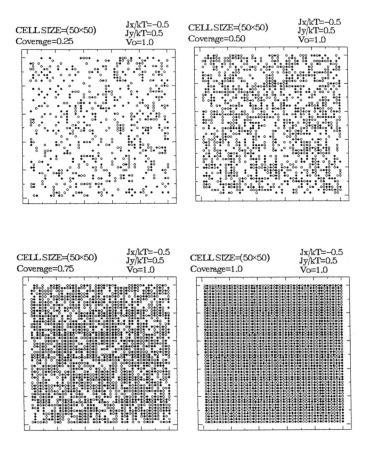

Figure 13. Atomic structure evolution of CuPt-type surface.

MBE or OMVPE growth with the highly supersaturation ratio and with the relatively low growth temperature. As is shown in Figure 15, LRO parameter increases drastically with increasing substrate temperature by the enhancement of atomic ordering caused by the cooperation of adsorption and diffusion processes at elevated temperatures. Then, it saturates at the maximum value ($S_{\mathrm{LRO}} \approx 0.6$ in this case). LRO value again decreases drastically with increasing the temperature and get to be almost zero around $k_B T / |J| = 1.0$. This disordering is naturally caused by the much enhanced diffusion and adsorption at excited-energy sites at elevated temperatures. Beyond the transition temperature of LRO parameter, SRO parameters still remain finite[30]. Systematic calculation is necessary in order to understand the behavior of LRO in the light of the nonequilibrium nature of dynamic growth process of epitaxy. For example, in the real growth process, many atomic jump

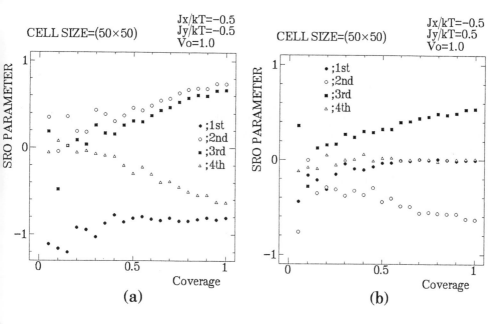

Figure 14. Change in the Warren-Cowley SRO parameters as a function of the surface coverage of chalcopyrite-type surface (a) and CuPt-type surface (b)

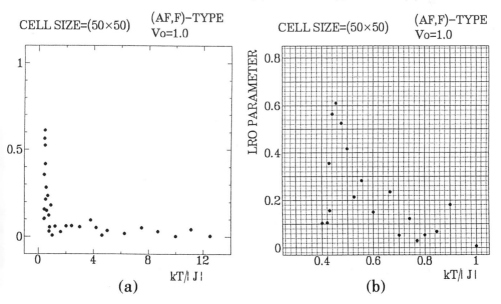

Figure 15. Change in the LRO parameters as a function of the normalized temperature. (b) is the enlarged part of (a), where characteristic feature of the order-disorder transition is demonstrated. See the presence of the peak in the LRO parameter at around $k_B T/J = 0.45$.

182

events at different sites should occur between the interval of atom impinging. However, it is worth noting that dynamic epitaxial processes should cause the characteristic order-disorder transition as is shown in Figure 15, whose feature is quite different from usual order-disorder transition in metallic alloys under (near) the thermal equilibrium condition. Finally, we would like to point out that the experimental data of the LRO parameter of InGaP//GaAs(001) showed such kind of order-disorder transition. It showed the peak value of LRO parameter at the substrate temperature around 940K and decreased drastically in both the lower and higher growth temperatures[20].

ACKNOWLEDGMENT

This work was supported in part by the Grant-in-Aid for Scientific Research on Priority Areas "Crystal Growth Mechanism in Atomic Scale" from the Ministry of Education, Science, sports and Culture. The authors would like to express sincere thanks to Prof. T. Nishinaga of the University of Tokyo for encouraging us during this work. This work was partly supported by the Photonics Materials Laboratory Project of the Graduate School of Science and Technology, Kobe University.

REFERENCES

1. H.Nakayama , S.Nakamura, E.Taguchi, H.Fujita, T.Yagi, T.Nishino and Y.Hamakawa; 1984 Abstract of 45th Fall Meeting of the Japan Society of Applied Physics, p.405.
2. H.Nakayama@and H.Fujita,in: GaAs and Related Compounds 1985, Inst.@Phys.@Conf. Ser.@No.79, Ed. M. Fujimoto (Inst. Phys., Bristol, 1986) p.289.
3. T.S.Kuan, T.F.Kuech, W.I.Wang and E.L.Wilkie, Phys. Rev. Lett., 54 (1985) 201.
4. D. De Fontaine, in Solid State Physics Vol.34 Eds. H. Ehrenreich, F. Seitz and D. Turnbull (Academic Press, New York,1979).
5. W.L.Bragg and E.J.Williams, Proc. Roy. Soc. A145 (1934) 699.
6. S. Yamazaki, A. Ushirokawa and T. Katoda, J. Appl. Phys. 51 (1980) 3722.
7. A.Gomyo, K.Kobayashi, S.Kawata, I.Hino, T.Suzuki and T.Yuasa, J.Cryst. Growth, 77 (1986) 367.
8. I.J.Murgatroyd, A.G.Norman, G.R.Booker, T.M.Kerr, Proc. 11th Int. Cong. on Electron Microscopy, Japan, 1986, p.1467, ed. by T.Imura, S.Maruse and T.Suzuki.
9. G.S. Chen, D.H.Jaw and G.B. Stringfellow, J. Appl. Phys. 69 (1991) 4263.
10. A.Zunger and S.Mahajan: Handbook of Semiconductors, Vol.3b, second. edition series ed. by T.S.Moss and volume ed. by S. Mahajan (Elsevier, Amsterdam, 1994) p.1399.
11. G.P. Srivastava, J.L.Martins and A.Zunger, Phys. Rev. B31 (1985) 2561.
12. T. Yagi, Y.Fujiwara, T.Nishino and Y.Hamakawa, Jpn. J. Appl. Phys. 22 (1983) L467.

13. A.Miller, A. MacKinnon and D. Weaire, in Solid State Physics Vol.36 Eds. H. Ehrenreich, F. Seitz and D. Turnbull (Academic Press, New York,1981).
14. H.R. Jen, M.J. Cherng and S.B. Stringfellow, Appl. Phys. Lett. 48 (1986) 1603.
15. A doubt about the structural assignment of famatinite-type InGaAs given by T.S. Kuan, Encyclopedia of Physical Science and Technology, Academic Press, 1989, is not correct in the same reason.
16. A. Gomyo, T. Suzuki and S. Iijima, Phys. Rev. Lett. 60, 2645 (1988).
17. T. Nishino, J. Cryst. Growth 98,44 (1989).
18. T. Nishino, Y. Inoue, Y. Hamakawa, M. Kondow, and S.Minagawa, Appl. Phys. Lett. 53, 583 (1988).
19. A. Mascarenhas, S. R. Kurtz, A. Kibber, and J. M. Olson, Phys. Rev. Lett. 63, 2108 (1989).
20. T. Kanata-Kita, M. Nishimoto, H. Nakayama, and T. Nishino, Phys. Rev. B 45, 6637 (1992).
21. T. Kanata-Kita, M. Nishimoto, H. Nakayama, and T. Nishino, Appl. Phys. Lett. 63, 512 (1993).
22. G. S. Horner, A. Mascarenhas, R. G. Alonso, S. Froyen, K. A. Bertness, and J. M. Olson, Phys. Rev. B 49, 1727 (1994).
23. T. Kita, A. Fujiwara, H. Nakayama, and T. Nishino, Appl. Phys. Lett. 66, 1794 (1995).
24. T. Kita, A. Fujiwara, H. Nakayama, and T. Nishino, J. Electronic Materials 25, 661 (1996).
25. S. H. Wei, D. B. Laks, and A. Zunger, Appl. Phys. Lett. 62, 1937 (1993).
26. S. H. Wei and A. Zunger, Appl. Phys. Lett. 64, 1676 (1994).
27. H.Nakayama, M.Tochigi, H.Maeda and T.Nishino,Applied Surf. Sci. 82/83 (1994) 214.
28. H.Nakayama, M.Tochigi, H.Maeda and T.Nishino, J.Cryst.Growth 150 (1995) 168.
29. H.Nakayama, T.Takeguchi and T.Nishino, J. Cryst. Growth 163 (1996) 135.
30. H.Nakayama, T.Takeguchi and T.Nishino, to be published in Surf. Sci. (1996)
31. J.M.Cowley, Phys. Rev. 120 (1960) 1648, and Phys. Rev.138 (1965) A1384.

Advances in the Understanding of Crystal Growth Mechanisms
T. Nishinaga, K. Nishioka, J. Harada, A. Sasaki and H. Takei (Editors)
© 1997 Elsevier Science B.V. All rights reserved.

Monte Carlo simulation of microstructures in ordered III-V semiconductor alloys

N. Kuwano* and K. Oki

Department of Materials Science and Technology,
Interdisciplinary Graduate School of Engineering Sciences, Kyushu University,
Kasuga, Fukuoka 816, Japan

Understanding the formation mechanism of ordered structures in the epitaxial layers of III-V semiconductor alloys is a significant fundamental problem in materials science and engineering. We proposed a new simple model for epitaxial growth and used it in Monte Carlo simulations of the microstructures in ordered III-V alloys. In this paper, we briefly describe the simple model and present some simulation results. We discuss our findings regarding the microstructures.

1. INTRODUCTION

It is known that most III-V semiconductor alloys tend to undergo phase separation in the bulk state [1,2]. A mottled structure resulting from interfacial spinodal decomposition is frequently observed in III-V alloys grown by the liquid phase epitaxy (LPE) method [3–5]. On the other hand, a number of authors [6–8] have reported the existence of ordered structures in epitaxial layers formed by molecular beam epitaxy (MBE), the metallorganic vapor phase epitaxy (MOVPE) and other methods. The atomic ordering is an interesting phenomenon since the electronic properties of the alloys, such as the energy band gap and the electron mobility, depend on the degree of order [9,10]. It is also important from the engineering point of view to characterize the microstructures of ordered III-V alloys. The formation of ordered structures has been attributed to lattice strains around absorbed atoms [11], surface reconstruction [12,13], atomic steps [14] or a combination of these [15]. In order to analyze the formation mechanism of the ordered structures, the atomic layer growth process should be clarified. The formation process of an epitaxial layer can be simulated very accurately using molecular dynamics (MD). However, an MD simulation needs a lot of calculations with a supercomputer even for a short time process of layer growth. We proposed a new Ising-like model which is quite simple [16] and applied it to Monte Carlo simulations of the microstructures in epitaxial layers [17,18].

In the present paper, we review the Ising-like model and its application to Monte Carlo simulations, and summarize the simulation results.

---

*Corresponding author (E-mail: kuwa-igz@mbox.nc.kyushu-u.ac.jp)

2. GROWTH MODEL

We consider III-V alloys in the $(A_{1-x}B_x)C$ system with a zincblende structure. The zincblende structure is made of two interpenetrating fcc sublattices, one of which consists of A and B atoms (α-sublattice) and the other of which consists of C atoms (β-sublattice). A and B atoms form an ordered structure in the α-sublattice. Figure 1 shows two typical ordered structure which will be referred to later. Atomic ordering occurs on the surface during the crystal growth. Atoms will not move to other sites inside the bulk crystal because the diffusion rate is very small.

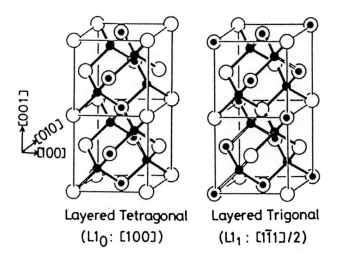

Layered Tetragonal
($L1_0$: [100])

Layered Trigonal
($L1_1$: [1$\bar{1}$1]/2)

Figure 1. Ordered lattices, $L1_0$ and $L1_1$, in a zincblende structure. The open and semiopen circles indicate α-sites occupied by A and B atoms, respectively, and the closed circles indicate β-sites occupied by C atoms.

It is assumed that epitaxial growth proceeds principally through the following steps.

Step-1 (adhesion process):

1. A and B atoms fall one by one onto the crystal surface.

2. The atoms diffuse over the surface and occupy particular preferred α-sublattice sites. Some atoms can leave the surface.

3. Eventually, the surface is fully covered with A and B atoms.

Step-2 (exchange process):

1. A and B atoms in the top surface layer exchange sites to reduce the configurational energy.

2. C atoms fall onto the surface and occupy β-sites on the layer consisting of A and B atoms. The configuration of the A and B atoms is then frozen.

During the actual growth, steps 1 and 2 proceed simultaneously. However, it is plausible that one process is dominant under certain growth conditions. In the "dynamic site correlation absorption process" proposed by Nakayama *et al.*[19], atomic ordering is assumed to occur in step 1. On the other hand, if the time available for atoms to exchange sites in step 2 is long, the atomic configuration is thought to be determined in step 2. Step-1 can then be ignored.

3. INTERACTIONS AND CONFIGURATIONAL ENERGIES

We first consider the case of (001) epitaxial growth. Since the α-sites in a zincblende structure form an *fcc* lattice, their positions can be presented by the vector

$$\mathbf{r} = n\mathbf{a} + m\mathbf{b} + l\mathbf{c}, \tag{1}$$

where $\mathbf{a}$, $\mathbf{b}$ and $\mathbf{c}$ are the primitive vectors

$$\mathbf{a} = [1\bar{1}0]/2, \quad \mathbf{b} = [110]/2, \quad \mathbf{c} = [10\bar{1}]/2, \tag{2}$$

and n, m and l are integers. When the origin of coordinates is at an α-site in the surface layer, l is zero for sites in the surface layer and greater than zero for those in the underlying layer. In order to describe the atomic configuration, an occupation operator $\gamma(\mathbf{r})$ is introduced. $\gamma(\mathbf{r}) = 1$, if the site at $\mathbf{r}$ is occupied by an A atom, and $\gamma(\mathbf{r}) = -1$ if it is occupied by a B atom.

The internal energy in the system can be divided into three components, (i) the absorption energy of A and B atoms, (ii) the mixing energy and (iii) the configurational energy. Since we assume that the exchange process is dominant, only the last component is taken into account. When the configurational energy E_c is assumed to be due only to the pair interactions, it can be given by

$$
E_c = -\frac{1}{2} \sum_{\mathbf{r}_1(l=0)} \sum_{\mathbf{r}_2(l=0)} V(\mathbf{r}_2 - \mathbf{r}_1) \gamma(\mathbf{r}_1) \gamma(\mathbf{r}_2)
$$
$$
- \sum_{\mathbf{r}_1(l=0)} \sum_{\mathbf{r}_2(l>0)} V(\mathbf{r}_2 - \mathbf{r}_1) \gamma(\mathbf{r}_1) \gamma(\mathbf{r}_2). \tag{3}
$$

The pair interactions (or the ordering energy) are defined as

$$V(\mathbf{r}_2 - \mathbf{r}_1) = E^{AB}(\mathbf{r}_2 - \mathbf{r}_1) - \frac{1}{2}\{E^{AA}(\mathbf{r}_2 - \mathbf{r}_1) + E^{BB}(\mathbf{r}_2 - \mathbf{r}_1)\}, \tag{4}$$

where $E^{ij}(\mathbf{r}_2 - \mathbf{r}_1)$ is the interaction energy (or the bonding energy) between the i- and j-atoms at $\mathbf{r}_2$ and $\mathbf{r}_1$, respectively. In the present model, only the atoms in the surface layer can exchange sites, and the interactions concerning these atoms govern the configuration of A and B atoms, that is, $l=0$ for $\mathbf{r}_1$. The symmetry $E^{ij} = E^{ji}$ is assumed. Unlike atom pairs have a lower configurational energy if $V(\mathbf{r})$ is negative and like atom pairs do if $V(\mathbf{r})$ is positive. E_c corresponds to the difference in internal energy between the ordered and random configurations.

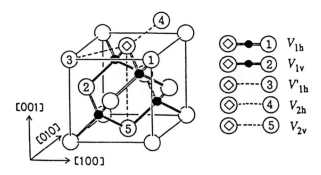

Figure 2. Atomic pairs in an α-sublattice and their interactions with an atom (marked $\diamond$) in the top surface layer. n in the subscript of V_{nh} and V_{nv} indicates the n-th nearest-neighbor shell and the letters h and v mean the intralayer and interlayer pairs, respectively.

Figure 2 shows atomic pairs up to the second-nearest neighbors for (001) growth. Here, the second-nearest neighbors are needed in order to account for the development of particular variants of the $L1_1$ or $L1_0$ structure in epitaxial layers. *Ab initio* calculations showed that the second-nearest neighbor interactions have significant magnitudes [20]. The atom marked $\diamond$ in the surface layer has three kinds of first-nearest neighbors. The pair formed with atom 1 is linked via a C atom, but the pair formed with atom 3 does not have such a linkage. The pair formed with atom 2 has this type of linkage, but atom 2 is in the underlying layer. Consequently, these pairs, in general, have different values for the pair interactions. Hereafter, pairs like the first two are called 'intralayer pairs' and those like the third are called 'interlayer pairs'. The second-nearest neighbor pairs formed with atoms 4 and 5 are an intralayer pair and an interlayer pair, respectively, and the values for the pair interactions of the pairs are different from each other. It should be noted that these atomic pairs can be distinguished from one another even if the surface is not reconstructed. The magnitudes of the pair interactions, however, strongly depend upon the surface reconstruction. Osòrio *et al.*[20] estimated the magnitudes by using the first principle theory and pointed out that they are much larger than those in the bulk. As far as the assumption of pair interaction mentioned above is accepted, the excess internal energy due to the atomic configuration can be expressed in the same manner as in the present, unless the reconstructed surface has a quite different symmetry from the unreconstructed one. Thus the effect of reconstruction is partly taken into account implicitly in the present model by using appropriate values for the pair interactions. When the surface has a more complicated reconstruction, the classification of the atomic pairs must be modified accordingly.

In the case of (110) growth, the positions of the α-sites are given by Eq.(1) with the primitive vectors

$$\mathbf{a} = [1\bar{1}0]/2, \quad \mathbf{b} = [00\bar{1}], \quad \mathbf{c} = [0\bar{1}\bar{1}]/2. \tag{5}$$

There are four kinds of first-nearest neighbor and two kinds of second-nearest neighbor. They are represented as

$$\mathbf{r} = \begin{cases} \mathbf{a}, \quad \mathbf{c}, \quad \mathbf{c} - \mathbf{a}, \quad 2\mathbf{c} - \mathbf{a} - \mathbf{b} & \text{for the first nearest neighbors} \\ \mathbf{b}, \quad 2\mathbf{c} - \mathbf{b} & \text{for the second-nearest neighbors} \end{cases} \tag{6}$$

Then, the excess internal energy for a given configuration can be deduced, if the pair interactions are properly determined.

4. MONTE CARLO SIMULATION

A Monte Carlo simulation of the microstructure in an epitaxial layer is carried out using the Kawasaki's dynamics [21,22] via the following procedure.

1. A and B atoms are distributed randomly at α-sites on the surface.

2. A first-nearest-neighbor pair is selected at random, and the atoms in the pair exchange with a following probability of

$$P(\Delta E, T) = \{1 - \tanh(\Delta E/2k_{\mathrm{B}}T)\}/2, \tag{7}$$

   where ΔE is the energy difference due to the exchange, k_{B} is Boltzmann's constant and T is the growth temperature. Trials are repeated over a given time interval $\Delta\tau$, which is measured in units of Monte Carlo steps (MCS). Here, one MCS corresponds to one trial per atom.

3. The atom configuration in the surface layer is fixed.

This procedure is for deposition of an atomic layer assuming the "exchange process ". An epitaxial layer can be grown by repetition of this procedure, or by a layer-by-layer process.

In the above procedure, the probability of exchange is assumed to be represented by a function of the energy difference between the states before and after the exchange, state 1 and state 2 (Figure 3). The actual probability of jumping from

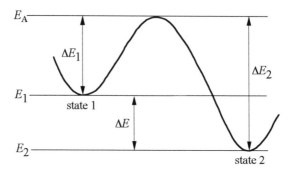

Figure 3. The energy path between state 1 and state 2. E_1, E_2 and E_A are the potential energies of state 1, state 2 and the activated state. When the system moves from state 1(or 2) to state 2 (or 1), it must pass through the activated state.

state 1 to state 2 should be represented in terms of the activation energy ΔE_1 instead of ΔE. However, if the reverse process from state 2 to 1 is also allowed, the state can revert to state-1 with a probability represented by ΔE_2. Then, the probabilities of staying in states 1 and 2 are approximately expressed by Eq.(7) when the jumping frequency is not very low. This means that local thermoeqilibrium is assumed to be reached. One trial in the present corresponds to a certain number of actual two-way exchanges between states 1 and 2. This approximation has the advantage that the exchange probabilities can be estimated without the activation energy-barriers which are usually difficult to be estimated. Besides, if the exchange process in the simulation was the same as that occurring in practice, a very large number of trials for jumping would be required in the simulation, which would consume long computing time.

In the simulation, the values of the pair interactions must be determined carefully. First, approximate values for the pair interactions are determined by considering the relationship between the pair interactions and the ground states [16]. A certain ordered structure is expected to correspond to a given set of interactions. Thus, the values of the pair interactions are limited to comparatively narrow ranges. Subsequently, one can determine the values more accurately by considering the values obtained by *ab initio* methods, such as the first principle method and the electron counting method, if available, and by seeking appropriate values within the limits so that they well reproduce the experimental results for the microstructures.

5. RESULTS AND DISCUSSION

We demonstrate results obtained so far for (001) growth.

5.1. L1₁ ordering during exact (001) growth

Typical simulation results for an "exact" (001) substrate are presented [17,23]. Figure 4 shows a (110) Fourier power spectrum and a (110) cross section of the simulated structure. In the spectrum, which corresponds to an electron diffraction pattern, a diffuse intensity distribution is observed with streaking along the growth direction [001]. The intensity maxima are at $hkl = \mp 1/2 \pm 1/2 \pm 1/2$ and $\pm 1/2 \mp 1/2 \pm 1/2$, indicating L1₁ ordering. It was confirmed that there is no such diffuse scattering in the $(1\bar{1}0)$ Fourier power spectrum, indicating that there are two variants of L1₁, formed in the epitaxial layer. The diffuse intensity distribution has a

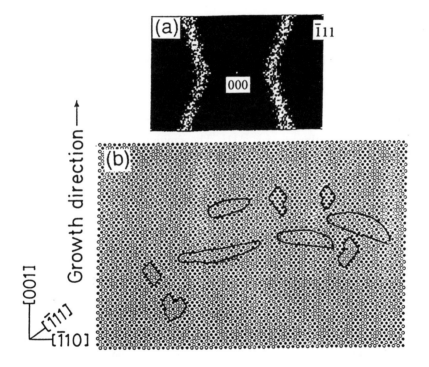

Figure 4. (110) Fourier spectrum and (110) cross section of the simulated structure grown on an exact (001) substrate.
$V_{1h} - k_B T$, $V'_{1h} = k_B T$, $V_{1v} = k_B T$, $V_{2h} = -0.5 k_B T$, $V_{2v} = -0.3 k_B T$,
$x = 0.5$, $\Delta \tau = 20$ MCS
The cross section is produced by superimposing two succesive (110) planes of α-sublattice. ○: A atom, ●: B atom. Some like-atom clusters and L1₁ domains are delineated.

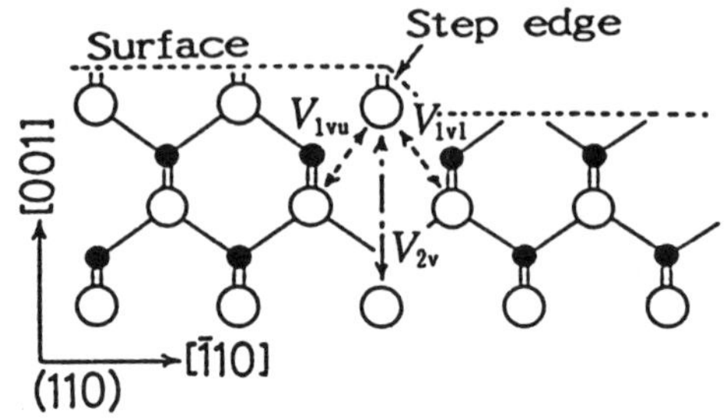

Figure 5. Pair interactions of an atom at a step edge.   The step edge is parallel to [110] and moves towards [$\bar{1}$10]. The large open circles denote α sites, and the small closed ones denote β sites.

wavy shape which bends towards the positions of the fundamental lattice reflections. These characteristics were observed in actual electron diffraction patterns [9,11–13]. The origin of this shape was controversial. The cross section in Figure 4(b) shows that the epitaxial layer is composed of small domains with L1$_1$ structure and small like-atom clusters, some of which are enclosed. It should be noted that L1$_1$ domains of ($\bar{1}$11)-variant have a "flying saucer" shape tilting towards [$\bar{1}$11] whereas those of (1$\bar{1}$1) tilts towards [1$\bar{1}$1]. The simulation reveals clearly that the intensity distribution results from both the presence of like-atom clusters and the shape of the ordered domains. From analyses using Warren-Cowley short-range-order (SRO) parameters, the shape of the intensity distribution is found to reflect the ordering of the interlayer first-nearest-neighbor pairs [23].

Epitaxial layers with a small $\Delta\tau$ contain mostly like-atom clusters and the diffuse scattering has the intensity maxima at around $hkl = -0.8, -0.8, 0$. In layers with large $\Delta\tau$, L1$_1$ domains are developed with few like-atom clusters. The progress of L1$_1$ ordering with a large $\Delta\tau$ was found to be contributed by the ordering of the interlayer second-nearest-neighbor pairs [23].

5.2.   Effect of miscut on L1$_1$ ordering during (001) growth

Bellon et $al.$[24] clearly showed that the miscut of the substrate has a strong effect on the L1$_1$ microstructure. One large variant of L1$_1$ develops with some slanting antiphase boundaries in an epitaxial layer grown on a (001) substrate inclined towards [$\bar{1}$10], while plate like domains of two variants are formed on the substrate inclined towards [110]. The miscut substrate has atomic steps on the surface, and under certain conditions, the crystal grows through moving of the steps (step flow mechanism). For simplicity, the steps are considered to be straight and parallel to [110] or [1$\bar{1}$0] for substrates miscut towards [1$\bar{1}$0] and [110], respectively. We assume that atoms which make the edge-line can exchange their sites with each other. The growth process is the same as for the layer-by-layer process, provided

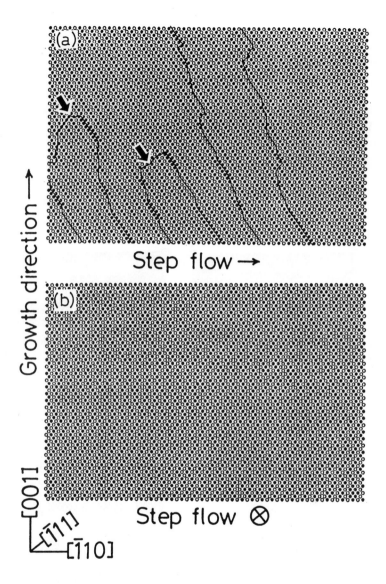

Figure 6. Projections of simulated structures grown on miscut substrates inclined towards [$\bar{1}$10] and [110].

$V_{1h} = -0.8k_BT,\ V'_{1h} = k_BT,\ V_{1vu} = 0.5k_BT,\ V_{1vl} = 0.9k_BT,$
$V_{2h} = -0.5k_BT,\ V_{2v} = -0.3k_BT,\ x = 0.5,\ \Delta\tau = 20\ \text{MCS}$

A pair of antiphase boundaries merge to each other to disappear; self-annihilation process (marked with arrows).

194

the term "surface" is replaced with "edge line" (line-by-line process)[25]. It is seen in Figure 5, that the pair interactions of the interlayer first-nearest-neighbor pairs have two different values, V_{1vu} and V_{1vl}, according to whether the second atom in the pair is located on the upper or lower side of the step-edge.

Figure 6 [26] shows simulated structures grown on (001) substrates miscut towards [$\bar{1}$10] and [110]. The same values for the pairwise interaction parameters were used in both cases. The structures have quite different morphologies, which depend on the direction of the step edge, that is, the direction of the miscut. The simulation quite successfully reproduces the microstructures observed by Bellon et $al.$ [24]

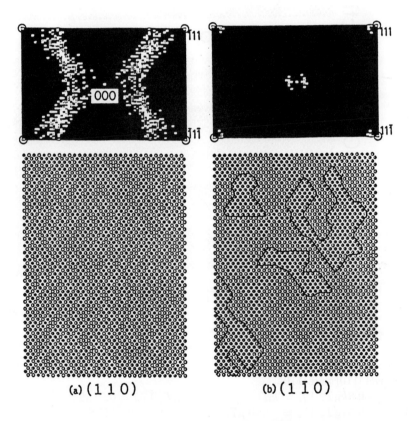

(a) (1 1 0) (b) (1 $\bar{1}$ 0)

Figure 7. Fourier power spectra and projections of the simulated structures in an epitaxial layer of $(A_{0.7}B_{0.3})C$ grown on a (001) substrate.
$V_{1h} = -k_B T,\ V'_{1h} = k_B T,\ V_{1v} = 0.7 k_B T,\ V_{2h} = -0.5 k_B T,\ V_{2v} = -0.2 k_B T,$
$x = 0.3,\ \Delta\tau = 1000$ MCS
(a) (110), (b) (1$\bar{1}$0).   Some $L1_1$ regions are contoured.

5.3. Effect of off-stoichiometry

We carried out a simulation for the microstructure with $x \neq 0.5$ grown on a (001) substrate by the layer-by-layer process[17,27]. Figure 7 shows the Fourier power spectra and the projected structures for $x = 0.3$. One can see wavy scattering in the (110) spectrum and satellite reflections around the fundamental lattice reflections in the ($1\bar{1}0$) spectrum. The diffuse scattering and satellite were observed experimentally in electron diffraction patterns for $GaAs_{1-x}Sb_x$ with a nonstoichiometric composition [12]. The diffuse scattering indicates the presence of two variants of $L1_1$, as shown in Figure 7(a). The structure projected onto ($1\bar{1}0$) reveals that the satellites are attributed to a modulated structure consisting of $L1_1$ ordered regions with a high concentration of B atoms and disordered regions with a low concentration of B atoms. It was also confirmed from a series of the simulation that $L1_1$ ordering proceeds in layers with $0 < x < 0.5$ and that the satellites are most intense at around $x = 0.25$ [27].

5.4. $L1_0$ ordering during (001) growth

Several investigators have reported the formation of $L1_0$ (CuAu-I) ordered structure in various III-V alloy layers grown on (001) and (110) substrates [28–30]. Kuan et al.[31] found, using the electron diffraction method, that two variants of $L1_0$, the c-axes of which are parallel to [100] and [010] (hereafter designated as [100] and [010] variants, respectively), are formed in an $Al_{1-x}Ga_xAs$ alloy grown on a (001) substrate. The appearance of the two variants is expected from the diagram for the ground states reported previously [16].

Figure 8 shows Fourier power spectra and a cross section of the simulated microstructure with $x = 0.5$ [32]. In the spectra, there are intensity maxima at $hkl = 100$ and 010 in (a) and $hkl = 101$ in (b), but not at $hkl = 001$ and 110. This indicates that the simulated microstructure contains only [100] and [010] variants of $L1_0$. The intensity maxima are accompanied by distinct streaks along [001], indicating the presence of (001) plane faults. The absence of [001] variant and the appearance of streaks have been observed experimentally[31].

In the (010) cross section, one can see two different atomic arrangements, a chessboard pattern and a vertical (parallel to [001]) stripe pattern. The former coincides with the [010] variant, and the latter with the [100] variant. If there were a [001] variant, regions with a horizontal stripe pattern would appear in the (010) cross section. The ordered domains are found to grow horizontally, resulting in streaks along [001] in the Fourier power spectrum. This tendency is due to the small absolute value for V_{2v} used in the simulation. We carried out a simulation for $A_{1-x}B_xC$ with $x = 0.25$, using the same values for the pair interactions, and found that the $L1_0$ structure develops even in a nonstoichiometric alloy. The microstructure consists of domains of highly ordered $L1_0$ embedded in a matrix of an AC compound [32].

5.5. Interfacial spinodal decomposition during LPE-growth

It is known that a compositional modulation is frequently formed in alloy layers grown by the liquid phase epitaxy (LPE) method. The modulation is due to the

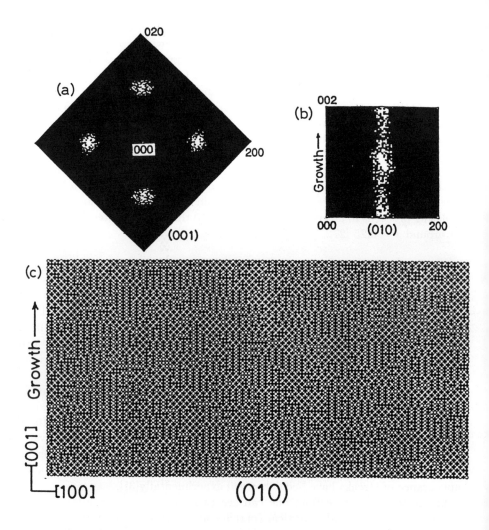

Figure 8. Fourier power spectra and cross section of a simulated L1$_0$ structure grown on a (001) substrate.
$V_{1h} = -k_B T$, $V'_{1h} = -k_B T$, $V_{1v} = -0.8 k_B T$, $V_{2h} = 0.5 k_B T$, $V_{2v} = 0.3 k_B T$, $x = 0.5$, $\Delta\tau = 20$ MCS
(a) (001), (b) (010), (c) (010) cross section.

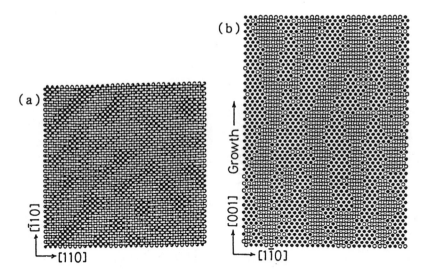

Figure 9. (a) (001) projection and (b) (010) cross section of the simulated structure in an epitaxial layer grown by LPE on a (001) substrate.
$V_{1h} = V'_{1h} = V_{1v} = k_BT$, $V_{2h} = V_{2v} = 0.3k_BT$, $x = 0.5$, $\Delta\tau = 100$ MCS
(a) (001) projection,   (b) (110) cross section.

"interfacial spinodal decomposition"[4]. We applied the present model to structures formed by LPE. It is believed that there is a thin intermediate phase on the liquid/solid interface. This phase has properties intermediate between those of solid and liquid. It is still liquid, but already has a structure like that of the solid phase. Therefore, we assume that surface reconstruction, which occurs due to dangling bonds on the surface, is undeveloped and that atoms are precipitated in the form of III-V molecules. Then, the pair interactions would have positive values as in the bulk. Figure 9 shows a (001) projection and a (110) cross section of the simulated microstructure formed by LPE. One can see that the compositional modulation is reproduced quite successfully[33]. The modulation exists along [100] and [010] in the (001) projection, and A-rich and B-rich regions are elongated along [001], forming a columnar texture in the (110) cross section. The orientation of the modulation has been explained in terms of the elastic anisotropy of the bulk crystal [34]. However, it should be emphasized that the morphology can be reproduced without considering the effects of the mechanical properties.

5.6. Preliminary attempt to simulate TP-A ordering

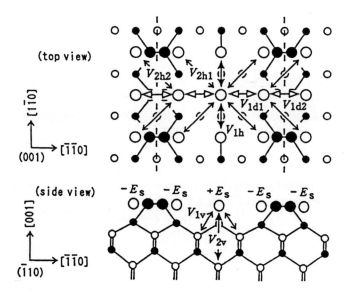

Figure 10. Pair interactions for α-sites on a (2×3) reconstructed surface.

Recently, a new type of ordered phase was discovered in (Al,In)As grown on a (001) InP substrate [35]. The ordered structure has a triple period in $(111)_A$ stackings (TP-A), where (111) layers have different concentrations of In and Al, whereas the $L1_1$ has a double period in (111) stackings. Gomyo $et\ al.$ [35] pointed out that the ordered structure results from the (2×3) surface reconstruction. The (2×3) reconstruction has quite different symmetry from the unreconstructed surface or those with other simply reconstructions, and the α-sites on the surface are not equivalent to one another. Therefore, the atomic pairs should be classified more minutely, as shown in Figure 10. In addition, there should be an energy difference among the α-sites. The internal energy is therefore given by

$$
E_c = -\frac{1}{2} \sum_{\mathbf{r}_1(l=0)} \sum_{\mathbf{r}_2(l=0)} V(\mathbf{r}_2 - \mathbf{r}_1)\gamma(\mathbf{r}_1)\gamma(\mathbf{r}_2)
$$

$$- \sum_{\mathbf{r}_1(l=0)} \sum_{\mathbf{r}_2(l>0)} V(\mathbf{r}_2 - \mathbf{r}_1)\gamma(\mathbf{r}_1)\gamma(\mathbf{r}_2)$$

$$- \sum_{\mathbf{r}_1(l=0)} E_s(\mathbf{r}_1)\gamma(\mathbf{r}_1), \qquad (8)$$

where $E_s(\mathbf{r})$ is the site correlation energy, which means that A (or B) atoms prefer particular α-sites at $\mathbf{r}$. Figure 11(a) shows the simulated microstructure grown on a (2×3) reconstructed surface. The TP-A structure is clearly seen. However, the microstructure is not the same as that observed experimentally. The simulated structure consists of well developed plate like regions of TP-A, whereas the actual alloy contains small TP-A domains embedded in a matrix[36]. These discrepancies can be reduced to some extent by adopting appropriate values for the pair interactions and by modifying a little the conditions used in the simulation. We carried out the simulation under the assumption that the (2×3) reconstruction contains some antiphase boundaries [37]. This simulated microstructure was found to be more similar to the observed one, as shown in Figure 11(b).

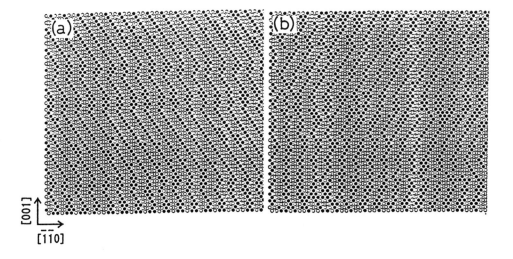

[001]

[$\bar{1}$10]

Figure 11. ($\bar{1}$10) projections of simulated TP-A structures grown on (001) substrates.
$V_{1h} = k_BT$, $V_{1d1} = -k_BT$, $V_{1d2} = k_BT$, $V_{1v} = k_BT$, $V_{2h1} = -0.5k_BT$,
$V_{2h2} = 0.5k_BT$, $V_{2v} = -0.3k_BT$, $E_s = k_BT$, $x = 0.48$, $\Delta\tau = 500$ MCS
(a) Perfect (2×3) reconstructed surface, (b) Reconstructed surface with antiphase boundaries.

6. CONCLUDING REMARKS

It has been demonstrated that the characteristics of the microstructures of epitaxially grown layers can be successfully reproduced by Monte Carlo simulations based on our simple model. By comparing actual microstructures with those simulated under various conditions, we can investigate the formation mechanism of ordered structures and the relationship between the microstructure and the growth conditions. These results are useful for determining the growth conditions for alloys with preferable microstructures.

However, the deposition process in the present model is over-simplified. We need to modify the model in order to take into account more effectively the effects of adhesion process and of complicated surface reconstructions. Then, we will be able to analyze more complex microstructures.

ACKNOWLEDGMENTS

This work was supported in part by Grants-in-Aid for Scientific Research on Priority Areas "Crystal Growth Mechanism in Atomic Scale" (Nos. 03243107, 04227107) and for Cooperative Research (A) "Structural Dynamics of Epitaxy and Quantum Mechanical Approach" (No. 07305001) from the Ministry of Education, Science, Sports and Culture of Japan. The authors are grateful to Dr. M. Ishimaru, Dr. S. Matsumura and Mr. Y. Kangawa for their cooperation in this study.

REFERENCES

1. K. Onabe, Jpn. J. Appl. Phys. 21 (1982) L323.
2. G. B. Stringfellow, J. Cryst. Growth 58(1982) 194.
3. O. Ueda, S. Isozumi and S. Komiya, Jpn. J. Appl. Phys. 23(1984) L241.
4. P.Henoc, A. Izrael, M.Quillec and H. Launois, Appl.Phys. Lett. 40(1982) 963.
5. M. Quillec, H. Launmois and M. C. Jourcour, J. Vac. Technol. B1(1983) 238.
6. T. S. Kuan, T. F. Kuech, W. I. Wang and E. L. Wilkie, Phys. Rev. Lett. 54(1985) 201.
7. H. R. Jen, K. Y. Ma and G. B. Stringfellow, Appl. Phys. Lett. 54(1989) 1154.
8. M. A. Shahid, S. Mahajan, D. E. Laughlin and H. M. Cox, Phys. Rev. Lett. 58(1987) 2567.
9. A. Gomyo, T. Suzuki, K. Kobayashi, S. Kawata and I. Hino, Appl. Phys. Lett. 50(1987) 673.
10. S. H. Wei, D. B. Laks and A. Zunger, Appl. Phys. Lett. 63(1993) 1937.
11. T. Suzuki, A. Gomyo and S. Iijima, J. Cryst. Growth 93(1988) 396.
12. I. J. Murgatroyd, A. G. Norman and G. R. Booker, J. Appl. Phys. 67(1990) 2310.
13. G. S. Chen, D. H. Jaw and G. B. Stringfellow, J. Appl. Phys. 69(1991) 4263.
14. T. Suzuki and A. Gomyo, J. Cryst. Growth 111(1991) 353.
15. B. A. Philips, A. G. Norman, T. Y. Seong, S. Mahajan, G. R. Booker, M. Skowrowski, J. P. Harbison and V. G. Keramidas, J. Cryst. Growth 40(1994)

249.

16. S. Matsumura, N. Kuwano and K. Oki, Jpn. J. Appl. Phys. 29(1990) 688.

17. S. Matsumura, K. Takano, N. Kuwano and K. Oki, J. Cryst. Growth 115(1991) 194.

18. N. Kuwano and K. Oki, J. Cryst. Growth 163(1996) 122.

19. H. Nakayama, M. Tochigi, H. Maeda and T. Nishino, J. Cryst. Growth 150(1995) 168.

20. R. Osòrio, J. E. Bernard, S. Froyen and A. Zunger, Phys. Rev. B45(1992) 11173.

21. K. Binder and D. W. Heerman, in Monte Carlo Simulation in Statistical Physics, An Introduction, Springer, Berlin, 1988.

22. K. Kawasaki, in Phase Transition and Critical Phenomena, C. Dom and M. G. Green (eds) Academic Press, NY, vol.2.

23. M. Ishimaru, S. Matsumura, N. Kuwano and K. Oki, Phys. Rev. B52(1995) 5154.

24. P. Bellon, J. P. Chevalier, E. Augande, J. André and G. P. Martin, J. Appl. Phys. 66(1989) 2388.

25. M. Ishimaru, S. Matsumura, N. Kuwano and K. Oki, J. Cryst. Growth 128(1993) 499.

26. M. Ishimaru, S. Matsumura, N. Kuwano and K. Oki, Phys. Rev. B51(1995) 9707.

27. M. Ishimaru, S. Matsumura, N. Kuwano and K. Oki, J. Appl. Phys. 77(1995) 2370.

28. O. Ueda, Y. Nakata and T. Fujii, Appl. Phys. Lett. 58(1991) 705.

29. O. Ueda, Y. Nakata and T. Fujii, J. Cryst. Growth 115(1991) 375, 504.

30. H. R. Jen, M. J. Cherng, M. J. Jou and G. B. Stringfellow, in Ternary and Multinary Compounds, S. K. Deb and A. Zunger (eds), MRS, Pittsburgh, 1987, p.353.

31. T. S. Kuan, T. F. Kuech, W. I. Wang and E. L. Wilkie, in Ternary and Multinary Compounds, S. K. Deb and A. Zunger (eds), MRS, Pittsburgh, 1987, p.325.

32. M. Ishimaru, S. Matsumura, N. Kuwano and K. Oki, Phys. Rev. B54(1996) (in presss).

33. M. Ishimaru, S. Matsumura, N. Kuwano and K. Oki, in Computer Aided Innovation of New Materials II, M. Doyama, J. Kihara, M. Tanaka and R. Yamamoto (eds), Elsevier Sci. Pub. B.V., 1993, p.367.

34. O. Ueda, S. Isozumi and S. Komiya, Jpn. J. Appl. Phys. 23(1984) L241.

35. A. Gomyo, K. Makita, I. Hino and T. Suzuki, Phys. Rev. Lett. 72(1994) 673.

36. D. Shindo, A. Gomyo, J-M. Zuo and J. C. H. Spence, J. Electr. Micros. 45(1996) 99.

37. Y. Kangawa, N. Kuwano and K. Oki, Proc. MRS-J Symposium R; Novel Semiconducting Materials, Makuhari Messe, H. Oyanagi (ed), MRS-Japan, 1996, p.26.

Advances in the Understanding of Crystal Growth Mechanisms
T. Nishinaga, K. Nishioka, J. Harada, A. Sasaki and H. Takei (Editors)
© 1997 Elsevier Science B.V. All rights reserved.

Control of rotational twins on heterointerface of fluoride and semiconductors

K. Tsutsui [a], S.Ohmi [a], K. Kawasaki [a] and N. S. Sokolov [b]

[a]Interdisciplinary Graduate School of Science and Engineering, Tokyo Institute of Technology,
4259 Nagatsuta, Midoriku, Yokohama 226, Japan

[b]Ioffe Physico-Technical Institute, Russian Academy of Science, 194021 St. Petersburg, Russia

Rotational-twin-free CaF_2 films (type A) can be grown on Si(111) by the 2-step growth method where the first thin layer is grown at a temperature lower than 300°C and the succeeding layer is grown at 600-750°C. It was found that more than 8 ML of the initial layer was necessary to obtain the type A film. The epitaxial relation of an initial layer thinner than 4 ML was found to rotate from type A to type B with respect to the Si substrate upon increasing the temperature from 200°C to 600°C. A model is proposed wherein the type A interface is in a metastable state and can transform to type B if the initial layer is thin enough to overcome the potential barrier of the transition. Photoluminescence study revealed that the type A CaF_2 films have less strain and higher structural defect density than type B films. Atomic structures of type A and type B interfaces are proposed on the basis of X-ray CTR analysis.

1. INTRODUCTION

Heteroepitaxial growth of group-IIa fluorides and silicides which have the fluorite structure on semiconductors of cubic structure has been investigated extensively. Growth on (111)-oriented substrates is preferable because the surface energy of these materials is extremely low on the (111) surface. However, in the case of (111)-oriented heteroepitaxy of fluoride on Si, there is a problem of the generation of rotational twins [1]. The epitaxial relation in which the crystallographic orientation of the grown film is the same as that of the substrate is referred to as "type A," while that in which the orientation of the grown film is rotated 180°C around the ⟨111⟩ surface normal axis, that is, the rotational twin, is referred to as "type B," as shown in Fig. 1. Both types of epitaxial relations have often been observed in many combinations of fluoride and semiconductor materials [2]. However, what determines which type appears has not yet been clarified. Coexistence of the two epitaxial relations in the epitaxial film, which often occurs [3], is undesirable. Even a uniform epitaxial film of type B is also undesirable because type B growth intro-

(a)

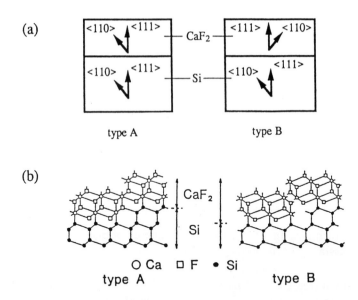

type A   type B

(b)

O Ca   □ F   • Si

type A   type B

Figure 1. (a) Epitaxial relations of type A and type B in the CaF_2/Si(111) heterostructures. (b) Atomic arrangements for type A and type B at CaF_2/Si(111) interface.

duces defects in the epitaxial film at steps on the substrate, as shown in Fig. 1(b). Hence, it is necessary to control the epitaxial relation of fluoride on Si(111) so that it is uniformly type A; that is, rotational twins should be suppressed at the interface. Among group IIa fluorides, CaF_2 is considered to be a good candidate for growth on Si because its lattice constant is close to that of Si. However, CaF_2 on Si(111) always shows type B growth when the conventional molecular beam epitaxy (MBE) method is used. Cho et al. reported the growth of type A structure on Si substrates [4]. They obtained the type A structure by increasing the substrate temperature gradually from 100°C to 600°C during the growth and then maintaining it at 600°C until the end of growth. However, the growth condition dependence and the reason why this method is so effective have not been studied sufficiently.

In this work, we propose a two-step growth method in which type A CaF_2 is grown on Si(111), and discuss the effects of the initial layer at an atomic layer level. We suggest a model wherein the type A interface is in a quasi-stable state compared to the type B interface. In addition, the interface structures are studied.

2. CONTROL OF ROTATIONAL TWINS ON CaF_2/Si(111) BY 2-STEP GROWTH METHOD

2.1. Experimental procedure

A CaF_2/Si heterostructure was grown by MBE. Si(111) wafers were used as substrates. These substrates were chemically cleaned and loaded into the MBE system. Prior to growth, flash heating was carried out at 900°C for 30min to obtain a clean surface. CaF_2 films were grown by the 2-step growth method as well as by the conventional single-step growth method. In the 2-step growth method, the first layer was grown at a low temperature and the succeeding layer, whose thickness was 300-500 nm, was grown at 600°C. The growth temperature and the thickness of the initial layer were varied in the range from 100°C to 600°C and 0.22 nm to 25 nm, respectively. The growth rates were 1 nm/min for the initial layer and 4 nm/min for the succeeding layer. The epitaxial relation between the Si substrate and CaF_2 film was observed by X-ray diffraction using the (115) plane. After the (333) diffraction peak of CaF_2 film was obtained, θ was scanned around the $\langle 110 \rangle$ axis with a fixed 2θ. Since the (115) diffraction should be observed at either small θ or large θ depending on crystal orientation, the epitaxial relation can be observed from relative positions of the (115) peaks of the Si substrate and CaF_2 film on the θ axis.

2.2. Effect of growth temperature of the initial layer

The growth temperature of the initial layer was varied when its thickness was fixed at 25 nm. The observed θ-rocking curves are shown in Fig. 2 [5]. In the case of temperatures lower than 300°C, the (115) peaks of both CaF_2 and Si are observed in the same region (small θ), which means that CaF_2 grew with type A structure. On the other hand, in the case of temperatures higher than 400°C, the (115) peaks of CaF_2 are observed in the opposite region (large θ), which means that the CaF_2 grew with type B structure. If type A and type B domains coexist in the CaF_2 film, (115) peaks appear in both regions. Figure 2 shows no evidence of coexistence. Since the initial layer grown at 200°C was confirmed to have type A structure by observation of the electron channeling pattern, it was considered that the whole film grew with type A from the interface for the case of initial growth temperature of 200°C and 300°C.

2.3. Effect of thickness of the initial layer

If the rotational twin cannot be generated during CaF_2 growth, in other words, the epitaxial relation dose not change in the film, the orientation of the film should be determined at the initial growth stage. Therefore, the first monolayer of CaF_2 should determine the epitaxial orientation of the film, depending on the initial growth temperature. We count one monolayer (ML) of CaF_2 as one unit of the F-Ca-F triple atomic layer. Figure 3(a) shows X-ray rocking curves for various thicknesses of the initial growth layer, where the initial growth temperature was 200°C. From this figure, it can be seen that the type B diffraction disappears and that only type A is observed when the initial layer is more than 8 ML thick. On the other hand, it is found that only type B diffraction is observed when the initial layer

206

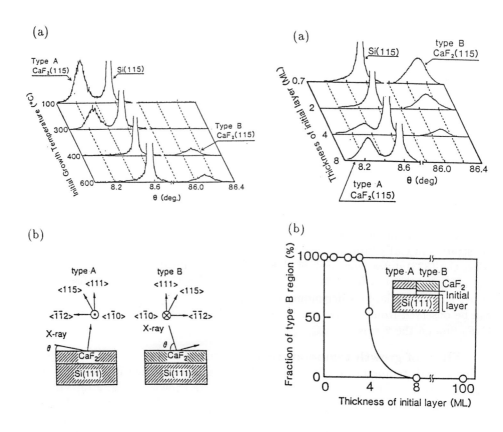

Figure 2. (a) X-ray rocking curves for CaF$_2$ films grown by the 2-step growth method in which growth temperature of the initial layer was varied from 100°C to 600°C. (b) Setup for the measurement.

Figure 3. (a) X-ray rocking curves for CaF$_2$ films grown by the 2-step growth method in which thickness of the initial layer was varied. (b) Thickness of the initial layer vs amount of type B region in the CaF$_2$ film grown by the 2-step growth method.

is less than 2 ML thick. Figure 3(b) shows the relationship between the thickness of the initial layer and amount of type B region. The mixture ratio is evaluated by comparing the area under each (115) peak [6]. Based on these results, it can be concluded that the initial layer should be thicker than 8 ML in the 2-step growth method in order to obtain type A, that is, rotational-twin-free, CaF$_2$ films.

This result is different from the previous prediction. Therefore, the behavior of the 2-3 ML-thick initial layer was investigated by electron diffraction, Rutherford backscattering and X-ray diffraction [7], and the results are summarized as follows.

(1) The initial layer grown at 200°C showed a smooth surface at an atomic level and covered the Si surface uniformly. (2) The initial layer grown at 200°C was type A. (3) After the initial layer was annealed at 600°C in vacuum, its thickness and uniformity were preserved. (4) The initial layer after annealing was type B. From these results, it was concluded that an initial layer with thickness less than 4 ML was rotated from type A to type B structure during the increase of substrate temperature from 200°C to 600°C.

In order to consider this phenomenon, the structures shown in Fig. 4 were grown. It was found that the epitaxial relation of the structure in Fig. 4(a) was type A, whereas that of the structure in Fig. 4(b) was type B. This implies that the type B

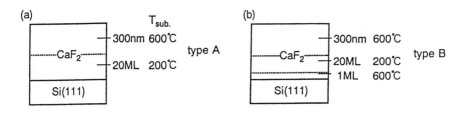

Figure 4. Schematic diagram of growth sequence indicating thickness and temperature of each layer and observed epitaxial relations.

CaF$_2$ layer is caused only by the first 1 ML, and that rotation from type B to type A structure does not occur during the decrease of substrate temperature from 600°C to 200°C. However, as we discussed above, the thin layer grown at 200°C tends to rotate to type B structure upon heating to 600°C. Based on these results, we propose the relationship between interface energy and epitaxial relation shown in Fig. 5. That is, the type B interface is a final stable state while that of type A is a metastable state. For the transition from type A to type B structure induced by incleasing the temperature, there is a potential barrier which corresponds to the energy of rearrangement of atomic sites in the CaF$_2$, and the potential barrier becomes higher as the thickness increases. Thus, the potential barrier can be overcome so that transitions from type A to type B structure occur during heating from 200°C to 600°C if the grown film is thin, while the type A structure is maintained, even after the temperature has been elevated, if the grown film is thick. This transition can be realized by rotation of F ions around the Ca ions by 60° in the surface plane or by site exchange between the upper-layer F and lower-layer F, as shown in Fig. 6. It is considered that such transitions can occur more easily in a two-dimensional thin layer than in a three-dimensional thick layer.

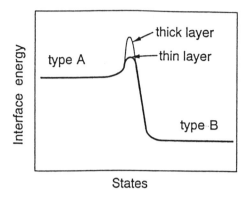

Figure 5. Potential diagram representing the concept of transition between type A and type B structures.

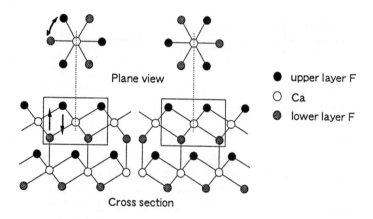

Figure 6. Schematics of atomic rotation and exchange for the transition between type A and type B structure.

3. STRUCTURE OF FILMS AND INTERFACES

3.1. Properties of CaF_2 films characterized by photoluminescence

Photoluminescence of Sm^{2+}-doped CaF_2 revealed the difference in residual strain and crystallinity between type A and type B films [8,9]. Figure 7 shows spectra of type B $CaF_2:Sm^{2+}$ films (a and b) and that of type A film (c). Peak heights are normalized to equal values in this figure since relative comparison of peak intensity is difficult, and the dashed line represents the peak from bulk CaF_2. The film of

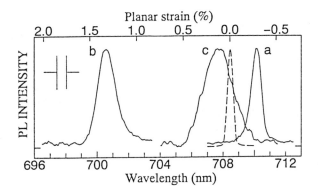

Figure 7. Photoluminescence spectra at 30K of CaF$_2$:Sm^{2+} layers on Si(111): (a) 15 nm-thick layer grown by the 2-step method; 2 ML at 770°C followed by growth at 200 °C (type B), (b) 15 nm-thick layer grown at 770°C (type B), (c) 10 nm-thick layer grown at 200 °C (type A). The dashed line is for the bulk CaF$_2$.

curve (a) in this figure was grown with decreasing temperature from the initial 770°C to 200°C, so as to achieve pseudomorphic growth to the Si substrate. The shift of the peak from the bulk reference shows 0.3 % compressive strain of the CaF$_2$ layer, which corresponds to lattice mismatch between CaF$_2$ and Si at 30 K. In the case of curve (b), CaF$_2$ film grown at a constant temparature of 770°C was relaxed at the high temperature, thus, tensile strain which corresponds to the difference in thermal expansion coefficients was observed as peak shift in the opposite direction. In comparison to these type B films, type A film grown at 200°C (c) showed smaller peak shift and a broader spectrum, which means that this film has less strain and degraded crystallinity. From these results, it is inferred that crystallinity of type A CaF$_2$ is inferior to that of type B CaF$_2$, and that strain generated in the film is relaxed due to the higher density of defects. From the viewpoint of the model shown in Fig. 1, the density of defects generated at steps on the substrate surface should be decreased. However, it is considered that the effect of other defects resulting from low-temperature growth becomes more significant than the effect of suppressing of defects from the type B interface.

3.2. Interface structure observed by X-ray CTR

Type A and type B interface structures were analyzed by the X-ray crystal truncation rod (X-ray CTR) method. The thickness of both type A and type B CaF$_2$ was 9 ML. The type A film was grown at 200°C, and the type B film was grown with the initial 2 ML growth at 740°C followed by a successive growth of 7 ML at 200°C. Figure 8 shows the observed intensity profiles for CTR scattering near the 111 Bragg point, and fitted curves calculated from the structure parameters of Si and CaF$_2$ and fitting parameters of d, the spacing at the interface, and Q, the amount of contribution from the CaF_2 film, which is referred to as crystalline quality of the film. The structure parameters were defined considering the strain observed in the photoluminescence spectra in this calculation. If all the CaF_2 deposited on the Si substrate is coherently aligned without any displacement, Q will be 1; otherwise it will be less than 1. It was found that, for the type A sample, normalized distance at the interface was $d/a_o = 1.2 \pm 0.05$ where a_o is the interlayer spacing between Si(111) planes and the value of Q was 0.4. On the other hand, for the type B sample, $d/a_o = 1.46 \pm 0.02$ and $Q = 0.9$ were obtained.

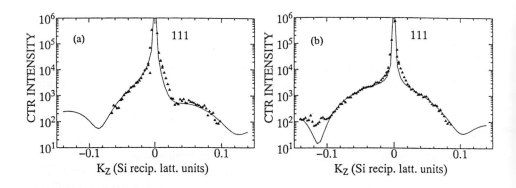

Figure 8. Integrated intensity of CTR scattering obtained for (a) type A CaF_2 grown at 200°C and (b) type B CaF_2 grown at 740-200°C.

The lower value of Q for the type A sample than that for the type B sample shows degraded crystallinity, which is consistent with the result obtained from the photoluminescence. The normalized distance at the interface for the type B sample is close to the value reported by Lucas and coworkers [10,11] who suggested the interface structure shown in Fig. 9(a). Thus, our type B sample is considered to have the same structure. Further studies of the type A sample by analysis of 220 CTR scattering were carried out in order to obtain information about Ca occupation sites in the lateral direction. As a result, it was speculated that Ca atoms of the first layer occupy T-sites on the Si(111) surface, as shown in Fig. 9(b). It is known that one atomic layer of F is desorbed from the interface between CaF_2 and Si when CaF_2 is grown at high temperature or when the epitaxial film is postannealed. As shown in Fig. 9, the type B structure whose interface is formed at high temperature corresponds to the F desorbed structure, whereas the type A structure whose interface is formed at low temperature preserves the interface F layer. Thus it is speculated that the epitaxial relation is related to the composition of the first monolayer.

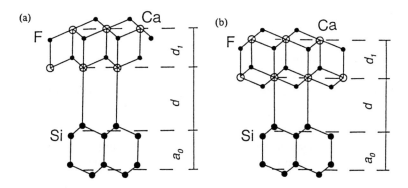

Figure 9. Schematic representation of the interface between CaF$_2$ and Si(111) inferred from X-ray CTR for (a) type B and (b) type A structures.

4. CONCLUSIONS

The heteroepitaxy of CaF$_2$/Si was taken as a typical case in which rotational twins are always generated in the conventional epitaxial method, and the control method of twin generation and its mechanism were investigated. The generation of the rotational twin at the heterointerface is equivalent to growth of CaF$_2$ with the epitaxial relation of type B.

The 2-step growth method in which the initial growth of CaF$_2$ is carried out at a temperature lower than 300°C is effective for obtaining the type A epitaxial relation which is the rotational-twin-free structure. It was also found that the initial layer should be thicker than 8 ML in order to obtain type A films, otherwise type B domains were generated even if the 2-step growth was employed. This phenomenon is explained as follows. The epitaxial relation is determined during the growth of the first monolayer of CaF$_2$: type A for low-temperature growth and type B for high-temperature growth. However, the type A interface is metastable compared to the type B interface, and the type A initial layer transforms into a type B layer as temperature increases if it is thin.

Photoluminescence and X-ray CTR studies revealed that the crystallinity of the type A CaF$_2$ was inferior to that of the type B film. Some irregularity generated by interface formation at low temperature might remain, which degrades the succeeding growth. It will be possible to improve the crystallinity of type A film by optimization of the growth process. Analysis of the interface structure by the X-ray CTR method showed that Ca atoms occupy T-site on the Si(111) surface, and stoichiometry of CaF$_2$ was preserved from the first monolayer in the type A film, whereas one atomic layer of F was desorbed from the interface in the type B film.

5. ACKNOWLEGEMENT

The authors are very grateful to the late Professor S. Furukawa for fruitful discussion and encouragement throughout this work. They also thank Professor J. Harada of X-ray Research Laboratory, Rigaku Corporation, Dr. I. Takahashi of Kwansei Gakuin University and Dr. Y. Itoh for help with X-ray CTR observation.

REFERENCES

1. H. Mizukami, K. Tsutsui and S. Furukawa, Jpn. J. Appl. Phys. 30 (1991) 3349.
2. K. Tsutsui, H. Ishiwara, T. Asano and S. Furukawa, Mater. Res. Soc. Symp. Proc. 27 (1985) 93.
3. T. Asano, H. Ishiwara and N. Kaifu, Jpn. J. Appl. Phys. 22 (1983) 1474.
4. C. C. Cho, H. Y. Liu, B. E. Gnade, T. S. Kim and Y. Nishioka, J. Vac. Sci. and Technol. 10 (1992) 770.
5. H. Mizukami, A. Ono, K. Tsutsui and S. Furukawa, Mater. Res. Soc. Symp. Proc. 237 (1992) 505.
6. A. Ono, K. Tsutsui and S. Furukawa, Jpn. J. Appl. Phys. 31 (1992) 3812.
7. S. Ohmi, K. Tsutsui and S. Furukawa, Jpn. J. Appl. Phys. 3 (1994) 1121.
8. N. S. Sokolov, J. C. Alvarez and N. L. Yakovlev, Appl. Surf. Sci. 60/61 (1992) 421.
9. N. S. Sokolov, T. Hirai, K. Kawasaki, K. Tsutsui, S. Furukawa, I. Takahashi, Y. Itho and J. Harada, Jpn. J. Appl. Phys. 33 (1994) 2395.
10. C. A. Lucas and D. Loretto, Appl. Phys. Lett. 60 (1992) 2071.
11. C. A. Lucas, G. C. L. Wong and D. Loretto, Phys. Rev. Lett. 70 (1993) 1826.

Advances in the Understanding of Crystal Growth Mechanisms
T. Nishinaga, K. Nishioka, J. Harada, A. Sasaki and H. Takei (Editors)
© 1997 Elsevier Science B.V. All rights reserved.   213

Growth of Ag crystallites on Mo(110) substrate observed by "in-situ" SEM

Y.Gotoh[a] , A.Horii[a], H.Kawanowa[a], M.Kamei[a], H.Yumoto[a], T.Gonda[b]

[a]Department of Materials, Science and Technology, Science University of Tokyo, Noda, Chiba 278, Japan

[b]Division of General Education, Aichi Gakuin University, Nisshin, Aichi 470-01, Japan

We have studied the growth of vapor-deposited Ag crystallites on a Mo(110) substrate heated at 400 ℃ using a method of "in-situ" observation by scanning electron microscopy (SEM). The shape of almost all Ag crystallites was a quasi-Wulff cubo-octahedron truncated by (100) planes, but a small number of multiply twinned particles of a regular icosahedron were observed. (111) Trigonal bipyramids truncated by (100) planes were observed in rare cases. Moreover, it was revealed that temporary morphological change to round shape and crystallite rotation took place during the growth.

1. INTRODUCTION

Studies on the initial stage of growth of thin films have been carried out widely using several techniques such as transmission electron microscopy, low-energy electron diffraction, reflection high-energy electron diffraction (RHEED), Auger electron spectroscopy, and thermal desorption spectroscopy. During the formation of thin films, three kinds of growth modes are observed: Volmer–Weber, Frank–van der Merwe, and Stranski–Krastanov growth modes[1,2].
The growth of Ag crystallites deposited on a Mo(110) substrate takes place in accordance with the Frank–van der Merwe growth mode at room–temperature condensation and with the Stranski–Krastanov growth mode at high–temperature condensation from 400 to 600 ℃ [3,4]. The Ag crystallites grown in the Stranski–Krastanov mode were observed by SEM. Ag crystallites having polyhedral shape grow epitaxially on the substrate surface.
In order to investigate the growth process of such polyhedral Ag crystallites, an "in-situ" SEM observation during the growth is useful. The shape of the Ag crystallites is, in many cases, a cubo-octahedron truncated by the (111) crystallographic plane parallel to the substrate surface. However, a small number of crystallites have the shape of a cubo-octahedron with the (100) upper plane, and multiply twinned particles [5,6] are also observed.
In the present study, special growth processes are described, in which a temporary

morphological change to round shape on the substrate, rotation of the Ag crystallite on the substrate and formation of singly twinned particles with bipyramidal shape occur.

2. EXPERIMENTAL PROCEDURE

A Hitachi S-4000 SEM with an electron gun of the cold field emission type was used in the present study. The resolution of the SEM is 15 Å. The total pressure of the specimen chamber was of the order of 10^{-6} Torr during the deposition. A Knudsen–type evaporating cell was built into the specimen chamber of the SEM. The vapor flux was controlled by an electric current passing through a Mo heater in the Knudsen cell. Normally, a vapor flux of 1.5 Å/min was used. The substrate was a Mo(110) single crystal with a size of 3 mm × 15 mm × 0.2 mm. The crystallographic orientation of the Mo single crystal was determined by the Laue method. The deviation from the (110) plane was less than 1°. The Mo surface was cleaned by heating at 1400 ℃ in a different vacuum apparatus, and then the Mo crystal was reheated in the SEM just before Ag deposition. The normal of the Mo substrate was inclined +30° for the electron beam and −60° for the Ag vapor flux. Therefore, the SEM image during "in-situ" observation is shortened in the longitudinal direction by a factor of 0.866. The substrate temperature was measured using a Pt–PtRh thermocouple attached to the Mo substrate and a calibration curve of the temperature versus direct current was obtained prior to the experiment.

In order to examine the electron beam effect on the growth of Ag crystallites on the Mo substrate, irradiated region and a nonirradiated region were compared. No difference in growth features, such as nucleation density and crystal shape was recognized in the two regions at substrate temperatures higher than 300 ℃. Therefore, in the present experiment, electron beam irradiation at a substrate temperature of 400 ℃ negligibly affects the growth of Ag crystallites on the Mo(110) substrate.

3. EXPERIMENTAL RESULTS AND DISCUSSION

3.1. General form of Ag crystallites

After opening the shutter between the evaporation source and the substrate, the substrate was observed by SEM. Small particles were nucleated nearly uniformly on the substrate after a few seconds of deposition, and one crystallite was chosen for successive observation of crystal growth.

The Ag crystallites are polyhedra with triangular or hexagonal shape, which are composed of (111) and (100) faces. In many cases, they exhibit the crystal form of a cubo-octahedron truncated by the (111) plane on the substrate, which we hereafter call a Wulff polyhedron. These crystallites have the epitaxial orientation relationship of [1$\bar{1}$0] Ag // [1$\bar{1}$1] Mo and (111) Ag // (110) Mo. A few crystallites have the relationship of [1$\bar{1}$0] Ag // [1$\bar{1}$1] Mo and (100) Ag // (110) Mo.

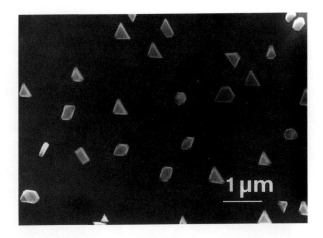

Figure 1. SEM image of as-deposited Ag crystallites on the Mo (110) substrate. The mean deposition thickness of Ag crystallites is 180Å, and the substrate temperature is 400 ℃.

Figure 1 shows the SEM image of as-deposited Ag crystallites on the Mo (110) substrate of 400 ℃. The mean deposition thickness of Ag crystallites is 180 Å and the deposition rate is 1.5 Å/min. Most of the crystallites are triangular or hexagonal with the epitaxial relationship of (111) Ag // (110) Mo. At the left-hand side of the image, a square crystal which shows the relationship of (100) Ag // (110) Mo is observed. Such square crystals were observed throughout a wide area.

Figure 2 shows the morphological change of Ag crystallites during the growth. The mean thickness of deposition varies from 13 to 270 Å. As the deposition thickness increases, the crystallites with nearly constant shapes grow on the substrate, but the growth does not occur isotropically. The crystallographic (111) plane develops, indicating that the growth rate of neighboring $\{1\bar{1}1\}$ and $\{100\}$ planes is higher than that of the (111) plane, but inverse growth takes place; that is, neighboring planes develop. Thus, the crystallites increase in width and thickness by changing the growth rate of each $(1\bar{1}1)$ and (100) plane. Therefore, the Ag crystallites observed in the present experiment do not have the exact Wulff polyhedron or equilibrium shape, but "the growth form".

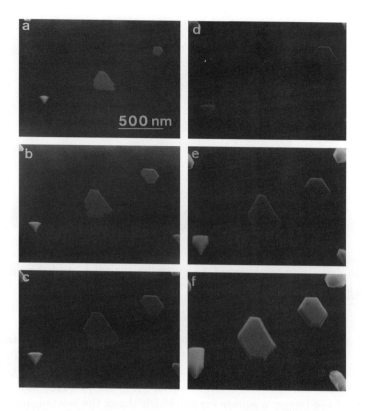

Figure 2. Morphological change of Ag crystallites during the growth. The mean deposition thickness is (a) 13Å, (b) 22Å, (c) 40Å, (d) 66Å, (e) 113Å and (f) 270Å.

3.2. Temporary morphological change to round shape

A series of micrographs of the growth process of a Ag crystal is shown in Figure 3. In Figure 3b, the lateral $(1\bar{1}1)$ and (100) facets of the hexagonal plate are vanished and the upper (111) plane also becomes rounded near the edges. The shape of the plate is rounded. In Figure 3c, the lateral and upper surfaces are reformed into a hexagonal shape. The new polyhedral crystal has smaller area of the upper (111) facet and larger areas of the lateral (100) and $(1\bar{1}1)$ facets compared with the former crystal. As shown in Figures 3d and 3e, the upper (111) plane begins to elongate to the right. The growth rate of the facets is not uniform. The crystal initially grows preferentially in lateral direction, and then, the preferred growth gradually changes to direction of the thickness. In Figures 3a and 3d, the edge directions are slightly different. This means that the orientation relationship between the crystal and the substrate is adjusted. As described above, the epitaxial relationships are $[1\bar{1}0]$ Ag

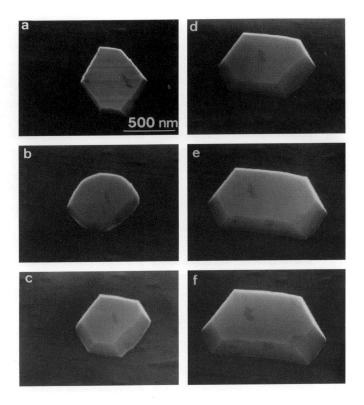

Figure 3. Temporary morphological change of a faceted Ag crystal during the growth. The mean deposition thickness is (a) 15Å, (b) 30Å, (c) 45Å, (d) 75Å, (e) 120Å and (f) 165Å.

// [1$\bar{1}$1] Mo and (111) Ag // (110) Mo. However, a few crystals do not have these relationships and are rotated slightly. This suggests that the growing crystals with a metastable orientation break into round shapes in order to ultimately attain the most stable epitaxial orientation by surface diffusion.

3.3. Rotation of the Ag crystals on the substrate

Figure 4 shows successive image of a growing Ag crystal with rotation. The crystal has the epitaxial relationship of (100) Ag // (110) Mo. The observation was started from the deposition thickness of 10 Å. As shown in Figure 4a, the crystal grows first as a Wulff polyhedron with the (100) upper plane. In Figure 4b, the diameter of the crystal increases without change of the external shape, that is, the crystal grows nearly isotropically. As shown in Figure 4c, the (1$\bar{1}$1) plane appears at the upper left-hand side of the (100) plane. The shape of the (1$\bar{1}$1)

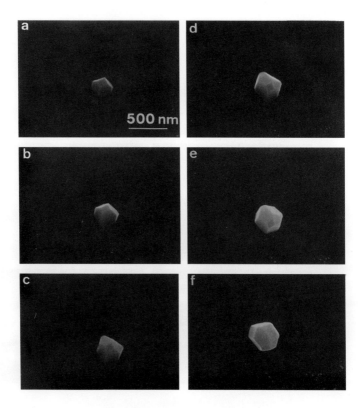

Figure 4. Successive micrographs of a growing Ag crystal with rotation. The mean deposition thickness is (a) 10Å, (b) 31Å, (c) 47Å, (d) 53Å, (e) 60Å and (f) 77Å.

facet is distinctly hexagonal (Figure 4d). The square (100) facet gradually moves downward. The Ag crystal continues to rotate during the growth (Figure 4e), and finally the (111) plane is supplanted by the ($\bar{1}\bar{1}\bar{1}$) plane (Figure 4f).

The rotation seems to cccur due to increasing adhesive force between the crystal and the substrate. The binding energy of the (100) plane to the substrate is small compared with that of the (111) plane, so that the adhesive plane of the crystal shifts from the (100) plane to the (111) plane, which is stable in combination with the substrate. Métois, Heinemann and Poppa[7] reported that deposited crystallites moved on the substrate. In the present study, it was found that the crystal moved by rotation.

Figure 5 shows a series of successive micrographs of a growing Ag crystal under the condition of a substrate temperature of 400 ℃ and a deposition rate of 1.5 Å/min. The change in the crystal shape during the growth is shown schematically in Figure

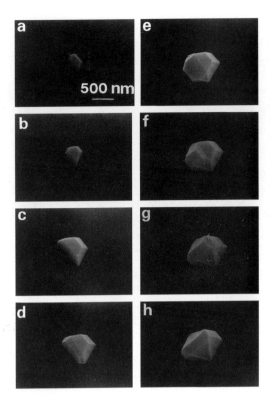

Figure 5. Successive micrographs of a growing Ag crystal. The mean deposition thickness is (a) 14Å, (b) 19Å, (c) 54Å, (d) 75Å, (e) 90Å, (f) 120Å, (g) 138Å and (h) 150Å.

6. Distinct habit planes were observed on the Ag crystal, as shown in Figure 5a. This crystal grew in lateral and thickness directions, but not isotropically (Figures 5b and 5c). The shape of the Ag crystal composed of the {100} and {111} planes is illustrated in Figure 6a, which shows a fundamental Wulff polyhedron truncated by the (111) plane. Unexpectedly, a new crystal having different contrast grows on the (11$\bar{1}$) face of the mother crystal, as shown in Figure 5d. The dark contrasts correspond to the ($\bar{1}\bar{1}$1) and small (1$\bar{1}$1) planes of the new crystal (Figure 6b). We will call the original mother crystal grown epitaxially on the Mo(110) surface crystal A and the new crystal with different contrast crystal B. From the morphological change of the growing crystal, it is found that crystal B itself is surrounded by the {100} and {111} planes. Crystal B grows toward the left with a wide B (001) surface (Figures 5e and 6c).

Crystal B increases in thickness as shown in Figure 5f, and the ($\bar{1}$11) and (1$\bar{1}$1)

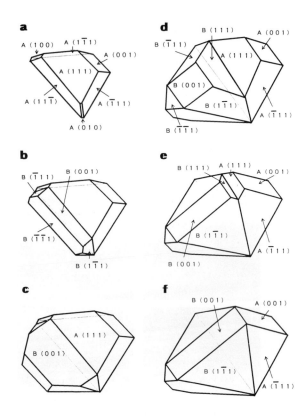

Figure 6. Schematical diagrams of the change in crystal shape during the growth.

facets develop as side faces and the (001) face shrinks (Figure 6d). The total crystal is a symmetrical bicrystal with a twin boundary. The bicrystal has a V–shaped groove at the twin boundary; the angle between the A(111) plane and B(111) facet is calculated to be 141 ° (Figure 5f). The groove is gradually filled (Figures 5g and 6e). As a result, the A(111) and B(111) planes disappear and A(001) and B(001) facets are formed. On the basis of the crystal shape, the growth due to the re-entrant corner mechanism profably takes place. Because the re-entrant corner acts as a step source, the growth rate of the (111) plane is much higher than that of other faces where two-dimensional nucleation is necessary.

After the disappearance of the (111) facets of crystals A and B, A(001) and B(001) facets come into contact with each other (Figure 5h). This is schematically shown in Figure 6f. The angle between A(001) and B(001) facets is calculated geometrically to be 109.5 ° . The grown crystal is a bipyramid with the {111} and {100} planes.

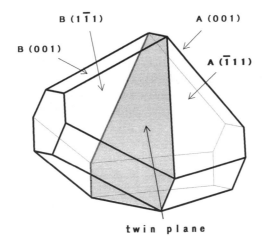

B (1$\bar{1}$1) A (001)

B (001) A ($\bar{1}$11)

twin plane

Figure 7. Ideal shape of the singly twinned particle in equilibrium in free space.

The shape of the singly twinned particle in equilibrium in free space is schematically shown in Figure 7. This is a bi–tetrahedron truncated by the (111) and (100) planes, which has well-developed (111) facets and a (111) twin boundary at the central part. Actually, Ueda [8] found such a singly twinned particle in Ag particles produced by evaporation in inert gas at low pressure in free space.

4. CONCLUSION

The growth process of Ag crystallites on the Mo(110) substrate was investigated by "in-situ" SEM observation. The Ag crystal does not grow isotropically on the Mo substrate. It was found that temporary morphological change to a round shape and rotation of the crystal took place during the growth. The growth process of singly twinned particles was clarified.

ACKNOWLEDGMENT

This work was supported by a Grant-in-Aid for Scientific Research on Priority Areas (no.04227101) from the Ministry of Education, Science and Culture of Japan.

REFERENCES

1. J.W. Matthews, in: Epitaxial Growth, Ed. J.W. Matthews (Academic Press, New York, 1975) p.599
2. R. Kern. G. LeLay and J.J. Métois, in: Current Topics in Materials Science, Ed. E. Kaldis (North-Holland, Amsterdam, 1979) p.131
3. Y.Gotoh and E.Yanokura, Surface. Sci. 269/270 (1992) 707.

4. Y.Gotoh and E.Yanokura, J. Crystal Growth 99 (1990) 588.
5. S.Ino, J. Phys. Soc. Jap. 21 (1996) 346.
6. S.Ino and S.Ogawa, J. Phys. Soc. Jap. 22 (1967) 1365.
7. J.J.Métois, K.Heinemann and H.Poppa, Phil. Mag. 35 (1977) 1413.
8. K.Ueda, Oyo-Butsuri 44 (1975) 611.

Advances in the Understanding of Crystal Growth Mechanisms
T. Nishinaga, K. Nishioka, J. Harada, A. Sasaki and H. Takei (Editors)
© 1997 Elsevier Science B.V. All rights reserved.

Induction time of electrical nucleation in supercooled sodium acetate trihydrate aqueous solution

Y. Iba and T. Ohachi*

Faculty of Engineering, Doshisha University,
Kyoto Tanabe 610-03, Japan

The induction time of the electrical nucleation in a supercooled sodium acetate trihydrate ($NaCH_3COO\cdot 3H_2O$) aqueous solution was measured by using a silver electrode as a function of the electrode potential. The relation between the induction time and the electrode potential was obtained by applying a step DC voltage to the electrode in the range of 0.15~0.5 V versus a saturated calomel electrode. It was found that a linear relationship is established between a logarithm of the induction time and the electrode potential. Experimental results of the induction time was confirmed by the ZFK(Zeldovich-Frenkel-Kashchiev) theory taking into account the relation between the electric field in a diffusion double layer and the electrode potential. When one compares the theory with experimental results, it becomes clear that the induction time of the electrical nucleation is greatly affected by the dissolution rate of silver and the incorporation rate of solute molecules into a nucleus.

1. INTRODUCTION

Atomic scale discussion for nucleation can be started by getting quantitative data on induction time, which is one of the important physical properties in studying nucleation kinetics [1]. It is well-known that the induction time is defined as a period between the time when a system was in a supersaturated or supercooled state and the time when the first nucleus appeared. Although many studies of the induction time have been reported [2–5], only a few have dealt with the influence of the electric field on the induction time [6–9]. Here we present a study of the induction time of electrical nucleation in a supercooled sodium acetate trihydrate ($NaCH_3COO\cdot 3H_2O$) aqueous solution [10–13].

Sodium acetate trihydrate is a very attractive material because its large supercooling is stable and spontaneous nucleation is not observed even at supercooling of about 50 °C or more. In this system, the induction time was defined as the period between the time when the voltage was applied to the electrode and the moment when the first nucleus was detected. It is easy to detect a nucleus because its rapid growth under large supercooling makes it quickly visible to the eye. [10].

---

*Corresponding author.

In industrial fields, sodium acetate trihydrate is useful as a heat storage material because of its high heat of fusion (264 J/g) and suitable melting temperature (58.3 °C) .

The electrical nucleation of sodium acetate trihydrate can be observed by using Cu-Hg electrodes [14] or Ag electrodes [15]. Evidences of the electrical nucleation are the onset of nucleation by the application of potential to the electrode and polarity independence using Cu-Hg electrodes [10]. In the case of the silver electrodes, positive potential is necessary for the electrical nucleation, because the dissolution of silver ions into the solution forms a diffusion double layer in which a large electric field would be created for the electrical nucleation.

A kinetic theory on homogeneous nucleation related to the induction time, which we call the ' ZFK(Zeldovich-Frenkel-Kashchiev) theory ' described by Zeldovich [17], Frenkel [18] and Kashchiev [6], is able to explain the present results in regard to the induction time. In the ZFK theory, Kashchiev has described the influence of the electric field on induction time. Incorporating the relation between the electric field in the diffusion double layer and the electrode potential into the ZFK theory, we have been able to explain the experimental results for the induction time of the electrical nucleation.

2. INDUCTION TIME OF ELECTRICAL NUCLEATION

In his seminal study, ' Kinetik der Phasenbildung ' [16], Volmer has pointed out that electric field (E) makes a square contribution (E^2) upon a driving force of nucleation. This is due to the well-known formula of electrostatic energy w in a system owing to permittivity ϵ i.e. $w = (1/2)\epsilon E^2$. Kashchiev [6] has advanced Volmer's approach to derive nucleation rate or induction time by including the effect of an electric field.

In considering the effect of the electric field in a system under constant temperature and pressure, the free energy change $\Delta G(n)$ required for the formation of a cluster containing n molecules is given by

$$\Delta G(n) = \Delta G_0(n) - \zeta kT E^2 n, \tag{1}$$

where $\Delta G_0(n)$ is the free energy change in the absence of electric field, k is the Boltzmann constant and T is the absolute temperature. The term ζ is a constant given by

$$\zeta = \frac{v_c \epsilon_m}{2kT} \frac{1 - \xi}{2 + \xi}, \quad \xi = \epsilon_c/\epsilon_m, \tag{2}$$

where v_c is the volume per molecule, ϵ_c and ϵ_m are the permittivity of cluster and mother phase, respectively, assuming that a spherical dielectric droplet crystallizes into a spherical nucleus. $\zeta > 0$ or $\xi < 1$ is required for the electrical nucleation, for in this case the induction time t_i is inversely proportional to increases in the electric field E , i.e. electrical nucleation will be promoted.

Kashchiev succeeded in solving Zeldovich-Frenkel's equation in detail [19] and derived the relation between the induction time t_i and the electric field E which is given by [6]

$$t_i = \frac{8\sigma^*}{kT\beta(s_0 + \zeta E^2)^2}, \tag{3}$$

where σ^* is the surface free energy density of a nucleus , β is the number of solute molecules joining the unit area of nucleus per unit time, which we call the 'incorporation rate' . The term s_0 shows the thermodynamic driving force. Although Kashchiev's formulation was for vapor condensation, we could use the following equation as s_0 in the case of the supercooled liquid or melt,

$$s_0 = \frac{L\Delta T}{kTT_m}, \tag{4}$$

where L is the heat of fusion per molecule which will be released at crystallization, T_m is the melting temperature, $\Delta T = T_m - T$ is the supercooling.

In order to obtain the relation between the electric field in the diffusion double layer and the electrode potential, we propose a simple model concerning a one-dimensional concentration distribution of ions in the diffusion double layer. If the concentration distribution of dissolved ions from the electrode is followed by

$$C(x) = C_s \exp(-\frac{x}{\delta}), \tag{5}$$

where $C(x)$ is the concentration of ions at the distance x from the electrode , C_s is the concentration of ions in the solution at the surface of the electrode , δ is the width of the diffusion double layer (assumed not to depend on the electrode potential), the electric field distribution can be determined by solving a well-known one-dimensional Poisson's formula. The largest electric field is obtained at the surface of the electrode E_0, which is assumed to be the most effective for the electrical nucleation , and is given by solving the Poisson's formula taken into account (5) at $x = 0$:

$$E_0 = -\frac{e\delta}{\epsilon_m}C_s, \tag{6}$$

where e is the elementary electric charge. The term C_s must be proportional to the Boltzmann factor which allows for the activation energy to remove an atom from the surface of the electrode to the solution, i.e.,

$$C_s = C_0\nu_0 \exp(-\frac{\Delta G_a}{kT}), \tag{7}$$

where C_0 is the number of atom per unit area of the electrode, ν_0 is the fundamental jump frequency at the surface of the electrode which is the order of kT/h (h is the Planck constant) and ΔG_a is the activation energy necessary to remove an atom from the electrode to the solution [20–22]. The activation energy taken into accout the electrode potential V with respect to the reference electrode is given by

$$\Delta G_a = \Delta G_{eq} - \alpha z eV, \tag{8}$$

where ΔG_{eq} is the activation energy in the equilibrium state, that is, at $V = 0$, α is the transport coefficient ($0 \leq \alpha \leq 1$) and z is the valence of ion [21]. If one combines (7) with (8) , C_s is given by

$$C_s = C_0 \nu_0 \exp\left(-\frac{\Delta G_{eq}}{kT}\right) \exp\left(\frac{\alpha z e}{kT}V\right). \tag{9}$$

If one inserts (9) into (6) , relation between the electric field and the electrode potential can be derived as

$$E_0 = R_p \exp\left(\frac{\alpha z e}{kT}V\right), \tag{10}$$

where $R_p = -(e\delta/\epsilon_m)(C_0\nu_0) \exp(-\Delta G_{eq}/kT)$ is a constant. Finally, we can obtain a relation between the induction time and the electrode potential by inserting (10) into (3) :

$$t_i = \frac{8\sigma^*}{kT\beta\{s_0 + \zeta R_p^2 \exp(\frac{2\alpha z e}{kT}V)\}^2}. \tag{11}$$

3. EXPERIMENTAL

3.1. Preparation of $NaCH_3COO\cdot 3H_2O$ aqueous solution

Sodium acetate trihydrate ($NaCH_3COO\cdot 3H_2O$) aqueous solution was prepared by dissolving sodium acetate ($NaCH_3COO$) in water. Sodium acetate can dissolve into water very easily and immediately turn to sodium acetate trihydrate due to the hydration. The concentration of the solution was controlled by changing a weight percent of sodium acetate in water. A 45 ~ 55 wt.% sodium acetate trihydrate aqueous solution of about 20 ml was used for measurements.

3.2. Three-electrode cell

The electrical nucleation was performed in a three-electrode cell as shown in Figure 1. A silver electrode was used as a working electrode and a platinum electrode was used as a counter electrode. A saturated calomel electrode (SCE) was used as a reference electrode , which was connected to the cell through a salt bridge (mixing a saturated KCl solution with agar powder). All potentials were measured and reported against SCE.

A polycrystalline silver electrode (about 20 mm^2) and single crystals of {110}, {100}, {111} silver electrode (about 14 mm^2) were used for measurements. The surface area of the working electrodes was adjusted by using a Teflon tape. In order to enhance the efficiency of the current flow , we adopted a platinum coil (0.8 mm, 5 turns) as the counter electrode to gain its electrode area. The current flow through the cell is defined as positive if the flow is into the working electrode and out of the counter electrode.

The temperature of the solution was controlled by a hot-plate and measured by a copper-constantan thermocouple. Uniformity of the solution was kept by using a stirrer. The solution was cooled down in 1 °C/min from 10 °C above the melting temperature ($T_m = 58.3$ °C) to 20 °C or 25 °C supercooling, where the experiments of the electrical nucleation were performed.

3.3. Electrical nucleation and induction time

The electrical nucleation was induced by applying a step DC voltage to the working electrode situated in a supercooled sodium acetate trihydrate solution. In-

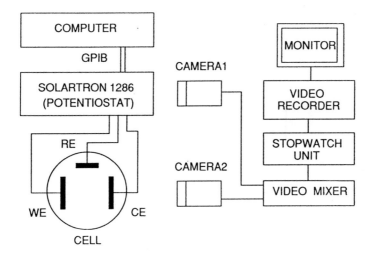

Figure 1. Schematic diagram of experimental setup.

situ observations of experiments were performed by using the video system which is constituted of two video cameras, video mixer, stop-watch unit, two video recorders and monitor. Using two video cameras, we recorded the three-electrode cell and the panel of the electrochemical interface in one frame in the video recorder to observe the nucleation process.

The induction time was measured with the aid of the video system, i.e. a stop-watch display panel was superimposed on the video frame. The moment of the electrical nucleation was easily recognized from the recorded video tape because a nucleus can grow into detectable size (by eye) very rapidly.

A detection limit of about 33 msec was determined by the frame interval of the video recorder. This detection limit is negligible because the measured induction time of the electrical nucleation is basically more than 1.0 sec.

4. RESULTS AND DISCUSSION

4.1. Electrode potential dependence of induction time

Figure 2 shows a relation between the induction time and the electrode potential. Each measurement was performed in a supercooled solution ($\Delta T = 20$ °C) by using a polycrystalline silver electrode. It is clear that the induction time decreases when one increases the electrode potential. This shows that the electric field in the diffusion double layer is closely related to the driving force of the electrical nucleation.

The solid line in Figure 2 denotes the theoretical fitting curve by using (11). The convergence indicates that it is possible to explain the mechanism of the electrical nucleation by using the ZFK theory.

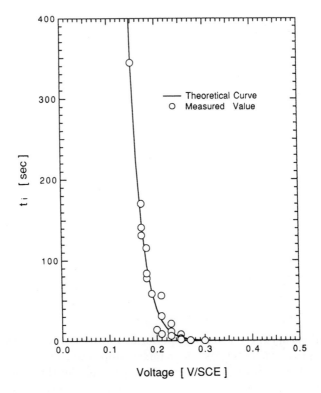

Figure 2. Relation between the induction time of the electrical nucleation and the electrode potential obtained by using a polycrystalline silver electrode in 55 wt.% sodium acetate trihydrate aqueous solution at $\Delta T = 20$ °C.

The best fitting was achieved by setting parameters to the following values : $T = 311.3$ [K], $v_c = 1.35 \times 10^{-27}$ [m^3], $L = 5.98 \times 10^{-20}$ [J], $\epsilon_m = 2\epsilon_0 = 2.66 \times 10^{-11}$ [F/m], $\xi \approx 0.6$, $C_0 = 1.38 \times 10^{19}$ [m^{-2}] using the value for $\{111\}$ of Ag, $\alpha = 0.38$ (from the slope of Figure 3 as mentioned in the next section). Under the assumption of $\sigma^* \approx 0.1$ [J/m^2] and $\Delta G_{eq} \approx 20kT$ [J], we can obtain the value of the incorporation rate β and the width of the diffusion double layer δ from the result of fitting : $\beta = 3.0 \times 10^5$ [m$^{-2} \cdot$ s^{-1}] , $\delta = 2.62 \times 10^{-4}$ [m].

4.2. Logarithm of induction time and electrode potential

The induction time can change widely as shown in Figure 2. Therefore we attempted to plot a logarithm of the induction time $(\ln t_i)$ against the electrode potential as shown in Figure 3. A linear relationship was found between $\ln t_i$ and the electrode potential. This means that there is an exponential relation between the induction time and the electrode potential. This linearity may be interpreted

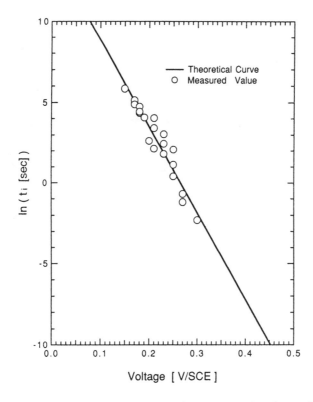

Figure 3. Plot of a logarithm of the induction time versus the electrode potential obtained by using a polycrystalline silver electrode in 55 wt.% sodium acetate trihydrate aqueous solution at $\Delta T = 20\ ^\circ C$.

as follows. In the case of the electrical nucleation, the supercooling is not the dominant driving force of the nucleation because spontaneous nuclei do not appear even in the large supercooling as mentioned before. Therefore, we can eliminate s_0 from (11) because $s_0 \ll \zeta E^2$ is established (for example, typical orders of s_0 and ζE^2 are 10^{-1} and 10^3, respectively). Thus a logarithm of (11) yields the following linear relationship

$$\ln t_i = K_1 - \frac{4\alpha e}{kT}V. \tag{12}$$

where $K_1 = \ln(8\sigma^*/kT\beta\zeta^2 R_p^4)$, $z = 1$ for Ag. The straight line in Figure 3 was obtained by fitting (12) to measured data, resulting in $K_1 = 14.2$ and $4\alpha e/kT = 53.5$. One can see that the theoretical fitting line is in good agreement with measured data. The value of α can be obtained from the slope of the fitting line in Figure 3. In this case, $\alpha = 0.38$, which is similar to generally reported values for silver disso-

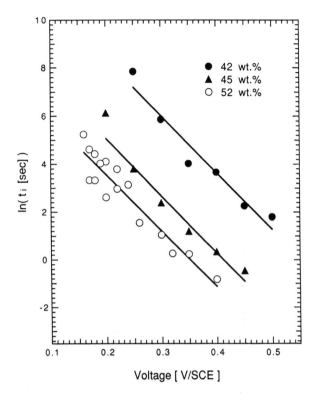

Figure 4. Plots of ln t_i versus electrode potential varying the concentration of the sodium acetate trihydrate aqueous solution at $\Delta T = 25$ °C measured by using a single silver {110} electrode.

lution in cyanide solution [23] $(\alpha = 0.46)$, was obtained. The transport coefficient α $(0 \leq \alpha \leq 1)$ is one of factors to determine a reaction rate of silver dissolution. Therefore, we can say that the slope of the straight line in Figure 3 is related to the kinetic process of the silver dissolution from the electrode. This means that the induction time corresponds to a period required for the establishment of the electric field appropriate for the electrical nucleation by the dissolution of silver.

4.3. Influence of incorporation rate on nucleation kinetics through concentration change

Figure 4 shows a relation between ln t_i versus the electrode potential varying the concentration of the solution. This measurement was performed by using a single crystal of {110} silver electrode of which surface atom density is smallest among {111}, {110} and {100}. It was found that the induction time increases with decreasing the concentration of the solution. However, the slopes of the fitting

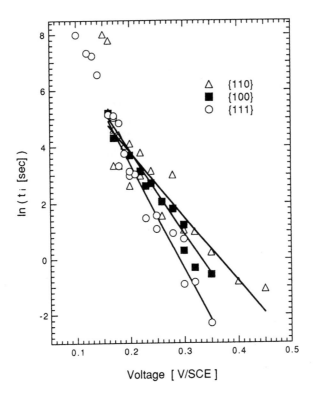

Figure 5. Plots of ln t_i versus electrode potential varying the crystallographic orientation of a single crystal silver electrode in 52 wt.% sodium acetate trihydrate aqueous solution at $\Delta T = 25$ °C.

lines do not change. This experimental result can be interpreted by using (12). Constant slopes mean that the value of $4\alpha e/kT$ is a constant which is related to a coefficient of the electrode potential in (10). This shows that an electric field is constant in spite of the concentration change of the solution. The increase of the induction time corresponds to the increase of K_1 in (12). The value of K_1 is determined by four factors; σ^*, β, ζ and R_p as seen in (11) and (12). σ^* and ζ are constant against the concentration of the solution. R_p is also almost constant because electric field described in (10) is constant. Consequently the incorporation rate β is an operative factor on a change of K_1. In other words, the increase of K_1 corresponds to the decrease of β. Therefore, the reason for the increase in the induction time is interpreted as a decrease in the incorporation rate β resulting from the decrease of the concentration of the solution.

4.4. Effect of crystallographic orientation of electrode

To study effects on nucleation at the atomic level, we investigated the effect of crystallographic orientation of an electrode on kinetics of the electrical nucleation. Figure 5 shows a relation between $\ln t_i$ and electrode potential with respect to {110} , {100} , {111} faces of silver electrode. A change of slopes on a linear relationship was observed in Figure 5 , where its tendency was {111} > {100} > {110} and the extracted values of α were 0.28, 0.21, 0.17 for {111}, {100}, {110}, respectively. We attempted to interpret these results by using (12). The change of slopes corresponds to change of α. The transfer coefficient α is defined as dependency of electrode potential on current , or discharge process through so-called electrical double layer. Therefore α is closely related to the surface free energy of the electrode which determines net activation energy of dissolution. In the case of a face centered cubic (for Ag) , the surface free energy changes with {111} > {100} > {110}. This tendency is parallel to surface density or electronic work function. Moreover Hamelin [24] reported that potential of zero charge (PZC) of the electrode changed with the surface free energy, i.e., the value of PZC is {111} > {100} > {110}. Therefore actual electrode potential increases also with {111} > {100} > {110} because a rest potential (potential of electrode under no external voltage to the electrode) changes with the PZC and the electrode potential is determined against the rest potential. Consequently rate of dissolution is changed as {111} > {100} > {110} , resulting in the change of the induction time : {111} < {100} < {110}.

5. CONCLUSION

Kinetic mechanism of the electrical nucleation of the sodium acetate trihydrate was investigated in terms of the induction time. From the experimental result that the induction time decreases with increasing the electrode potential, we concluded that the electric field is a dominant driving force of the electrical nucleation. Good agreement of the transport coefficient between the value extracted from the experimental curve with the reported value indicates that the dissolution of silver ion plays crucial role in determining the induction time of the electrical nucleation. The result of the induction time which varies with the concentration of the solution can be interpreted by the ZFK theory and it is clear that the induction time of the electrical nucleation is greatly affected by the incorporation rate of the solute molecules into a nucleus.

ACKNOWLEDGEMENT

The authors would like to express sincere gratitude to Professor Ichiro Taniguchi in Doshisha Univ. and Professor Hiroshi Komatsu in Tohoku Univ. for their valuable suggestions and discussions, Mrs. Shuji Kano and Yoseke Hisamori for their experimental help. This research is supported by Grant-in Aid for Scientific Research on Priority Areas "Crystal Growth Mechanism in Atomic Scale"

No.03243102 and 04227101.

REFERENCES

1. J. W. Mullin, *Crystallization*, (Butterworths , 1972) p.174.
2. S. Toschev and I. Markov, J. Crystal Growth 3/4 (1968) 436.
3. K. Wojciechowski and W. Kibalczyc, J. Crystal Growth 76 (1986) 379.
4. H. E. Lundager Madsen, J. Crystal Growth 80 (1987) 371.
5. J. Torrent, R. Rodriguez and J. H. Sluyters, J. Crystal Growth 131 (1993) 115.
6. D. Kashchiev, J. Crystal Growth 13/14 (1972) 128.
7. A. G. Crowther, J. Crystal Growth 13/14 (1972) 291.
8. R. Dhanasekaran and P. Ramasamy, J. Crystal Growth 79 (1986) 993.
9. T. Shichiri and T. Nagata, J. Crystal Growth 54 (1981) 207.
10. T. Ohachi, M. Hamanaka, H. Konda, S. Hayashi, I. Taniguchi, T. Hashimoto and Y. Kotani, J. Crystal Growth 99 (1990) 72.
11. T. Ohachi, Y. Iba and I. Taniguchi, Proc. of the 6th Topical Meeting on Crystal Growth Mechanism (Awara , 20-22 January 1993) ed. T.Nishinaga, p.89.
12. Y. Iba, T. Ohachi and I. Taniguchi, Proc. of the 5th Meeting on Crystal Growth Mechanism (Hokkaido Univ. , June 1993) ed. T.Nishinaga, p.236.
13. Y. Iba, Y. Hisamori, S. Kanou, T. Ohachi and I. Taniguchi, Proc. of the 7th Topical Meeting on Crystal Growth Mechanism (Atagawa , 19-21 January 1994) ed. T.Nishinaga, p.117.
14. Y. Kotani and T. Hashimoto, J. Japan. Assoc. Crystal Growth 9 (1982) 31 (in Japanese).
15. T. Chida and K. Morimoto Proc. of the 4th meeting on Energy and Resource Association (1985) 43 (in Japanese).
16. M. Volmer, *Kinetik der Phasenbildung*, (Steinkopff, Dresden Leipzig, 1939) p.193.
17. J. Zeldovich, J.Exp.Theor.Phys. 12 (1942) 525.
18. J. Frenkel, *Kinetic Theory of Liquids*, (Oxford , 1946) p.366.
19. D. Kashchiev, Sur. Sci. 14 (1969) 209.
20. D. Turnbull, *Solid State Physics*, Vol. 3, (Academic Press , 1956) p.268.
21. K. J. Vetter, *Electrochemical Kinetics*, (Academic Press , 1967) p.115.
22. J. C. Brice, *Crystal Growth Process*, (John Wiley and Sons , 1986) p.72.
23. J. Li and M. E. Wadsworth, J. Electrochem. Soc. 140 (1993) 1921.
24. A. Hamelin , L. Stoicoviciu , L. Doubova and S. Trasatti, Sur. Sci. 201 (1988) L498.

Advances in the Understanding of Crystal Growth Mechanisms
T. Nishinaga, K. Nishioka, J. Harada, A. Sasaki and H. Takei (Editors)
© 1997 Elsevier Science B.V. All rights reserved.

Growth mechanism of smoke particles

C. Kaito[a], S. Kimura[a] and Y. Saito[b]

[a]Department of Physics, Ritsumeikan University,
1916 Noji-cho, Kusatsu-shi, Shiga-ken 525, Japan

[b]Department of Electronics and Information Science, Kyoto Institute of
Technology, Matsugasaki, Sakyo-ku, Kyoto 606, Japan

In this paper we describe the mechanism of the production and growth of ultra-fine particles by the evaporation technique. The first step involved the production of ultrafine particles by this method and the analyses of the morphology of the particles produced and their growth mechanisms. The second step involved clarification of the nucleation and growth processes. Important fields of application of these particles include star formation (planetary science) and mesoscopic physics from the material science point of view.

1. INTRODUCTION

One of the most advanced methods for producing ultrafine particles is the "gas evaporation technique", in which a material is heated in an atmosphere of inert gas. Vapor from the material is subsequently cooled and condensed in the inert gas atmosphere, resulting in a smoke which resembles the flame of a candle. Metallic and compound particles in the smoke, with size of 100 nm order, showed characteristic crystal habits. Results of systematic studies on the structure and morphology of ultrafine metallic particles have been summarized by Uyeda [1,2]. Production of oxide and sulfide particles has been done systematically by us [3,4]. Applications of these particles to planetary science have been done intensively using the advanced gas evaporation method (AGEM) developed by us [5,6]. The particle formation mechanism, size control method and advanced gas evaporation methods which were developed in order to clarify the mechanism of coalescence have been summarized in the book "Kemurino Himitsu" in Japanese [7]. We introduce here the mechanism of the growth of smoke particles and their development in astrophysics and quantum dots.

2. MECHANISM OF PRODUCTION OF SMOKE PARTICLES

Smoke particles were produced using a vacuum chamber developed by us. After evacuating the chamber, inert gas was introduced in the pressure range of 1 ~ 100 Torr. Evaporation was done in a closed system. The practical design of the vapor source is based on the need to concentrate thermal energy at the source

236

Table 1
Temperature at a vapor pressure of 1 Torr for various materials.

Material	Temperature at vapor pressure of 1 Torr ($^\circ$C)	Material	Temperature at vapor pressure of 1 Torr ($^\circ$C)
Te	520	Cu	1617
Fe	1783	MoO_3	737
Au	1867	CdS	1125
Ag	1786	WO_3	1300

Convection

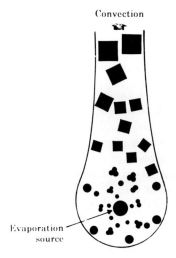

Evaporation source

Figure 1. Schematic diagram of the particle formation process using the gas evaporation technique.

material and to avoid excessive heating of the surroundings, similar to the case of vacuum evaporation. Power is transmitted to the evaporant by several methods, the simplest and most widely used of which is direct resistance heating of a small boat, strip, wire, or basket of a refractory material (W, Mo, Pt, Ta) in which a small amount of evaporant is placed.

The vacuum evaporation method involves the growth of film on a substrate by condensation from the vapor phase. Therefore the interaction between condensate and substrate is important. On the other hand, the gas evaporation technique involves condensation in free space due to the collision with an inert gas, leading to the formation of three-dimensional particles. The particles produced in space are transported by the convection flow produced by the evaporation source and stationary gas flow is observed as smoke with the aid of grown particles. When the vapor pressure of the evaporant exceeds 1 Torr, smoke is observed. Temperature at a vapor pressure of 1 Torr for various materials are shown in Table 1. Heating temperatures that exceed those given in Table 1 are necessary for the production of ultrafine particles. Figure 1 shows a schematic diagram of the particle formation

Table 2
Crystal structure and growth habits of particles.

Structure	Basic external shape	Examples
fcc metal	octahedron	Au,Ag,Ni,Al
bcc metal	cubic, rhombic dodecahedron	Fe,Cr
hcp metal	plate	Mg,Cd,Zn
NaCl type	cubic	MgO,CdO,PbSe,PbTe
layer structure	plate	MoO_3,PbO,SnO,SnS_2
correction of MO_6	octahedron	WO_3,In_2O_3,In_2O_3,SnO_2
II-VI compounds	plate, sphere, tetrapod	ZnO,CdS,CdSe,CdTe

Figure 2. Smoke production apparatus for our experiments.

process. Generated vapor collided with inert gas to produce clusters. These clusters collided with each other to produce ultrafine particles. The particles produced near the evaporation source were spherical. In contrast, particles collected at a distance from the evaporation source showed characteristic crystal habits. The crystal habits of particles due to their crystal structure are summarized in Table 2. Particles generated from metal with the fcc structure are icosahedral, pentagonal or decahedral and their size is larger [2,3] than that observed for particles generated in the initial stage of vacuum evaporation. In II-VI compounds, tetrapod crystals grew from cubic nuclei [8]. These findings may be attributed to the fact that particles were generated in a gas atmosphere with no effects from a substrate.

Our work chamber was a glass cylinder 17 cm in diameter and 30 cm in height, covered with a stainless steel plate and connected to a high-vacuum exhaust through a valve at its bottom, as shown in Figure 2. Electrodes for evaporation were set on the stainless steel plate. The state of the smoke was clearly observed by the naked eye. Temperature distribution was measured using a chromel-alumel thermocouple of 20 μm diameter which was previously developed by us to elucidate the growth of smoke particles. A tungsten V-boat (50 mm long, 1 mm wide and 2 mm high) was used as the evaporation source at an Ar gas pressure of 100 Torr, and the temperature distribution around the heater is shown in Figure 3 [9]. The convection flow of Ar gas, which can be visualized by metallic or oxide smoke of materials evaporated from the heater, is shown at the right of the figure. The size of particles increased with increasing distance of collection from the evaporation source. As can be seen in Figure 3, atmospheric temperature is lower than the

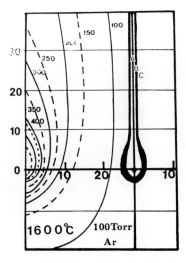

Figure 3. Typical temperature distribution and shape of smoke at an Ar gas pressure of 100 Torr.

source temperature even near the evaporation source. Vapor from the evaporant exists within a few mm of the evaporation source. Therefore the size increment took place without vapor, i.e., coalescence is an important process in the gas evaporation technique. Evidence for, and the mechanism involved in the coalescence of smoke particles have been elucidated in a series of experiments [10–13]. Coalescence takes place within $10^{-3} \sim 10^{-2}$ sec. The coalescence of smoke particles was classified into two stages: surface melting coalescence and liquid-like coalescence [13]. These two stages depend on the particle size and temperature. The morphology, particle size, and crystal structure are controlled by the mass density and the temperature of the smoke, and also by the type of atmospheric gas and gas pressure. If the density of the produced particle is ρ and particles radius is r, the growth rate of the particles can be expressed as

$$\frac{dr}{dt} = \frac{mk}{4\pi\rho},$$

(1)

Table 3
Coalescence temperature of particles

Ultrafine particles	Temp.(°C)	Ultrafine particles	Temp.(°C)
Al	200	MoO_3-Bi_2O_3	350
Fe	250	WO_3-Bi_2O_3	400
MoO_3	200	Fe-Fe_2O_3	350
Fe_2O_3	300	Zn-Te	200
PbO	200	Cu-Ag	300
SnO	200	Fe-Ni	200
SnO_2	200	Fe-S	100
CdS	250		

where m is the mass density expressed as the total mass of particles per unit volume of smoke and k is the coalescence probability factor which depends on the temperature of the particles. Since m and k increase with increasing source temperature and gas pressure, the particle size also becomes large. The coalescence temperature has been used in the growth of film on substrate. In the case of smoke, the coalescence took place in free space, accompanying the collision process between particles. Since the interaction between the film and the substrate is very large, the coalescence in smoke predominantly took place than that in the case of film growth.

The coalescence temperature, which is defined as the temperature below which no coalescence takes place, is an important factor when taking into account the growth and physical properties of smoke particles. Coalescence temperatures for various smoke particles are presented in Table 3.

Growth of ultrafine particles by the gas evaporation technique proceeds as follows.

i)   Material evaporation occurs in inert gas.

ii)   Collision between evaporated vapor and inert gas takes place.

iii) The evaporated vapor cools down rapidly near the evaporation source, because the mean free path of the evaporated vapor becomes less than 1 μm in the gas evaporation technique.

iv) Since the vapor pressure at the evaporation source is more than 1 Torr, many nuclei of the evaporant appear near the evaporation source.

v)   By heating the evaporant, stationary gas flow takes place. All the evaporant and nuclei are transported by this convection flow. Since all the nuclei produced around the evaporation source are transported by convection flow, as shown in Figure 3, the convection flow can be observed as smoke. As shown in Figure 3, the temperature around the heater becomes lower than that of the evaporation source. Therefore vapor is hardly observed at a height exceeding a few mm from the evaporation source. However, particle size increases, i.e., coalescence takes place.

3. A NEW APPROACH FOR DETECTING THE INITIAL STAGE OF NUCLEATION PROCESS AND THE PRODUCTION OF CARBON-COATED QUANTUM DOTS

With increasing interest in meso- or nanophysics, the production of particles less than 10 nm in size becomes important. In the gas evaporation experiment, particle whose size corresponds to that of a perfect three-dimensional quantum dot are hardly produced, and most of the particles form chains due to coalescence. We recently developed a method for producing carbon-coated particles or silicon oxide coated-particles with size ranging from 2 nm to 20 nm, using one of the AGEMs. Since the formation of particles takes place in the gas, in situ observation of particles, similarly to that in the vacuum evaporation technique, could not be accomplished. Thus, the development of experimental equipment for observing the initial stage of particle formation becomes very important. One method of obtaining small particles is to avoid astriction of the evaporant. Experiments under low-gravity conditions have been planned and some experiments have been conducted by a group from Nagoya University in order to eliminate convection. However, with increasing particle size, its speed decreases; therefore, coalescence among particles could not be avoided, as was revealed from our experiment on the production of organic particles [14].

The carbon electrode shown in Figure 4 was used as the evaporation source in the AC arc-discharge evaporation of carbon at an Ar gas pressure of about 13kPa. The end of electrode B was pushed with a spring. When an electric current is passed through electrodes A and B, the temperature of electrode A in contact with electrode B becomes extremely high, and metallic powder placed on the pared-off part of the tip of electrode A can evaporate together with carbon. Therefore evaporation of metal took place in a carbon atmosphere. Figure 5(a) shows an example of Ag particles produced by the ordinary gas evaporation technique from a tungsten heater in Ar gas at 13kPa. Particles connect with each other. Figure 5(b) shows the particles produced by the method shown in Figure 4. Each particle was covered with an amorphous carbon layer which appears as an area of weak contrast surrounding individual particles. Figure 6(a) also shows an example of carbon-coated Ag particles with size of $2 \sim 30$ nm [15]. A similar example of Ag particles covered with a SiO_2 layer is shown in Figure 6(b). Detailed analysis of quantum dots may reveal the initial stage of growth of smoke particles. A particle of ~ 1 nm size near the heater is composed of about 30 atoms. Particles of 20 nm size,

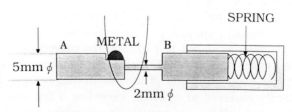

Figure 4. Schematic representation of method of producing metallic clusters.

which is the typical size of fcc metallic particles, accumulated near the heater in the ordinary gas evaporation technique, resulting in the coalescence of about 8000 particles of 1nm size. By further coalescence of particles of 20 nm size in smoke, particle size increases to about 50 nm. In this case, about 16 particles (20 nm) are necessary. Therefore, coalescence frequency becomes high in the process, i.e., the nucleus size to particle growth. In the process of nucleation and growth, a vapor-liquid-solid process may take place. Further studies are under way.

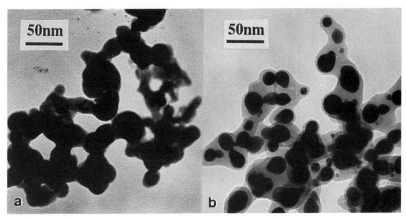

Figure 5. Example of (a) Ag particles and (b) Ag particles covered with amorphous carbon.

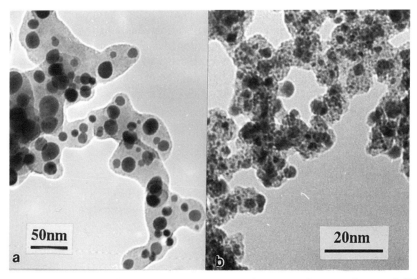

Figure 6. Example of (a) Ag particles covered with amorphous carbon and (b) Ag particles covered with SiO_2.

4. PRODUCTION OF METALLIC COMPOUND PARTICLES

Using stationary convection flow of smoke and coalescence of particles, compound particles were produced. These methods were first used to give corroborative evidence of the coalescence among particles in smoke. A schematic representation of the AGEM is presented in Figure 7. We have reported the systematic investigation of the morphology and growth of metallic oxide particles. Oxide particles such as ZnO and MgO are formed by the oxidation of metallic vapor in a gas mixture. The morphology and structure of some oxide particles showed their dependence on oxygen partial pressure.

When we use a molybdenum or tungsten heater as the evaporation source in a mixture gas, the partial pressure of oxygen in a mixture gas should not exceed 130Pa, otherwise tungsten or molybdenum oxide particles formed on the heater would be evaporated. Many oxide particles were prepared by using a nichrome boat or wire which was stable in an atmosphere of oxygen gas. This heater cannot be used as an evaporation source at temperatures higher than 1200°C. Since the melting point of most metallic oxides and metals is higher than 1200°C, astrophysically interesting particles such as magnetite, alumina and silicate cannot be produced by the conventional resistance heating method. On the other hand, almost all metallic particles can be produced using the tungsten or molybdenum heater. The AGEM methods involve the use of the resistance heating method. The mechanism and application are also shown in Figure 7. Direct evaporation of metallic compounds is not always achieved using the same components of the evaporant, i.e., decomposition took place. In spite of increasing interest in particle production and optical properties of metallic oxides or sulfides, decomposition and the kind of evaporation source are some of the key points in the production of compound particles. As shown in Figures 7 (f) and (g), we proposed two methods: one is the use of a molybdenum silicide evaporation source at temperatures below 1700°C in air [16], and the other is the spontaneous change into metallic oxide by introducing oxygen gas in a metallic smoke in order to produce the compound particles [17]. Details of the method for the production of oxides and of more advanced control methods are under investigation.

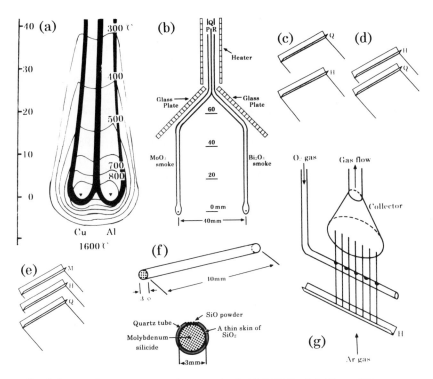

Figure 7. Advanced gas evaporation method (AGEM). (a) Two different types of smoke were mixed near heaters; this method is applied to the production of alloy and compound oxides. (b) Two different types of smoke were mixed after the coalescence near heaters was finished. The coalescence of different particles took place at the heater. Evidence of coalescence is given and coalescence temperature can be determined. (c) Heater H produces the stationary flow of the gas. The temperature produced by heater H at position Q was used as the evaporation temperature of the substance put on boat Q. Ultrafine particles of organic material and high-vapor-pressure material can be obtained. (d) Heater H is the evaporation source of low-vapor-pressure material and produces the stationary flow of inert gas. High-vapor-pressure material such as sulfur is put on boat Q. Sulfur vapor is transported by convection flow and reacts with the metallic gas from boat H. (e) Convection flow and evaporation source charged material (high vapor pressure) in boats M and Q are achieved using heater H. II-VI compounds are easily produced. (f) Molybdenum silicide evaporation source. (g) Metallic particles produced using heater H were changed by introduction of O_2 gas.

244

REFERENCES

1. R. Uyeda, in : Morphology of Crystals, Part B, Ed I. Sunagawa (Terra, Tokyo, 1987), p.369.
2. R. Uyeda, in : Progress in Materials Science, Vol.35 (Pergamon, Oxford, 1991), pp.1 - pp.96.
3. C. Kaito, K. Fujita, H. Shibahara and M. Shiojiri, Jpn. J. Appl. Phys., 16, (1977), 697.
4. C. Kaito, Y. Saito and K. Fujita, J. Cryst. Growth, 94, (1989), 967.
5. C. Kaito and Y. Saito, Proc. Japan Acad., 65, Ser.B, (1989), 125.
6. C. Kaito, H. Nakamura, T. Sakamoto, S. Kimura, N. Shiba, Y. Yoshimura, Y. Nakayama, Y. Saito and C. Koike,
7. C. Kaito, Kemurino Himitsu, (Kyoritsu, Tokyo, (1991)) (In Japanese).
8. C. Kaito, K. Fujita and M. Shiojiri, J. Cryst. Growth, 7, (1982), 199.
9. C. Kaito and K. Fujita, Jpn. J. Appl. Phys., 25, (1986), 496.
10. C. Kaito, Jpn. J. Appl. Phys., 17, (1978), 601.
11. C. Kaito, J. Cryst. Growth, 55, (1981), 273.
12. C. Kaito, Jpn. J. Appl. Phys., 23, (1984), 525.
13. C. Kaito, Jpn. J. Appl. Phys., 24, (1985), 261.
14. C. Kaito, Hyomen Kagaku, (J. Surf. Sci., Japan), 8, (1987), 144.
15. C. Kaito, T. Sakamoto, D. Ban, T. Izuta, Y. Kitano and Y. Saito, J. Cryst. Growth, 142, (1996), in press.
16. C. Kaito, R. Shoji and K. Fujita, Jpn. J. Appl. Phys., 6, (1987), L965.
17. C. Kaito, N. Shiba, A. Sakagami, S. Kimura, N. Suzuki, N. Tsuda, C. Koike and Y. Saito, Jpn. J. Appl. Phys., Vol.35, (1996), No.9A, in press.

PART III

Observations of Growth Surfaces and Interfaces

Advances in the Understanding of Crystal Growth Mechanisms
T. Nishinaga, K. Nishioka, J. Harada, A. Sasaki and H. Takei (Editors)
© 1997 Elsevier Science B.V. All rights reserved.

Thermally oxidized layers on Si-wafers
–Surface X-ray scattering and field ion microscopy–

J.Harada[a], I.Takahashi[b], T.Shimura[c], and M.Umeno[c]

[a]X-ray Research Laboratory, Rigaku corporation
3-9-12 Matsubara-cho, Akishima, Tokyo, 196 Japan

[b]School of Science, Kwansei Gakuin University,
Uegahara, Nishinomiya, 662 Japan

[c]Department of Material and Life Science, Graduate School of Osaka University,
2-1 Yamadaoka, Suita, Osaka 565, Japan

Surface X-ray scattering techniques that we have developed so far are briefly surveyed. Photography using Imaging Plate in combination with a synchrotron-radiation source is advocated as being one of the most promising probes for the ex-situ and in-situ characterization of the surface and the interface of epitaxially-grown crystal systems. A particular application, the study of oxide films on Si wafers, is described in detail. The diffraction evidence to show the existence of epitaxially grown oxide crystallites on Si(001) wafer and its structure model proposed by Iida et al. [Surf. Sci. 258 (1991) 235] and Takahashi et al. [J. Phys. Cond. Matt. 5 (1993) 6525] are reviewed. The change of the growth of those epitaxial crystallites under different oxygen atmospheres, such as O_3, afterglow of microwave plasma of O_2 and dry and wet O_2 and also additional recent X-ray diffraction results obtained from oxide layers on Si(111) and Si(110) wafers are presented.

The field ion microscope (FIM) is also known as a useful tool for the surface structure analyses, especially, of metallic materials. However, the observation of an atomic image of the semiconductor Si is not so easy as in the case of metallic materials, for several reasons. As we recently succeeded to observe crystalline SiO_2 on Si surface by FIM, some results of such direct observations of Si surface are presented. By comparing the results obtained by different techniques we discuss a general aspect of the existence of crystalline SiO_2 phase at the interface of oxide Si wafers.

1. INTRODUCTION

Because of the extremely weak interaction of X-rays with materials compared with electrons, X-ray scattering techniques would not seem to be as effective a prove for the characterization of crystal surfaces, as LEED and RHEED). However, the following several advantages should be considered. X-ray scattering provides us quantitative structural information on an atomic scale with respect to the surface and interface, as the data analysis can be carried out directly, on the basis of kinematical theory of scattering, which is free from multiple scattering seen in the cases of LEED and RHEED. Furthermore, X-ray scattering is a non-destructive technique. Recently, such an ineffectiveness of X-ray prove has been removed by using a synchrotron radiation source and may be improved in both precision and range of application by the appearance of the next-generation facilities.

On this occasion it would be meaningful to promote a better understanding of a number of fundamental problems still remaining in X-ray scattering from crystal surfaces and interfaces, and also to determine the applications in the characterization of epitaxial films grown by MBE, OMVPE and other methods. In this paper we briefly review our X-ray scattering techniques so far employed for the characterization of growth surfaces and interfaces [1-5] and then we present our studies on the structure of the amorphous oxide layer on Si wafer. The observation of the diffraction pattern from the oxide film on Si wafer clearly demonstrated the existence of crystalline SiO_2 in the amorphous oxidized layer [6]. More surprisingly the analysis of the data clearly exhibited that the crystalline scatterers were not predominantly located close to the interface, but were percolated through the entire oxide layer and retaining an epitaxial relation with the substrate [7]. This is a quite remarkable result, which may provide a key for further understanding of the SiO_2/Si system. In this paper the essential results of the studies on Si(001) are reviewed [6-8] and additional results for the structural models of the oxide layer on the Si(111) and Si(110) surfaces are presented [9-11].

The field ion microscope (FIM) is also known as a useful tool for surface structure analyses, especially, of metallic materials. The FIM can image individual surface atoms on a screen with a high magnification determined simply by the geometrical relations between the tip radius and the distance from the screen. Because of such an imaging property in real space, the FIM has the following advantages: 1) an atomic arrangement can be directly obtained without any analyses; 2) the FIM has a unique potentiality for atom probe chemical analysis; 3) the three-dimensional atomic structure of samples or a buried interface can be observed, if a field evaporation technique is incorporated. Thus the FIM is another promising method to investigate the structure of Si/SiO_2 interfaces on an atomic scale. In general, the observation of an atomic image of the semiconductor Si is not so easy as in the case of metallic materials of high melting points. In spite of many trials in recent decades only a few works have succeeded in obtaining the atomically resolved FIM patterns from Si whiskers [12]. However, we found that the atomic images of Si tips could be obtainable, when the field evaporation was conducted after preliminary thermal oxidation in the FIM chamber. In this paper, therefore, by presenting some results of our direct observations obtained for SiO_2/Si interfaces we discuss the existence of crystalline SiO_2 phase on oxide Si surface in connection with X-ray studies.

2. OBSERVATION OF CTR SCATTERING

Our main X-ray scattering technique is photography using Imaging Plate in combination with a synchrotron-radiation source [3]. The imaging plate is a two-dimensional detector which has been developed by Fuji Photo Film Co. It has several advantages such as high detective quantum efficiency and a wide dynamic range of intensity, about 10^5, and so on. It will be referred to as IP, hereafter. We use the system developed by Sakabe [13] for protein crystallography without making any modification. It is installed at BL-6A$_2$ of Photon Factory, KEK (Tsukuba), and consists of a bent mirror, a bent Si(111) monochromator and an oscillation camera. The original papers [13] and cited in reference [3] should be referred to for the details of the system. We have used the Fuji BA100 system for read-out of image data from an exposed IP so that the pixel size of read-out data is 0.1×0.1 mm^2 [3].

The X-ray diffraction photography is a very simple technique compared with other diffractometry but important fundamental aspects of scattering phenomena have been revealed by this technique: the validity of kinematic diffraction theory has been proved for CTR scattering, furthermore it became possible to estimate the coverage of epitaxial layer on any substrate crystal, as the CTR scattering from the substrate can be used as the reference scattering [5]. In addition it has been revealed that the analysis of observed CTR scattering provides surprisingly detailed structural information about the interface: as demonstrated by the studies of the AsH$_3$ exposed InP/InPAs/InP single heterostructure [14,15]. We also use a 4-circle diffractometer of vertical-type with Si(111) double-crystal monochromator and a Si(111) analyzer crystal, which is installed at BL-4C of Photon Factory, KEK for high-precision measurements. It is recommended to refer to [1] for the resolution of this diffractometer.

3. STRUCTURE OF OXIDE FILM ON Si(001) WAFERS

3.1. Oxidation of samples

The Si substrates were boron-doped (001) Czochralski-grown wafers with resistivity in the range 8.5-11.5 Ωcm. After a conventional cleaning procedure including chemical etching and rinsing with deionized water, the Si(001) wafers were placed in an atmospheric furnace equipped with an in-situ ellipsometer for measuring the thickness of the oxidized film. For high-temperature oxidation samples were prepared by heating wafers at about 960–1000°C in dry- and wet-O$_2$ atmosphere, while for low-temperature oxidation the atmosphere of either 97.5 : 2.5 mixture gas of dry-O$_2$ and O$_3$ at 1 atom (abbreviated as O$_3$ oxidation), or the afterglow of 2.45 GHz micro wave plasma of 270 Pa (abbreviated as AGMP oxidation). The oxidation temperature was 650°C in both cases.

3.2. X-ray diffraction from Si(001) wafer

If there exists some epitaxial crystalline material in the oxide layer on Si-wafer, it is expected that Bragg peaks would appear on the needle-shape CTR scattering from the substrate Si. As a result of their search of the intensity distribution along all the CTR scatterings observable with CuKα radiation for the oxide surface of a Si(001) wafer by using conventional 4-circle diffractometer, Iida et al. [6] revealed the existence of a very weak extra peak on the low-angle side of the CTR scattering from the 111 Bragg point and also its equivalent points in reciprocal space. The positions at which the extra peaks have been observed in the reciprocal lattice are represented schematically in Figure 1. The fact that such peaks are exactly on the CTR scattering in the reciprocal space can be confirmed by an oscillation photograph, as shown in Figure 2. The characteristic features of the extra peak revealed by further precise measurement are summarized as follows [7].

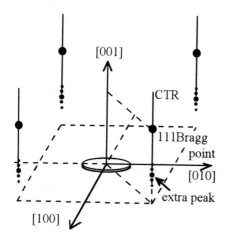

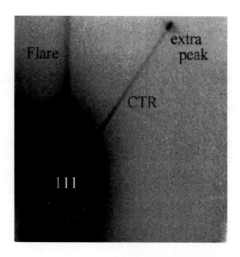

Figure 1. Schematic representation of the location of extra peaks in the Si reciprocal lattice

Figure 2. Oscillation photograph showing the extra peak located just on the CTR scattering elongated from the 111 Si Bragg point

1. The extra peak is located at the position 1 1 ∼0.45 in reciprocal space as seen in Figure 1, while the higher-order reflections such as 2 2 ∼0.9 are not observed. Here ∼0.45 means about 0.45 since it is slightly sample dependent.

2. The extra peak is exactly located on the lower-angle side of the CTR scattering, as seen in the oscillation photograph of Figure 2, and not correlated with any other scatterings, such as the Debye-Scherrer ring from some crystallites or the halo pattern from an amorphous phase.

3. The intensity of the extra peak depends on the thickness of the oxide layer and the peak does not appear for the specimen where the oxide layer is etched

off with HF solution.

4. The width of the peak perpendicular to the CTR scattering is the same order of magnitude as that of the CTR scattering from the substrate as shown in Figure 3.

5. The intensity profile along the CTR scattering is not a simple oscillatory fringe but shows a Laue-function-like fringe pattern, so that the width of the main peak is twice as much as that of sub-peaks and the period corresponds to roughly the inverse of the film thickness as shown in Figure 4.

6. The extra peak is not observed on the 00ℓ line (the specular reflection from the (001) crystal) while a simple fringe pattern is observed due to interference between the oxide surface and the interface boundary.

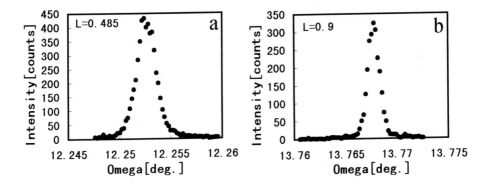

Figure 3. Intensity profiles perpendicular to the CTR scattering. a) for the extra peak at 1,1,0.485 and b) for near the Bragg point at 1,1,0.9

3.3. The origin of the extra peak

It is easily recognized that the extra peak originates from the oxide layer because it disappears when the oxide is removed. The possibility that the peak is caused by the lattice distortion of the Si substrate induced by the oxidation is small, because the intensity profile along the CTR scattering shows a clear fringe pattern, which occurs as an interference effect between two distinct boundaries, and the lattice distortion usually has no clear boundaries inside the crystal. In addition, the extra peak does not vary with the annealing in N_2 atmosphere at 950°C for 1 hour after

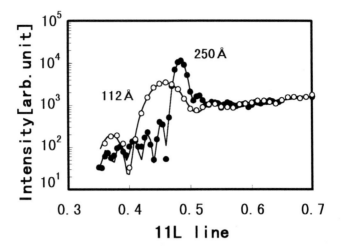

Figure 4. Intensity distributions along the CTR scattering around the extra peak. Solid and open circles represent the observed points for the samples of oxide thickness of 250 and 112 Å, respectively. Solid curves show the model calculations.

the oxidation. Therefore, it is hardly possible to consider that the extra peak is due to the lattice distortion of the substrate.

All these experimental facts are in rather good agreement with the diffraction pattern expected from an epitaxial thin crystal on some substrate. It is, therefore, appropriate to conclude that the extra peaks are the Bragg scatterings from crystallites in the oxide layer on the Si wafer and having an epitaxial relation with it. Thus, we can carry forward this idea.

Generally speaking the inverse of the peak width corresponds to the size of the crystal, so we can estimate the average size of the crystalline phase. We see from Figure 3 that the lateral width of the extra peak is very narrow and nearly the same order of magnitude as that of the CTR scattering and also of the Bragg scattering from the substrate, while the profile along the CTR scattering shows a Laue-function-like oscillation with a period corresponding to the inverse of the film thickness. This means that the crystallites extend all over the substrate surface with an ideal epitaxial relationship with it in the lateral direction and terminating at the top of the oxide film.

The idea of considering a large solid epitaxial crystal is not realistic and is also inconsistent with many other experiments on oxide film. In addition, the observed intensity of the peak is very weak compared with calculation based on such a large solid epitaxial crystal. It is thus confirmed that the epitaxial crystal phase is not one entity but SiO$_2$ crystallites or segments that are distributed in the oxide film and retain an epitaxial relationship with the substrate. In this model the number of coherent scatterers will be reduced and could be estimated from the observed intensity. This is our final interpretation of the extra peak observed. All the data we have examined are explained by this model by simply adjusting the distribution of crystallites. We have an image that the SiO$_2$ crystallites or its segments are probably linked together in the oxide film, forming a percolated structure as seen in a spin glass alloy [16].

3.4. Atomic structure of the SiO$_2$ crystallites

What structure do the epitaxial crystallites possess in the stable form in the oxide layer? This is the most interesting subject for us to study. We regard the extra peaks observed as the Bragg scatterings from the epitaxial oxide layer on the Si wafer, for a number of reasons. However, the number of non-equivalent (independent) Bragg reflections observed is only one, so that it is almost impossible to do a conventional structure analysis. Iida et al., therefore, attempted to differentiate among the structure models so far proposed for the crystalline transition layer between the amorphous SiO$_2$ and the Si(001) surface: the tridymite structure proposed by Ourmazd et al. [17], the β-cristobalite model of Foss et al. and Hane et al. [18,19], and pseudo-cristobalite structure of Hattori et al. [20]. Among them, only the pseudo-cristobalite structure proposed by Hattori et al. produces Bragg peaks just at the position observed and furthermore satisfies the extinction rule that the intensities of any other Bragg reflections are extremely weak compared with the one we observed. Exceptions are the 004 and the 220 reflections. For those higher-order reflections a strong intensity reduction due to the static Debye-Waller factor, $\exp\{-8\cdot\pi^2\cdot <\Delta r^2> (\sin\theta/\lambda)^2\}$ is conceivable as the static displacement of atoms from average sites is possibly very large, as will be discussed later. If we accept this model the structure is represented by space group $I41/amd$ of the tetragonal system, with a unit cell with $a = a_{Si}/\sqrt{2}$ and $c = 2\cdot a_{Si}$. From the symmetric arrangement of observed extra peaks in the reciprocal lattice of the Si substrate, it is easy to see that there is a four-fold symmetry around the [001] axis. In Figure 5 the structural model is reproduced.

3.5. Distribution of crystalline SiO$_2$ in the amorphous oxide layer

In Figure 4 the Laue function-like fringe pattern observed is reproduced. Its profile should be given by the square of the Fourier transform of a density distribution of scatterers. Thus, this is a inverse problem; the density distribution is obtained from the inverse Fourier analysis of the observed intensity profile. As the Bragg reflection from the crystalline SiO$_2$ is just on the line of CTR scattering from the Si substrate due to having an epitaxial relationship between them, the interference effect should be taken into account between the CTR scattering from the Si substrate and the Bragg reflection from the epitaxial crystalline SiO$_2$. Thus the intensity $I(K)$ along the 11ℓ is given as

$$I(K) = |F_{Si}(K) + F_{SiO2}(K)\cdot\exp(i\cdot d\cdot\ell)|^2 \qquad (1)$$

254

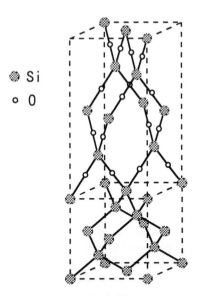

● Si

○ O

Figure 5. Pseudo-cristobalite structure of the epitaxially aligned crystalline SiO$_2$ on the Si(001) substrate. Oxygen atoms are located between the Si atoms. The lattice spacing along the normal to the interface is approximately twice that of Si, while the basal plane is common with that of Si. It belongs to space group $I41/amd$ of the tetragonal system.

where $F_{Si}(K)$ and $F_{SiO2}(K)$ are the scattering amplitude of CTR scattering from the Si substrate and the scattering amplitude from the crystalline SiO$_2$, respectively, and d is the atomic interface spacing between the substrate and the epitaxial crystal [21]. In $F_{SiO2}(K)$ the distribution function $\rho(z)$ is included as

$$F_{SiO2}(K) = F_{SiO2} \cdot \sum \cdot \rho(z) \cdot \exp(i \cdot K \cdot c \cdot z) \tag{2}$$

where the summation $\sum$ is taken from the interface to the top of the oxide film and F_{SiO2} is the structure factor for the crystalline SiO$_2$. If $\rho(z)$ is constant, the well known simple Laue function is obtained. The result based on this simple model is compared with the observation in Figure 6 a) where the solid curve represents the calculation for the sample of the thickness 752 Å. Agreement between the calculations performed and observation is rather poor. Several trial calculations by modifying the density distribution revealed that the profile is very sensitive to changes of this distribution. It was found that the profile is reproduced very well if a modified-exponential distribution function is assumed [8]:

$$\rho(z) = \begin{cases} \rho_0 + \rho_1 & (z = 1) \\ \rho_1 \cdot \exp(-z/\xi) & (z = 2, 3, ... Z_{max}), \end{cases} \tag{3}$$

where z is an integer, representing the level from the interface in units of c and ξ is a parameter representing effective depth of ρ. Thus, we can analyze all the intensity profile on the basis of such a modified-exponential distribution function by a least-squares fitting procedure and obtain several parameter values; ρ_0, ρ_1, ξ and Z_{max} for any samples obtained under several different oxidation conditions. Those parameters were found not to be correlated with each other in the least-squares fitting procedure. Comparison of the calculation with observation is reproduced in Figure 6 b) for the sample of thickness 250 Å where the parameters refined are $\rho_0 = 0.060(4)$, $\rho_1 = 0.062(1)$, $x = 21(1)$, $Z_{max} = 21$ and interface roughness parameter $< \Delta z^2 > = 0.074(7)$ Å^2.

What we found from this analysis is that the concentration of SiO_2 crystallites is only 15-18 % of the SiO_2 oxide and these are not homogeneously distributed but predominately reside in the interface region, (about 2/3 of the crystallites; and the number decreases with increasing distance from the substrate.) Another interesting point is Z_{max} which is given always to be smaller than the thickness of oxide layer by a few layers. It means that there is no crystalline SiO_2 near the surface of the oxide layer.

3.6. Effects of oxidation condition on the crystalline SiO_2 distribution

In addition to the normal high-temperature oxidation in dry and wet atmosphere (abbreviated as HTO), we examined two other kinds of oxidization at low temperature: oxidation under the gas mixture of O_3 and dry O_2 (O_3) and oxidation under afterglow of microwave plasma of 270 Pa in dry O_2 (abbreviated as AGMP-). In the case of O_3-oxidation with a 13 nm thick oxide layer (abbreviated as O_3-13), we determined that the crystallites are distributed in the layer in a similar way to the case of HTO, while the number of crystallites is about half of that for HTO-11. On the other hand , in the case of AGMP-7 (7nm thick SiO_2 film) we determined that these are very few crystallites and that they are only localized in the region of the $SiO_2/Si(001)$ interface. By comparing the results obtained from the HTO-, O_3- and AGMP- samples, the following trends were noticed [10] .

1. Just below the surface of the SiO_2 layer, the crystallites do not exist , since the parameter Z_{max} is always given to be smaller than the thickness of the SiO_2 film .

2. Interface roughness parameter $< \Delta z^2 >$, which represents the degree of roughness at the $SiO_2/Si(001)$ interface [10], has a relationship,

$$<\Delta z^2(\text{HTO})> \; < \; <\Delta z^2(\text{O}-13)> \; < \; <\Delta z^2(\text{AGMP})> .$$

3. The distribution function $\rho(z)$ has a relationship,

$$\rho(z)(\text{HTO}) > \rho(z)(\text{O}-13) > \rho(z)(\text{AGMP}).$$

256

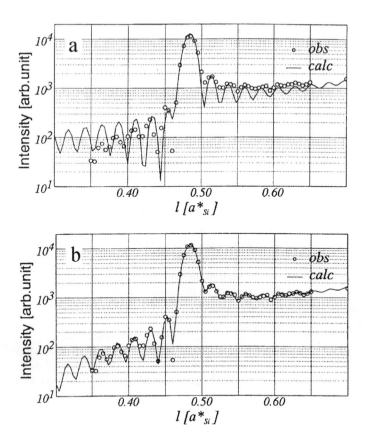

Figure 6. a) b) Profile fittings to the observations on the basis of density distribution of SiO_2 crystallites a) for uniform distribution, b) for modified exponential distribution model given by eq.(2) in the text.

In addition to the above experiments we examined various other oxide layers prepared with several different conditions: different wafer materials such as CZ or FZ and p or n-type wafers and different oxidization such as dry-O_2 or wet-O_2 atmospheres and high and low oxidation temperatures. Essentially similar extra peaks are observed for all the samples, while some differences are noticed in the position and the intensity of the peak. We, therefore, conclude that the crystalline SiO_2 is formed more or less in all cases of thermal oxide Si(001) wafers, while its amount and the distribution of crystallites in the oxide layer depends on the oxidation condition.

4. OXIDE LAYER FOR WAFERS OF OTHER ORIENTAIONS

4.1. Oxide layer of vicinal Si(111) surfaces

We have been concerned so far with the oxide layer on Si(001). It is also helpful for our understanding of the oxidation mechanism, to investigate oxide layers on Si-wafers with other orientations. We examined vicinal Si(111) surfaces which are tilted from the [111] by 4° towards the [11$\bar{2}$] direction and confirmed that the extra peaks do exist just on the CTR scattering from the 111 Bragg point. The extra peaks show similar features to those for the SiO_2/Si(001) samples. The oscillation diffraction photograph taken from the 111 Bragg point to the origin of the reciprocal lattice is reproduced in Figure 7 a). The direction of the CTR scattering is not along the [111] axis, because of the 4° mis-orientation of the surface. Thus the CTR scattering from the 111 Bragg point does not coincide with the reflectivity line or the CTR scattering from the 000 reciprocal lattice point as seen in the photograph. The position of the extra peak was confirmed to be ∼0.42 ∼0.42 ∼0.51 in the units of the Si reciprocal lattice.

The intensity distributions of the extra peaks also show a similar fringe pattern to that of the Si(001) wafers, suggesting existence of SiO_2 crystallites in the oxide layer. Our present model, achieved for this peak, is shown in Figure 8: a two-dimensional atomic arrangement of the interface structure projected along the [1 $\bar{1}$ 0] direction for the Si(111) wafer. This is essentially the same structure as that of the (001) wafer. We put O atoms between Si atoms and make the lattice spacing along the surface normal approximately twice that of Si, while the lateral lattice spacing is kept to that of the substrate. The lattice plane of the oxide layer is considered to be slightly inclined to the substrate as shown in Figure 8, because the longitudinal lattice spacing of the crystalline SiO_2 at the interface could not change so abruptly at the step edge. This epitaxial atomic arrangement is constructed by making reference to the model proposed by Nagai (1974) and Auvray (1989) [22,23] for the heteroepitaxial layers on the vicinal surface of GaAlAs/GaAs. The difference between the two cases is the mismatch between the lattice constants: much larger in the present case. The simulated diffraction pattern obtained by taking the numerical Fourier transform of the model of Figure 8 is shown in Figure 7 b). The extra peak is reproduced on the CTR scattering from the 111 Bragg point, at the same position as that obtained by experiment. The small peak seen beside the 111 Bragg spot in Figure 7 b) is not recognized in the photograph of

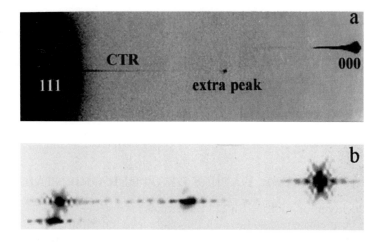

Figure 7. a) Oscillation photograph of vicinal $SiO_2/Si(111)$ interface, $4°$ off from the (111) plane. Two CTR scatterings, one from the 000 and the other from the 111 Bragg points, do not coincide. The extra peak is located at the tip of CTR scattering from the 111 point. b) Fourier transform of the atomic structure for vicinal $SiO_2/Si(111)$ interface shown in Figure 8.

Figure 7 a), because this peak is concealed behind the thermal diffuse scattering (TDS) in the experiment.

4.2. Oxide film on Si(110) surfaces

Similar extra peaks were also observed on the low-angle side of the 111 CTR scattering for thermally oxidized Si(110) wafers. The position of the observed extra peak is ~0.45 ~0.45 1 in the units of the Si reciprocal lattice. This was also confirmed by an oscillation photograph taken around the 111 Bragg point. The extra peaks showed again similar features to those of Si(001) and Si(111) wafers, indicating the existence of the epitaxial crystalline SiO_2 phase in the oxide layers on Si(110) wafers. For more detailed results the reader is referred to reference [11].

5. DIRECT OBSERVATION OF CRYSTALLINE SiO_2 WITH FIM

5.1. Apparatus

The FIM apparatus used is composed of the main chamber, which can be evacuated to up to 2×10^{-10} Torr and is equipped with the specimen stage cooled by a cold finger connected to a cryo-generator and with a micro-channel plate imaging system, and of the sub-chamber connected to the main chamber by a gate valve.

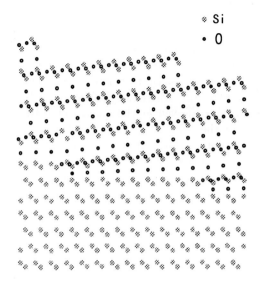

Figure 8. Schematic illustration of the structural model for vicinal $SiO_2/Si(111)$ interface. O-atoms are placed between Si atoms and the lattice is elongated along the surface normal direction by about twice the (111) spacing of Si. The lattice spacing of the oxide layer at the interface gradually change to accommodate terrace edges.

Sample tips were prepared from n-type (001) CZ wafer with the specific resistivity of 0.005 to 0.02 Ωcm and from p-type (110) CZ wafer of 6 to 10 Ωcm. Sample tips for the FIM observations were prepared from the square rods obtained from the wafers by the method described elsewhere [24]. The Si tips so prepared were firstly annealed at 1000K for 5 minutes and field evaporated until some net plane rings were observed. The thermal oxidation of Si tips were done by exposing to 300L of dry O_2 at 1000K in the FIM chamber and the stepwise field evaporation was conducted on the tips. FIM images were taken at every stage of the field evaporation using 3.4×10^{-4} Pa of Ne as the imaging gas.

5.2. Observation of Si(011) surfaces

Figure 9 shows the FIM images of [011] tips at different stages of field evaporation after oxidation. Just after the thermal oxidation the image is characterized by the randomly distributed bright spots as shown in Figure 9 a), showing the existence of a non-crystalline phase, i.e. a-SiO_2. When the field evaporation was performed, the image did not change greatly for a while, but when the oxide surface approached to the interface, the FIM image changed suddenly to the pattern characterized by dense bright spots and field evaporation became severe and then dark spaces appeared at 010 and 001 in the image as shown in Figure 9 b). The layer imaged as dense bright spots was nearly two or three atomic layers and is considered to be the transition SiO_2 layer at the Si/SiO_2 interface. The dark area in the image corresponds to the substrate Si where the ionization of the imaging gas is low. Figure 9 c) shows the image taken at the stage where the oxide remained only at 111 and $\bar{1}11$ axes of the substrate. By a careful observation of the pattern circular arrangements of bright spots are discerned at 111. By the successive field evaporation the pattern changed into the FIM image of substrate Si as shown in Figure 9 d). The net plane rings of 111 and $\bar{1}11$ are clearly seen at the same place where the oxide image with the circular arrangement of spots are seen in Figure 9 c).

5.3. Observation of Si(001) surfaces

Figure 10 shows an example of a FIM image of a Si/SiO_2 interface of a (001) tip. Figure 10 a) shows the pattern where both images of SiO_2 and substrate Si are appearing. The oxide remains at 112 and $11\bar{2}$ showing also a ring-like arrangement of bright spots as is enlarged in Figure 10 b). The existence of a ring arrangement of atoms means that the oxide at the interface has a crystalline structure. When the oxide is perfectly removed, a nearly perfect FIM image of Si was observed as shown in Figure 10 c) where the main low indexed axes of 001, 011 and 111 are clearly seen. From this pattern the radius of the tip was calculated to be 40 nm and using this value the evaporation field of Si and SiO_2 were determined to be 40 V/nm and 34 V/nm, respectively. The lower evaporation field strength of SiO_2, compared to that of Si, enabled us to obtain a highly resolved FIM image of Si by field evaporating a preliminary oxidized Si tip. Once a smooth interface was formed by thermal oxidation, the field evaporation stops when substrate Si appears if an appropriate electric field is applied for the field evaporation.

6. DISCUSSION

6.1. Comparison of the results obtained by two techniques

In the X-ray diffraction patterns from the thermal oxide layers on (001), (111), and (110) silicon surfaces, the extra peaks are always observed on the low-angle side of the CTR scattering from the 111 Bragg points. They show similar characteristic fringes, indicating the existence of the epitaxial crystalline SiO_2 in the oxide layer. In the FIM image of interfacial SiO_2 layer, on the other hand, net plane rings related to the atomic arrangement of substrate are observed, indicating also the existence of a thin crystalline SiO_2 layer on the low-indexed crystal planes of Si. Thus, the results obtained by two different techniques are in rather good agreement with each other, in regard to the existence of epitaxial crystalline SiO_2 on Si substrate.

However, the crystalline phase was observed only at the very vicinity of the interface and not in the main oxide layer using the FIM, whereas the CTR study suggested a wide distribution of the crystalline phase in oxide layer. There are two

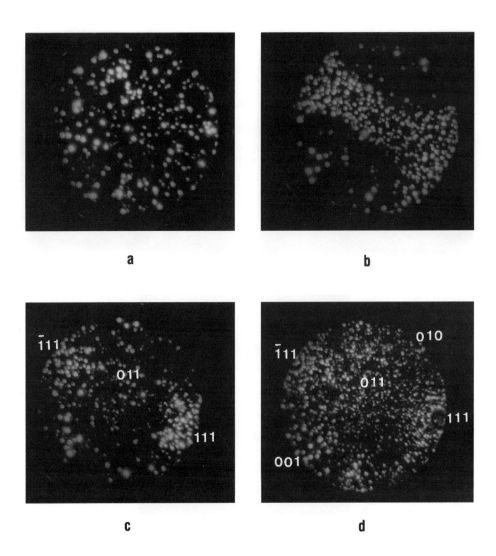

Figure 9. FIM images of [110] tips at different stages of field evaporation after oxidation. a) oxide image just after the thermal oxidation, b) transition layer at the interface c) oxide remaining only at 111 and $\bar{1}11$ axes of the substrate and d) the FIM image of substrate Si

262

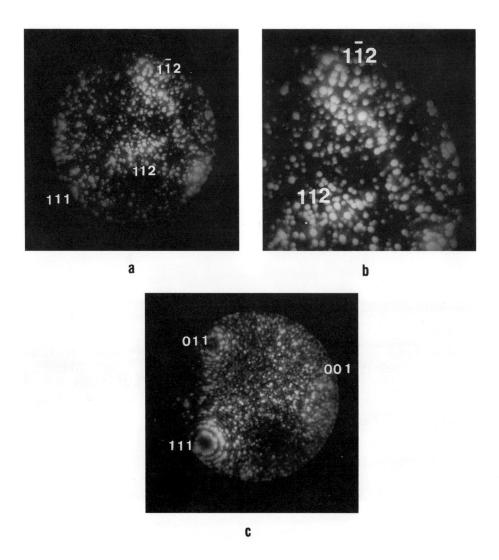

Figure 10. FIM image of a Si/SiO$_2$ interface of a (001) tip. a) both the images of SiO$_2$ and substrate Si, b) remained oxide at 112 and 1$\bar{1}$2, showing a ring-like arrangement, c) enlargement of b).

reasons we can offer for this discrepancy: one from the difference of oxide formation in the samples used. For the FIM observation the oxide is formed on the apex of a sharp Si tip in a dilute oxidizing atmosphere, so the amorphization of the oxide firstly formed in crystalline phase occurs immediately in the succeeding oxidation. On the other hand the oxide samples for CTR study are formed on wafer surfaces with an oxidation procedure usually used in the LSI process, where amorphization occurs gradually when the intrinsic stress in the oxide layer exceeds a critical value. Another reason for the discrepancy is as follows: most of the crystalline SiO_2 is considered to be predominately localized on the interface and only one third of the crystallites are percolated in the amorphous oxide layer. The crystallites are very isolated in the amorphous layer and the number of crystallites is very small, as estimated to be less than a few per cent for each layer. Thus, it would be very hard to observe the crystallites in the amorphous oxide layer by FIM, even if they exist.

6.2. Crystalline state in amorphous oxide layer

From the present study we found that the SiO_2/Si interface is partially crystallized for all the three cases, $SiO_2/Si(001)$, $SiO_2/Si(111)$ and $SiO_2/Si(110)$. Their interface structures are slightly different to each other but it is possible to say that they are more or less similar in respect of the lattice spacing along the normal to the surface being roughly twice the original spacing, while the lattice spacing perpendicular to the surface remain unchanged. The structure of pseudo–cristobalite shown in Figure 5 is proposed as a typical example of the interface structures.

As has been discussed in a previous paper [7], the structure of pseudo–cristobalite consists of unusually small and large bond angles of O–Si–O such as about 65° and 135°, respectively and is unfamiliar in crystallography. Thus, we imagined that the crystalline SiO_2 phase is only stabilized under some high stressed circumstances near the SiO_2/Si interface. The formation of such an epitaxial layer is, therefore, supposed to depend on the process of oxidization of Si wafers. Thus, even the unit cell size along the c axis of each crystallite may not be well defined, although the lateral size of unit cell should be matched to the period of the substrate. Therefore, the unit cell size along the c axis may vary from place to place in the amorphous oxide layer but the average value may be well-determined for each sample. Actually we found that the position of the extra peak in the CTR scattering changes from sample to sample, supporting the above argument. These facts are also consistent with our speculation as to the reason why the higher-order reflections are not observed: that is, the intensity is diminished by the large effect of the static Debye-Waller factor.

6.3. Oxidation process of Si interface

If we look carefully at the refined parameters ρ_0, ρ_1, ξ, for the samples oxidized by different processes, we notice that there is an interesting systematic change among them. If the distribution function $\rho(z)$ is reproduced using the parameters listed in Table 1 such systematic changes are visualized. In general the density of the crystallites at the interface (=coverage of the crystalline phase) , $\rho(1) = \rho_0 + \rho_1$ is

	Sample A (250Å)	Sample B (112Å)
R factor	0.040	0.062
Weighted R factor	0.065	0.079
Number of intensities	74	47
k	1.23(1)	1.37(2)
$<\Delta z^2>$	0.074(7)	0.10(1)
ρ_0	0.060(4)	0.067(9)
ρ_1	0.062(1)	0.067(4)
ξ	21(1)	14(4)
Z_{max}	21	7
Δc	0	0
$s(=1/(1-q))$	2.077(1)	2.226(5)
u_0	0	0

Reprinted from: I.Takahashi, T.Shimura, and J.Harada, J.Phys.: Condes. Matter 5 (1993) 6525

about several percent. This is roughly twice as much as the value of $\sum_z \rho(z)$ (for $z > 1$) for all the samples , while individual values of ρ_0 and ρ_1 are fairly different from one another, depending on the oxidation process. (the more precise discussion is given in ref. [8], [9]). Among the several trends we can find we discuss here the reason why the distribution is well described by above modified exponential function $\rho(z) = \rho_0 + \rho_1 \exp(-z/\xi)$, because it is closely related to the growth process of the crystallites at interface. As we believe that the oxidation proceeds at the interface, we consider that oxygen (possibly in the form of molecules) is constantly supplied to the interface through the diffusion into the amorphous SiO_2 layer. Substrate Si atoms will be continuously oxidized in a way that a kink or a step edge which may exist at the $SiO_2/Si(001)$ interface plays an important role for making the interface smooth. It is something like a layer by layer growth process seen often in crystal growth [9]. In such an oxidation process some of the oxygen molecules will react with substrate Si to form crystallites and some will contribute to the the amorphous material. Then, the interface will be partially covered by crystallites. When the interface itself goes one step further into the interior of the substrate by further oxidation, some of the crystallites remain as they were but some others may change to the amorphous state because of the instability of their under unfavorable conditions. This is the scenario we can propose, from the form of the distribution function and the flatness of the interface.

ACKNOWLEDGEMENTS

The authors express their thanks to Y.Iida of Toyota Automobile Co. Ltd., K.Nakano, Mr.S.Samata and Mr.Y.Matsushita of Toshiba Co. Ltd. for their collaboration at an early stage of the present study, and to Mr. H.Misaki of Osaka University for his cooperation for the study of $SiO_2/Si(111)$. We also thank Professor T.Sakabe, Dr. A.Nakagawa, N.Watanabe, Dr. S.Kishimoto, and Associate

Professor Y.Murakami of KEK for their assistance in using the facilities at Photon Factory in KEK. The authors also express their sincere thanks to A.W.Stevenson of MDST, CSIRO, Australia for his kind critical reading of this manuscript. The major part of the experiments were done under proposal Nos. 90-086 and 94G105 of PF in KEK. This work was supported partly by Grants-in-Aid for Scientific Research Nos. 03243105, 04227105, 07555337, and 07750038

REFERENCES

1. N. Kashiwagura, Y.Kashihara, M. Sakata, J. Harada, S.W. Wilkins and A.W. Stevenson, Jpn J. Appl. Phys. 26 (1987) L2026.
2. J. Harada, Acta Cryst. A48 (1992) 764.
3. T. Shimura and J. Harada, J.Appl.Cryst. 26 (1993) 151.
4. J. Harada , Ultramicroscopy 52 (1993) 233.
5. T. Shimura and J. Harada, Physica B 198 (1994) 195.
6. Y.Iida, T.Shimura, J.Harada, S.Samata, and Y.Matsushita, Surf. Sci. 258 (1991) 235.
7. I.Takahashi, T.Shimura, and J.Harada, J. Phys.: Condens. Matter 5 (1993) 6525.
8. I.Takahashi, K.Nakano, J.Harada, T.Shimura, M.Umeno, Surf. Sci. 315 (1994) L1021.
9. T.Shimura, H.Misaki, M.Umeno, I.Takahashi, and J.Harada, J. Cryst. Growth, in press.
10. I.Takahashi and J.Harada: *SEMICONDUCTOR SILICON*, H.R.Huff, W.Bergholz and K.Sumino (ed.) ,The Electro Chemical Soc. Inc., Pennington N. J., (1994) 1147.
11. T.Shimura, I.Takahashi, J.Harada and M.Umeno, *The Phys. and Chem. of SiO$_2$ and the Si-SiO$_2$ Interface 3*, H.Z.Massoud, E.H.Poindexter, and C.R.Helms (ed.) Proc. vol.96-1 (1996) 456.
12. T. T. Tsong, H.M.Liu and D.L. Feng, Phys. Rev. B36 (1987) 4446.
13. N.Sakabe: Nucl. Instrum. Methods A246 (1991) 572.
14. Y. Takeda, Y. Sakuraba, K. Fujibayashi, M. Tabuchi, I. Takahashi, J.Harada and H. Kamei: Appl. Phys. Lett., 66 (1995) 332.
15. M. Tabuchi, Y. Takeda, Y. Sakuraba, T. Kumamoto, K. Fujibayashi, I.Takahashi, J. Harada, and H. Kamei: J. Cryst. Growth, 146 (1995) 148.
16. K. Ohshima, N. Iwao and J. Harada, J. Phys. F: Met. Phys. 17 (1987) 1769.
17. A.Ourmazd, D.W.Tylor, J.A.Rentschler, and J.Bevk, Phys. Rev. Lett. 59 (1987) 213.
18. P.H.Fuoss, L.J.Norton, S.Brennan, and A. Fischer-Colbrie, Phys. Rev. Lett. 60 (1988) 600.
19. M. Hane, Y.Miyamoto, and A.Oshiyama, Phys. Rev. B41 (1990) 12637 .
20. T.Hattori, T.Igarashi, M.Ohi, and H Yamaguchi, Jpn.J.Appl.Phys. 28 (1989) L1436.
21. I.K.Robinson, W. K.Waskiewicz, and R. T.Tung, Phys. Rev. Lett. 57(21) (1986) 2714.

22. H.Nagai, J.Appl.Phys. 45 (1974) 3789.
23. P.Auvray, M.Baudet, and A.Regreny, J. Cryst. Growth 95 (1989) 288.
24. J.Suwa, N.Mori, M.Tagawa, N.Ohmae and M. Umeno: Prod. 7th Topical Meeting on Crystal Growth Mechanisms (1994) 227.
25. F.Rochet, M.Froment, C.D'Anterroches, H.Roulet, and G.Dufour, Phil. Mag. B59 (1989) 339.

Advances in the Understanding of Crystal Growth Mechanisms
T. Nishinaga, K. Nishioka, J. Harada, A. Sasaki and H. Takei (Editors)

Atom exchange process at the growth front in OMVPE revealed by X-ray CTR scattering measurement

Y. Takeda and M. Tabuchi

Department of Materials Science and Engineering, School of Engineering
Nagoya University, Nagoya 464-01, Japan

In the OMVPE growth process of III-V compound heterostructures with different group-V elements, there is always atom exchange of group-V elements at the growth front. This process is a clear obstacle to obtain an abrupt heterointerface in composition, which is required for high performance of devices with heterostructures. We conducted X-ray CTR scattering measurement, for samples prepared by exposing the grown InP surface to AsH_3 and capped by InP, to reveal the As/P exchange process during the growth of the heterointerface. By the analysis of the CTR spectra for samples with different exposure times, we obtained the As atom distribution profiles in the samples on the monolayer scale. It was found that considerable amounts of As atoms existed in all the samples, though the InP surface was only exposed to AsH_3, and that the P atoms on the surface are quite quickly exchanged with the As atoms but the exchange stops at the As composition of 0.3, probably due to formation of a surface reconstructed structure. The long tail of As atom distribution in the cap layer was tentatively attributed to the memory effect of the As source on the reactor wall.

1. INTRODUCTION

In the organometallic vapor phase epitaxy (OMVPE) of III-V compounds with different group-V elements, it is difficult to achieve the abrupt change of composition at the heterointerface which is essential for obtaining well-defined heterostructures [1-5]. For example, $InP/In_{0.53}Ga_{0.47}As/InP$, which is an essential structure for long-wavelength optoelectronic devices and ultrahigh-frequency electronic devices, is a typical case. When a thick (over several hundred Å) $In_{0.53}Ga_{0.47}As$ layer is grown on InP, no lattice-mismatch is detectable by double-crystal X-ray diffraction. However, when multiple quantum-well (MQW) structures or superlattices are grown with thin alternating $In_{0.53}Ga_{0.47}As$ and InP layers, a noticeable lattice mismatch is observed[1]. An example of X-ray diffraction measurement is shown in Figure 1, where fifty periods of 100Å-thick $In_{0.53}Ga_{0.47}As$ and 150Å-thick InP layers were grown on InP by OMVPE. This has been attributed to atom exchange[1,3], atom contamination from reactor wall[2,4], or atom interdiffusion across the heterointerface[5].

Since the width of quantum wells is as thin as 20Å in many cases, even a fluctuation of 1 monolayer (ML, 2~3Å) in the layer growth results in a 10% fluctuation in thickness. Thus, to control, or at least to determine, the layer thickness at the 1ML level, we need techniques that can reveal the interface structures at the atomic level. For this purpose and to investigate the exchange of As and P atoms at the growth front, we applied X-ray crystal truncation rod (CTR) scattering measurement to several samples which were grown under various conditions described below. During the OMVPE growth of InAs on InP and InP on InAs, atom exchange was observed by optical reflection technique such as SPA (surface photo-absorption)[6]. This is a powerful technique for *in situ* observation of such atom exchange. We want to know the final structure at the interface which is formed during the growth of the $In_{0.53}Ga_{0.47}As$ layer and also after capping it by the InP top layer. This can be done only by X-ray CTR scattering measurement nondestructively on the monolayer scale.

X-ray CTR scattering is a rod that appears around a Bragg diffraction spot in **k**-space[7]. It is caused by the abrupt truncation of a crystal at the surface. A spectrum from an InP wafer is shown in Figure 2(a) where a broad and symmetric tail around a Bragg peak is displayed. Because of this tail, perturbations in the crystal structures, if any, can be superimposed on the spectrum. A simulated spectrum for InP with such perturbations is shown in Figure 2(b). In the spectrum, the surface roughness affects the damping of the tail, the layer thickness and the difference in lattice constants in the case of a heterostructure change the period

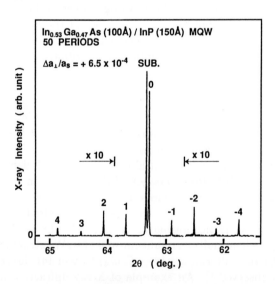

Figure 1. X-ray diffraction pattern from MQW of 50 periods of 100Å-thick $In_{0.53}Ga_{0.47}As$ and 150 Å-thick InP. A noticeable lattice mismatch for the 0th peak of $+6.5\times10^{-4}$ is detected. Diffraction peaks at higher order indicate the high quality of the MQW.

and the phase of oscillation, respectively, and the interface roughness affects the amplitude of the oscillation. These modulations in the spectrum are sensitive to the perturbations on a monolayer or even sub-monolayer scale since the X-ray wavelength ($\sim$1Å) is comparable to the lattice spacing (2$\sim$3Å).

We have demonstrated that X-ray CTR scattering is a powerful technique for revealing the layer structure of heteroepitaxially grown samples on the atomic scale[8-12].

In this work, we apply X-ray CTR scattering measurement to reveal the exchange process of As and P atom in OMVPE growth.

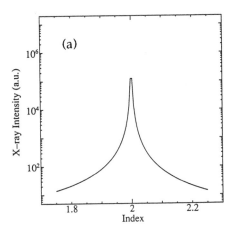 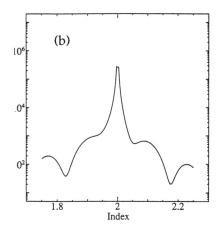

Figure 2. (a) shows a CTR spectrum from InP wafer with a smooth surface, and (b) shows a simulated spectrum for InP with several perturbations in crystal structure.

2. EXPERIMENTS

2.1. Sample preparation

In the OMVPE, AsH$_3$ and PH$_3$ were used as group-V sources and TMIn (trimethylindium) as the group-III source. The samples were prepared by exposing the grown InP surface to AsH$_3$ and capped by a 20Å InP layer in one continuous gas sequence, as shown in Figure 3. (001)-oriented InP was used as the substrate. The exposure time was varied from 0.5 to 30s. During the growth and exposure, the reactor pressure was kept at 76Torr and the growth temperature at 600°C. These growth conditions were previously found as the optimum conditions for obtaining the high-quality InP/In$_{0.53}$Ga$_{0.47}$As/InP quantum wells[1].

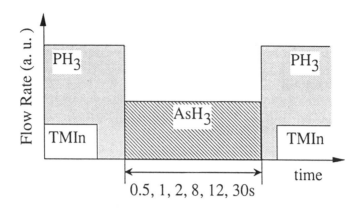

Figure 3. Gas sequence for the preparation of AsH$_3$-exposed InP surface to reveal the As/P exchange process at the interface. After growing InP, the surface was exposed to AsH$_3$ for between 0.5 and 30s and capped by InP.

2.2. X-ray CTR scattering measurement

The X-ray CTR scattering measurement was conducted using synchrotron radiation at the Photon Factory in the National Laboratory for High Energy Physics at Tsukuba. Beam line BL6A$_2$ was used and the X-ray wavelength was set at 1.000Å.

The diffraction spots and the CTR scattering were recorded using a Weissenberg camera with an IP (imaging plate). The recorded patterns on the IP were read out optically and stored as digital data. The diffraction spot and CTR around the 002 peak were used in this work. By subtracting the background X-ray diffuse scattering from the measured X-ray intensity spectra around the 002 Bragg point, we obtained the CTR spectra. It is essential to subtract the background X-ray diffuse scattering to obtain correct CTR spectra, in order to discuss the interface structures both qualitatively and quantitatively as described in the introduction. The strong X-ray beam from synchrotron radiation is necessary to obtain good-quality signals in the tail part of the spectra where most of the CTR signals are superimposed. The X-ray CTR scattering intensity varies widely from 10^2 to 10^7 as for the spectrum of the 0.5s-exposed sample shown in Figure 4. Other spectra are shifted upward by one order each for clarity.

Index ℓ which is the abscissa of Figure 4 means the index of $\mathbf{k}$-space. We measured the X-ray CTR scattering along the (00ℓ) direction which is normal to the surface of the samples. The shoulders observed in Figure 4 at ℓ around 2.1~2.2 are induced by the inclusion of As atoms in InP. Therefore, these modified spectra reflect information of As atoms contained in InP.

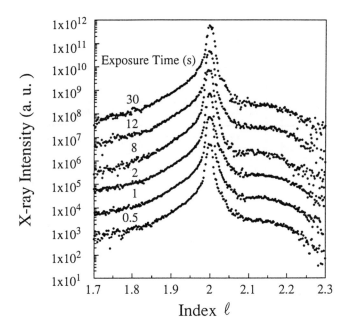

Figure 4. Measured CTR spectra. AsH$_3$-exposure time is shown at the left end of each spectrum. Data for a 0.5s-exposed sample is plotted on the real scale and other spectra are shifted upward by one order each for clarity.

3. MODEL STRUCTURE

To analyze the CTR scattering data, we need a model structure to generate the CTR spectrum with several fitting parameters. We assumed the model structure shown in Figure 5[12]. This model contains parameters such as n_c, n_h, x_h, d_c, d_b, $<\Delta z^2>$, and c/a. Here, n_c and n_h are the thicknesses of the cap layer and the heterolayer, respectively. x_h denotes the As composition at the heterolayer which is assumed to be formed by the As and P atom exchange during the AsH$_3$-exposure period. d_c and d_b denote the distributions of As atoms in the InP cap layer and in the InP buffer layer, respectively. The As distribution is assumed to be represented by the formula $x = x_h \exp(-n/d)$ where n in units of monolayer [ML] is the distance from the upper or lower interface, and x is the As composition of InP$_{1-x}$As$_x$ in the layer at the distance n[ML] from the interface. d is equal to d_c in the cap layer and equal to d_b in the buffer layer. $<\Delta z^2>$ is the mean square deviation which represents the roughness of the surface. c/a is the ratio of the lattice constants normal(c) and parallel(a) to the surface. Coherent growth is assumed and then a is constant. When the ratio is not equal to 1, there is a tetragonal distortion of the lattice in the InP$_{1-x}$As$_x$ layer. The c/a values at $x=1$ are listed in Table 1.

272

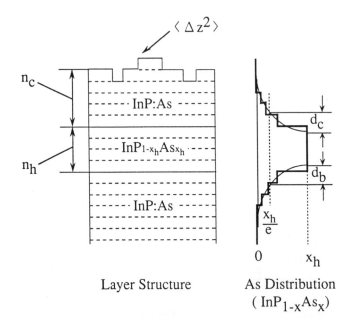

Layer Structure   As Distribution
                              ($InP_{1-x}As_x$)

Figure 5. A model for the CTR spectrum calculation. The parameters d_c and d_b are used to determine the As distribution in the cap and buffer layers as the As composition $x = x_h \exp(-n/d)$. $d = d_c$ or d_b.

Kinetic theory was used for the CTR spectrum calculation[13]. Comparing the calculated spectrum and the measured data, we determined the best fit values of the above parameters at the lowest R-factor (residual error ratio).

4. RESULTS AND DISCUSSION

Figure 6 shows an example of curve fitting. Data points close to the 002 Bragg point were not used for curve fitting since the kinetic theory is incorrect when the diffraction is too strong.

The parameters for the heterostructures obtained at the best fit are listed in Table 1 together with the lowest R-factors at each best fit.

4.1. Layer thicknesses and surface roughness

Most of the n_c values (the cap layer thicknesses) are 7ML which is quite close to the designed InP cap layer thickness of 20Å. The n_h values are 1 or 0. The difference is not very meaningful in the analysis of the As distribution. It slightly changes the shape near the peak As composition. The $< \Delta z^2 >$ values are much less than $1ML^2$. This means that even if there are steps of 1ML height, the smooth terraces predominate dips or plateaus. The c/a values are distributed between

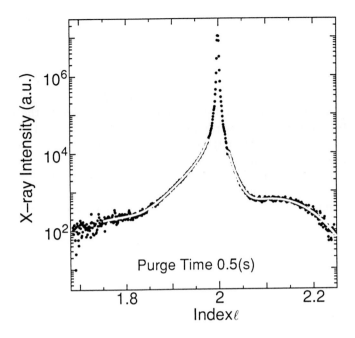

Figure 6. An example of curve fitting. Data points near the 002 Bragg point were not used for curve fitting since the kinetic theory is incorrect when diffraction is too strong.

Table 1. Values of fitting parameters which give the best fit to the measured CTR spectra of AsH$_3$-exposed samples.

	0.5s	1s	2s	8s	12s	30s
n_c (ML)	7 (20.5Å)	7	7	7	8 (23.5Å)	7
n_h (ML)	1	0	0	1	0	1
x_h (ML)	0.291	0.275	0.294	0.279	0.318	0.210
d_c (ML)	0.525	1.02	1.41	3.10	2.15	3.10
d_h (ML)	0.442	0.393	0.497	1.40	0.791	0.720
$\langle\Delta z^2\rangle$(ML^2)	0.088	0.084	0.155	0.269	0.00375	0.212
c/a	1.087	1.091	1.078	1.061	1.087	1.073
R-factor	0.0145	0.00905	0.00927	0.0205	0.0133	0.0157

1.06 and 1.09. c/a in the coherently grown InAs between InP layers is calculated to be 1.068 using a simple elastic model[14] with stiffnesses $C_{11}(=8.65\times10^{11}$[dyne cm^{-2}]) and C$_{12}$ (=4.85×10^{11} [dyne cm^{-2}]) of InAs. In the curve fitting, c/a value has little influence on the R-factor. However, the obtained values at the best fit are quite close to the calculated one.

4.2. As atom distribution

The important parameters for discussion of atom exchange are x_h, d_c and d_b which describe the amount and the distribution of As atoms due to the AsH$_3$ exposure of the InP surface. The distributions of As atoms calculated using these three parameters and the cap layer thicknesses (n_c) are shown in Figure 7. Considerable amounts of As exist in each sample.

In Figure 7, two features are observed. Firstly, the As compositions at each peak are almost the same, $i.e.$, 0.3 in all the AsH$_3$-exposed samples in spite of very different exposure times. There must be some mechanism that stops the exchange at that composition, $e.g.$, the formation of a surface reconstructed structure. This can be confirmed by measuring grazing incidence X-ray scattering in $situ$ as demonstrated by D.W. Kisker et $al.$[15]. Secondly, the As atoms distribute deeply in the cap layer with increasing AsH$_3$-exposure time. The amounts of As atoms in the buffer layer, which are equal to (As composition)×(layer thickness in units of ML), are about 0.3ML and independent of the exposure time. On the other hand, the

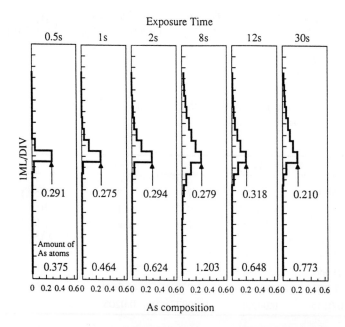

Figure 7. Profiles of As atom distribution in each AsH$_3$-exposed sample. As compositions at each peak are almost the same, $i.e.$, 0.3. From the area of the profile the amount of As atoms in units of ML is calculated.

amount of As atoms in the cap layer increases steeply for the first few seconds and appears to saturate after about 10s.

4.3. As/P atom exchange process

When the PH_3 flow is switched to the AsH_3 flow after the InP growth, P atoms on the top surface of InP must be removed since there is no P over pressure in the $[AsH_3 + H_2]$ flow. Then, As atoms readily occupy the P vacancies on the top surface. From Figure 7, it appears that this exchange process stops when the As composition reaches about 0.3 which is probably due to the formation of a surface reconstructed structure. This process is quite quick and is completed within 0.5s. Once the surface reconstructed structure is formed, the exchange process does not seem to proceed to the layer beneath the reconstructed layer. The solid phase diffusion of As atoms is apparently quite slow in this process.

To explain the long tails of As distribution in the cap layer, two processes are considered; one is the memory effects of As source gas which remains in the gas phase or adsorbed on the reactor wall, and the other is due to the excess As atoms adsorbed on the reconstructed surface.

The memory effect due to the remaining As source gas in the gas phase is not likely to occur because at the shortest exposure time of 0.5s the As distribution in the cap layer is as abrupt as that in the buffer layer as shown in Figure 7. It means that AsH_3 and PH_3 are switched quickly in the gas phase.

It is not easy, based on the present experiments, to separate the effect of the adsorbed As source on the reactor wall and that of the excess As atoms adsorbed on the reconstructed surface. If the desorption of the excess As from the surface in the PH_3 or P atmosphere is as quick as the As/P exchange at the P surface of InP in the AsH_3 flow, no excess As atoms should remain when the growth of InP starts with the TMIn flow. Indeed, even the As atoms once bonded to In atoms in InP surface, which formed 2ML of $InP_{0.72}As_{0.28}$ by the 4s exposure of AsH_3, are removed by the PH_3 purge immediately after the AsH_3-exposure as shown in Figure 8. Therefore, the adsorbed excess As atoms should be removed very quickly from the InP surface.

If the excess As atoms remain longer even in the PH_3 flow and these As atoms are incorporated into the growing InP, then the thickness of the heterolayer of the As composition 0.3 should increase with increasing AsH_3-exposure time. However, Figure 7 shows only 1ML of such layer at all the exposure times. Therefore, from the above discussion we tentatively conclude that the long tail of the As atom distribution in the cap layer is due to the memory effect of the As source on the reactor wall, though we still do not have direct evidence for this.

5. CONCLUSIONS

X-ray CTR scattering measurement was conducted for samples prepared by exposing the grown InP surface to AsH_3 and capped by InP to reveal the As/P exchange process during the growth of the heterointerface. By the analysis of the CTR spectra for samples with different exposure times, we obtained the As atom

276

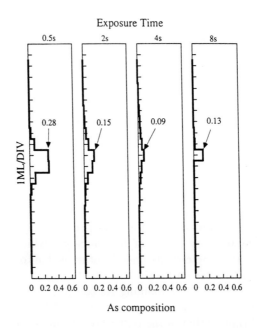

Figure 8. Profiles of As atom distribution in AsH$_3$-exposed and PH$_3$-purged sample. AsH$_3$ was supplied for 4s and then PH$_3$ was supplied for between 0.5 and 8s.

distribution profiles in the samples.

It was found that considerable amounts of As atoms existed in all the samples, though the InP surface was only exposed to AsH$_3$, and that the P atoms on the surface are quite quickly exchanged by the As atoms but the exchange stops at the As composition of 0.3, probably due to formation of a surface reconstructed structure.

The long tail of As atom distribution in the cap layer was tentatively concluded to be due to the memory effect of the As source on the reactor wall.

ACKNOWLEDGMENT

This work was performed as a part of the project (Proposal No. 93G195) accepted by the Photon Factory Program Advisory Committee. The authors would like to thank Prof. J. Harada and Dr. I. Takahashi at Nagoya University for their helpful discussion and assistance in the X-ray CTR measurement, N. Yamada and K. Fujibayashi at Nagoya University for their technical assistance, and H. Kamei at Sumitomo Electric for the preparation of samples. This work was supported in part by Grants-in-Aid for Scientific Research on Priority Areas "Crystal Growth Mechanisms in Atomic Scale" No. 04227106 and No. 05211101 from the Ministry of Education, Science and Culture.

REFERENCES

1. H. Kamei and H. Hayashi, J. Cryst. Growth **107** (1991) 567.
2. A.R. Clawson, T.T. Vu, S.A. Pappert and C.M. Hanson, J. Cryst. Growth **124** (1992) 536.
3. K. Streubel, V. Harle, F. Scholz, M. Bode and M. Grundmann, J. Appl. Phys. **71** (1992) 3300.
4. A.R. Clawson, X. Jiang, P.K.L. Yu, C.M. Hanson and T.T. Vu, J. Electron. Mat. **22** (1993) 155.
5. J.P. Wittgreffe, M.J. Yates, S.D. Perrin and P.C. Spurdens, J. Cryst. Growth **130** (1993) 51.
6. Y. Kobayashi and N. Kobayashi, Jpn. J. Appl. Phys. **31** (1992) 3988.
7. I.K. Robinson, Phys. Rev. B **33** (1986) 3830.
8. Y. Takeda, Y. Sakuraba, K. Fujibayashi, M. Tabuchi, T. Kumamoto, I. Takahashi, J. Harada and H. Kamei, Appl. Phys. Lett. **66** (1995) 332.
9. M. Tabuchi, Y. Takeda, Y. Sakuraba, T. Kumamoto, K. Fujibayashi, I. Takahashi, J. Harada and H. Kamei, J. Cryst. Growth **146** (1995) 148.
10. Y. Fujiwara, N. Matsubara, J. Yuhara, M. Tabuchi, K. Fujita, N. Yamada, Y. Nonogaki, Y. Takeda and K. Morita, Inst. Phys. Conf. Ser. No. **145** (1996) 149.
11. M. Tabuchi, K. Fujibayashi, N. Yamada, K. Hagiwara, A. Kobashi, H. Kamei and Y. Takeda, Inst. Phys. Conf. Ser. No. **145** (1996) 227.
12. M. Tabuchi, N. Yamada, K. Fujibayashi, Y. Takeda and H. Kamei, J. Electron. Mat. **25** (1996) 671.
13. J. Harada, Acta Cryst. **A48** (1992) 764.
14. E. Estop, A. Izrael and M. Sauvage, Acta Cryst. **A32** (1976) 627.
15. D.W. Kisker, G.B. Stephenson, P.H. Fuoss, F.J. Lamelas, S. Brennan and P. Imperatori, J. Cryst. Growth **148** (1992) 1.

Advances in the Understanding of Crystal Growth Mechanisms
T. Nishinaga, K. Nishioka, J. Harada, A. Sasaki and H. Takei (Editors)

279

Surface structures during silicon growth on an Si(111) surface

A. Ichimiya,[a] H. Nakahara[b] and Y. Tanaka[b]

[a]Department of Quantum Engineering, Nagoya University,
Furo-cho, Chikusa-ku, Nagoya 464-01 Japan

[b]Department of Applied Physics, Nagoya University,
Furo-cho, Chikusa-ku, Nagoya 464-01, Japan

Atomic structures during silicon growth on a Si(111)7×7 surface are investigated by reflection high-energy electron diffraction (RHEED) and scanning tunneling microscopy (STM). RHEED intensity rocking curves during growth are analyzed by dynamical calculations. From the analysis we conclude that backbonds of adatoms on the dimer-adatom-stacking-fault (DAS) structure are broken by two adsorbed silicon atoms at an initial stage of the deposition. Subsequently the structure is reconstructed into a pyramidal cluster type one. The formation of the metastable structure, such as the pyramidal cluster one, promotes successive epitaxial growth accompanied with stacking fault dissolution at the dimer-stacking-fault framework. Comparing RHEED pattern and oscillations we conclude that the 5×5 DAS structure is formed on terraces during growth, and the 7×7 DAS structure grows from step edge. In STM images of isolated silicon islands formed on the Si(111)7×7 surface, we observed stable (long lifetime) shapes of 5×5 islands, and found magic numbers of the 5×5 units in the islands. For 7×7 islands, however, it is hard to find stable ones in STM images. We discuss relation between formation of the 5×5 DAS structure during growth and the stability of the isolated 5×5 islands.

1. INTRODUCTION

It is well known that a clean Si(111) surface has a 7×7 structure. For the 7×7 structure, Takayanagi's dimer-adatom-stacking-fault (DAS) model [1] is widely accepted by several experimental and theoretical studies. This structure includes a stacking fault region in the unit cell. When silicon layers are grown on the silicon surface, it is expected that the surface structure during growth is also the DAS structure. Therefore epitaxial growth of silicon on the Si(111)7×7 surface requires successive dissolution of the stacking fault regions during growth, because it is considered that the surface intends to form the dimer-adatom-stacking fault(DAS) structure. In experiments on homoepitaxial growth of Si(111), intensity oscillation periods of reflection high-energy electron diffraction (RHEED) are irregular at the initial stage of the deposition called incubation region [2–6]. Altsinger et al. [7] showed by intensity analysis of LEED that double height islands nucleate at this

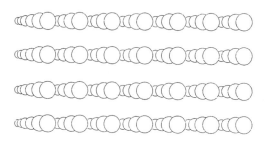

Figure 1. Projection of a crystal lattice at the one-beam condition.

region, and have suggested that such double height nucleation stimulates destruction of stacking fault layer in the 7×7. Such islands were also observed by STM experiments [8]. During homoepitaxial growth on Si(111) with RHEED intensity oscillations, the 5×5 structure is observed with the 7×7 one [6]. By microprobe RHEED system, Ichikawa and Doi [4] observed the 5×5 structure only at wide terrace region and disappearance of this structure by annealing at 700°C. Since step flow growth takes place with the 7×7 structure, it seems that the 5×5 structure promotes destruction of the stacking fault in terrace region during growth [6]. Using rocking curves at a one-beam condition of RHEED at which the incident electron beam direction is several degrees off from a symmetric crystal axis [9,10], Nakahara and Ichimiya [11] have found a new metastable structure during homoepitaxial growth on the Si(111) surface.

We describe at first a method of analysis of the growing surfaces by the one-beam approximation of RHEED briefly. Then we show results of structural analysis of Si(111) surface during homoepitaxital growth by RHEED, and scanning tunneling microscopy (STM) observations of thermal decomposition processes of isolated islands formed on a Si(111)7×7 at various substrate temperatures. Comparing above results, we discuss surface structures during growth and growth mechanism of silicon on Si(111)7×7.

2. ONE-BEAM APPROXIMATION OF RHEED

Since fast electrons are scattered dominantly in a forward direction by atoms, dynamic diffraction of high energy electrons in a crystal mainly occurs in a forward direction. Therefore it is considered that RHEED intensities depend on a lattice arrangement of a projection of crystal into the direction of the electron incidence, but scarcely depend on displacements on the incident direction. When a crystal is rotated around the surface normal axis, incident electrons see the crystal as shown in Figure 1. At this condition, called the one-beam condition at which the main diffraction beam is simply the specular one, a rocking curve of the specular reflection intensity is a function of surface normal components of atomic positions and layer atomic densities, but scarcely depends on lateral components of the atomic

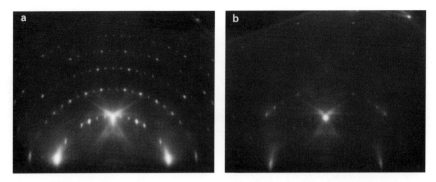

Figure 2. RHEED patterns with [11$\bar{2}$] incidence at glancing angle of 3.8°: (a) Si(111)7×7 and (b) δ7×7 pattern after 0.7BL deposition.

positions. Therefore surface normal components of atomic positions and atomic densities of surface layers are determined by dynamical calculation analysis of a one-beam rocking curve with short computation times. It is easy to determine surface structures during adsorption processes, epitaxial growth and phase transition processes by the one-beam RHEED analysis [12,13,6].

3. INITIAL STAGE OF SILICON GROWTH AT ROOM TEMPERATURE

The DAS structure of a clean Si(111)7×7 surface consists of adatoms, dimers and stacking faults. Therefore destruction of the DAS structure is required during epitaxial growth. At low temperatures chemical reactions between substrate atoms and adsorbates are very slow. Therefore metastable structures formed during epitaxial growth are able to be observed at several intermediate stages of the growth. At an initial stage of silicon deposition on the Si(111)7×7 surface, a RHEED pattern by the DAS structure changes into another 7×7 structure called δ7×7 [14]. Figure 2a shows a RHEED pattern of the 7×7 surface before deposition. A RHEED pattern of a silicon deposition of 0.7BL (BL:bilayer, 1BL=1.56×10^{15} atoms/cm^2) is shown in Figure 2b. This pattern is similar to those from the Si(111) surface adsorbed by hydrogen, lithium, sodium and some metal atoms at room temperature [15].

The RHEED intensities of the (00) and (3/7,3/7) rods changes during deposition as shown in Figure 3. The (00)-rod intensity was taken under the one-beam condition [10] with a glancing angle of 1.1°; the (3/7,3/7) rod one was measured for the [11$\bar{2}$] incidence at a glancing angle of 3.8°. Just after the deposition, the (3/7,3/7) rod intensity decreased rapidly and the almost disappeared at 0.25BL deposition. Since the (3/7,3/7) spots are mainly due to periodicity of the adatoms of the DAS structure, disappearance of the spots indicates disarrangement of the adatoms. Since the coverage of 0.25BL correspond to twice the adatom density of the DAS structure, this reveals that two adsorbed silicon atoms break backbonds of one adatom at the initial stage.

282

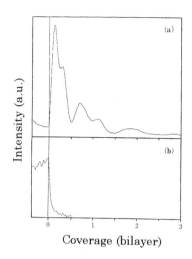

Coverage (bilayer)

Figure 3. Changes of the RHEED intensity during silicon deposition at room temperature; (a) for (00) rod and (b) for (3/7,3/7) rod. Curve (a) was measured under the one-beam condition with a glancing angle of 1.1° and curve (b) for [11$\bar{2}$] incidence with glancing angle of 3.8°.

In order to analyze the surface structure during growth, rocking curves at the one-beam condition was measured for various coverages of the silicon overlayer. Figure 4a shows rocking curves during silicon deposition at room temperature. The rocking curves were analyzed by the one-beam RHEED dynamical calculations [10] with the parameters shown in Figure 5. For the calculations we assumed that the fraction of the DAS region was proportional to the coverage of the adatoms. The surface normal components of the atomic positions of the DAS region were taken the same as the values obtained for the clean surface [9]. For the $\delta 7 \times 7$ region, it was assumed that layers 0 and 1 in Figure 5 included dimers and vacancies. The dimer positions were taken the same as the positions of the DAS structure [9]. In the $\delta 7 \times 7$ region, the positions of layers 0 and 1 were taken at the bulk values referred to the $\delta 7 \times 7$ structure of hydrogen adsorbed Si(111) [16]. For the calculations the parameters shown in Figure 5, $\theta_a, \theta_2, \theta_3, d_2$ and d_3, were changed widely for each deposition coverages. The calculated rocking curves shown in Figure 4b are in the best agreement with the experimental ones shown in Figure 4a. The values of d's and layer coverages during growth obtained by the rocking curve analysis are shown in Table 1.

The above result is consistent to the results of intensity variation of (3/7,3/7) during growth. From this analysis it is concluded that backbonds of adatoms are broken by adsorbed atoms and randomly distributed pyramidal clusters are formed subsequently during deposition on the dimer-stacking-fault framework as shown Figure 5. In STM images of a silicon surface after growing on Si(111)7×7 at room temperature, it is observed that growth nuclei of silicon atoms distribute randomly on the 7×7DAS unit cell, mostly on faulted half parts, at initial stage as

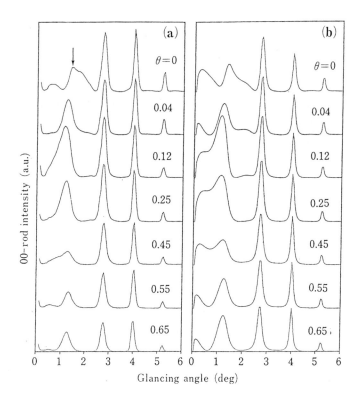

Figure 4. Rocking curves during the deposition at room temperature for various deposition coverages, θ, shown in the figure; (a) experimental ones and (b) calculated ones. Parameter used in the calculations are shown in Table 1 and Figure 3.

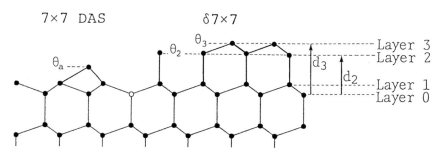

Figure 5. [01$\bar{1}$] side view of geometric arrangements for the calculations in Figure 4b. d and θ are parameters of layer positions and layer coverages respectively. The parameters $d_2, d_3, \theta_a, \theta_2$ and θ_3 correspond to those in Table 1. Layers 0 and 1 are fixed at the bulk position. Positions of dimers, adatoms and atoms of the DAS structure regions are taken from the clean 7×7 DAS structure.

Table 1
Parameters for calculated rocking curves with the best fit shown in Figure 4b. Errors were estimated to be about ±0.1Å for positions and about ±0.05BL for coverages.

	Coverage(BL)			Positions(Å)	
Total	Adatom	Layer 2	Layer 3	Layer 2	Layer 3
θ	θ_a	θ_2	θ_3	d_2	d_3
0.00	0.12	0.00	0.00	–	–
0.04	0.10	0.06	0.00	3.1	–
0.12	0.06	0.18	0.00	3.1	–
0.25	0.03	0.28	0.06	3.1	4.0
0.45	0.00	0.43	0.14	3.3	3.7
0.55	0.00	0.43	0.24	3.3	3.7
0.65	0.00	0.43	0.34	3.3	3.7

less than 0.025BL deposition [17]. Upon further deposition the nucleated clusters cover the whole and do not cover the dimer rows, because lines of the dimer rows of the DAS structure are clearly seen [18]. Therefore it is concluded that coalescence of the nuclei does not occur crossing over the dimer rows at room temperature.

These RHEED and STM results suggest that silicon atoms adsorb randomly at adatom sites and induce no destruction of the dimer-stacking-fault framework at room temperature. Upon further deposition, amorphous silicon layer is formed on the $\delta7\times7$ structure because the rocking curve profile changes periodically with damping intensity oscillation [10]. This means that stacking fault and dimers remain at the interface between the crystal and the amorphous silicon layer. These results are consistent with those of X-ray diffraction and cross-sectional transmission electron microscopy (TEM). Values for d's and layer coverage (statistical occupancies for each layer) for the RHEED analysis are compared with X-ray diffraction results by Robinson et al. [18]. The values of atomic positions are in very good agreements each other. The values of coverage θ for the layers d_2 and d_3 are, however, very large in comparison with those of the X-ray results. Since the one-beam RHEED analysis is insensitive to lateral displacements of atoms, it is considered that the coverage values obtained by the present analysis include density of disordered atoms.

4. METASTABLE STRUCTURES DURING HOMOEPITAXIAL GROWTH

At substrate temperature higher than about 280°C, RHEED intensity oscillation is observed during homoepitaxial growth on a Si(111)7×7 surface as shown in Figure 6. For the beginning 5BL deposition at temperature up to 500°C, the oscillation amplitudes are irregular, but after that the oscillation become stable and periodic. At the periodic region, profiles of rocking curves changed periodically

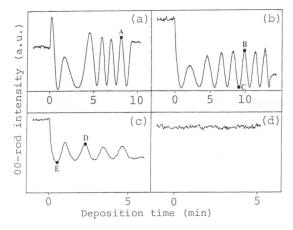

Figure 6. RHEED intensity oscillation during epitaxial growth of silicon on Si(111); (a) at 300°C, (b) at 400°C, (c) at 600°C and (d) at 700°C.

with coverage [5]. Therefore it has been concluded that defects and stacking faults remain scarcely in the epitaxial layers. In RHEED patterns during oscillations, 5×5 spots were observed with the 7×7 ones [6]. At low temperature up to 400°C, the spots of the 5×5 and the 7×7 are diffuse and streaky. At higher temperature above 500°C, the 5×5 spots were clearly seen in the patterns as shown in Figure 7. Therefore the size of the 5×5 and the 7×7 domains increases with increase of temperatures [6,19]. When RHEED intensity oscillation disappeared at about 700°C, only 7×7 spots were observed in a RHEED pattern. It is understood that RHEED oscillation is observed when an epitaxial layer-by-layer growth with 5×5 nucleation on terraces take place, and that intensity oscillation is scarcely observed when the growth proceeds by the 7×7 growth from step edges as step flow mode [6,4].

At the maximum intensity of the oscillations, one-beam rocking curves at higher than 400°C are very similar to the curve from the Si(111)7×7 DAS structure as shown in Figure 8. Therefore it is concluded that the structures of the 5×5 and the 7×7 during growth at this temperature region are nearly the same as the DAS structure. At a substrates temperature of about 300°C, the one-beam rocking curve is very different from those of the DAS structure as shown in Figure 8. The rocking curve is similar to that from δ7×7 structure [6,11]. Since RHEED intensity oscillation is very stable at 300°C as shown in Figure 6a, it seems that there are no defects remaining in the growing layers excepting growth front at surface. This suggest that the stacking fault regions are dissolved into the normal stacking with stimulation by further deposition of atoms because of existence of the metastable structure such as the pyramidal cluster type structure shown in Figure 9. Atomic positions of the structure are determined by dynamical diffraction analysis of the one beam rocking curve in Figure 8a.

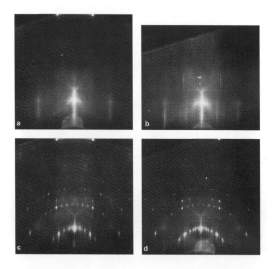

Figure 7. RHEED patterns during homoepitaxial growth on the Si(111)7×7; (a) at 280°C, (b) at 380°C, (c) at 500°C and (d) at 600°C.

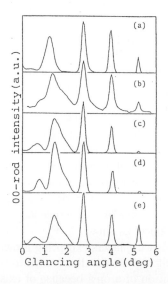

Figure 8. RHEED intensity rocking curves at the one-beam condition during homoepitaxial growth; (a) at 300°C, (b) at 400°C, (c) at 600°C, (d) at 700°C and (e) the clean surface at room temperature.

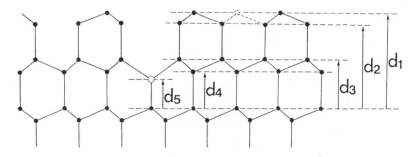

Figure 9. [01$\bar{1}$] side view of the pyramidal cluster model. Atomic positions shown by broken lines are added for the $\delta 7 \times 7$ structure. The values of d's are $d_1 = 6.1$Å, $d_2 = 5.43$Å, $d_3 = 3.135$Å, $d_4 = 2.34$Å and $d_5 = 2.25$Å.

During homoepitaxial growth on the Si(111)7×7, reconstruction of the stacking fault layer into the normal stacking by deposition of silicon atoms is assisted via breaking backbonds of adatoms and dimer bonds by coexistence of the 5×5 and 7×7 phases. Homoepitaxial growth on the 7×7 surface takes place at lower temperatures than the temperature of the phase transition, while the reconstruction into the normal stacking, such as phase transition from the 7×7 to "1×1", occurs at high temperature of about 830°C for the clean 7×7 surface [13]. The stacking fault layer is reconstructed by the transition from the $\delta 7 \times 7$ structure to a $\sqrt{3} \times \sqrt{3}$ one by adsorption of several metals at about 400°C. In this case it is suggested that $\delta 7 \times 7$ structure by metal adsorption promotes dissolution of the stacking fault [15]. Therefore it is concluded that the pyramidal cluster type or the nucleated cluster type structure promotes epitaxial growth as a precursor state at an intermediate stage at low substrate temperatures, and that adatom bond breaking makes a trigger of the dissolution of the stacking fault layer leading to the epitaxial growth at higher temperatures. Successive epitaxial growth at low temperature is promoted by instability of the metastable structures.

5. APPEARANCE OF 5×5 ISLANDS DURING HOMOEPITAXIAL GROWTH

It is well known that the 5×5 DAS structure is energetically rather unstable than the 7×7. There is a question why the 5×5 islands prefer to appear during homoepitaxial growth on the Si(111)7×7 surface. In order to find an answer of the question, we observed thermal instabilities of isolated silicon islands by STM at various temperatures. Figure 10 shows a thermal decomposition process of a silicon island formed by an STM tip at 520°C. Figure 11 shows a time dependence of island size measured by STM images. Alphabets indicated in the decomposition curve correspond to the islands shown in Figure 10a, b and c. The initial island in Figure 10a consists of two domains of the 7×7 and 5×5. The 7×7 part has round edges and the 5×5 part has straight edges. The 7×7 area decreases faster than

288

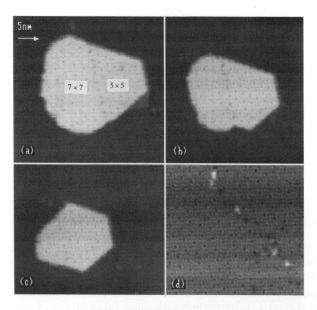

Figure 10. STM images during decomposition of a island at 520°C: (a) Just after deposition, (b) after 140 sec and (c) after 220 sec. The right-hand side of the island is the 5×5 and the left-hand side is the 7×7.

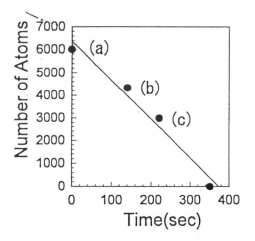

Figure 11. A time dependence of number of atoms in the island shown in Figure 10. The alphabets correspond to the Figure 10a, b and c.

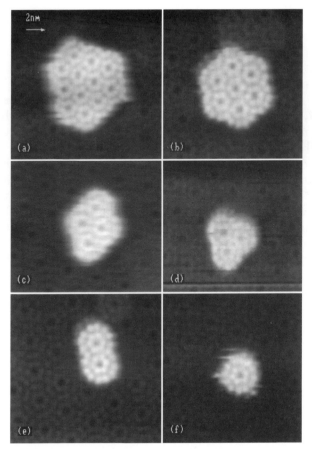

Figure 12. 5×5 islands with long lifetimes: 5×5 units of (a) 18.5 units, (b) 12 units, (c) 9 units, (d) 6.5 units, (e) 5 units and (f) 3 units.

the 5×5 one. Subsequently the 5×5 island remains with a characteristic shape. Fluctuations of the island images are observed due to moving atoms at the edges, while such fluctuations are hardly observed in the terrace and the substrate. Since the time dependence of the decomposition is nearly linear, it is considered that the island decomposition is due to atomic detachment at step edges [20]. The quick decomposition of the 7×7 area of the island is owing to instability of the round edge. 7×7 islands are mostly observed with round edges. It is hard to find 7×7 islands with completely straight edges, but 7×7 islands have one or two straight edges with a round side. Most islands are mixed two phases of 7×7 and 5×5. These islands are finally reduced into 5×5 with characteristic shapes. After formation of 5×5 islands, some islands have long lifetimes during decomposition process. Figure 12 shows long lifetime (hereafter called stable) islands found during STM observations at 400 to 600°C. These islands have all the 5×5 structure. The smallest one shown in Figure 12f consists of 3 units of the 5×5 DAS structure. The next one in Figure

290

12e has 5 units. The islands in Figure 12d, c, b and a have 6.5, 9, 12 and 18.5 units, respectively. These values are magic numbers of the 5×5 units for the stable shape of islands. For other numbers of the 5×5 units, kinks are formed at step edges, and lead to instability of the edge. In Figure 12a and d, step edges at faulted halves are dominantly observed.

These results reveal that the step edge energy of 5×5 structure is lower than that of the 7×7, because stable 7×7 islands are hardly observed, while the 7×7 structure is the most stable on the Si(111) surface. The 5×5 structure appears on terraces of small islands which are energetically predominant at the 5×5 step edge structure. Difference between the 7×7 step edge and the 5×5 one is number of dimers and corner holes for unit length; 1/5 holes and 2/5 dimers for the 5×5 and 1/7 holes and 3/7 dimers for the 7×7. Therefore it is considered that the 5×5 step edge is energetically predominant due to the less dimers and more holes than that of the 7×7. From above results of STM observations, we believe that appearance of the 5×5 islands during growth is due to the long life time of the 5×5 islands, although the chemical potential at the step edges of the islands are different from those of islands during growth.

ACKNOWLEDGMENTS

This work was carried out under the support of a Grant-in-Aid to the Scientific Research on Priority Areas by the Japanese Ministry of Education, Science and Culture (Nos. 03243219 and 04227216).

REFERENCES

1. K. Takayanagi, Y. Tanishiro, M. Takahashi and S. Takahashi, Surf. Sci. 164 (1987) 367.
2. T. Sakamoto, N. J. Kawai, T. Nakagawa, K. Ohta and T. Kojima, Surf. Sci. 174 (1986) 651.
3. J. Aarts and P. K. Larsen, Surf. Sci. 188 (1987) 391.
4. M. Ichikawa and T. Doi in: Reflection High Energy Electron Diffraction and Reflection Electron Imaging of Surfaces, Eds. P. K. Larsen and P. J. Dobson (Plenum, New York, 1988) p. 343.
5. H. Nakahara and A. Ichimiya, J. Cryst. Growth 95 (1989) 472.
6. H. Nakahara and A. Ichimiya, Surf. Sci. 241 (1991) 124.
7. R. Altsinger, H. Busch, M. Horn and M. Henzler, Surf. Sci. 200 (1988) 235.
8. U. Köhler, J. E. Demuth and R. J. Hamers, J. Vac. Sci. Technol. A7 (1989) 2860.
9. A. Ichimiya, Surf. Sci. 192 (1987) L893.
10. A. Ichimiya, Structure of Surfaces III (1990) 162.
11. H. Nakahara and A. Ichimiya, Surf. Sci. 242 (1991) 162.
12. S. Kohmoto, S. Mizuno and A. Ichimiya, Appl. Surf. Sci. 41/42 (1989) 107.
13. S. Kohmoto and A. Ichimiya, Surf. Sci. 223 (1989) 400.
14. H. Daimon and S. Ino, Surf. Sci. 164 (1985) 320.
15. A. Ichimiya, Mater. Res. Soc. Symp. Proc. 208 (1991) 3.

16. A. Ichimiya and S. Mizuno, Surf. Sci. 191 (1987) L765.
17. A. Ichimiya, T. Hashizume, K. Ishiyama, K. Motai and T. Sakurai, Ultramicroscopy 42-44 (1992) 910.
18. I. K. Robinson, W. K. Waskiewicz and R. T. Tung, Phys. Rev. Lett. 57 (1986) 2714.
19. M. Horn von Hoegen, J. Falta and M. Henzler, Thin Solid Films 183 (1989) 213.
20. A. Ichimiya, Y. Tanaka and K. Ishiyama, Phys. Rev. Lett. 76 (1996) 4721.

Advances in the Understanding of Crystal Growth Mechanisms
T. Nishinaga, K. Nishioka, J. Harada, A. Sasaki and H. Takei (Editors)
© 1997 Elsevier Science B.V. All rights reserved.

REM studies of surfactant-mediated epitaxy

H. Minoda and K. Yagi
Department of Physics, Tokyo Institute of Technology,
2-12-1 Oh-okayama, Meguro, Tokyo, 152, Japan.

In this chapter, some examples of the surfactant-mediated epitaxy studied by REM-RHEED are presented. Changes of the surface activity by surfactants modified the growth processes drastically. Another important factor for the heteroepitaxial systems in the surfactant-mediated epitaxy is a change of the surface energy of the overgrown films.

1. INTRODUCTION

It is well known that there are three different growth modes in thin film growth[1]. (I) A monolayer-by-monolayer (ML-by-ML) mode which is called Frank-van der Merwe (FM) growth mode in which only two-dimensional (2D) growth occurs. (II) An islanding mode which is called Volmer-Weber (VW) growth mode in which 3D islands nucleate, grow and coalesce. (III) Stranski-Krastanov (SK) growth mode in which growth starts in FM mode and VW growth starts at some stage. Figure 1 schematically illustrates these three growth modes. The growth mode depends on combination of substrate and overgrowth materials in the Molecular Beam Epitaxy (MBE). Thus, the growth mode can be changed by only change the combination of substrate and overgrowth materials.

Surfactant-mediated epitaxy (SME) is one of the techniques to obtain thin film with flat surface[2]. As described above, the growth mode is determined when the combination of substrate and overgrowth materials is fixed. In the SME, however, it is possible to modified the growth mode because the surfactant modifies surface properties of substrate and overgrowth materials. The processes of SME are as follows. At first, surfactant material is predeposited on a substrate surface. By the predeposition of the surfactant material, surface properties such as the surface activity and the surface energy can be modified. Then, overgrowth materials are deposited on the surface. If interatomic exchange occurs between the surfactant and the overgrown materials during the growth, the surfactant always on the topmost surface. Thus, the overgrowth material is deposited always on the surfactant and a film of overgrown material is formed directly on the bare substrate surface not on the surfactant by this rising process. In this way the growth mode can be modified in the SME, by changing the combination of substrate and overgrowth materials effectively.

In the case of growths of Si and Ge on Si substrate, the surfactant is classified into two different types [3]. Some materials such as As, Sb and Bi make the

294

surface active and overgrown materials (in this case Si and Ge) easily react with
the substrate surface[4-10]. The activation energies of surface diffusion of Si and
Ge atoms are higher and/or critical sizes for nucleation of Si and Ge atoms are
smaller on the surface covered by the surfactant than those on the bare Si surface.
Thus, in the case of heteroepitaxy, high density of small 3D islands are formed and
surface roughness of Ge films is suppressed on the surface mediated. However, it is
not possible to suppress the formation of 3D islands in this system. Degree of the
surface roughness can be controlled by selecting the optimum growth condition.

The other materials such as Ga and In make the surfaces inactive and Si and
Ge atoms do not easily react with the substrate surface [11-14]. The activation
energies of surface diffusion of Si and Ge atoms are lower and/or critical sizes for
nucleation of Si and Ge atoms are larger on the surfaces covered by such materials
than those on the bare surface. The surface energies of substrate and overgrown
materials are also reduced by the predeposition of the surfactants. Thus, in the
case of heteroepitaxy, 3D islands formation itself is suppressed and Ge films with
flat surface are formed on the surface mediated with this type of the surfactants.
Of course, if the reduction of the surface energy is not sufficient, 3D islands starts
to form. However, since 3D island formation itself is suppressed in this system,
it can be expected that Ge films with almost perfectly flat surface is obtained by
selecting the optimum growth condition.

It is well known that there is anisotropy in the growth on Si(001)2×1 surfaces
[15-16]. The predeposition of the surfactant modifies not only the surface activity
but also the surface anisotropy on the Si(001) surfaces.

Domain structures are formed on some surfaces. The domain structures affect the
thin film growth processes[17-18]. SME has been studied not only in the growths
of semiconductors on the Si substrate but also the growths of metal on metal
substrates[19]. In the metal on metal system modification of diffusion barrier at
the step edges by adsorption of the surfactants was reported in addition to above
mentioned effects of surfactant.

In this chapter, some examples of the SME which were studied by reflection
electron microscopy and diffraction (REM-RHEED) are presented.

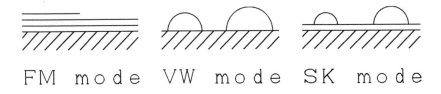

Figure 1. Diagrams illustrate three growth modes of films on a substrate.

2. GENERAL REMARKS IN REM

An ultra-high vacuum electron microscope was used in this study[20]. Imaging geometry of REM is illustrated in Figure 2[21-22]. The geometry of REM is similar to a RHEED geometry. RHEED pattern, with use of a conventional transmission electron microscope (CTEM), is at the back focal plane of the objective lens. The corresponding REM image is formed by selecting the RHEED spots by an objective aperture at the back focal plane. The REM image is foreshortened along the incident electron beam direction due to the small incident glancing angle. In this study, the foreshortening factor in the REM images is about 1/50 for Si(111) surfaces and about 1/40 for Si(001) surfaces. In all the REM images shown in this chapter, the direction of the incident electron beam (the direction of the foreshortening) is vertical.

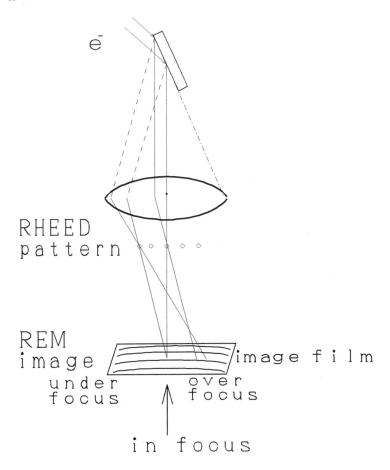

Figure 2. A schematic diagram of image formation in REM with CTEM.

Because the RHEED intensities depend on the surface crystallographic structure, the image contrast is sensitive to the surface structure (Bragg contrast)[23]. Dark and bright contrasts seen on the $Si(001)\sqrt{26}\times 1$-Au surfaces (Fig. 9) correspond to the Bragg diffraction contrast. They are due to a formation of different orientational domains on the surfaces. There is another contrast called Fresnel diffraction contrast[24]. It is seen in out of focus regions of the image. Contrast of the surface steps is due to the Fresnel diffraction contrast. Thus, the surface steps are clearly seen in the REM images.

3. RESULTS AND DISCUSSION

3.1. Surface segregation of the surfactant

One of the most important points in the SME processes is that interatomic exchange between the surfactants and overgrown materials occurs and the surfactants segregate to the topmost surface during film growth. Thus, it is important to investigate whether, the surfactants exist on the surface or not during the growth of Si and Ge. In this study the surface segregation of the surfactants was concluded from RHEED observations.

Figure 3 reproduces RHEED patterns taken before (a) and after (b) the deposition of Si on the $Si(111)"5\times 2"$-Au surface[24]. The superlattice reflections indicated by arrows correspond to those from the "5×2" structure. Weak diffuse streaks seen in (a) which divide the fundamental reflections are also from the "5×2" structure. The RHEED pattern did not change after the deposition of Si as shown in (b). This means that the Au atoms segregated to the topmost surface during the growth of Si to form the Au adsorbed "5×2" structure. The surface segregation of Au atoms was observed at temperatures between 400-730°C[15]. We found that In atoms also segregated to the topmost surface during the growth of Si. And the surface segregation of In atoms was noted at temperatures between 300-500°C[12]. At above temperature ranges, the surfactant materials sublimated during the growth. The sublimation temperatures of Au and In atoms from Si(111) surfaces were noted to be higher than those during the deposition of Si. Thus, the sublimation temperature is reduced during the deposition of Si due to the interatomic exchanges between the surfactant and overgrown materials.

RHEED observations were also performed in the case of heteroepitaxial growth to study surface segregation of surfactants. Figure 4 reproduces RHEED patterns taken before (a) and after (b) the deposition of Ge on Au-predeposited "5×2" structure[25]. Direction of incident electron beam was parallel to the [112] direction of Si substrate crystal. The "5×2" structure seen in (a) transformed to the $\sqrt{3}\times\sqrt{3}$ structure after the deposition of Ge at 500°C in (b). The $\sqrt{3}\times\sqrt{3}$ structure reversibly transformed to the 1×1 structure at 600°C. It should be noted that the $\sqrt{3}\times\sqrt{3}$ structure was formed by the deposition of Au on a Ge film of 2ML formed on a Si(111) surface. This $\sqrt{3}\times\sqrt{3}$ structure also reversibly transformed to the 1×1 structure at 600°. Thus, the $\sqrt{3}\times\sqrt{3}$ structure seen in Fig. 4(b) correspond to the Au-adsorbed Ge(111) structure. It was also reported that the $\sqrt{3}\times\sqrt{3}$ structure is

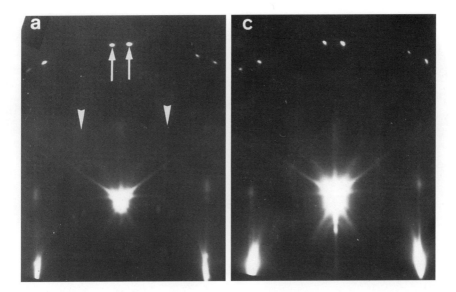

Figure 3. RHEED patterns (a) before and (b) after the deposition of Si of 3ML on a Si(111)"5×2"-Au surface. It is noticed that the RHEED pattern did not change indicating that Au atoms segregated to the topmost surface during deposition of Si.

formed by the deposition of Au on a Ge(111) surface[25]. Thus, the $\sqrt{3}\times\sqrt{3}$ structure seen in (b) is due to the surface segregation of Au atoms to the topmost surface during the growth of Ge. We found that In atoms also segregated to the topmost surface to form In-adsorbed Ge(111) structure during the growth of Ge independent of the surface structures (the Si(111)$\sqrt{3}\times\sqrt{3}$-In and Si(111)4×1-In) before the growth of Ge.

3.2. Suppression of 2D island formation

Growth features of Si and Ge on Si(111) surfaces were drastically changed by the predeposition of Au and In. In this section, examples of effects of adsorption of surfactant on the growth of Si and Ge on Si surfaces are presented. Figure 5 reproduces the REM images taken during the deposition of Si on Si(111) clean surface (a) and the In-predeposited Si(111) $\sqrt{3}\times\sqrt{3}$ surface (b)[12]. Growth conditions in (a) and (b) were the same. The substrate temperature was 420°C and the deposition rate was 0.3ML/min and the total amount of deposited Si was 0.5ML. Wavy lines seen in the images correspond to single height steps of the Si(111) surface and steplike marks in (a) and (b) show the sense of the steps. Dark fine images seen on the terraces correspond to nucleated 2D islands. Island size is very small in (a)

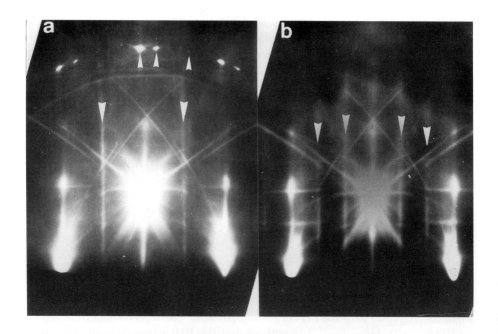

Figure 4. RHEED patterns taken (a) before and (b) after the deposition of Ge on a Si(111)"5×2"-Au surface. Transformation of the RHEED patterns from the "5×2" structure to the $\sqrt{3}\times\sqrt{3}$ structure is seen. The $\sqrt{3}\times\sqrt{3}$ structure is one of the Au-adsorbed Ge(111) structure, which shows surface segregation of Au atoms during the growth of Ge.

and almost all the surface is covered by islands. Low density of 2D islands are seen in (b) and their sizes are larger than those in (a). Surface terraces are not completely covered with 2D island in the both images. Bright areas indicated by arrow heads in (a) and (b) correspond to the flat areas without 2D island formation (denuded zone). Width of the denuded zone related to the surface diffusion distance under this growth condition. Thus, it is obvious that the surface diffusion distance on the $\sqrt{3}\times\sqrt{3}$ structure is longer than that on the 7×7 surface.

Changes in density of nuclei in the heteroepitaxy due to the predeposition of surfactants were also observed. Figure 6 compares growth features of Ge on a Si(111)7×7 (a) with that on a Si(111)"5×2"-Au surfaces at 500°C[25]. Deposition rate was about 0.5ML/min and the total amount of deposited Ge was 0.5ML in both cases. High density of 2D nuclei on the Si(111) surface is noted in (a). The 2D islands are not seen on narrow terraces and their density is very low on wide terraces in (b).

2D island nucleation of Ge is also suppressed on the In-adsorbed surfaces. An effect of domain boundaries of the "5×2" observed on the growth of Si [15] was not observed in the case of growth of Ge on the "5×2" structure. This is due to the transformation of the surface structure during the growth of Ge by surface segregation of Au atoms. As described above, the surface structure transformed from the "5×2" to the $\sqrt{3}\times\sqrt{3}$ structure in this system and the domain structure of the "5×2"structure does not remain on the surface during the growth.

3.3. Suppression of 3D island formation

It is well known that Ge grows in SK growth mode on Si surfaces. In the section 3.2, the effects of the surfactant on formation of 2D islands are described. The surfactant also affects formation of 3D islands of Ge. Formation of 3D islands of Ge was suppressed on the Au-adsorbed Si(111) surface[25].

Figure 7 compares growth features at later stages of Figure 6: Ge of 6ML was deposited on the Si(111) 7×7 (a) and on the Si(111)"5×2"-Au surfaces at 500°C. Dark images in (a) are 3D islands of Ge. The surface is not covered with 2D islands

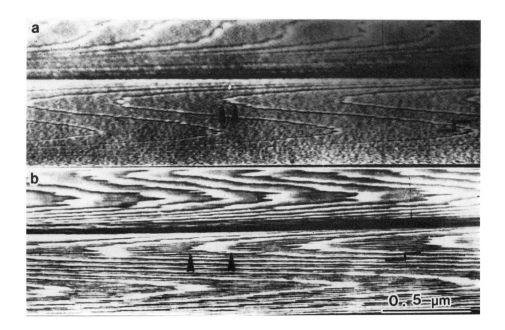

Figure 5. REM images comparing growth of Si on a 7×7 clean surface (a) with that on a $\sqrt{3}\times\sqrt{3}$-In surface at 420°C. Deposition rate and deposited amount of Si are 0.3ML/min and 0.5ML, respectively. Denuded zone width with bright contrast in (a) is smaller than that in (b).

Figure 6. REM images comparing features of 2D island formation of Ge on a 7×7 clean surface (a) and "5×2"-Au (b) surface at 500°C. High density of 2D islands are seen in (a) and relatively flat terraces are seen in (b).

in (b) showing suppression of 2D nucleation on the "5×2" structures. 3D islands were not formed on flat terraces at this stage in (b). This means that the 3D island nucleation as well as 2D nucleation is suppressed on the Au-adsorbed surfaces.

3D island nucleation was found to be suppressed also on a $\sqrt{3}\times\sqrt{3}$-In and a 4×1-In structures[11]. Figure 8 compares growth features of Ge on the 7×7, the $\sqrt{3}\times\sqrt{3}$-In and the 4×1-In surfaces. Growth conditions were almost the same in three cases. Total amount of deposited Ge was 3ML in all the images. Arrowheads seen in (b) and (c) correspond to the position of the step edges. In (a), dark images correspond to the 3D islands of Ge. On the other hand 3D islands of Ge were not seen in (b) and (c). Only 2D islands were formed on the $\sqrt{3}\times\sqrt{3}$ and the 4×1 structures. These facts mean that the formation of the 3D islands of Ge was suppressed on the In-adsorbed surface and the critical thicknesses for 3D islands on the In-adsorbed surfaces were thicker than that on the clean surface. The critical thickness on the $\sqrt{3}\times\sqrt{3}$ surface was about 6ML and that on the 4×1 surface was about 8ML under this growth condition. Coverage of In of the $\sqrt{3}\times\sqrt{3}$ structure is 1/3ML and that of the 4×1 structure is 1ML, respectively. Thus, the difference in the critical thickness on these two In-adsorbed surface structures should not be due to the difference of the surface structure during the growth of Ge. The difference in the critical thickness is caused by the difference in the surface energy due to the difference in the atomic density of In[11, 26]. Due to the lower surface energies, the surfaces of thin films tends to keep the (111) surfaces and islands nucleation is suppressed by the predeposition of In.

The corresponding RHEED patterns of (b) and (c) indicated the 1×1 structure which is one of the surface structures on the In-adsorbed Ge(111) surface[27]. And the surface structure did not change by the further deposition of Ge on the both In-adsorbed Si(111) surface. The 1×1 structure which was formed after deposition of Ge on the $\sqrt{3} \times \sqrt{3}$ is disordered structure because coverage of In is 1/3ML. On the other hand the 1×1 structure which was formed after deposition of Ge on the 4×1 is ordered structure because coverage of In is 1ML.

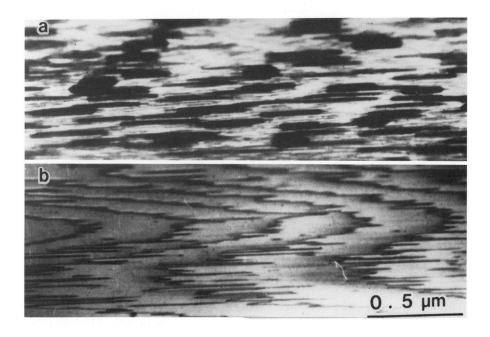

Figure 7. REM images comparing features of 3D island formation at latter stages of Fig. 6: Ge deposition on the 7×7 clean surface (a) and on the "5×2"-Au surface at 500°C. Amount of deposition is 6ML in both cases. High density of 3D islands are seen in (a) and relatively flat terraces with no islands are seen in (b).

Figure 8. REM images taken after the deposition of Ge of 3ML on the 7×7 (a), the $\sqrt{3}\times\sqrt{3}$ (b) and the 4×1 structures(c). The formation of 3D islands in (a) but no island formation in (b) and (c) are noted.

3.4. Anisotropy of growth

The Si(001)2×1 structure is anisotropic. By predeposition of Au atoms anisotropy of the surface diffusion and capturing rate of Ge atoms at the 2D island edges are modified[18].   Figure 9 reproduces a series of REM images taken during growth of Ge on a Si(001)$\sqrt{26}$×1-Au surface at 400°C[28-29]. The total amount of deposited Ge in (a)-(c) are 0, 1, and 2ML, respectively. Bright and dark contrast domains of the $\sqrt{26}$×1 structure are seen in (a). The domain boundaries (DBs) are along $\langle110\rangle$ directions and the domains are rectangular, although they are seen as oblique shape due to foreshortening and small misalignment of the incident electron beam from the exact [110] direction. Dark and bright contrast domains correspond to $\sqrt{26}$×1 and 1×$\sqrt{26}$ domains, respectively. DBs are clearly seen even after the deposition of Ge. 2D islands of Ge indicated by small arrows which elongate to the vertical direction were formed on the $\sqrt{26}$×1 domains (the dark domains in (a)). Taking into account of foreshortening factor of the REM image, a typical size of

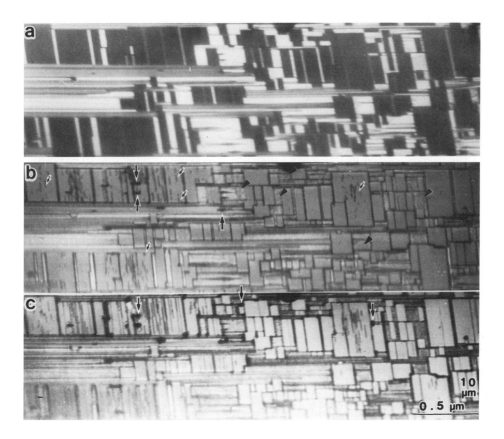

Figure 9. REM images taken during Ge deposition on a Si(001)$\sqrt{26}\times 1$-Au surface at 400°C. Coverages of Ge in (a)-(c) are 0, 1 and 2ML, respectively.

islands in the $\sqrt{26}\times 1$ domain is $1000\times 10\text{nm}^2$ and the islands have more anisotropic shapes than apparent shapes seen in the REM image. It is noted that the islands are always away from the vertical DBs (existence of denuded zone of nucleation) and 2D island nucleation did not occur in small domains. On the other hand, it is seen that some of the islands touch the horizontal DBs.

The 2D islands which elongate to the horizontal direction in the $1\times\sqrt{26}$ domains (the bright domains in (a)) are also seen in (b) as indicated by arrow heads. Density of the islands in the $1\times\sqrt{26}$ domains is seemingly higher than that in the $\sqrt{26}\times 1$ domains. Edges of almost all the islands in the horizontal direction touched to the vertical DBs. The denuded zones in the $1\times\sqrt{26}$ domain are difficult to observe due to the foreshortening. 3D islands nucleated preferentially at the DBs as indicated by large arrows in (b). After 2ML deposition the DBs are still clearly seen in (c).

The islands in the $\sqrt{26}\times1$ domains increased in size and density. The islands expanded mainly in the vertical direction and their width have not changed from (b) to (c). These facts suggest that the capturing rate of Ge in the vertical direction is high while that in the horizontal direction is low.

Figure 10 is a diagram which schematically illustrates anisotropy of surface diffusion and capturing rate of Ge in the $1\times\sqrt{26}$ domains (A) and the $\sqrt{26}\times1$ domains (B). Thick vertical and horizontal solid lines represent DBs. The domains are quite anisotropic in shape due to the heating current effect[28]. Rectangles on each domain show typical shapes of the 2D islands formed in the two domains. The shapes of the islands formed in the $\sqrt{26}\times1$ domains is very anisotropic due to anisotropy of capturing rate of Ge. On the other hand the island shapes in the $1\times\sqrt{26}$ domains are illustrated differently from those in the $\sqrt{26}\times1$ domains (less anisotropic in shape) because the islands could not elongate sufficiently in the domains due to their small horizontal domain size. Widening of the 2D islands might occur (rectangles in the $1\times\sqrt{26}$). Arrows put near the 2D islands show the directions of large capturing rate of Ge atoms in each domain.

The distances between islands indicated by P~S and their nearest vertical DBs are larger than those between the islands and the horizontal DBs. These distances are associated with the surface diffusion distances in two directions. Difference of the distances in the vertical and the horizontal directions shows the anisotropy in the surface diffusion of Ge atoms. The islands indicated by S touches to the horizontal DBs because of the slower surface diffusion in the vertical direction. The vertical direction in the image is the parallel to the 1-fold direction in the $\sqrt{26}\times1$ domains. (The 1-fold direction is not perpendicular to the $\sqrt{26}$-fold direction. However, in this paper, the direction perpendicular to the 1-fold direction is presented as $\sqrt{26}$-fold direction.) On the other hand, in the $1\times\sqrt{26}$ domains the vertical distances between islands are larger than the distances between islands and their nearest vertical DBs. This fact also shows anisotropy of the surface diffusion. In this way, the anisotropy of surface diffusion of Ge is concluded as follows: the surface diffusion in the $\sqrt{26}$-fold direction is faster than that parallel to it in the both domains. Large arrows at the bottom show the faster surface diffusion directions of Ge atoms in the domains.

It is seen that the apparent coverage of Ge depends on the domain (difference in area covered with 2D islands of Ge in each domain). The apparent coverages of Ge in the $1\times\sqrt{26}$ domains is higher than those in the $\sqrt{26}\times1$ domains. The difference in the apparent coverage of Ge in the two domains can be explained as follows. Due to the anisotropy of surface diffusion and shape of the domain, Ge atoms deposited on the $\sqrt{26}\times1$ domains diffuse in the $\sqrt{26}$-fold direction and many Ge atoms can diffuse to the vertical DBs. Thus, small amount of deposited Ge atoms are captured by 2D islands in the $\sqrt{26}\times1$ domains. On the other hand, since deposited Ge atoms diffuse in the $\sqrt{26}$-fold direction and the $1\times\sqrt{26}$ domains are elongated in the $\sqrt{26}$-fold direction, most of the deposited Ge atoms are captured by the 2D islands in the $1\times\sqrt{26}$ domains. Ge atoms captured by the DBs cause formation of 3D islands at the DBs.

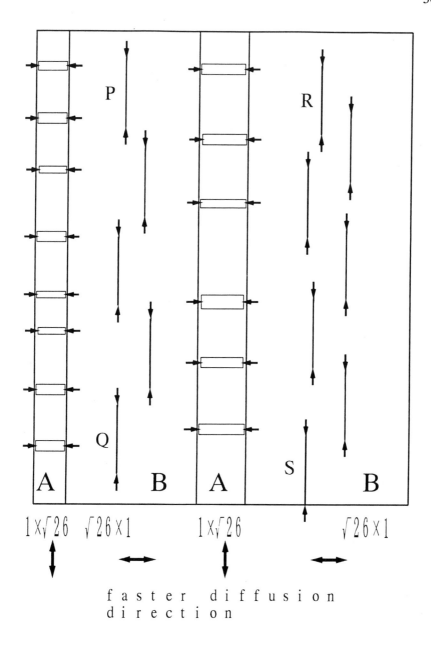

Figure 10. A diagram which schematically shows anisotropy of surface diffusion and capturing rate at step edges of Ge islands on the $\sqrt{26} \times 1$ domains.

4. CONCLUSION

In this chapter, some effects of surfactants on the crystal growth of semiconductor on the Si surfaces studied by in situ REM and RHEED are presented. The surface segregation of the surfactants to the topmost surface are noted and this is one of the most important factors in the SME. The effects of the surfactant are as follows. (I) the change of the surface diffusion of Si and Ge was noted. This correspond to the change of activity of the surface by the predeposition of the surfactant. The formation of the 2D islands is suppressed due to the reduction of the surface activity. (II) the change of the critical thickness of the formation of 3D islands of Ge was noted. And the critical thickness for 3D island formation depends on the coverage of the surfactants (In atoms). This is due to the change of the surface energy. (III) the change of the anisotropy of the growth of Ge was noted on the Si(001) surface.

REFERENCES

1. E. Bauer, Z. Kristallogr. 110 (1958) 423.
2. M. Copel, M. R. Reuter, E. Kaxiras and M. R. Tromp, Phys. Rev. Lett. 63 (1989) 632.
3. B. Voigtländer, A. Zinner, T. Weber and H. P. Bonzel, Phys. Rev. B52. (1995) 7583.
4. G. Mayer, B. Voigtländer and N. M. Amer, Surf. Sci.272 (1992) L541.
5. A. Kawano, I. Konomi, H. Azuma, T. Hioki and S. Noda, J. Appl. Phys.74(1993)4265.
6. S. Higuchi and Y. Nakanishi, Surf. Sci., 254 (1991) L465.
7. K. Sakamoto, K Kyoya, K. Miki, H. Matsuhata and T. Sakamoto, Jpn. J. Appl. Phys., 32 (1993) L204.
8. M. Copel, M. R. Reuter, M. H. von Hoegen and M. R. Tromp, Phys. Rev., B42 (1990) 11682.
9. H. J. Osten, J. Appl. Phys.74 (1993) 2507.
10. G. Mayer, B. Voigtlander and N. M. Amer, Surf. Sci. 272 (1992) L541.
11. H. Minoda, Y. Tanishiro, N. Yamamoto and K. Yagi, Surf. Rev. Lett. 2 (1995) 1.
12. H. Minoda, Y. Tanishiro, N. Yamamoto and K. Yagi, Surf. Sci. 287/288(1993)915.
13. S. Iwanari and K. Takayanagi, J. Cryst. Growth 119 (1991) 229.
14. H. Nakahara, M. Ichikawa and S. Stoyanov, Ultramicroscopy 48 (1993)417.
15. Y. W. Mo, B. S. Swartzentruber, R. Kariotis, M. B. Webb and M. G. Lagally, Phys. Rev. Lett 63(1989) 2393.
16. Y. W. Mo, B. S. Swartzentruber, R. Kariotis, M. B. Webb and M. G. Lagally, J. Vac. Sci. Technol. B8(1990) 232.
17. H. Minoda, Y. Tanishiro, N. Yamamoto and K. Yagi, Appl. Surf. Sci., 60/61 (1992)107.
18. H. Minoda, Y. Tanishiro, N. Yamamoto and K. Yagi, Surf. Sci., 331-333 (1995).

19. H. A. van der Vegt, H. M. Pinxteren, M. Lohmeier, E. Vlieg and J. M. C. Thornton, Phys. Rev. Lett. 68 (1992) 3335.
20. P. K. Larsen and P. J. Dobson, Reflection High-Energy Electron Diffraction and Reflection Electron Imaging of Surfaces, Plenum, New York, (1988).
21. K. Yagi, Appl. Cryst. 20 (1987) 147.
22. K. Yagi, RHEED and REM, in: Electron Diffraction Techniques II, J. M. Cowley1 ed. Oxford University Press, Oxford, (1993).
23. N. Osakabe, Y. Tanishiro, K. Yagi and G. Honjo, Surf. Sci. 102 (1981) 424.
24. H. Minoda and K. Yagi, Ultramicroscopy 48 (1993) 371.
25. G. Le Lay, M. Manneville and J. J. Metois: Surf. Sci. 123 (1982)117.
26. D. J. Eaglesham, F. C. Unterwald, D. C. Jacobson, Phys. Rev. Lett. 70 (1993) 966.
27. T. Ichikawa,Surf. Sci. 111(1981) 227.
28. A. Yamanaka, Y. Tanishiro and K. Yagi, Surf. Sci., 264 (1992) 120.
29. K. Oura, Y. Makino and T. Hanawa: Jpn. J. Appl. Phys., 15 (1976) 737.

Advances in the Understanding of Crystal Growth Mechanisms
T. Nishinaga, K. Nishioka, J. Harada, A. Sasaki and H. Takei (Editors)

Oscillations of the intensity of scattered energetic ions from growing surface

K. Nakajima, Y. Fujii*, K. Narumi, K. Kimura and M. Mannami

Department of Physical Science and Mechanics, Kyoto University,
606-01 Kyoto, Japan

At glancing-angle scattering of energetic ions from a surface under epitaxial growth, we first observed that the intensity of scattered ions oscillates with the period corresponding to the growth of one mono-molecular layer on the surface; sinusoidal oscillations for the MBE growth of GaAs(100) surface at 3 keV He scattering and skewed oscillations for vapour growth of PbSe(111) surfaces at 500 keV proton scattering. We have shown from the simulation of ion trajectories that the the oscillations are caused by ion scattering at steps which are formed during layer-by-layer growth of crystal surface. The oscillations can be used as a convenient monitoring method of epitaxial growth, since the ions from the ion gun, which is often equipped for surface cleaning in growth chamber, can be used for this purpose.

1. INTRODUCTION

Epitaxial growth of single crystal is one of the important methods of the controlled crystal growth for sophisticated materials fabrication. Studies of phenomena which are sensitive to the surface topography during growth have been central to advances in our understanding of the epitaxial growth processes.

One of the growth modes of vapour-phase epitaxy is the layer-by-layer growth. By the lateral growth at surface steps after the nucleation of two dimensional nuclei on a low index surface, one layer of atoms/molecules on the surface is completed. During the completion of the layer, the density of surface steps increases and then decreases. Thus the phenomenon sensitive to the surface step-density oscillates during the layer-by-layer growth. For molecular beam epitaxy (MBE), several growth monitoring methods have been contrived, which use the phenomena sensitive to the surface topography. The most well-known and well-established method is the intensity oscillations of scattered electrons of RHEED (Reflection High-Energy Electron Diffraction) [1]. In the method, it is shown that the period of the intensity oscillations of scattered electrons corresponds to the completion of a layer of atoms/molecules on the growing surface.

Similar periodic change during the growth has been known for Auger electron

---

*Present address: Department of Mechanical Engineering, Kobe University, 657 Kobe, Japan

line shape [2,3], angular dependence of ellipsometry [4], secondary electron emission at RHEED observation [5], He atom diffraction intensity [6], photo-electron emission from GaAs by irradiation of continuous light source [7], x-ray reflectivity and diffraction [8] and the x-ray intensity of crystal truncation rod [9].

We found that the intensity of scattered ions at glancing-angle incidence of energetic ions on a low index crystal surface oscillates when the surface is growing by layer-by-layer growth mode [10]. In the present paper, firstly we review briefly our studies on the ion-surface interaction at glancing-angle surface scattering of fast ions ($v > v_F$) with emphasis on the angular distribution of scattered ions, and then present our studies of epitaxial growth processes using the glancing-angle scattering of energetic ions.

2. GLANCING-ANGLE SCATTERING OF ENERGETIC IONS AT CRYSTAL SURFACE

2.1. Ion trajectories

At glancing-angle incidence of a fast ion on an atomically flat surface, the ion is reflected if the angle of incidence is smaller than a critical angle when the beam direction is not along low index axis. Angular distribution of the scattered ions has a maximum at scattering angle $2\theta_i$, where θ_i is the angle of incidence to the surface, and this is called specular reflection.

Consider the Cartesian coordinates with the x-axis perpendicular to the surface and the z-axis parallel to the surface plane with the origin on the surface atomic plane. A semi-infinite crystal is at $x < 0$ and the vacuum is $x > 0$. The motion of ion in the xz-plane is rewritten in the one-dimensional motion along x-axis, and the trajectory projected along the x-axis becomes

$$2E\frac{d^2x}{dz^2} = -\frac{dU(x)}{dx}, \tag{1}$$

where E is the energy of the ion and $U(x)$ is the ion-surface scattering potential.

There are two contributions to the potential $U(x)$; the continuum planar potential, which is the planar average over the yz-plane of ion-atom interaction potentials [11] and the dynamic image potential [12]. It must be noted that the trajectory calculated by eq. (1) applies to the cases where $\theta_i < \theta_c$, where the critical angle θ_c is approximately given by $E \sin^2 \theta_c = U(a_{TF})$. a_{TF} is the Thomas-Fermi screening distance and is of the order of 10^{-2} nm. Thus the critical angle is of the order of 10 mrad for MeV He ion at the (001) surface of SnTe.

The reflection of the ions from an actual surface is affected by surface defects and the scattering by individual electrons and thermally vibrating atoms. Thus the angular distribution of the scattered ions is broad centered at the scattering angle for specular reflection. Figure 1 shows an example of the angular distribution of scattered ions at the incidence of 500 keV protons with glancing-angle $\theta_i = 3.1$ mrad from the (111) surface of PbSe. A bright spot at the bottom is residual incident beam. The distribution has a maximum at the scattering angle of the specular reflection, i.e., $\theta_s = 6.2$ mrad and $\phi_s = 0$ mrad, where the scattering angle

θ_s and the azimuthal angle ϕ_s are defined in Figure 1. In addition, more than 50% of the incident ions penetrate the surface at step edges even at a carefully prepared flat surface. These ions which once penetrated the surface hardly reappear outside the surface, but only a small fraction of the ions appears from the step edges after suffering from larger energy losses. Such ions are easily identified as having larger energy losses [13].

If the incident beam direction is nearly parallel to one of the low index axes parallel to the surface plane, the ions are scattered by an atomic row potential to form a broad semicircle as observed downstream the target. This is called surface channeling and the trajectory of ion depends on the impact parameter of collision of the ion with the row of atoms [14–16].

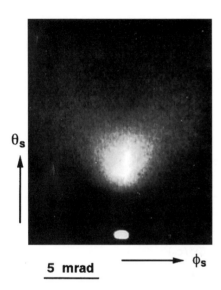

Figure 1. Two-dimensional angular distribution of scattered protons from the (111) surface of PbSe at the incidence of 500 keV protons with an angle of incidence 3.1 mrad relative to the surface. The bright spot at the bottom shows the residual incident beam.

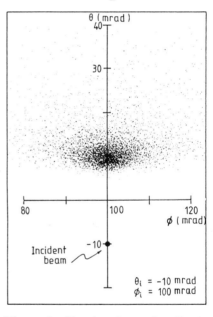

Figure 2. Simulated angular distribution of reflected ions at glancing-angle indicence of 0.7 MeV He$^+$ ions on a flat (001) surface of SnTe. The angle of incidence $\theta_i = 10$ mrad and the azimuthal angle relative to the [100] is 100 mrad.

2.2. Effects of surface steps on glancing-angle scattering

To study the effects of surface steps on the glancing-angle scattering of ions, we simulated ion trajectories at glancing-angle scattering at stepped (001) surfaces of SnTe [17]. The details of the simulation will be described later and only the results are shown here. The simulated angular distribution of reflected ions shows a broad peak centered at the angle for specular reflection, as shown in Figure 2, which reproduces the characteristic features of the observed ones. The origin of the broadening of the distribution is the scattering at the steps where the ions are deflected away from the angle for specular reflection.

For the simulation of ion trajectories, we assumed random arrays of steps on the (001) surface where only the mean separation D_s is given. Upward and downward steps, as viewed from the incident ions, were also introduced randomly. Total yield of scattered ions (the reflectivity) and the energy spectrum of ions at the angle for specular reflection were calculated. The angular distribution becomes broader, the energy spectrum has lower energy components and the reflectivity decreases as the step density increases. It was found from the comparison of the calculated results that the yield of ions at the angle for specular reflection is most sensitive to the step density. Figure 3 shows the relation between the yield of ions and the step density on the (001) surface of SnTe.

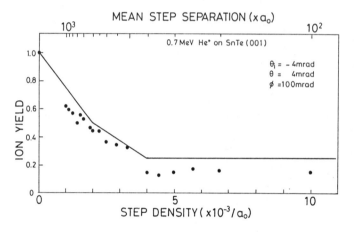

Figure 3. Calculated dependence of the yield of He ions scattered at specular reflection angle on the step density $1/D_s$. The detector window is 1×1 mrad^2. The glancing and azimuthal angle of the incident 0.7 MeV He ions are 4 mrad and 100 mrad respectively. The solid line shows the ion yield calculated with a simple mirror reflection.

The yield of ions scattered at the mirror reflection angle can be derived with a simple optical model. Four configurations of two neighbouring steps with separation D_s are possible; they are up-step·up-step, up-step·down-step, down-step·up-step and down-step·down-step pairs, as the up-steps and down-steps are equally probable. For the incidence of ions on the pairs with angle of incidence θ_i and the step height a_0, it is assumed that the terrace areas which are exposed to the ions reflect the ions at the mirror reflection angle and that the ions incident at the step edges penetrate inside the crystal. Then the yield $Y(D_s)$ of ion is expressed as

$$Y(D_s) = \begin{cases} 1 - a_0/(D_s|\theta_i|) & \text{for } D_s > 2a_0/|\theta_i| \\ 3/4 - a_0/(D_s|\theta_i|) & \text{for } 2a_0/|\theta_i| \geq D_s \geq a_0/|\theta_i| \\ 1/4 & \text{for } D_s < a_0/|\theta_i| \end{cases} \quad . \tag{2}$$

The yield $Y(D_s)$ is independent of the mean step separation D_s at step separation smaller than $a_0/|\theta_i|$, which shows that the incident ions are reflected only at the terraces surrounded by the up-step·down-step pairs and that the terraces of other pairs are not exposed to the incident ions. The yield $Y(D_s)$ is drawn by the solid line in Figure 3, which agrees well with the simulation.

The yield of ions scattered at the angle for specular reflection is the quantity that can be observed easily and quickly by an ion detector. Thus the ion yield at the angle for mirror reflection can be used as the monitor of surface step density.

3. MEASUREMENT OF ANGULAR DISTRIBUTION OF IONS

A system for measuring the angular distribution of ions at glancing-angle incidence on a surface of crystal was constructed. It is composed of a fluorescence screen combined with micro-channel plates (MCP) set in the vacuum so that the angular distribution of scattered ions can be observed by naked eyes through the view port of the vacuum chamber. With a CCD camera connected to a PC and video recorder, we could continuously record the angular distriubtion of scattered ions. Since the exposure time of a picture frame of video camera is less than 1/30 seconds, noise to signal ratio was improved by adding many picture frames. From the angular distribution thus prepared, intensity of scattered ions was determined for a chosen area of the distribution. In most of the studies here we measured the area surrounding the angle for specular reflection.

4. EXPERIMENTAL RESULTS AND ANALYSES

4.1. MBE growth of GaAs(001)
4.1.1. Experimental
The experiments were performed in an MBE growth chamber which is connected to a 3 keV ion source [10]. This MBE growth chamber had a conventional MBE system and a 30 keV electron gun for RHEED study. Ga and As beams were supplied from standard effusion cells. A clean (001) surface of GaAs was prepared by MBE growth of an undoped GaAs layer of 0.15 μm thick on a GaAs substrate which has been polished in a H_2SO_4-H_2O_2-H_2O solution and dissipated in situ by the thermal etch at about $600°C$ under an arsenic vapor pressure of 10^{-7} Torr. The surface showed 2×4 reconstructed structure under RHEED observation. Growth of GaAs layers on the prepared surface was performed by sending a Ga beam on the surface kept at $580°C$ under As vapor pressure 9×10^{-8} Torr. The growth rate was 0.4 nm/min.

A beam of 3 keV neutral He or He^+ ions from the ion source was collimated by apertures to a diameter less than 0.2 mm and to a divergence angle less than 0.8 mrad. The typical beam current on the target crystal was 3 pA. The GaAs crystal was mounted on a five-axis goniometer, and the beam of ions was incident on the crystal with an angle less than 20 mrad measured from the surface plane. Most of scattered He ions were neutral and the fraction of He^+ ions was estimated to be less than a few %.

4.1.2. Results and discussion

The intensity of the ions scattered at the angle for specular reflection is shown as a function of the growth time in Figure 4. When the As-stabilized surface was exposed to a constant Ga flux, the intensity showed periodic oscillations. The oscillations faded out during the growth. Ga flux was stopped at the time indicated by an arrow in Figure 4, and then the intensity recovered quickly. The period of the oscillations was equal to that of RHEED oscillations with the same growth condition, and corresponds to the growth time of one mono-molecular layer of GaAs.

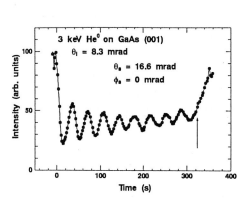

Figure 4. Intensity oscillations of the scattered ions at glancing angle incidence of 3 keV He^0 on the (001) surface of GaAs during the growth.

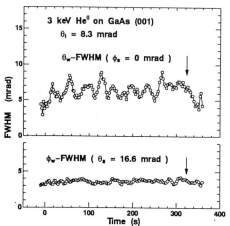

Figure 5. Dependence of FWHM of θ_s-distribution, θ_w, and FWHM of ϕ_s-distribution, ϕ_w, of ions scattered from the (001) surface of GaAs at glancing angle incidence of 3 keV He^0 during the growth.

The full width at half maximum of θ_s-distribution, θ_w, of scattered ions and that of ϕ_s-distribution, ϕ_w, are shown as a function of the growth time in Figure 5. The θ_s-distribution means angular distribution of scattered ions along θ_s-axis perpendicular to the surface and the ϕ_s-distribution means that along ϕ_s-axis parallel to the surface. θ_w shows periodic oscillations with the same period and the opposite phase to the intensity oscillations, while ϕ_w does not change.

The intensity oscillations are explained as due to the periodic change in the density of surface step accompanied by the layer-by-layer growth of GaAs.

In order to understand the θ_w-oscillations shown in Figure 5, the effect of steps on the ion trajectory has to be considered. When an ion travel near a step, the ion is deflected away from the angle for specular reflection. An upward step makes the ion scattered at a larger angle, while a downward step makes the ion scattered at a smaller angle. Thus the angular distribution of scattered ions in θ_s-direction becomes broader when the density of steps is large, and as the result θ_s showed periodic oscillations with the opposite phase to the intensity oscillations.

4.2. Homoepitaxial growth of PbSe(111)
4.2.1. Experimental

A single crystal of PbSe(111) was prepared by epitaxial growth in situ by vacuum evaporation on a cleaved surface of BaF_2(111) at 340°C, which was mounted on a high precision goniometer in an UHV chamber (base pressure 3×10^{-10} Torr) [21]. The chamber was connected to the Tandetron accelerator via a differential pumping system. A beam of 500 keV protons was collimated by a series of slits to less than 0.1×0.1 mm^2 and to a divergence angle less than 0.3 mrad. The typical beam current was less than 1 pA. The effect of surface channeling was avoided by choosing the incident azimuthal angle 28 mrad from the [11$\bar{2}$] axis.

The angular distribution of the scattered ions at glancing-angle incidence of 500 keV protons was observed during the homoepitaxial growth of PbSe at 220°C. At this temperature, the RHEED intensity oscillations were observed showing that the growth mode is the layer-by-layer mode [19]. The growth rate was about 0.3 nm/min, which was monitored with a calibrated quartz microbalance. Before each observation of the homoepitaxial growth of PbSe at 220°C, the temperature of the PbSe(111) was raised up to 340°C and more than 10 nm PbSe was evaporated on the surface. Because the growth mode at 340°C is known to be step flow mode [19], the step density on the surface was expected to become small by this procedure.

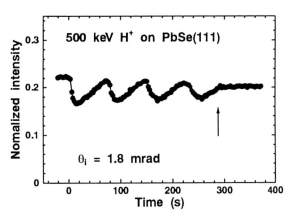

Fugure 6. The change of the intensity of the ions scattered into an area 3×3 mrad^2 around the position of the specular reflection at the incidence of 500 keV protons during the growth of PbSe(111) at 220°C. The intensity has been normalized by the the intensity of total ions detected by the MCP.

316

4.2.2. Experimental results

The intensity of the ions scattered into an area of 3×3 mrad^2 around the angle for the specular reflection (to be referred to as the intensity of specular reflection hereafter) was derived from the observed distribution. The change of the intensity of specular reflection during the homoepitaxial growth of PbSe is shown in Figure 6, normalized by the total yield of scattered ions detected by the MCP in order to avoid the effect of the beam fluctuation. The normalization procedure cancels out the beam fluctuation and the intensity oscillations are clearly seen. The growth started at $t = 0$ s and stopped at $t = 285$ s. The period of the oscillations is about 70 s, which corresponds to the growth time for one mono-molecular layer of PbSe. The shape of the oscillations is very different from a sinusoidal curve. The first intensity minimum occurs at $t = 15$ s, which is about a quarter of the oscillation period.

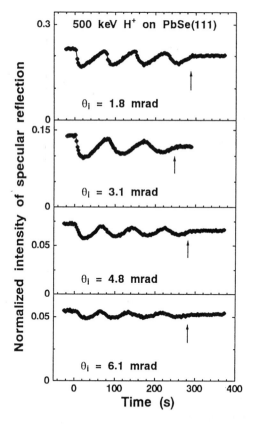

Fugure 7. Oscillations of the normalized intensity at various incident angles. The amplitude of the oscillations decreases with increasing incident angle. The arrows indicate the time at which the growth stops.

Figure 7 shows the normalized intensity of specular reflection observed during the growth at various incident angles. The oscillations can be seen at any incident angle. The period of the oscillations always corresponds to the growth time for one mono-molecular layer irrespective of the incident angle and the phases of the oscillations are the same. With increasing incident angle, the amplitude of the oscillations becomes smaller and the position of the intensity minimum moves to the right.

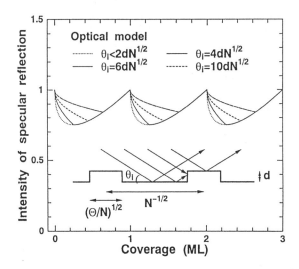

Figure 8. Intensity oscillations of the specularly reflected ions calculated with the optical model. The inset shows the ion trajectories.

4.2.3. Discussion - optical model

The observed result can be qualitatively discussed using the optical model for the specular reflection of fast ions [17]. For convenience, we assume (1) the growth mode is a perfect layer-by-layer mode, (2) the shape of the two-dimensional islands is square with one side parallel to the incident beam direction, (3) the density of the island, N, does not change during the growth, (4) all the islands have the same size and (5) the islands are arranged regularly with a separation of $N^{-1/2}$ between adjacent islands. With these assumptions, the intensity of specular reflection is written by

$$I = \begin{cases} 1 - 2a_0\sqrt{\Theta N}/\theta_i & \text{for } \sqrt{\Theta} < 1 - 2a_0\sqrt{N}/\theta_i \\ 1 - \sqrt{\Theta} + \Theta & \text{for } \sqrt{\Theta} \geq 1 - 2a_0\sqrt{N}/\theta_i \end{cases}, \tag{3}$$

where a_0 is the step height and Θ is the coverage of the two-dimensional islands. Figure 8 shows the calculated intensity of the specular reflection as a function of Θ at various incident angles. At small incident angle, the intensity decreases rapidly with increasing coverage and has a minimum at $\Theta = 0.25$. With increasing θ_i, the coverage of intensity minimum becomes larger and the amplitude of the oscillations becomes smaller. The calculated result reproduces the characteristic features of the observed result.

4.2.4. Simulation of ion scattering on a modelled surface

A computer program was developed for simulation of ion scattering at glancing-angle incidence on a (111) surface of NaCl-type crystal. The basic function of the program is to trace the energies and the trajectories of ions. The trajectory of an ion is decided as follows. An ion starts at a distance of $7d$ from the surface atomic layer, where d is the interplanar distance between adjacent atomic layers and gradually approaches to the surface by a given angle of incidence. For an ion moving in vacuum at the distance larger than $3d$ from the surface atomic layer, the interaction is described by the continuum surface planar potential. The dynamic image potential is neglected. For an ion moving in the crystal or in vacuum at the distance closer than $3d$ from the surface, the interaction is reduced to a series of isolated binary collisions of the ions with the atoms. The angle of deflection in a binary collision is derived with the use of the impulse approximation. For the ion-atom interaction potential, the analytical Molière approximation was employed [20]. The correlation of the thermal displacements of atoms from their equilibrium positions is neglected and the displacement of each atom is derived with the use of the isotropic Gaussian distribution defined by the mean square displacement calculated from the Debye temperature, for which 144 K is chosen.

The interaction between the ions and target electrons is approximated by an empirical energy loss function which depends on the distance of the ion to the surface atomic plane [13], and the ion deflection due to electron scattering is also calculated from the energy loss with the use of the impulse approximation in ion-electron collision. Thus the calculated ion trajectory is made up of a series of linear hops. More than 10^4 trajectories were calculated for a set of incident ions and surface structure to obtain an angular distribution.

The growth model of the PbSe(111) surface used in our simulation is very simple. The surface before growth is perfectly smooth with no steps. At the moment when the homoepitaxy of PbSe starts, two-dimensional nuclei of the growth with a height of one molecular layer are formed randomly on the surface to the given density. They grow into two-dimensional islands in the shape of circle and no more nucleus is generated until a new molecular layer is completed. When the two-dimensional islands overlap each other during the growth, the parts of overlap are eliminated. Eventually, the two-dimensional islands cover the whole surface and a new mono-molecular layer is completed.

4.2.5. Results of simulation and discussion

Angular distributions of protons scattered from the growing surface modelled above were calculated during the completion of one monolayer on the surface, i.e., as a function of the coverage of growing layer on the initial surface. From a calculated angular distribution, the number of protons scattered in the angular window 3×3 mrad^2 centered the angle for specular reflection was derived and divided by the number of calculated trajectories. We call this as the normalized intensity of scattered protons. Figure 9 shows the dependence of the calculated normalized intensities on the coverage of the surface at the incidence of 500 keV protons on the (111) surface of PbSe. Four broken curves show the normalized intensity for different densities of two-dimensional nuclei on the surface from 10 to $10^4/\mu m^{-2}$. The calculated normalized intensities do not change sinusoidally and the minima are at coverage about $\Theta = 0.25$. The coverage at which the minimum of the intensity appears is larger for smaller density of two-dimensional nuclei.

The experimental normalized intensity is shown by the solid line in Figure 9, which is the first period of the oscillations shown in Figure 6. For comparison with the calculated intensities, the experimental intensity multiplied by 2 is plotted in the Figure. Characteristic features of the experimental intensity change are well reproduced. From the comparison of the profiles of experimental and simulated intensities, it may be concluded that the number of nuclei formed on a flat (111) surface of PbSe is of the order of $N = 10^2/\mu m^2$.

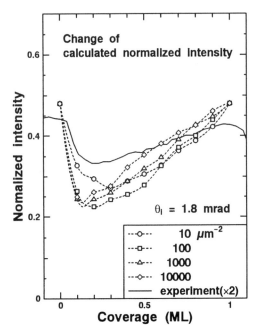

Figure 9. Changes of the normalized intensity as a function of coverage calculated by the simulation at the incidence of 500 keV protons on growing (111) surface of PbSe with various densities of two-dimensional nuclei. A solid curve denotes the first period of the normalized intensity oscillations observed in the experiments, multiplied by 2.

Quantitative agreement of the experimental and simulated intensities is poor if we compare the intensities of the first period of oscillations. This may be due to the assumed initial flat surface; there may be many steps on the surface which exist as the surface is slightly vicinal to the crystallographic (111) plane. The fading out of the oscillations after a long growth-run is also understood as the result of the steps due to the uncompleted lateral growth.

5. DISCUSSION

With the computer simulation of trajectories of energetic ions at glancing-angle incidence on stepped surfaces, we showed that the ions are scattered at steps and penetrate the surface at step edges. Thus the yield of scattered ions from surface depends on the surface topography. Our experimental finding that the yield of reflected ions oscillates during the vapour growth of low index crystal surface is fairly explained by the periodic change in the step density on growing surface. Different from the techniques using diffraction phenomena, we can derive the "oscillations" of yield of reflected ions with trajectory simulation and thus can discuss the ion scattering from a surface with complex topography.

We showed that the optical model of ion scattering at stepped surface is convenient for deriving the yield of scattered ions. In the model we neglect the contribution of the ions scattered at step edges and of the ions scattered inside the crystal after penetrating the surface at step edges. The former ions are scattered away and have little contribution to the yield at the angle for specular reflection. The latter ions are only a small fraction of ions which penetrate the surface since most of these ions travel along the planar channels parallel to the surface plane. These ions have less chance of being suffered from multiple scattering and reappear in the vacuum only at the step edges. However, they form a rather narrow distribution centered at the angle for specular reflection as they are confined in the planar channels. Since the contribution of these ions is neglected in the optical models, we must be careful to use the model for ion reflection from surface with higher step density, where the fraction of ions which reappear after penetrating the surface is large.

In the present experiments, we measured only the angular distributions of scattered ions, because these are observed within a short interval (30 frames per second with a conventional video camera) with the ion detection system mentioned in section 3. The ions travelled in crystal are suffered from energy loss larger than those of ions reflected by the topmost atomic layer. Thus the detailed information of the surface steps will be obtained from the energy spectrum of ions as in our previous study of the growth of PbSe on SnTe(001) [18] and the ion reflectivity of the surface. It is desirable for this purpose to construct an ion detecting system where both angle and energy of scattered ions can be measured simultaneously.

REFERENCES

1. J.J. Harris and B.A. Joyce, Surf. Sci. Lett. 108 (1981) L90.

2. S.S. Chao, E.-A. Kanbbe and R.W. Vook, Surf. Sci. 100 (1980) 581.

3. Y. Namba, R.W. Vook and S.S. Chao, Surf. Sci. 109 (1981) 320.

4. L.V. Sokolov, M.A. Lamin, V.A. Markov, V.I. Mashanov, O.P. Pchelyakov and S.I. Stenin, JETP lett. 44 (1985) 357. Pis'ma Zh.Eksp. Teo. Fiz. 44 (1986) 278.

5. L.J. Gòmez, S. Bourgeal, J. Ibàñez and M. Salmeròn, Phys. Rev. B31 (1985) 2551.

6. L.P. Erickson, M.D. Longerbone, R.C. Youngman and B.E. Dies, J.Crystal Growth 55 (1987) 81.

7. J.N. Eckstein, C. Webb, S.-L. Weng and K.A. Bertness, Appl. Phys. lett. 51 (1987) 1833.

8. E. Vlieg, A.W. Denter van der Gon, J.F. van der Veen, J.E. MacDonald and C. Norris, Phys. Rev. Lett. 61 (1988) 2441.

9. P.H. Fuoss, D.W. Kister, F.J. Lamelas, G.B. Stephenson, P. Imperatori and S. Brennan, Phys. Rev. Lett. 69 (1992) 2791.

10. Y. Fujii, K. Narumi, K. Kimura, M. Mannami, T. Hashimoto, K. Ogawa, F. Ohtani, T. Yoshida and M. Asari, Appl. Phys. Lett. 63 (1993) 2070.
11. J. Lindhard, K. Dans. Vid. Selsk. Mat. Fys. Medd. 34 no. 14 (1965) 1.
12. P.M. Echenique, F. Flores and R.H. Ritchie, Solid State Physics, vol. 44, Ed. H. Ehrenreich and D. Turnbull, Academic Press, Inc., 1990.
13. K. Kimura, M. Hasegawa and M. Mannami, Phys. Rev. 36 (1987) 7.
14. A.D. Marwick, M.W. Thompson, B.W. Farmery and G.S. Harbinson, Radiat. Eff. 15 (1972) 195.
15. E.S. Mashkova and V.A. Molchanov, Medium-energy Ion Reflection from Solids, North-Holland, Amsterdam, 1985.
16. Y. Fujii, K Kishine, K. Narumi, K. Kimura and M. Mannami, Phys. Rev. A47 (1993) 2047.
17. M. Mannami, Y. Fujii and K. Kimura, Surface Sci. 204 (1988) 213.
18. Y. Fujii, S. Fujiwara, K. Kimura and M. Mannami, Radiat. Eff. 116 (1991) 111.
19. J. Fuch, Z. Feit and H. Preier, Appl. Phys. Lett. 53 (1988) 894.
20. G. Molière, Z. Naturforsch, A2 (1947) 133.
21. Y. Fujii, K. Nakajima, K. Narumi, K. Kimura and M. Mannami, Surf. Sci. Lett. 318 (1994) L1225.

Scanning tunneling microscopy study of solid phase epitaxy processes on the Si(001)-2×1 surface

T. Yao[a] [b], T. Komura[a], K. Uesugi[c] and M. Yoshimura[d]

[a]Institute for Materials Research, Tohoku University, 2-1-1, Katahira, Aoba-ku, Sendai 980, Japan

[b]Joint Research Center for Atom Technology, National Institute for Advanced Interdisciplinary Research, Tsukuba 305, Japan

[c]Research Institute for Electronic Science, Hokkaido University, Sapporo 060, Japan

[d]Toyota Technological Institute, Hisakata, Tenpaku-ku, Nagoya 468, Japan

Atomistic processes of solid phase epitaxy of amorphous Si layers on the Si(001)-2×1 surface are studied using a scanning tunneling microscope (STM) at elevated temperatures. Amorphous layers are prepared on the Si surface either by vacuum evaporation or Ar^+-ion sputtering at room temperature. The amorphous layers prepared by vacuum deposition consist of two-dimensional clusters with 1 dimer length. Anisotropic island formation is observed on the Si surface during annealing at 250–300°C. Well developed step and terrace are observed by annealing at 400–500°C. Line defects consisting of missing dimers are observed with almost equal distance between neighboring line defects on the surface annealed at higher temperatures. High temperature annealing above 1000°C produces more perfect surface with reduced surface defects. Amorphous layers prepared by sputtering the Si(100)-2×1 surface consist of three-dimensional fine clusters. Coalescence of clusters is observed at around 250°C and partly crystallized layers appear after prolonged annealing. Higher temperature annealing shows similar evolution of the surface morphology as that of vacuum evaporated layers. The recovery processes of crystallinity are investigated by in-situ STM observation during annealing.

1. INTRODUCTION

Because of technological and scientific importance, the solid phase epitaxy (SPE) of amorphous Si (a-Si) has been intensively investigated [1–7]. From the technological point of views, it enables low temperature growth of large area. In fact it has been established that the controlled epitaxial regrowth of amorphous layers fabricated by Si ion implantation onto single-crystal Si substrates occurs in a low-temperature range around 600°C [1]. During thermal annealing, the reordering of a-Si is characterized by a uniform translation of the amorphous-to crystalline interface towards the top surface. However, details of this process have not been elucidated at least from an atomistic point of views.

Amorphous Si layers have been formed by ion implantation [1,2], vacuum evaporation [3–5], and chemical vapor deposition (CVD) [6]. In the case of Si(100) and (110) substrates, a linear regrowth in time is observed for isothermal anneal, and the resultant layers are relatively defect free [7]. The SPE growth processes have been investigated by transmission electron microscopy (TEM) [2,3,5,6], scanning electron microscopy (SEM) [1,3], Nomarski optical microscopy [3], transmission electron diffraction (TED) [3,6], and Rutherford back scattering (RBS) [1,2,4–6]. Since most of those techniques give information on rather microscopic scale, there have been strong demand for atomistic investigation on solid phase epitaxy processes which should elucidate details of the processes on an atomic scale.

Very recently, we have performed microscopic investigation of the regrowth processes of amorphous layers prepared by vacuum evaporation, Ar^+ ion sputtering, or P^+-ion implantation on Si(001) surfaces using a scanning tunneling microscope (STM) [8–12]. The atomic scale regrowth processes during annealing have been elucidated. The purpose of this paper is to review our recent study on atomistic processes of solid phase epitaxy of Si amorphous layers on Si (001) surfaces with special attention placed upon the SPE of vacuum-evaporated layers and Ar^+-ion sputtered Si surfaces.

2. EXPERIMENTAL

We have used an ultra-high vacuum STM system consisting of an STM chamber, a preparation chamber equipped with a vacuum evaporation source or an Ar^+-ion sputtering gun, and a load-lock chamber. This STM can be operated at elevated temperatures. The tip used was an electrochemically etched tungsten wire. Most of the STM image was taken at a sample bias of -2 V.

The amorphous Si layers investigated in the present study were prepared by vacuum evaporation and Ar^+-ion sputtering. The respective sample preparations were as follows: (a) An amorphous Si layer of 0.03–4 monolayer (ML) thickness was deposited on the clean Si(001)-2×1 surface at room temperature; (b) the clean Si(001)-2×1 surface was amorphized by Ar^+-ion sputtering to a dose of (0.2–5) $\times 10^{15}$ cm^{-2} in the preparation chamber. Before being loaded into the load-lock chamber, the implanted specimen was etched in a buffered HF solution to remove an oxide layer.

3. SOLID PHASE EPITAXY PROCESS OF Si LAYERS PREPARED BY VACUUM DEPOSITION

The Si(001)-2×1 surface was covered with tiny Si clusters by the vacuum evaporation of Si at room temperature. The average size of the Si cluster observed on the surface was 0.72 nm for 0.03 ML coverage with a dispersion of 0.153 nm. This fact suggests that most of the evaporated Si adatoms form single dimers at extremely low coverage (0.03 ML) even at room temperature. With increasing coverage, sin-

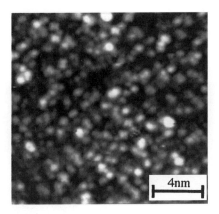

Figure 1. As-prepared amorphous Si surfaces by vacuum evaporation at room temperature. $(6.8 \times 10^{14} \ Si/cm^2)$

gle dimers tend to coalesce and form larger clusters. For instance, the average size of the Si cluster for 1ML coverage was 0.77 nm with a dispersion of 0.23 nm. Some of them were narrow rectangular in shape and form dimer strings, which suggests that deposited Si atoms migrate on the surface even at room temperature to form epitaxial islands. Increase in deposition results in formation of three-dimension clusters as can be seen in Figure 1.

When the Si amorphous layer is annealed, the surface migration of adatoms is enhanced to form dimer strings, which are running perpendicular to the dimer row in the atomic plane of the substrate surface. This phenomenon is observed even at 250°C as shown in Figure 2 (a), where a deposited 1ML-thick Si layer was annealed for 150 min at 200°C. We note that most of Si adatoms form dimer strings. However there still remains Si clusters. As the annealing temperature increases these clusters dissolved to form dimer strings. When annealing time increases much more, these residual clusters dissolved and dimer strings develop as shown in Figure 2(b), where the layer was annealed for 500 min at 200°C. The formation and development of the dimer strings are due to anisotropic diffusion of Si adatoms on the 2×1 reconstructed surface [15]. Further increase in annealing temperature results in the development of the cluster size both along and perpendicular to surface dimer rows as seen in Figure 2(c), where the annealing temperature is 300°C. As a result, the aspect ratio of the islands decrease. By comparison with Figure 2(b), we note that both S_A and S_B steps of the clusters develop with preferential growth along S_B step. It should be noted that similar anisotropic development of Si clusters are observed at the initial stage of molecular beam epitaxy [16]. Hence the initial stages of Si epitaxy both for MBE and SPE are characterized by the anisotropic development of clusters with preferential direction parallel to the S_A step, i.e. the S_B step develops preferentially.

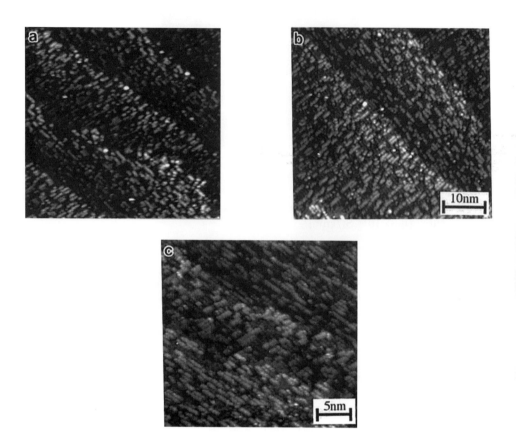

Figure 2. STM images during annealing at 250°C for (a) 150 min and (b) 500 min and (c) at 300°C.

Figure 3 shows how the dimer string develops with annealing time and temperature. Here, (i,j) denotes a dimer string with i dimer unit (du) width and j atomic unit (au) length (1 du = 0.768 nm, 1 au = 0.384 nm). Short dimer strings (1, 1–3) are prominent on the surface during annealing at 250°C for 150 min, while both the length and width of the island increase with prolonged annealing (500 min). Additional high temperature annealing up to 300°C develops the islands both in length and width at the expense of short dimer strings. This is because the critical cluster size becomes larger as temperature increases, where clusters smaller

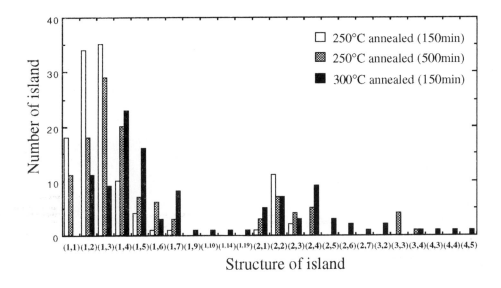

Figure 3. The evolution of island structures on the vacuum-deposited Si layers with 1 ML thickness during annealing. The annealing sequence was 250°C (150 min) ⇒ 250°C (500 min) ⇒ 300°C (150 min).

than the critical size become unstable. Consequently, smaller clusters dissolve to produce Si adatoms which migrate to be incorporated into larger stable clusters.

Most of dimer strings disappear by further annealing up to 400°C as shown in Figure 4 (a). Monatomic height surface steps develop and results in wide surface terraces. Monatomic height two-dimensional islands are observed on the terrace, while small number of clusters are still observed. Prolonged annealing results in decrease in number of clusters and more development of islands, which can be seen in Figure 4(b). These phenomena indicate that small clusters are dissolved into Si adatoms to be incorporated into surface steps and large clusters on the surface. The former process promotes the development of monatomic height surface steps, while the latter favors the formation of two-dimensional islands on the terraces.

A close look at those STM images reveal anti-phase boundaries on islands as indicated by arrows in Figures 4(a) and (b). Figure 4(c) is a magnified image at around an anti-phase boundary. As clearly seen in the figure, the coalescence of two islands consisting of surface dimers of different phases by 180° results in the formation of an anti-phase boundary [16]. It is noted that a cluster is formed as to cover the anti-phase boundary. Similar growth features are observed at the initial stage of MBE. It is argued that the anti-phase boundary acts as effective nucleation site at the initial stage of epitaxy in MBE, which is the case for SPE. Although the atomic arrangement below the cluster has not been elucidated, we believe that the atoms are arranged so as to stabilize the anti-phase boundary. No anti-phase boundary is observed on the surface annealed at higher temperatures (see for example Figure 5). This implies that the formation of an anti-phase boundary is limited by kinetics and the anti-phase boundary disappears through reconstruction of the dimers at high temperature.

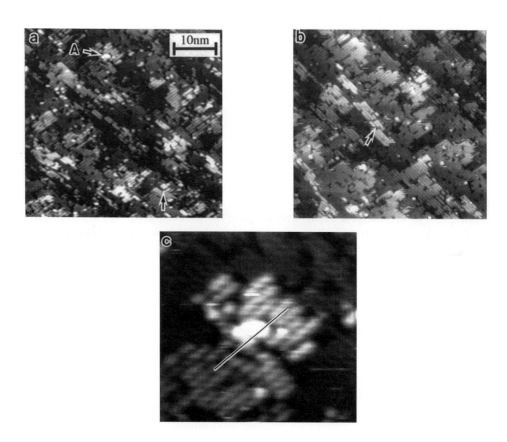

Figure 4. The evolution of the surface structure during annealing at 400°C for (a)175 min and (b) 275 min. The arrows indicate anti-phase domain boundary. (c) A magnification of the anti-phase domain labeled as "A" in Figure (a).

By further annealing up to 500°C, the islands on the terrace disappeared and a 2×1 reconstructed surface developed with many A-type defects [16] on the terrace as shown in Figure 5. These A-type defects fluctuated along the dimer rows during annealing. However, the defect does not move across the dimer row. It is interesting to see how these defects move with annealing time, which is shown in Figures 5(a) and (b), where (a) was observed during annealing for 175 min at 500°C, while (b) during annealing for 275 min. When the annealing time is short, the defect distributes almost in a random fashion. With prolonged annealing, the defects on adjacent dimer rows tend to come close to each other as shown in Figure 5(b). This fact suggests the existence of attractive interaction between A-type defects on the adjacent dimer rows. Similar fluctuation of A-type defects were observed in the annealing process of Ar^+-sputtered surface at 500°C [11].

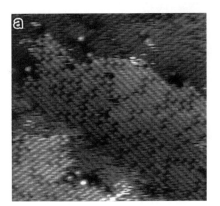 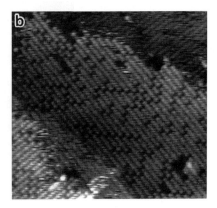

Figure 5. The evolution of the surface structure during annealing at 500°C for (a) 175 min and (b) 275 min.

A high temperature annealing results in well developed surface steps and terraces with a 2×1 reconstructed surface. Figure 6 (a) shows an STM image of the surface of a 4 ML thick Si layer after annealing at 800–900°C. It is interesting to note that the A-type defects are aligned perpendicular to the dimer rows [14]. In other word, line defects of missing dimer are formed on the surface. Figures 6(b) and (c) show the variations of the length of the line defect and the distance of the line defect, respectively. The distance of the line defect is 4.4 nm on average, which results in a 2×11 reconstructed surface. The length of the line defect is 7.8 nm on average, which suggests that the line defect extends over 10 dimer rows at this temperature. Figure 6(b) and (c) indicate that the distributions of both the length of the line defect and the distance between the line defects are small. The formation of the line defects apart almost equal distance each other suggests that there exists repulsive interaction between A-type defects on the same dimer row, while attractive interaction between A-type defects on the adjacent dimer rows. Both interactions should originate from strain interaction between A-type defects on the dimer rows [17]. It should be noted that similar line defects are observed on Ar$^+$-ion sputtered surface during annealing process at 600°C. In this particular case, the missing dimer defects are separated approximately 5 nm apart, which resulted in a 2×13 reconstructed surface [18]. It is natural to consider that the surface lattice is more relaxed to reduce the repulsive strain energy at higher temperature. Hence, the missing dimer defects develop more at higher temperature which results in smaller distance between the missing dimer defects.

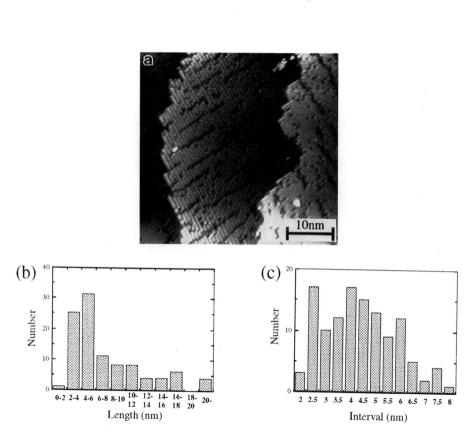

Figure 6. (a) The surface structure of a 4ML-thick Si layer after annealing at 800°C observed at a sample bias of −1.7 V. (b) Distribution of the length of the line defects observed in Figure (a). (c) Distribution of the separation of line defects.

Annealing at much higher temperature produces more perfect surface with less number of surface defects. Figure 7 shows an STM image after annealing at 1050°C. There still remains line defects. However both its number and length across dimer rows are greatly reduced. The length is typically 2–3 dimer long, i.e., 1.5–2.3 nm. By comparing Figure 6(a) and 7, we note that the number of surface defects are reduced tremendously. Based on the real time observation of the fluctuation of missing dimer defects as described later, the dimer defects disappear in the following processes by high temperature annealing: Fluctuation of missing dimer defects is enhanced at high temperature, which results in the dissociation of line defects into segments of line defects consisting of missing dimer defects. Those segments move longer distance on the surface than the longer line defects and some of them reach step edges from which the segments disappear, resulting in decrease in number of missing dimer defects. Since the missing dimer defect moves along the dimer row, the defects should disappear at S_B step edges. This sweeping-out process of dimer defects becomes more frequent as the annealing temperature is raised. Therefore, the high temperature annealing produces more perfect crystals, which has been experimentally observed.

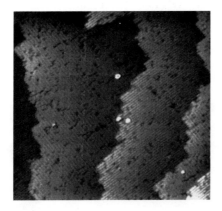

Figure 7. The surface structure of a 4 ML-thick Si layer after annealing at 1050°C observed at −1.7 V.

4. SOLID PHASE EPITAXY PROCESS OF AMORPHOUS Si LAYERS PREPARED BY Ar⁺-ION SPUTTERING

Figure 8 shows typical STM images of the amorphous Si surfaces prepared by Ar⁺-ion sputtering at 3 keg to a dose of 2×10^{14} cm^{-2}. The surface is covered with homogeneous circular particles, which is similar with that of vacuum-deposited layers. The average size of the Si cluster observed on the surface was 0.72 nm for 0.03 ML coverage with a dispersion of 0.153 nm. This fact suggests that most of the evaporated Si adatoms form single dimers at extremely low deposition (0.03 ML) even at room temperature. With increasing coverage, single dimers tend to

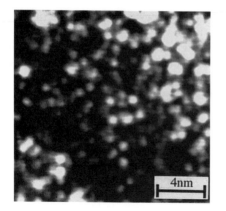

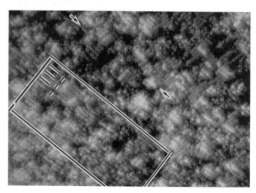

Figure 8. As-prepared Ar$^+$-ion sputtered surface.

Figure 9. Ar$^+$-ion bombarded Si(001) surface to a dose of 1.9×10^{14} cm^{-2} during annealing at 245°C. ($V_s = -0.9$ V).

Figure 10. Ar$^+$-ion bombarded Si(001) surface to a dose of 1.9×10^{14} cm^{-2} during annealing at 500°C. ($V_s = -1.2$ V).

coalesce and form larger clusters. For instance, the average size of the Si cluster for 1ML coverage was 0.77 nm with a dispersion of 0.23 nm. Some of them are narrow in shape and form dimer strings, which suggests that deposited Si atoms at room temperature migrate on the surface to form epitaxial island.

The surface of the Ar^+-ion sputtered surfaces is smoothed by annealing. Figure 9 shows an STM image of an Ar^+-ion bombarded Si(001) surface to a dose of 1.9 $\times$ 10^{14} cm^{-2} during annealing at 245°C [10]. The as bombarded surface consisted of 0.63–1.6 nm diameter grains and the density of the grain was about 2 $\times$ 10^{15} cm^{-2}. By comparing Figure 8 and Figure 9, it is noted that the surface became smooth by annealing, although the granular surface morphology was still observed. The grain grew in size by annealing: typical size of the grain was 2–3.6 nm in diameter at 245°C, for instance. This is presumably due to the coalescence of fine grains during annealing. We found that prolonged annealing even at this temperature promoted crystallization of the amorphous surface. Eventually, 2×1 surface reconstruction was found on relatively smoothed areas on the surface as shown in Figure 10. Although the surface consists of reconstructed regions and grains, atomic steps were noted as indicated by arrows. In this figure, dimer rows running along a [011] direction are clearly observed in the areas indicated by a square. The dimer rows are observed in other areas. Hence, that part of the surface shows a 2×1 reconstruction. We note that the dimer rows on the terrace (dashed lines) are running perpendicular to the underlying dimer rows (solid lines), which indicates the onset of the growth of a monatomic layer on the terrace. The Si atoms which formed the layer were supplied from the grains, since the surface roughness was eventually reduced with annealing temperature and time. Hence, the surface reconstruction changed from 2×1 to 1×2 or vice versa, as the annealing time was prolonged. Since the formation of a dimer on the surface indicates that the underlying layer is crystallized, these facts indicate that the crystallization process initiates at the crystal/grain interface and extends towards the surface. The crystallization process proceeds inhomogeneously due to the presence of grains until the crystal/amorphous interface reaches the surface. Once a part of the surface is crystallized, the crystallization proceeds in a layer-by-layer mode, in which Si atoms are supplied from grains. Thus, the surface morphology is greatly improved through annealing.

As the annealing temperature is raised, the reconstructed regions develop and a mixture of 2×1 and 1×2 reconstruction is observed, where the amorphous layer epitaxially crystallizes up to the topmost surface. However, the surface was defective and contains many point defects as well as line defects. Figure 10 shows an STM image of the surface at 500°C. Although the terrace and step structures do not develop at this temperature, both S_A and S_B steps are clearly observed.

Although the amorphous layer epitaxially crystallizes up to the topmost surface at annealing temperature below 500°C, the surface is defective and contains many monatomic-height islands. When the annealing temperature increases to 500°C, smoothing of step edges is observed in situ on the surface. Figure 11 shows successive images during annealing at 500°C taken every 9 s. The S_A and S_B stems of an

334

island (A) are observed, and the smoothing of a kink (B), "step-flow growth", is observed in real time. The surface contains many point defects as well as line defects (C and D) which fluctuates along the dimer rows by annealing. The observed line defects consist of A-type vacancies and are perpendicular to the dimer row. The defect (E) is annealed out with time. However, the missing dimers at the antiphase boundary (F) do not move. Figure 12 shows schematics of the movement of the defect (C) in Figure 7. The arrows indicate the movement of the defects in the corresponding image. It is observed that A-type defect frequently fluctuate along the dimer rows, each of which are connected with A-type defects on the adjacent dimer rows. In contrast, neither B-type defects nor a triple-dimer vacancy (D) do not fluctuate at all.

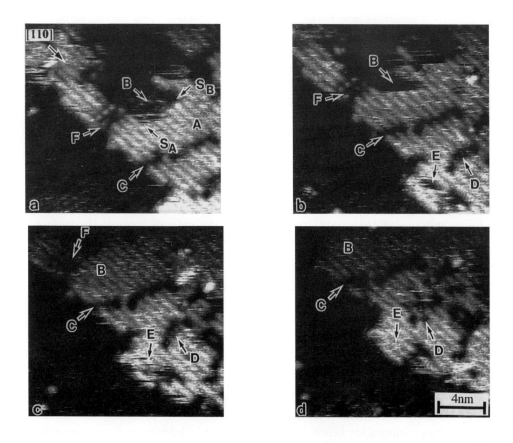

Figure 11. Successive STM image of Ar$^+$-ion bombarded surface during thermal annealing at 500°C taken every 9 s. ($V_s = +1.6$ V).

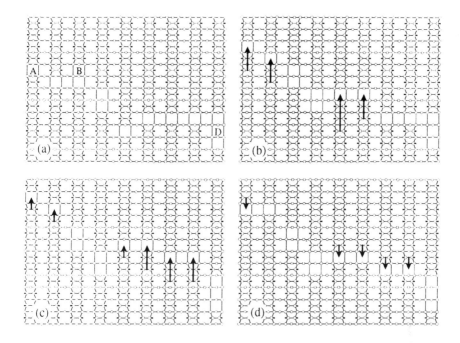

Figure 12. Schematic illustration of the fluctuation of defects observed in Figure 11.

5. CONCLUSIONS

SPE processes of amorphous Si(001) surfaces prepared by vacuum evaporation and Ar$^+$-ion sputtering have been investigated in situ using STM operated at elevated temperatures. Amorphous layers prepared by vacuum evaporation consist of two-dimensional clusters with 1 dimer length, while the Ar$^+$-ion sputtered surface consists of three-dimensional fine clusters. Anisotropic island formation is observed on the vacuum evaporated Si surface during annealing at 250–300°C. Well developed steps and terraces are observed on the surface annealed at 400–500°C. The formation of anti-phase boundary by coalescence of two dimensional islands is observed together with the formation of cluster on the anti-phase boundary. Line defects consisting of missing dimer defects are formed on the surface annealed at higher temperatures, which suggest repulsive interaction between missing dimer defects on the same dimer row and attractive interaction on the adjacent dimer rows. High temperature annealing above 1000°C produces more perfect surface with reduced surface defects. In the case of an Ar$^+$-ion sputtered surface, coalescence of clusters is observed at around 250°C and partly crystallized layers appear after prolonged annealing. Higher temperature annealing results in similar evolution as that of vacuum evaporated layers. Based on "in situ" annealing observation, the recovery processes of crystallinity is elucidated in terms of the seeping out of missing dimer defects from S_B step edges at high temperature. This recovery process dominates both in for SPE of vacuum-deposited layers and Ar^+-ion sputtered surfaces.

REFERENCES

1. Csepregi, E.F. Kennedy, J.W. Mayer, and T.W. Sigmon, J. Appl. Phys. 49 (1978) 3906.
2. S.S. Lau, S. Matteson, J.W. Mayer, P. Revesz, J. Gyulai, J. Roth, T.W. Sigmon, and T. Cass, Appl. Phys. Lett. 34 (1979) 76.
3. H. Ishiwara, H. Yamamoto, and S. Furukawa, Appl. Phys. Lett. 43 (1983) 1028.
4. S. Saitoh, T. Sugii, H. Ishiwara, and S. Furukawa, Jpn. J. Appl. Phys. 20 (1981) L130.
5. L.S. Hung, S.S. Lau, M. von Allmen, J.W. Mayer, B.M. Ullrich, J.E. Baker, and P. Williams, Appl. Phys. Lett. 37 (1980) 909.
6. Y. Kunii, M. Tabe, and K. Kajiyama, Jpn. J. Appl. Phys. 21 (1982) 1431.
7. L. Csepregi, J.W. Mayer, and T.W. Sigmon, Phys. Lett. A54 (1975) 157.
8. K. Uesugi, T. Yao, T. Sato, T. Sueyoshi, and M. Iwatsuki, Appl. Phys. Lett. 62 (1993) 1600.
9. K. Uesugi, M. Yoshimura, T. Sato, T. Sueyoshi, M. Iwatsuki, and T. Yao, Jpn. J. Appl. Phys. 32 (1993) 6203.
10. T. Yao, K. Uesugi, M. Yoshimura, T. Sato, T. Sueyoshi, and M. Iwatsuki, Appl. Phys. Sci. 75 (1994) 139.
11. K. Uesugi, M. Yoshimura, T. Yao, T. Sato, T. Sueyoshi, and M. Iwatsuki, J. Vac. Sci. Technol. B12 (1994) 2018.
12. K. Uesugi, T. Komura, M. Yoshimura, and T. Yao, Appl. Surf. Sci. 82/83 (1994) 367.
13. R.M. Feenstra and G.S. Oehrlein, Appl. Phys. Lett. 47 (1985) 97.
14. I.H. Wilson, N.J. Zheng, U. Knipping, and I.S.T. Tsong, J. Vac. Sci. Technol. A7 (1989) 2840.
15. Z. Zhang and H. Metiu, Surface Science (1991) pp. 175–232.
16. R.J. Hamers and U.K. Koher, J. Vac. Sci. Technol. A7 (1989) 2854.
17. A. Natori, private communication.
18. K. Uesugi, M. Yoshimura, and T. Yao, private communication.

Advances in the Understanding of Crystal Growth Mechanisms
T. Nishinaga, K. Nishioka, J. Harada, A. Sasaki and H. Takei (Editors)
© 1997 Elsevier Science B.V. All rights reserved.

Crystal Growth of Polymers in Thin Films

K. Izumi[a], Gan Ping[a], M. Hashimoto[a*], A. Toda[a†], H. Miyaji[a], Y. Miyamoto[b]
and Y. Nakagawa[c]

[a]Department of Physics, Faculty of Science,
Kyoto University, Kyoto 606-01 Japan

[b]Department of Fundamental Sciences, Faculty of Integrated Human Studies,
Kyoto University, Kyoto 606-01 Japan

[c]Toray Research Center, Inc., 3-3-7 Sonoyama, Otsu 520 Japan

Crystal growth of isotactic polystyrene in thin amorphous films has been investigated by transmission electron microscopy (TEM) and atomic force microscopy (AFM). The single crystals grown above 200°C are hexagonal plates parallel to the (001) planes with {110} facets. The crystals become round due to a kinetic roughening transition at about 195°C. At temperatures below 170°C two dimensional spherulites grow. Crystals with spiral overgrowth terraces grow to be thicker than the original films. The amorphous parts surrounding the planar crystals are thin and are observed as 'bright haloes' in bright field images of TEM.

In situ observations showed that the bundle of lamellae with a 'halo' grows first and the planar crystals grow from the end/ends of the bundle. AFM observation has shown that the crystals are covered with amorphous layers and have a hollow dislocation at the center. The growth rate of crystals in thin films is 70% and for the overgrowth terraces is 40% of that in the bulk.

1. INTRODUCTION

The crystallization of linear polymer is a process in which random coils of long chain molecules as long as 1 μm in a liquid state are organized into a crystal lattice with the three-dimensional translational symmetry. Crystal growth is generally controlled by the well-known three factors: 1) the diffusion of materials to the growth front, 2) the kinetics of organization of the materials into the crystal, and 3) the removal of latent heat, i.e. the diffusion of heat.

In the melt growth of polymers here, the last process of diffusion of heat is ignored on account of the small growth rate of polymer crystals. The temperature

---

*Present address: Department of Polymer Science and Engineering, Faculty of Textile Science, Kyoto Institute of Technology, Kyoto 606 Japan
†Present address: Faculty of Integrated Arts and Sciences, Hiroshima University, Higashi-Hiroshima 724 Japan

variation along the growth direction is described by an exponential function characterized by the diffusion length defined by $2D/G$ (D = thermal conductivity/specific heat/density and G = linear growth rate). This length amounts to a value of more than $0.1\,m$ for polyethylene [1] and $25\,m$ for isotactic polystyrene [2] respectively and much larger than the lateral dimension of crystals (less than $1\,mm$), and hence the whole materials including crystals and liquid are in an isothermal state. For the diffusion of polymers, the WLF equation, the Eyring equation or the reptation model has been successfully applied. Thus the kinetics at the interface between the crystal and melt is the main issue in polymer crystallization and is investigated by the observation of morphology of grown crystals and the growth rate.

There are three morphologies in polymer crystals revealed by electron microscopy and small angle x-ray scattering: extended chain crystals, folded chain crystals and nodular crystals. Extended chain crystals are lamellar crystals thicker than $0.1\,\mu m$ along the chain axis [3] and are observed under an extreme condition that the grown crystal is in a disordered mobile phase, such as the hexagonal phase of polyethylene at a high pressure and high temperature, and the paraelectric phase of several ferroelectric fluoride polymers above the transition temperature; polymer chains can move along the chain direction easily even in the crystal [4]. Nodular crystals are small crystallites of a few tens of nm in three dimensions and grow under another extreme condition of very large supercoolings near the glass transition temperature [5]; polymer chains are immobile not only in the crystal but also in the amorphous so much as in the melt or solution. Folded chain crystals, which are discovered by A. Keller in 1957 [6], are thin lamellar crystals a few tens of nm in thickness along the chain axis, and hence a polymer chain $1\,\mu m$ long has to be folded back and forth in the lamella. This lamellar crystal is universally observed for all linear flexible polymers in the usual crystallization condition for both of solution and melt crystallizations; polymer chains can move in the liquid state but essentially cannot move once organized in the crystal. Thus the folded chain crystal is the most common morphology in polymer crystals. Therefore the main issue in polymer crystallization has been the folding process of long chain molecules at the growth plane. As such, the nucleation theories [7] and rough-surface growth theory [8] have been developed for the growth of the folded chain crystals.

In this paper we investigate the growth of the folded chain crystals in thin films of isotactic polystyrene (iPS) with rather large side groups of phenyl groups; owing to the slow growth rate (several $\mu m/h$) and high glass transition temperature (90°C), it is possible to freeze a growing state by quenching and to observe the morphology of a growing crystal by transmission electron microscopy (TEM) and atomic force microscopy (AFM) at room temperature. Further, polystyrene has a high resistance against the radiation damage by electron beam, and hence *in situ* electron microscopy has been successfully performed for the crystal growth from the melt. The crystal structure of iPS is illustrated in Figure 1. The chains have the 3_1 helical conformation and arrange to form the trigonal cell with both right-handed and left-handed helices [9].

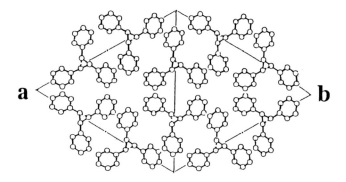

Figure 1. Crystal structure of isotactic polystyrene [9]. Projection on (001) plane

The shapes of iPS lamellar crystals are observed at various crystallization temperatures and are discussed on the basis of the kinetics of a nucleation control process at the growth face, taking account of the temperature dependence of lateral growth rate. Lamellar crystals are organized to form a spherulite through branching of lamellae. A possible mechanism of the branching is the multiplication of lamellae by the spiral growth through screw dislocations. Atomic force microscopy elucidates the specific features of the spiral growth and diffusion of chain molecules to overgrowth crystals through the thin films of the melt.

2. EXPERIMENTAL

Thin films of iPS (M_w = 5.9 × 10^5, M_w/M_n = 3.4 and M_w = 1.57 × 10^6, M_w/M_n = 6.4) 10 ∼ 30 nm in thickness were prepared by casting 0.1% xylene or cyclohexanone solutions on an evaporated carbon film for TEM observation and on a cleaved surface of mica for AFM observation.

For *in situ* electron microscopy, the thin film on a carbon film was melted, cooled to a growth temperature and observed. For AFM and TEM observations at room temperature, the specimens were melted, quenched to room temperature, heated up to a growth temperature, quenched again to room temperature and observed.

The morphology was investigated mainly by TEM. TEM is a powerful method for the observation of a small and thin crystal like a polymer single crystal because the diffraction contrast, the bright field and dark field images, and the electron diffraction patterns can be obtained easily. By the defocusing over several hundred nanometers, a clear phase contrast at crystal edges is imaged without shadowing even after heavy electron irradiation. Atomic force microscopy (AFM) was done to obtain three dimensional quantitative information with a sufficient resolution on the surface topology.

3. *IN SITU* ELECTRON MICROSCOPY

In situ transmission electron microscopy during the crystallization process is possible owing to the low vapor pressure and high viscosity of polymer melts, when the polymers of interest are relatively electron-irradiation resistant [10]. The sequence of isothermal crystallization process is shown by successive electron micrographs recorded at intervals of 20 minutes in Figure 2. The iPS thin film on carbon film was cooled to a crystallization temperature of 215°C in one minute after melting at 260°C on a heating holder in an electron microscope. A weak electron beam

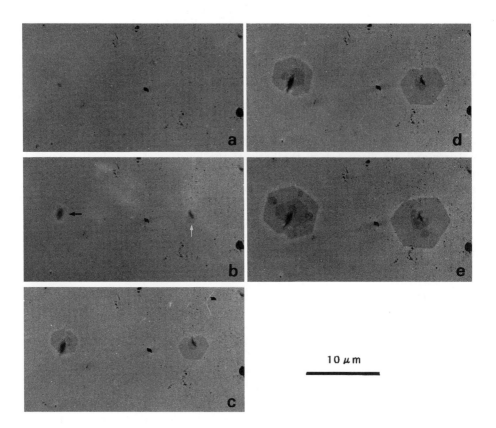

10 μm

Figure 2. *In situ* bright field images of iPS crystals taken at 215°C at an interval of 20 min successively. (a) After 5 min crystallization. Faint dark spots are indicated by the arrows. (b) A 'halo' and planar crystal are indicated by a black arrow and a white arrow respectively. (c) Planar crystals surround the bundle of lamellae. (d) Overgrowth terraces grow larger. (e) New overgrowth terraces emerge.

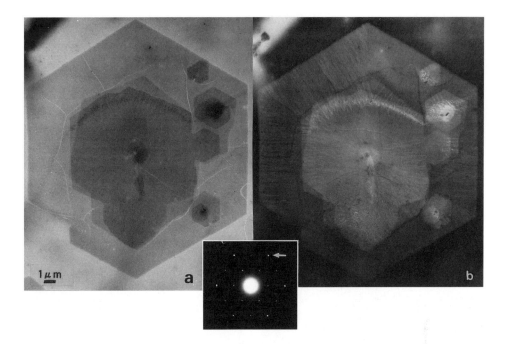

Figure 3. An iPS crystal grown at 220°C, 19.5 h., $M_w = 1.57 \times 10^6$. A part of the overgrowth terrace beneath the mother crystal does not show the flat facet by the deficit of the molecular supply. (a) Bright field image. Cracks generated by quenching from the growth temperature. (b) Dark field image by 220 reflection indicated by an arrow in the electron diffraction. The domains show a bright and dark contrast alternately. The domain boundaries are not straight.

irradiated the specimen only at a time of exposure for 2.6 s. Faint dark spots emerge from the melt in five minutes after quenching to 215°C (Figure 2a). These dark spots consist of lamellar crystals, the normals of which are parallel to the film surface, i.e., edge-on crystals, and grow into bundles of lamellae. A planar crystal (the lamellar normal is perpendicular to the film surface, a flat-on crystal) starts growing from one end of each bundle (Figure 2b). Planar crystals sometimes grow from both ends of a bundle. The bright periphery of each crystal can be seen at this stage as indicated by a white arrow in Figure 2b. This bright region in the supercooled melt, hereafter called a halo, is thinner than the outer region of the melt, and expands a few hundreds nanometers in width from the crystal-melt interface. The planar crystals display a hexagonal habit with the smooth and flat {110} facets (Figure 2c) and grow into hexagonal plates surrounding the bundle from which they started growing (Figures 2d and 2e). The electron diffraction reveals that these planar crystals are single crystals and that the chain axes are parallel to the lamellar normal. Spiral-growth terraces can be seen in Figures 2c to 2e, which indicate the existence of a screw dislocation at the center of planar crystal. Besides the overgrowth terraces from the central dislocation, the overgrown crystals at a distance from the center are observed in Figures 2d and 2e. They are aligned along the trace of the periphery of the crystal which is recorded in the preceding micrograph. The electron irradiation enhanced the induction of the screw dislocations.

The hexagonal habit and the well-developed {110} facets of planar crystals are similar to the crystals grown from the bulk [11–13]. The growth of planar crystal is linear with time, much as seen in the growth of spherulites and axialites in the bulk [14].

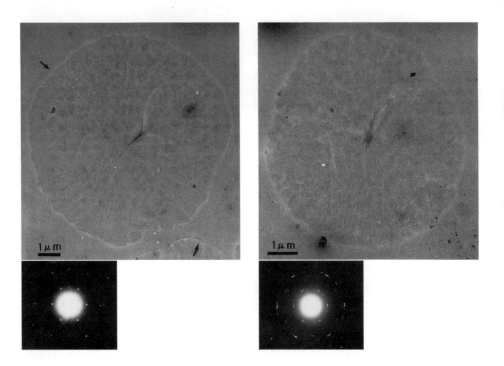

Figure 4. Shadowed image of a crystal of iPS grown at 180°C for 20 min. The 'halo' and fibrils are indicated by arrows.

Figure 5. Crystals of iPS grown at 170°C for 15 min. Electron diffraction pattern shows arcs.

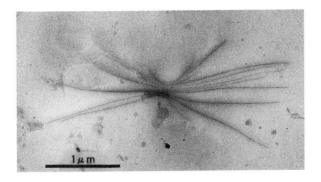

Figure 6. Spider like edge-on crystals grown at 200°C for 20 min

4. MORPHOLOGY

Morphology of growing crystal is an important aspect in crystal growth. Since the dimension of a polymer crystal along the chain axis is a few tens of nm, morphology is mainly concerned with the lateral shape of the crystal. We can estimate the kinetics of the lateral crystal growth from the morphology of the crystal; namely, facetted crystals are to be formed under a nucleation-controlled process, and rounded crystals are under rough surface growth. However, in polymer crystallization, we will meet a difficult problem.

4.1. Temperature dependence of morphology and growth rate

At crystallization temperatures above 200°C, the flat-on single crystals are hexagonal plates parallel to the (001) plane with the {110} facets; the molecular chains are perpendicular to the plate (Figure 3) [11–13]. They show stripes perpendicular to the growth plane in the dark field images of TEM (Figure 3b). Since the stripes are not straight (Figure 3b), they were introduced during the crystal growth and not by deformation after growth. The stripes are not imaged in the two sectors of which the {110} growth planes are parallel to the reflecting plane but imaged in other sectors. The displacement vector is therefore perpendicular to the stripe but not parallel to the c-axis. The stripes correspond to crystal domains elongated along the growth direction. The molecular chains tilt in the growth plane from domain to domain alternately in the opposite direction. At crystallization temperatures below 200°C, the facets of the flat-on crystal are rounded and show wavy edges (Figure 4) through a kinetic roughening transition temperature at 195°C ($M_w = 5.9 \times 10^5$) and 200°C ($M_w = 1.57 \times 10^6$). The diffraction patterns from the rounded crystals grown below 170°C show arcs (Figure 5); the arcs are attributed to many incoherent overgrowth terraces and branching of the flat-on crystal.

The edge-on crystals crystallized at high temperatures are observed as a bundle of lamellae in the center of the planar crystals (Figure 2) or also as "spider like" fibrils grown along the ⟨110⟩ direction (Figure 6). The edge-on crystals protrude from the film surface by a few hundred nanometers. The edge-on crystals crystallized at low temperatures are twisting fibrils branching away to form a two-dimensional spherulite (Figure 7). The flat-on crystals grow in the spaces between the edge-on crystals.

In thin films the lateral growth rate is reduced to about 70% of the one in the bulk in a range of crystallization temperatures studied (Figure 8). Further the overgrowth terraces grow more slowly. Since the temperature dependence of the growth rate is the same as that in the bulk state, the slowing down of the growth rate in thin films can not be attributed to the kinetics at the growth face but to the slowing down of the diffusion of molecules. While the morphology changes from the facetted to rounded crystal through the kinetic roughening temperature around 200°C with decreasing crystallization temperature, the growth rate does not show any transition; the nucleation controlled process in the growth of facetted crystals seems to work also for the rough surface growth. This is the mystery pointed out, for the first time, in the solution growth of iPS crystals by Tanzawa [15] and confirmed also for the growth in the concentrated solutions and melt [14], and hence yet to be investigated.

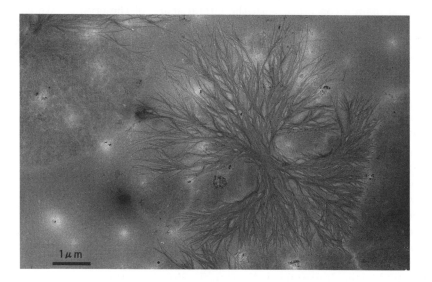

Figure 7. Two dimensional spherulitic iPS crystals grown at 150°C for 40 min. At the tips of the fibrils no 'halo' is observed. Planar crystals grown in the spaces between the bundles of fibrils show 'haloes'.

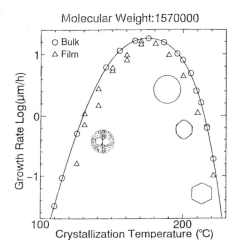

Figure 8. Growth rate of iPS crystals vs. growth temperature. Morphologies are shown at corresponding temperatures.

4.2. Overgrowth [16]

Most of the flat-on crystals are multilayered by the spiral growth through a screw dislocation with a Burgers vector with a magnitude of lamellar thickness. Overgrowth at high temperatures shows the {110} facets but the edges are not always parallel to the ones of the mother crystal even though they grow by the spiral growth through a screw dislocation (Figure 9).

The terraces grow both on the upper surface and below the lower surface of the mother crystal. The upper terraces are crushed to the substrate and deformed at the edges of lower terraces to give dull spiral traces in an AFM image (Figure 9).

The center of the spiral terraces is concave, and hence the screw dislocation is a hollow dislocation. The height of the crystal is 80 nm with five layers and corresponds to five times of the value of the long spacing obtained from small angle x-ray scattering [17]. The height profile shows that the edges of the steps are rounded, thus suggesting that the crystal is covered by an amorphous layer. In fact, removing the amorphous layer by etching, we observe the sharp edges of a crystal [17]. The surface tension of this thin amorphous layer covering the upper growth terraces may cause to collapse them down to the substrate. By the diffusion of polymer molecules through this amorphous layer, the overgrowth terraces on the mother crystal can grow.

The overgrowth terraces apart from the center also show the spiral patterns (Figures 2 and 3). The populations of overgrowth increase with decreasing growth temperature, and hence we eventually get the dense branching morphology, i.e.

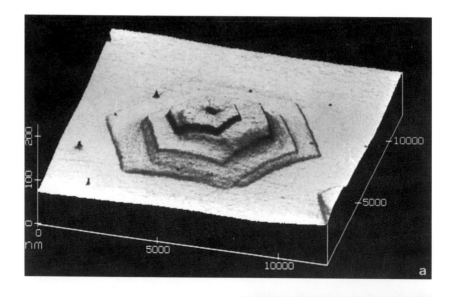

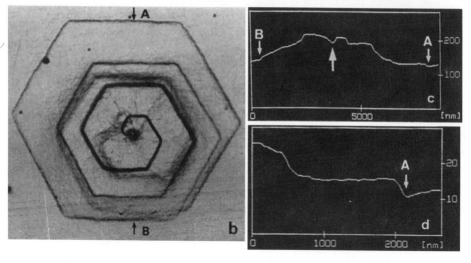

Figure 9. AFM (a) Shaded three dimensional perspective views of iPS crystal grown at 210°C for 3 h. The crystal has collapsed. A hollow screw dislocation is present at the center of the crystal. (b) Shaded top view. Overgrowth terraces show a spiral pattern. The edges are not parallel to each other due to a rotation by an angle of a few degrees due to the dislocation. The terraces under the mother crystal are observed as dull spiral traces. (c) Cross-sectional profiles along the line connecting points A and B. At the center indicated by the arrow, there is a dip corresponding to the hollow dislocation. (d) The magnified profile around point A. The edges of the steps are rounded. The 'halo' surrounding the crystals is concave, 2 nm in depth and 400 nm in width.

spherulites [18].

As a result of the overgrowth, the crystals protrude highly over the film and sometimes they can reach as high as a few hundred nanometers.

4.3. Halo

The haloes are the thin amorphous regions at the interface between the flat-on crystal and the melt. They are observed as a bright halo in the bright field image and as a dark one in the dark field image not only at the *in situ* TEM but also at room temperature after quenching (Figures 3a and 3b), and hence the quenching can freeze the growing crystals, melt and interface between them at a crystallization temperature into room temperature.

The height profile of AFM shows that a concave amorphous region corresponds to the halo and is 2 nm in depth and 400 nm in width (Figure 9). The values of the depth and width remain unchanged during growth, a few nanometers and a few hundred nanometers respectively, and are independent of molecular weight and crystallization temperature. It is to be noted that the haloes are not observed at the interface of the edge-on crystals and the melt. Since the haloes were also observed around growing flat-on crystals of polyoxymethylene in thin films (Figure 10), it is a general phenomenon that the concave region is formed at the interface between a crystal and the melt in the melt crystallization of polymers in thin films.

5. CONCLUSION

In the crystallization of iPS from the melt, the lateral growth planes are the {110} facets at high temperatures and become round with the decreasing of growth temperature; the kinetic roughening occurs. The roughening-transition temperature depends on the molecular weight. Both the edge-on and flat-on lamellar crystals

Figure 10. *In situ* bright field image of a growing polyoxymethylene crystal at 150°C. 'Halo' is indicated by arrows.

grow in thin films and protrude from the film surface over a few hundred nanometers. The molecules are supplied to the growing crystals through the thin amorphous layer covering the surface of the crystals. The haloes observed by TEM are the thin amorphous regions surrounding the growing flat-on crystals. Their widths are almost constant, a few hundred nanometers, during growth, and do not depend on growth temperature, molecular weight and morphology. The screw dislocation with the Burgers vector of the lamellar thickness is induced at the growth front during crystal growth, and the electron irradiation enhances the generation of the dislocations.

The halo, amorphous layers on the growing crystal and the reduction of the growth rates are general features for the crystal growth of polymers in thin films.

REFERENCES

1. A. Toda, Faraday Discuss. 95 (1993) 129
2. J. Brandrup and E.H. Immergut (eds.), Polymer Handbook, Third Ed., John Wiley & Sons, New York, 1989, p.V81
3. P.H. Geil, F.R. Anderson, B. Wunderlich and T. Arakawa, J. Polym. Sci., A2 (1964) 3707
4. M. Hikosaka, Polymer 31 (1990) 458
5. H. Miyaji and P.H. Geil, Polymer 22 (1981) 701
6. A. Keller, Phil. Mag. 2 (1957) 1171
7. J.D. Hoffman, G.T. Davis and J.I. Lauritzen Jr., Treatise on Solid State Chemistry, ed. N.B. Hannay, Plenum Press, New York, 1963, Vol.3 Chapt.7
8. D.M. Sadler, Polymer 24 (1983) 1401
9. G. Natta, P. Corradini and I.W. Bassi, Nuovo Cimento, Suppl. 15 (1960) 68
10. K. Izumi, Gan Ping, A. Toda, H. Miyaji and Y. Miyamoto, Jpn. J. Appl. Phys. 31 (1992) L626
11. D.C. Bassett and A.S. Vaughan, Polymer 26 (1985) 717
12. A.S. Vaughan and D.C. Bassett, Polymer 29 (1988) 1397
13. H.D. Keith and F.J. Padden Jr., J. Polym. Sci. B, Polym. Phys. 25 (1987) 2371
14. Y. Miyamoto, Y. Tanzawa, H. Miyaji and H. Kiho, Polymer 33 (1992) 2496
15. Y. Tanzawa, Polymer 33 (1992) 2659
16. K. Izumi, Gan Ping, A. Toda, H. Miyaji, M. Hashimoto, Y. Miyamoto and Y.Nakagawa, Jpn. J. Appl. Phys. 33 (1994) L1628
17. S.J. Sutton, K. Izumi, H. Miyaji and Y. Miyamoto, Polymer (in press)
18. N. Goldenfeld, J. Cryst. Growth 84 (1987) 601

Advances in the Understanding of Crystal Growth Mechanisms
T. Nishinaga, K. Nishioka, J. Harada, A. Sasaki and H. Takei (Editors)
© 1997 Elsevier Science B.V. All rights reserved.

Crystal growth and control of molecular orientation and polymorphism in physical vapor deposition of long-chain compounds

K. Sato[a], H. Takiguchi[a], S. Ueno[a], J. Yano[a] and K. Yase [b]

[a] Faculty of Applied Biological Science, Hiroshima University, Higashi-Hiroshima 739, Japan

[b] Department of polymer physics, National Institute of Materials and Chemical Research, Tsukuba, 305, Japan

Crystal growth mechanisms of vapor-deposited thin films of long-chain compounds have been studied with transmission electron microscopy, X-ray diffraction, Fourier-transform infrared (FT-IR) spectroscopy and atomic force microscopy. The controllable external conditions for the vapor deposition were type of substrate, substrate temperature, supersaturation and thermal annealing after deposition. In the physical vapor deposition of the long-chain compounds, the molecules are arranged either parallel (lateral growth mode) or normal (normal growth mode) with respect to a substrate surface. The type of molecular orientation strongly influences the physical properties of the thin films. In the present work, particular attention was paid to two aspects; (a) control of the orientation of the long-chain molecules with respect to the substrate, and (b) control of the occurrence of polymorphic modifications. It was confirmed that supersaturation was most determinative for controlling molecular orientation, and that substrate temperature was a critical factor in determining the formation of the polymorphic forms. We also found that surface diffusion, which was induced during thermal annealing after deposition, remarkably modified the crystallinity and molecular orientation of the thin films.

1. INTRODUCTION

There has been great interest in organic thin films which reveal multiple functional properties, such as insulator, lubricating coating, molecular electronic devices etc[1–4]. Two techniques have so far been applied for the formation of the organic thin films; vapor growth technique and dipping technique using molecular assembly adsorbed at an air/water interface. In order to utilize the organic thin films for newer functional materials, elucidation for crystal growth mechanism of thin film formation is pertinent, since the physical properties of the organic thin films are sensibly dependent on molecular orientation, polymorphic crystallization and transformation, twin formation, growth morphology, lattice defects, epitaxy etc. This is true for the vapor deposition techniques, since little is known in regards to thin film structures and deposition conditions. Recent studies on molecular beam

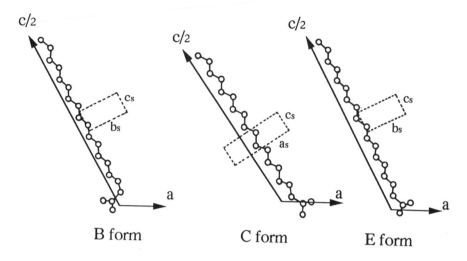

Figure 1. Molecular structures of B, C and E forms of stearic acid.

thin film structures and deposition conditions. Recent studies on molecular beam epitaxy of phthalocyanine molecules have enabled to form high-quality crystalline thin films[5,6].

The authors have dealt with the crystal growth mechanism in the physical vapor deposition of long-chain molecules, paying attention to the control of molecular orientation with respect to the substrate surface[7–10]. It has been found[11] that the most important factors for a better control of the molecular orientation are the rate of deposition, substrate temperature and type of substrate.

Since then, further research has been done on the effects of annealing of the as-deposited thin films at elevated temperatures on the transformation in molecular orientation from lateral to normal, and also on the crystallinity of the growth hillocks of the thin films[12–18]. For the molecular-level characterization of the thin films, atomic force microscopy (AFM) was recently applied and an excellent agreement with previous studies was confirmed[19]. Furthermore, it was recently found that the preferable occurrence of specific polymorphs of a long-chain fatty acid can be controlled by systematically varying the substrate temperature[20].

This article presents recent work on the control of the molecular orientation and polymorphic crystallization of long-chain fatty acids and their metal salts in the physical vapor deposition, which were examined by electron microscopy (EM), X-ray diffraction (XRD), Fourier-transform infrared (FT-IR) spectroscopy and AFM.

2. CRYSTAL STRUCTURES OF LONG-CHAIN FATTY ACIDS

Figure 1 illustrates molecular structures of three typical polymorphs, B, C and E forms of even-numbered saturated fatty acids, which have been determined by

	B	C	E
space group	$P2_1/a$	$P2_1/a$	$P2_1/a$
a (nm)	0.5587	0.936	0.5603
b (nm)	0.7386	0.495	0.7360
c (nm)	4.933	5.07	5.079
β (deg)	117.24	128.15	119.40
V (nm^3)	1.810	1.845	1.825
Z	4	4	4

Table 1
Unit cell parameters of B, C and E polymorphs of stearic acid

single crystal XRD[21–24], powder XRD[25] and FT-IR[26,27] and Raman techniques[28,29]. The three forms occur in homologous series of saturated fatty acids with even numbers of hydrocarbon atoms, n_c, 16 through 22. Unit cell parameters of B, C and E forms of stearic acid, which is the most typical even-numbered saturated fatty acid, are summarized in Table 1.

C and E forms contain straight hydrocarbon chains of stearic acid in the unit cell, yet *gauche* conformation reveals at the hydrocarbons closest to COOH group in B. As shown in Table 1, crystal density is largest for B and smallest for C. The differentiation of C form from other two forms can easily be done by observing crystal shape and powder XRD spectra. However, B and E cannot be differentiated by these convenient techniques since their crystal structures and thereby the crystal shapes are quite similar. Instead, an FT-IR technique using three specific absorption spectra of progression bands due to CH$_2$ wagging mode, $w(CH_2)$, in-plane O-C=O deformation band, $\delta(OCO)$ and out-of-plane O-H deformation band, $\sigma(OH)$, can facilitate identification of the two forms[26,27].

3. MATERIALS AND METHODS

The materials examined in the present study were arachinic acid and behenic acid, which are long-chain fatty acids with n_c of 20 and 22, respectively, and calcium stearate which is calcium salt of stearic acid (n_c =18). The melting points were 75.3°C (arachinic acid), 80.2°C (behenic acid) and 180°C (calcium stearate). All the materials with the purity of more than 99% were provided by Nippon Oil and Fats Co. The physical vapor deposition was done in an ambient vacuum of 5×10^{-7} Torr. The temperatures of furnace (T_f) and substrate (T_s) were carefully controlled separately. The substrate was heated at 150°C before deposition. It should be noted here that the differences in temperature between T_f and T_s correspond to supersaturation for deposition, which are determined by the ratio of equilibrium vapor pressures at T_f and T_s. Hence, $\Delta T = T_f - T_s$ correspond to driving force of the deposition.

The distance between the furnace and substrate was 2cm. The deposition time

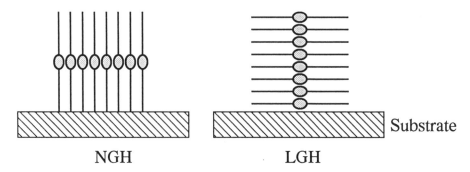

Figure 2. Lateral growth hillock (LGH) and normal growth hillock (NGH).

was changed from several minutes to several hours, depending on the technique of characterization. For example, the deposition time was 3 minutes for the observation of the initial stage of deposition examined by EM. A 5 hours deposition time was used for the determination of the polymorphic modification of the deposited films with XRD and FT-IR techniques.

The surface structures were observed by EM in replica and by AFM as well. As for the EM observation, the films were cooled to -10°C soon after deposition to prevent evaporation and replica films were made by germanium decoration with supporting carbon film. The EM replica image provided macroscopic structures of the thin films, yet no information of the surface structures on a molecular level was available.

Next, the AFM observation was attempted (Nano Scope II and III, Digital Instruments Co.), using Si_3N_4 tip with a radius of curvature of 40 nm and aspect ratio of 1.43:1 placed on a cantilever with a spring constant of 0.12N/m. The microscope was isolated from building vibrations by an antivibration table. For high-resolution images, the AFM was operated with repulsive forces on the order of 10^{-8}N and scan rate ranged from 2-61Hz.

The molecular orientation and polymorphic structure were evaluated by XRD (Rigaku, Roterflex) and FT-IR (Perkin Elmer Spectrum 2000) methods. The intensity ratio of XRD long spacing (small angle region) and short spacing spectra (wide angle region) is sensitive to the molecular orientation. This is because the intensity of the long spacing spectra increases relative to the short spacing spectra with increasing concentration of the normally oriented growth hillocks (see below).

4. MOLECULAR ORIENTATION AND GROWTH CONDITIONS

The molecular orientation in the thin films is dependent upon the orientation of the long-chain axes in the crystalline films with respect to the substrate surface. As illustrated in Figure 2, normal growth hillock (NGH) refers to the orientation normal to the substrate surface, and parallel orientation refers to lateral growth

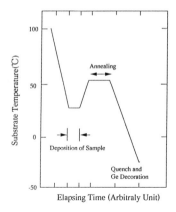

Figure 3. A schematic illustration of thermal annealing processes.

hillock (LGH). Actually, this illustration disregards inclination of the long-chain axes which are not exactly perpendicular against the substrate surface because of monoclinic crystal structures of the three polymorphic forms (Table 1).

The three most influential deposition conditions for better control of the molecular orientation[7–11] are as follows. As for the type of substrate, NGH is preferred on the hydrophobic substrate surfaces such as amorphous carbon. On the contrary, ionic substrates such as cleaved KCl crystal surface prefers LGH. As for the rate of deposition, LGH prevailed at the higher rates of deposition. Finally, high T_s preferred NGH, and vice versa.

5. THERMAL ANNEALING OF VAPOR-DEPOSITED THIN FILMS

It was confirmed that morphology, crystallinity, crystal size and molecular orientation of the as-deposited thin films were modified by thermal annealing[12–18], The thermal annealing was carried out at elevated temperature below the melting points of the deposited materials after the cessation of deposition. These effects have been observed in physical vapor deposition of the long-chain molecules for the first time in the present work. It is considered that the thermal annealing processes are controlled by surface diffusion of the molecules, which may occur within relaxation times of transient adsorption states of the molecules before desorption. Namely, the molecules would be detached from less stable growth hillocks and incorporated into the more stable growth hillocks after surface diffusion. The stability of this kind is still uncertain, yet it may be due to lattice defect density of the growth hillocks, interfacial energy between the growth hillocks and the substrate surface, occurrence of metastable polymorphic forms etc. Temperature variation of the deposition and thermal annealing performed in the present study is illustrated in Figure 3.

The furnace was sealed after the deposition was ceased, and T_s was then raised

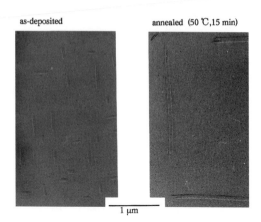

as-deposited   annealed (50 ℃,15 min)

1 μm

Figure 4. Lateral growth hillocks of calcium stearate deposited on KCl after thermal annealing.

to the programmed annealing temperature, kept there over certain period and subjected to replica EM observation and XRD. The following annealing behavior was observed.

5.1. Changes in morphology and size of lateral growth hillocks

The thermal annealing of the LGH at 30-50°C of calcium stearate deposited on KCl gave rise to a reduction in the number and an increase in the size of growth hillocks (shown in Figure 4). The annealing period was 15 minutes. As for morphological change, the length of LGH increased, yet no increase in width was detectable, relative to the as-deposited forms. This means that the thermal annealing did not induce the growth of the basal plane of LGH, but the growth of the lateral faces was induced.

It can be assumed that an inverse of the number of LGH per unit area, $1/n$, is proportional to the area, which is covered by the surface diffusion, and thereby $1/n^{1/2}$ is proportional to mean free path of surface diffusion. In addition, the increase in the length of LGH is proportional to the growth rate induced by the thermal annealing.

Based on this assumption, activation energy values for surface diffusion and growth rate for LGH of calcium stearate films deposited on KCl were calculated by measuring temperature dependence of $1/n^{1/2}$ and the length of LGH shown in Figure 5(a). The values were 44kJ/mol for surface diffusion and 87kJ/mol for growth rate.

5.2. Changes in morphology and size of normal growth hillocks

The thermal annealing of NGH deposited on amorphous carbon films at 40-60°C induced drastic morphological changes of the as-deposited films from dendrite patterns to polyhedral forms, as shown in Figure 6. At the same time, the size of

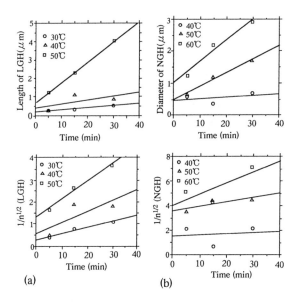

Figure 5. (a) Mean length and $1/n^{1/2}$ of lateral growth hillocks on KCl, and (b) mean diameter and $1/n^{1/2}$ of normal growth hillocks on amorphous carbon films of calcium stearate, at different elapsing time at different annealing temperatures.

NGH increased at the expense of their number per unit surface area. The stabilized crystal morphology after the thermal annealing corresponds well to the monoclinic crystal structure of calcium stearate[30]. The height of the NGH crystal in Figure 6(c) was about 6nm, which corresponds to the length of one-dimer. Hence, the growth hillocks shown in Figure 6 are two-dimensional crystals having the height of one unit cell.

The activation energy values for surface diffusion and growth rate of calcium stearate on amorphous carbon film induced by thermal annealing for the NGH were calculated, as shown in Figure 5(b); 100kJ/mol for surface diffusion, and 108kJ/mol for growth rate.

5.3. Transformation from lateral to normal growth hillocks

It was observed that thermal annealing induced the transformation of the molecular orientation from lateral to normal for the as-deposited calcium stearate films, when annealing was carried out at 50-90°C . This phenomenon was more easily observed by replica EM on the substrate surfaces of cleaved mica and amorphous carbon films, both of which prefer the NGH formation. However, the same transformation was also observed on KCl surfaces, although quite slowly compared to the former two substrates.

The transformation in the molecular orientation was also monitored by XRD for calcium stearate deposited on mica substrate, as shown in Figure 7. The diffraction

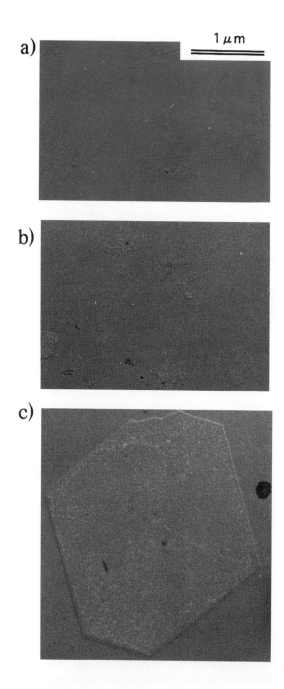

a)

1 μm

b)

c)

Figure 6. Electron micrographs of (a) as-deposited, (b) annealed at 30°C for 30 min and (c) annealed at 60°C for 35 min after (b), of calcium stearate deposited on amorphous carbon films.

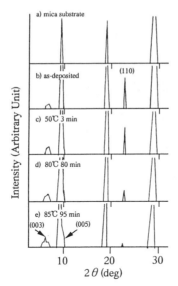

Figure 7. Schematic X-ray diffraction patterns of calcium stearate thin films deposited on cleaved mica annealed at high temperatures.

patterns of mica substrate were obtained before deposition(Figure 7(a)). After deposition at $T_s = 20°C$, a (110) spectrum was observed which is a characteristic spectrum of LGH (Figure 7(b)). As the thermal annealing was subjected to the as-deposited films at 50, 80 and 85°C, long spacing (003) and (005) spectra increased in strength at the expense of the (110) spectrum, as shown in Figures 7(c)-(e).

Since no deposition occurred during the annealing, actual processes involved in this transformation were nucleation and growth of NGH at the expense of LGH during the annealing, all of which are induced by surface diffusion. The replica EM observation revealed[15] that nucleation and growth of NGH occurred at the places in contact with as well as free from the LGHs. This means that the nucleation of NGH in the LGH-dominated films occurred through spontaneous surface nucleation, and through heterogeneous surface nucleation which was assisted by LGH.

The replica EM observations enabled the calculation of activation energies for the transformation in molecular orientation for the three substrates; 81kJ/mol (KCl), 16kJ/mol (mica) and 20kJ/mol (carbon film).

At present, there is no precise numerical evaluation method for the three types of activation energy for processes induced by thermal annealing. This task, an area of future research, may involve complicated elementary processes of surface diffusion, surface nucleation and crystal growth, all of which may be dependent on the molecular orientation and type of substrate.

358

| | | Single Crystal | | Thin film |
		XRD[a]	AFM[b]	AFM[b]
	d(10)	0.936	0.90	1.02
C	d(01)	0.495	0.53	0.56
	d(11)	0.437	0.49	0.49
	d(10)	0.560	0.58	0.59
E	d(01)	0.736	0.76	0.82
	d(11)	0.446	0.48	0.50

[a]stearic acid, [b]behenic acid

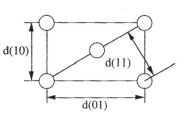

Table 2
Lattice parameters of two-dimensional orthorhombic subcells of stearic acid and behenic acid

6. CONTROL OF POLYMORPHIC OCCURRENCE IN PHYSICAL VAPOR DEPOSITION OF BEHENIC ACID

Polymorphism is always revealed in the long-chain compounds[31]. The occurrence of polymorphic modifications of the long-chain compounds from solution has been thoroughly examined. This has been done for stearic acid[32,33]. As for physical vapor deposition of stearic acid, the occurrence of a high-temperature stable form, C form, was observed by XRD and EM[7]. Since then, this form is believed to uniquely crystallize in the vapor deposition. However, the most recent studies using AFM and FT-IR has revealed[19,20] that the occurrence of E and C form are systematically controllable by varying T_s, as briefly explained below.

Figure 8 shows the AFM images of NGH of behenic acid deposited on mica. A low-magnification image of the as-deposited film with a deposition time of 20 minutes at $T_f = 43°C$ and $T_s = 23°C$ is shown in Figure 8(a). Island structures with widths ranging from 20nm to 1μm have a step height of 6nm which corresponds to a one-dimer length[19]. Figure 8(b) shows a top-view molecular resolution image of NGH grown at $T_f = 60°C$ and $T_s = 40°C$. The fine structure corresponds to a rectangular lattice of a typical subcell of orthorhombic perpendicular of the aliphatic chain packing (see Table 2)[34].

The two-dimensional lattice parameters are calculated from the Bragg peaks in Fourier transformation of Figure 8(b), as summarized in Table 2, together with those of bulk single crystals of stearic acid and behenic acid obtained by XRD and AFM. The image in Figure 8(b) correspond well to those of C forms of the bulk system. Figure 8(c) shows the same AFM image of NGH grown at $T_f = 50°C$ and $T_s = 10°C$, whose two-dimensional lattice parameters correspond well to those of E form of the bulk system as summarized in Table 2[20]. These results indicate the occurrence of the two forms in the physical vapor deposition, which was highly dependent on T_s as examined by FT-IR.

Figure 9 shows an FT-IR spectrum of out-of-plane O-H deformation band, $\sigma(OH)$ of LGH and NGH of behenic acid deposited on mica at $T_f = 75°C$ at different

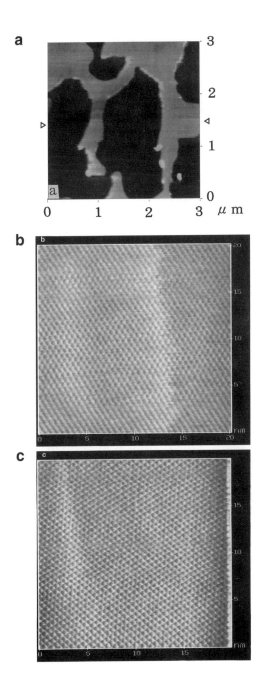

Figure 8. AFM images of normal growth hillocks of behenic acid, (a) low-magnification image, (b) top view of C form, (c) top view of E form.

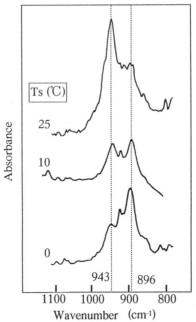

Figure 9. FT-IR spectra of out-of-plane O-H deformation band, $\sigma(OH)$ of behenic acid at different substrate temperatures.

substrate temperatures. At $T_s = 25°C$, a strong $\sigma(OH)$ spectrum appears at $943cm^{-1}$ which is typical of the C form. A weak $\sigma(OH)$ band appears at $896cm^{-1}$ which is typical of E form and B form[27]. The intensity ratio of the two spectra varied inversely with decreasing T_s from 25 to 0°C. The differentiation of E and B forms of the thin films was done by observing $w(CH_2)$ wagging band, which proved that the polymorphic form preferably crystallized at $T_s = 0°C$ in Figure 9 was E form,although not shown here. This indicates the preferable occurrence of E form at lower temperature of substrate.

The temperature dependence of the occurrence behavior of E and C forms of behenic acid can be qualitatively interpreted by taking into account of thermo-dynamic stability relationship. Figure 10 shows the relationship between Gibbs energy and temperature of the three monoclinic forms of even-numbered saturated fatty acids which was thoroughly examined in the case of stearic acid[11,33]. C form is stable at high-temperatures, B form at low-temperatures, and E form is metastable at all temperatures. The transformation in a solid state occurs uni-directionally from E to C, and from B to C upon heating. Transformations in the opposite direction did not occur on cooling because of steric hindrance. The crossing point of Gibbs energy values of B and C, T_{B-C}, is 32°C for stearic acid, as experimentally determined by solubility measurement[35]. Since E form is less stable than B form, T_{B-C} must be higher than T_{E-C}. As for the present case of

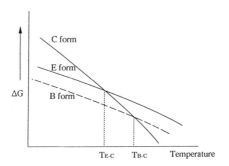

Figure 10. Temperature dependence of Gibbs energy values (ΔG) of B, C and E forms of even-numbered saturated fatty acids.

behenic acid, it is reasonable to conceive that the two crossing points are shifted toward higher temperature ranges by about 15°C .

This phase behavior of the three forms enables the occurrence of the three forms, which are sensitively dependent on supersaturation and surface energy of the polymorphic forms. Calculation of surface energy of the C and B forms of stearic acid for a crystal-vapor system showed that the surface energy values of C are smaller than those of B for all well-developed crystal faces[36]. This relationship can be applied to C and E forms of behenic acid, by making a homologous analogy between stearic and behenic acids, and between B and E as well. Therefore, a prediction of nucleation behavior of E and C forms can be described in the case of enantiotropic free energy relation like Figure 10 in what follows[37]. Nucleation of the high-temperature form C is manifest in a high temperature region, where two factors of surface energy and supersaturation prevail in C form. By contrast, E form can be nucleated at low temperatures and low supersaturation because of smaller values of Gibbs energy. The experimental observation shown in Figures 8 and 9 agrees well with this argument.

7. CONCLUSION

The present study deals with an attempt to fabricate high-quality crystalline thin films of long-chain compounds in which the molecular orientation and polymorphic forms were controlled by varying the deposition conditions of supersaturation and temperature. This trial attempts to meet a demand in molecular electronics which turns from a phase of searching functional substances to a phase of controlling structures and properties[38]. The elucidation of crystal growth mechanisms of molecular crystals which contain multiple facets of polymorphism, crystal habit, lattice defects and molecular orientation, is described in the present article for the specific examples of long-chain compounds.

362

ACKNOWLEDGMENTS

This work was supported by a Grant-in-Aid for Scientific research on Priority Area "Crystal Growth Mechanism in Atomic Scale" No.03243015 from the Ministry of Education, Science, Sports and Culture of Japan. The authors are indebted to Professor K.Inaoka, Yuge Mercantile Marine College for fruitful discussion.

REFERENCES

1. P.E.Burrows *et al.*, Mol. Cryst. Liq. Cryst. Sci. Thechnol. Sec. B. Nonlinear Optics 2 (1992) 193.
2. N.Ara-Kato, K.Kajikawa, K.Yase, M.Hara, H.Sasabe and W.Knoll, Mol. Cryst. Liq. Cryst., 280 (1996) 295.
3. K.Yase, Y.Yoshida, S.Sumimoto, N.Tanigaki, H.Matsuda and M.Kato, Thin Solid Films, 273 (1996) 218.
4. K.Yase, T.Saraya and K.Kudo, Mat. Res. Soc. Symp. Proc. 359 (1995) 417.
5. H.Tada, K.Saiki and A.Koma, Jpn. J. Appl. Phys. 30 (1992) L306.
6. H.Tada *et al.*, Appl. Phys. Lett. 61 (1992) 2021.
7. F.Matsuzaki, K.Inaoka, M.Okada, K.Sato, J. Cryst. Growth 69 (1984) 231.
8. K.Yase, T.Inoue, K.Inaoka and M.Okada, Jpn. J. Appl. Phys. 28 (1989) 872.
9. T.Inoue, K.Yase, M.Okada, S.Okada, H.Matsuda. H.Nakanishi and M.Kato, Jpn. J. Appl. Phys. 28 (1989) 2037.
10. K.Yase, T.Inoue, K.Inaoka and M.Okada, J. Electron. Microsc. 38 (1989) 132.
11. K.Sato and M.Kobayashi, in Crystals/Growth, Properties and Applications, vol.13, Organic crystals, ed.N.Karl, Springer-Verlag, Heidelberg, (1991), pp.65-108.
12. K.Yase, M.Yamanaka, T.Sasaki, K.Inaoka and M.Okada, J. Cryst. Growth 118 (1992) 348.
13. K.Mimura, M.Yamanaka, K.Yase, S.Ueno and K.Sato, in Proceedings of the 6th Topical meetings on Crystal Growth Mechanism, Awara, Jpn. 20-22 (1993) p.221.
14. K.Yase, M.Yamanaka and K.Sato, Appl. Surf. Sci 60/61 (1992) 326.
15. K.Yase, M.Yamanaka, K.Sato and M.Okada, Jpn. J. Appl. Phys. 31 (1992) L731.
16. M.Yamanaka, K.Sato, K.Inaoka and K.Yase, Jpn. J. Appl. Phys. 31 (1992) 1632.
17. M.Yamanaka, K.Mimura, K.Yase, K.Sato and K.Inaoka, J. Cryst. Growth 128 (1993) 113.
18. K.Yase, M.Yamanaka, M.Mimura, S.Ueno and K.Sato, Thin Solid Films 243 (1994) 228.
19. H. Takiguchi, M.Izawa, K.Yase, S.Ueno, M.Yoshimura, T.Yao and K.Sato, J. Cryst. Growth 146 (1995) 645.
20. H.Takiguchi, T.Nakata, S.Miyashita, J.Yano, S.Ueno and K.Sato, Collected Abstracts of 27th National Conference on Crystal Growth, Kusatsu, Japan, p.154.

21. M.Goto and E.Asada, Bull. Chem. Soc. Japan 51 (1978) 2456.
22. F.Kaneko, H.Sakashita, M.Kobayashi, M.Kitagawa and Y.Matsuura, Acta Crystallogr. C46 (1990) 1490.
23. F.Kaneko, H.Sakashita, M.Kobayashi, M.Kitagawa, Y.Matsuura and Y.Suzuki, Acta Crystallogr. C50 (1994) 245.
24. F.Kaneko, H.Sakashita, M.Kobayashi, M.Kitagawa, Y.Matsuura and M.Suzuki, Acta Crystallogr. C50 (1994) 247.
25. B.Malta, G.Celotti, R.Zannetti and A.F.Martelli, J. Chem. Soc. (B) 548.
26. F.Kaneko, M.Miyamoto, T.Shimofuku and M.Kobayashi, J. Phys. Chem. 96 (1992) 10554.
27. F.Kaneko, O.Shirai, H.Miyamoto, M.Kobayashi and M.Suzuki, J. Phys. Chem. 98 (1994) 2185.
28. M.Kobayashi, T.Kobayashi, Y.Ito and K.Sato, J. Chem. Phys. 80 (1984) 2897.
29. F.Kaneko, M.Kobayashi and H.Sakashita, J. Raman Spectrosc. 24 (1993) 527.
30. D.M.Small, The Physical Chemistry of Lipids, Plenum, (1986), Chapter 9, pp.285-343.
31. K.Sato, in Crystallizaiton and Polymorphism of Fats and Fatty Acids, ed. N.Garti and K.Sato, Marcel Dekker, New York, 1988, pp. 227-263.
32. K.Sato and R.Boistelle, J.Cryst.Growth, 66 (1984) 441.
33. F.Kaneko, H.Sakashita, M.Kobayashi and M.Suzuki, J.Phys.Chem., 98 (1994) 3801.
34. In ref.30, p.97.
35. K.Sato, M.Kobayashi and H.Morishita, J.Cryst.Growth, 87 (1988) 236.
36. W.Bechkmann and R.Boistelle, J.Cryst.Growth, 67 (1984) 271.
37. M.Kitamura, S.Ueno and K.Sato, in Crystallization Processes, Solution Chemistry Series, ed.H.Ohtaki, John Wiley & Sons Ltd., New York, in press.
38. A.J.Hiller and M.D.Ward, Science, 263 (1994) 1261.

Advances in the Understanding of Crystal Growth Mechanisms
T. Nishinaga, K. Nishioka, J. Harada, A. Sasaki and H. Takei (Editors)
© 1997 Elsevier Science B.V. All rights reserved.

Polymer crystallization approached from a new view point of chain sliding diffusion

M. Hikosaka
Faculty of Integrated Arts and Sciences, Hiroshima University,
Higashi-Hiroshima, 739 Japan

Linear chain polymers have an extreme topological nature, different from atomic systems, which results in characteristic morphology of lamellar crystals. In this study, two important long unsolved questions, "what is the origin of very thin lamellar crystals named folded chain crystals (FCCs) and thick ones named extended chain crystals (ECCs)?" and "what determines the lamellar thickness?" are answered by introducing a new point of view of chain sliding diffusion within crystals or nucleus. The first answer is that all polymers will form ECCs when they crystallize from the melt into the mobile phase, where chain can slide easily, so the lamellar can thicken easily, whereas all polymers will form FCCs when they crystallize from the melt into the immobile phase where chain can not slide easily, so the lamellar can not thicken significantly. The second answer is that the lamellar thickness of ECCs is determined by apparent stoppage of lamellar thickening growth with the mobile metastable hexagonal to immobile stable orthorhombic phase transition which can be observed widely in polymer systems.

1. INTRODUCTION

Linear chain polymer molecules is one of the typical soft materials and has particular tendency that they prefer to construct spontaneously a lot of kinds of anisotropic and complicated structures, textures and/or morphologies, the scale and class of which covers very wide, from microscopic (nm) to macroscopic (mm), scale. This means that self-organization and class-formation are very remarkable character (or potential) to polymer chain systems. The ultimate example is seen in life systems constructed of biopolymers. It is interesting for both science and technology to make clear the mechanism of the self-organization and class-formation processes, because this requires new scientific ideas, quite different from those used in conventional hard materials such as metals, silicones etc. and this is inevitable to control the structures and morphologies of polymer materials.

Study in this field starts developing recently by introducing a new view point of topological nature of chain molecules, such as chain sliding diffusion presented by the present author, [1], [2] which will be shown below. In this study, polymer crystallization will be investigated which is a typical example of the self-organization processes.

1.1. Lamellar crystals of polymers

Polymer crystals usually show a characteristic morphology of lamellar shape which means plate like crystals.[3] Polymer chains are aligned nearly perpendicular to the end surface of the lamellae. The lamellae are elemental morphology within complicated super structures such as spherulite. Lamellar thickness (l) changes at large scale, i.e., from the order of nm to μm.

Thin lamella of the order of nm thick is named an folded chain crystal (FCC) first identified by Keller,[4] while thick one thicker than μm is named an extended chain crystals (ECCs) first observed by Wunderlich and Arakawa.[5] The FCCs are so thin that they are thermodynamically unstable and their structures and properties are imperfect, while the ECCs are stable and perfect. It has been well known that the polymer crystals are crystallized into unstable FCCs in many cases, whereas they are crystallized into stable ECCs in rather small cases. This is paradoxical to thermodynamics. Therefore "what is the origin of FCCs and ECCs?" has been one of the most basic unsolved problem in polymer science.

The l is one of the most important parameter in polymer crystals, because l strongly controls physical properties of polymer solid, such as melting temperature, mechanical properties, ferroelectric properties and so on. It has been interesting unsolved problem for long time to solve "what determines the lamellar thickness of lamellar crystal?" It must be closely related to crystallization kinetics of chain molecules.

1.2. Distinction between lamellar thickening and thickening growth

It is well known that the l of FCCs with stacked lamellae increases with crystallization temperature (T_c) or annealing temperature (T_a), which is well known lamellar thickening, on which many studies have been done.[6] Hoffman et al proposed a nucleation theory and explained the temperature dependence of l of FCC [7] and Sadler did an entropic controlled theory to explain it.[8] But in these theories the role of topological nature of chains has not been taken in consideration.

We have to distinguish between the lamellar thickening and the new lamellar thickening growth which will be explained in 2.3, in spite of their apparent close resemblances. The former is an increasing process of l of stacked lamellar system, whereas the latter is that of an isolated single crystal. The former is essentially an annealing process which succeeds after formation of stacked lamellae, that is one of so called ripening process, while the latter is a crystal growth process. In another word, in the former case, materials are not supplied from outside of the lamellae, whereas in the latter case, they can be supplied from outside.

1.3. Purpose

The purpose of this chapter is to propose answers briefly to the above two unsolved problems, what is the origin of FCCs and ECCs? and what determines the lamellar thickness of lamellar crystal? In this study, the important role of chain sliding diffusion which is due to topological nature of polymer chains and that of mesophase such as hexagonal or liquid crystalline phases will be stressed.

2. CHAIN SLIDING DIFFUSION THEORY [1] [2]

The author recently presented a chain sliding diffusion theory to explain the origin of FCCs and ECCs. The theory showed that the origin of ECCs and FCCs is related to the respective ease and difficulty of chain sliding diffusion within the nucleus and/or crystals, respectively. The concept of sliding diffusion along the chain axis reflects the topological nature of long chain molecules, as shown in the next subsection, which is quite different from that in atomic systems or systems of low molecular weight molecules.

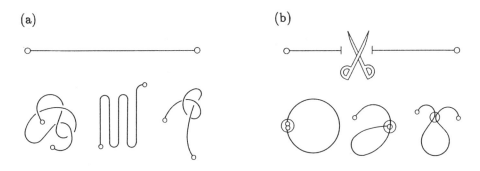

Figure 1. Schematic illustration explaining the topological nature of a linear chain. (a) any chain is topologically equivalent if it changes its conformation without cutting or healing due to chemical reaction, while (b) any chain can not be topologically equivalent if it is cut or healed within the chain.

2.1. Topological nature of polymer chains

A linear polymer chain can take various conformation without any cutting or combining as illustrated in Figure 1(a). In this case, they are called that they are topologically the same with each other, while if a chain is cut or combined as in Figure 1(b), then they are not topologically the same. The topological nature takes an important role in polymer crystallization, because the chains has to change its conformation without any cutting or combining within the chain during crystallization process.

2.2. Two competing factors in crystal growth

It is general view point in a classical nucleation theory that crystal growth is controlled by two competing factors, one is a thermodynamic one and the other is a kinetic one related to diffusion.[9] It is obvious that the thermodynamic factor requires to minimize the free energy of the system, so thermodynamics predicts that any crystal grows into three dimensional shape so as to minimize the ratio of its surface area to bulk volume. The kinetic factor, on the other hand, modifies the crystal shape in the case that the activation energy of diffusion is too high for some direction of growth.

2.3. Lateral growth and thickening growth

From the thermodynamic point of view, a polymer single crystal grows into three dimensional shape, i.e., into two directions, nearly perpendicular and parallel to the chain axis as is illustrated in Figure 2. The former is well known lateral growth and the latter is lamellar thickening growth, respectively. The latter is newly found by the author.[12] Therefore polymer crystals will be formed via coupling of the two different growths. It is to be noted that in both growth processes, chains

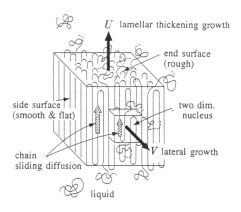

Figure 2. Model of a polymer single crystal grwoing into two directions via lateral and thickening growths with growth rates V and U, respectively.

have to slide along their chain axes within crystals or nuclei in order to satisfy the above topological requirement (see Figure2). The effect of the chain sliding is formulated as a diffusion factor, henceforth the theory is named "chain sliding diffusion theory". Thus the sliding diffusion is the kinetic factor in the case of polymer crystal growth.

2.4. Origin of FCCs and ECCs [1] [2]

The sliding diffusion theory succeeded to formulate the nucleation rate or crystal growth rate of polymers introducing the new kinetic factor related to the chain sliding diffusion. The activation free energy of which is denoted as ΔE. The sliding diffusion theory proposed a prediction illustrated in Figure 3 which is our answer to the question, "what is the origin of FCCs and ECCs?" All polymers will form ECCs when they crystallize from the melt into the mobile phase, because in the mobile phase, chain can slide easily, i.e., ΔE is rather small, so the lamellar thickening growth occurs significantly, which gives formation of ECCs. Here the

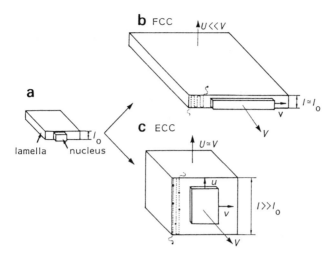

Figure 3. Model for the origin of FCC and ECC. (a) initial FCC type lamella or nucleus, (b) FCC type growth for large ΔE whre chain sliding diffusion is difficult and the thickening rate U is much smaller than the lateral growth rate V, (c) ECC type growth for small ΔE where chain sliding dussusion is easy and U is comparable to V.

mobile phases include the hexagonal phase, liquid crystalline phase and any other disordered phases with high chain mobility and entropy which are widely seen in chain-molecule system at a rather high temperature. The close relation between the mobile phase and formation of ECCs was first suggested by Bassett.[10]

All polymers will form FCCs, on the other hand, when they crystallize from the melt into the immobile phase, because in the immobile phase, chain can not slide easily, i.e., ΔE is rather large, so the lamellar thickening growth could not occur significantly, which gives formation of FCCs. The immobile phases include ordered phases with low entropy.

The prediction has been confirmed recently on several polymers, in addition to PE, such as polychrolotrifluoroethylene (PCTFE), polytrans 1,4 butadiene (PT1,4BD), vinilidenefuloride/ trifluoroethylene copolymers (P(VDF/TrFE)) (Figure 4) and also in liquid crystalline polymers. [12] [11]

2.5. Correspondence between phase and morphology

A good example of confirmation of the prediction is shown on PE using a P-T phase diagram shown in Figure 5. [13] There are three phases, liquid, orthorhombic (o) and hexagonal (h) phases. The h and o phases are mobile and immobile phases, respectively. The triple point is indicated by Q. The $P = 0.5$ GPa is the triple point pressure (P_{tri}).

T^0 is an equilibrium melting temperature. The solid lines indicates real melting

370

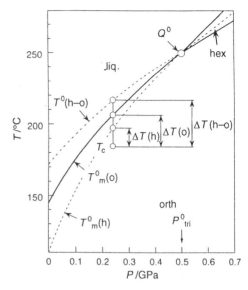

Figure 4. Transmission electron micrograph showing thick lamellae of ECCs of P(VDF/TrFE) crystallized at 80MPa. $\Delta T = 13$K. Scale bar= 1 μm.

Figure 5. Equilibrium pressure (P)-temperature (T) phase diagram of PE constructed from Figure 7 in reference 8. Q^0 indicates the triple point and P_{tri} is the triple point pressure (0.5GPa). Definitions of $\Delta T(h)$, $\Delta T(o)$ and ΔT(h-o) are shown in the text.

or phase transition curves and broken lines does virtual curves extrapolated from the real curves. There can be three kinds of degree of supercooling temperature, i.e., that for melting of h and o crystals, ΔT(h), ΔT(o), respectively and that for o-h transition, ΔT(o-h).

It is found that crystallization from the melt into h phase and o phase give formation of ECCs and FCCs, respectively, which clearly confirmed the prediction.

2.6. Important role of the metastable mobile phase in crystallization

A much more remarkable fact was found that even below the triple point pressure, ECCs can be formed. This could be easily explained by sliding diffusion theory by finding a fact that a metastable mobile phase of h crystal was formed at first and then it transformed into a stable immobile o crystal after some waiting time. This is the reason why virtual broken lines are illustrated in the phase diagram. Universality of this finding was confirmed on other polymers mentioned above.[14]

This is one of Ostwald's rule of stage, [17] but the more detailed physical implication was recently developed combining with a new aspect of inversion of stable phase with crystal growth due to size effect.[18] This is an interesting topics interesting for not only chain molecular systems but also for atomic systems.

It should be stressed that another significant fact was found at the same time that the lateral growth actually stops with the h-o phase transition.[15] We will turn to this point in the next section to explain the thickness-determining mechanism.

3. WHAT DETERMINES THE LAMELLAR THICKNESS? [19]

Next we will answer to the second question "what determines the lamellar thickness?" by introducing a new stoppage mechanism of the lamellar thickening growth. Here we will limit our answer to the case of an extended chain single crystal (ECSC) of polyethylene (PE). This may be generally applicable to ECSCs of other polymers, while it is not certain whether this is directly applicable to FCCs or not, but this must give a good clue to solve the question of FCCs.

In this section, all ECSCs of PE were isothermally crystallized from the melt at a range of pressures (P) from 0.3 to 0.5 GPa, below the triple point pressure (Figure 5), where the metastable mobile h phase first crystallizes and then it transforms into the stable mobile o phase, as seen in the previous section. The mobile-immobile phase transition will play the main role in this study. As is mentioned in the

(a)

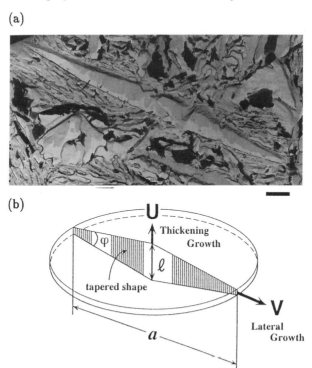

Figure 6. (a) Transmission electron micrograph of the linear tapered cross section of an isolated ECSC. Crystallized at 0.4GPa and $\Delta T = 2.6$K. Scale bar = 1 μm. (b) Schematic perspective of an ECSC.

previous section, an ECSC is formed via coupling of well known lateral growth and newly found lamellar thickening growth, as illustrated in Figure 6 which shows a perspective of an ECSC of PE. The transmission electron micrographs of cross section of the isolated ECSCs are shown in Figure 7. It is to be noted that the cross section of an ECSC in Figure 7 shows tapered shape, so the l is defined as the maximum thickest thickness at the center, as illustrated in Figure 6.

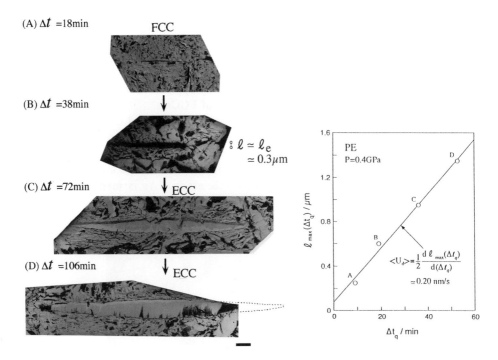

Figure 7. Transmission electron micrographs showing an evidence that the ECC formed from FCC. Scale bar $= 1\mu$m. Crystallized at 0.4GPa and $\Delta T = 2.6$K. Δt is crystallization time

Figure 8. lamaellara thickness l vs time t obtained on the same ECCs as shown in Figure 7

3.1. Direct evidence of lamellar thickening growth [20]

Direct evidence of the lamellar thickening is shown in Figure 7 which shows cross sections of isolated single crystals arranged in the order of increasing lateral size (a). A larger crystal arises after longer crystallization time (Δt). Therefore Figure 7. is a representation of the growing process for an ECSC with increasing Δt. It shows that the l increases with growth, which is quantitatively shown in Figure 8 where A, B, C and D correspond to ECSCs in Figure 7. This is direct evidence of lamellar thickening which is observed at the first time.

It is found from Figure 8 that the l increases linearly with crystallization time (t), thus l is given by

$$l = l^* + 2Ut \tag{1}$$

where l^* is l of a critical nucleus and U is defined by $(dl/dt)/2$.

It is also found that the U increases with increase of ΔT, which is just opposite to the negative ΔT dependence of lamellar thickening rate of lamellar stacked system.

3.2. ECC is formed from FCC via thickening growth [19]

It is apparent from Figure 7 that the l is small, only 0.15 μm, for a small single crystal growing for short Δt, while it increases up to a few μm for a large single crystal after long Δt. It is obvious that the former is thinner than the extended chain length (l_e), in this case $l_e = 0.3$ μm, whereas the latter exceeds the l_e. Consequently the small crystal in Figure 7(A) must be a folded chain single crystal (FCSC), whereas the large crystals in Figure 7(C) and (D) must be ECSCs.

It is therefore concluded that an ECSC is generated from an FCC and an FCC grows into an ECC via coupling of the newly recognized lamellar thickening growth and the long familiar lateral growth. Thus there appears to be no significant gap between the formation of an FCC and an ECC, i.e., the morphological change from an FCC to an ECC is a continuous process.

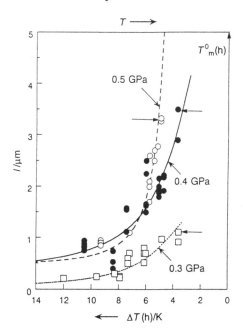

Figure 9. l vs. $\Delta T(h)$ or T at $P - 0.3$, 0.4 and 0.5 GPa. Points are observed values, curves are calculated as described in the text.

374

3.3. Significant increase of lamellar thickness with decrease of ΔT(h)

Figure 9 shows significant ΔT dependence of l of an isolated ECSC observed at $P = 0.3, 0.4$ GPa and 0.5 GPa. l increased with decrease of ΔT, (i.e., increase of T_c) which is similar to well known ΔT dependence of l of FCCs. The curves indicate the theoretical ones which will be explained later.

It is to be noted that wide distribution of l was observed on many single crystals within a sample crystallized at a given condition. Important point was that there exists a maximum lamellar thickness for a given crystallization condition, i.e., for a given ΔT(h) and P. Therefore it is defined in Figure 8 the maximum lamellar thickness as observed one.

3.4. Cessation of thickening growth with mobile-immobile phase transition[19]

It is found that the maximum lamellar thickness mentioned above was obtained usually on single crystals which have transformed from the metastable mobile h phase to the stable mobile o phase.

Based on this fact a theory is constructed by combining the idea that all ECSC which starts from metastable hexagonal crystal transforms into stable orthorhombic crystal after some "life time" of the hexagonal crystal and that the lamellar thickening growth nearly stops with the hex.-orth. transition, which is schematically illustrated in Figure 10(a). Figure 10(a) shows schematically that the actual lamellar thickness of an orthorhombic crystal will be nearly given by that at hex.-orth. transition, $l(t(h - o))$.

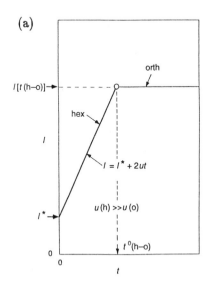
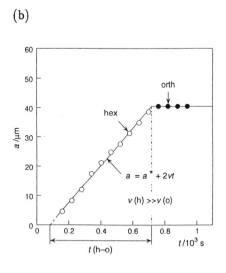

Figure 10. Stoppage of (a)lamellar thickening growth (schematic illustration) and (b) observed lateral growth of an ECSC with hex.-orth. transition at $P < P_{\text{tri}}$.

The assumption of cessation of thickening growth was suggested from the observed fact that the lateral growth nearly stops with the hex.-orth. transition as is shown in Figure 10(b). [15] From Figure 10(b) t(h-o) was estimated.

3.5. Formulation of lamellar thickness [19]

Thus the lamellar thickness can be formulated as a function of lamellar thickening growth rate of the hexagonal crystal $U(h)$ and the "life time $(t(\text{h-o}))$",

$$l = l^* + 2Ut(\text{h-o}) = 2Ut(\text{h-o}); \quad \text{for} \quad \Delta T(\text{h}) = \text{a few K}, \tag{2}$$

where l^* was neglected in this case because $l^* \ll l$. The life time will be also formulated as a function of nucleation rate of a orthorhombic primary nucleus (i_{h-o}) within the hexagonal crystal and the volume of the "mother" orthorhombic crystal (Φ),

$$t(\text{h-o}) \propto (i_{h-o}\Phi)^{-1}. \tag{3}$$

Φ is given by

$$\Phi V^2 U t^3(\text{h-o}) = U^3 t^3(\text{h-o}); \quad \text{for} \quad \Delta T(\text{h}) = \text{a few K} \tag{4}$$

where recent result that $U \propto V$ was used. Combination of equations 2, 3 and 4 gives the final formula of l,

$$l = 2Ut(\text{h-o}) \propto (i_{h-o}{}^{-1}U)^{1/4}; \quad \text{for} \quad \Delta T(\text{h}) = \text{a few K} \tag{5}$$

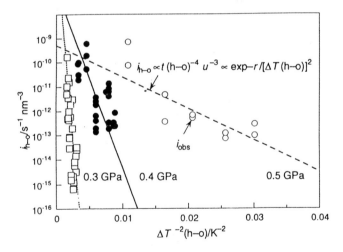

Figure 11. $\log[i_{h-o}]$ vs. $\Delta T^{-2}(\text{h-o})$ at 0.3 GPa, 0.4 GPa and 0.5 GPa. Points are observed ones and straight lines are calculated using equation 5. The linear relations suggest that the hex.-orth. transition is a primary nucleation controlled one.

3.6. Comparison with experimental results [19]

The observed nucleation rate $i_{h\text{-}o}$ obtained from observed life time $t_{obs}(h\text{-}o)$ was plotted against $\Delta T^{-2}(h\text{-}o)$ for $P=0.3$, 0.4 and 0.5 GPa in Figure 11. This showed that i_{h-o} vs. $\Delta T^{-2}(h\text{-}o)$ give straight lines, which supports the above assumption that the hex.-orth. transition is nucleation process of the primary nucleus.

Theoretical l (l_{th}) was obtained from inserting observed nucleation rate $i_{h\text{-}o}$ and thickening growth rate U into equation 5 which is shown by a solid curve in Figure 12. l_{th} explained the observed plot of l (l_{obs}) well.

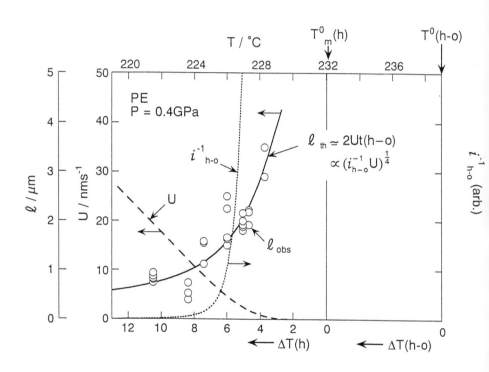

Figure 12. Typical l vs. $\Delta T(h)$ at $P=0.4$ GPa $< P_{tri}$. Points are observed values and the curves are calculated l_{th} from equation 5. U and i_{h-o} are also shown.

Thus we have an answer to the second question; the lamellar thickness is determined by apparent stoppage of lamellar thickening growth through the metastable hexagonal to stable orthorhombic phase transition.

4. SUMMARY

1. What is the origin of folded chain crystals (FCCs) and extended chain crystals (ECCs)? was answered by chain sliding diffusion theory. The theory predicted that all polymers will form ECCs when they crystallize from the melt into the mobile phase, because in the mobile phase, chain can slide easily, i.e., ΔE is rather small, so the lamellar thickening growth occurs significantly, whereas all polymers will form FCCs when they crystallize from the melt into the immobile phase, because in the immobile phase, chain can not slide easily, i.e., ΔE is rather large, so the lamellar thickening growth could not occur significantly. The prediction was confirmed on polyethylene (PE) and other polymers, such as polychrolotrifluoroethylene (PCTFE), polytrans 1,4 butadiene (PT1,4BD) and vinilidenefuloride/trifluoroethylene copolymers (P(VDF/TrFE)) were mentioned.

2. What determines the lamellar thickness? was answered in the case of ECC by applying the chain sliding diffusion theory; the lamellar thickness is determined by apparent stoppage of lamellar thickening growth with the mobile metastable hexagonal to immobile stable orthorhombic phase transition.

ACKNOWLEDGEMENT

I would like to express my special thanks to Prof. A. Keller of University of Bristol and Dr. S. Rastogi of Eindhoven Technische Universiteit, Prof. A. Toda of Hiroshima University and all collaborators in my laboratory for their collaboration in this work. This work was partly supported by Grant-in-Aid for Scientific Research on Priority Areas, No.04227105, that No.6651038 and that on KIBAN KENKYU (B) No.07455386 from Ministry of Education, Japan and by International joint research grant from New energy and industrial technology development organization (NEDO), Japan.

REFERENCES

1. M.Hikosaka, Polymer, 28   (1987) 1257.
2. M.Hikosaka, Polymer, 31 (1990) 4.
3. B.Wunderlich, Macromolecular Phys., Academic Press, New York, 1973, Chap.4.
4. A.Keller, Phil. Mag. 2 (1957) 1171.
5. B.Wunderlich and T.Arakawa, J. Polym. Sci., 2 (1964) 3697.
6. B.Wunderlich, Macromolecular Phys. , Academic Press, New York 1973, Chap.7.
7. J.D.Hoffman, G.T.Davis and J.I.Lauritzen Jr., Treatise on Solid State Chem., Plenum Press, New York 1976, Chap7.
8. D.M.Sadler, Nature, 326 (1987) 174.
9. D.Turnbul and J.C.Ficher, J.Chem. Phys. 17   (1949) 71.
10. D.C.Bassett, S.Block and G.J.Piermarini, J. Appl. Phys., 45 (1974) 4146.

378

11. M.Hikosaka, K.Sakurai, H.Ohigashi and T.Koizumi, Jpn. J. Appl. Phys., 32 (1993) 2029.
12. M.Hikosaka, S.Rastogi, A.Keller and H.Kawabata, J. Macromol. Sci. Phys., B31 (1992) 87.
13. M.Hikosaka, K.Tukijima, S.Rastogi and A.Keller, Polymer, 33 (1992) 2502.
14. M.Hikosaka, K.Sakurai, H.Ohigashi and A.Keller, Jpn.J.Appl.Phys., 33, No.1A, (1994) 214.
15. S.Rastogi, M.Hikosaka, H.Kawabata and A.Keller, Macromolecules 24 (1991) 6384.
16. S.Rastogi, M.Hikosaka, A.Keller and G.Ungar, Progress in Colloid and Polymer Science, 87 (1992) 42.
17. W.Ostwald, Z. Physik. Chem. 22 (1897) 286.
18. A.Keller, M.Hikosaka, S.Rastogi, A.Toda and P.J.Barham, J.Mat.Sci., 29 (1994) 2579.
19. M.Hikosaka, H.Okada, A.Toda S.Rastogi and A.Keller, J. Chem. Soc. Faraday Transactions, 91 (1995) 2573.
20. M.Hikosaka, K.Amano, S.Rastogi and A.Keller, Submitted to Macromolecules

PART IV

Mechanisms of Heteroepitaxy

Advances in the Understanding of Crystal Growth Mechanisms
T. Nishinaga, K. Nishioka, J. Harada, A. Sasaki and H. Takei (Editors)
© 1997 Elsevier Science B.V. All rights reserved.

Initial growth layer, island formation, and critical thickness of InAs heteroepitaxy on GaAs substrate

A. Sasaki *

Department of Electronic Science and Engineering, Kyoto University
Kyoto 606, Japan

We investigate the initial growth layer, island formation, and the critical thickness of InAs on GaAs from which heteroepitaxial process and growth mode can be revealed. The InAs layers are grown on the (001)-oriented GaAs substrates at 480°C by molecular beam epitaxy. They grow two-dimensionally at first and then begin to transit to three-dimensional growth at 1.8 mono-molecular layer(ML). The experimental results show that the critical thickness is 3ML beyond which misfit dislocations are generated. It can be made thicker to 5ML by growth on the vicinal GaAs substrate inclined to $[1\bar{1}0]$-direction by 3.5-5.0°. These are observed and measured by reflection high energy electron diffraction, atomic force microscope, electroluminescence, photocurrent spectroscopy, transmission electron microscope, and photoluminescence. The critical thickness is theoretically derived with taking account of the strain energy calculated by valence-force field method. The experimental result agrees well with the theoretical result. The heteroepitaxial process and the growth mode are discussed and described.

1. INTRODUCTION

Semiconductor heteroepitaxy and critical thickness have been considered important for the realization of quantum-effect materials and hetero-structured devices. Heteroepitaxial process and growth mode have to be revealed in an atomic scale to obtain a high quality of growth layer, but they become complex caused by the difference in the bond length. The initial growth layers of lattice-mismatched heteroepitaxy have been numerously investigated [1-16]. Although heteroepitaxial process and mechanism cannot be uniquely described, the InAs/GaAs has been chosen in our study to investigate them. The InAs/GaAs contains neither a reactive element such as aluminum nor a high vapor-pressure element such as phosphorus. Their lattice mismatching is about 7 % and the layer can be grown with a conventional growth technology.

This article is not a general review paper of the InAs/GaAs heteroepitaxy, but describes and summarizes experimental and theoretical results of that subject in-

---

*present address:Osaka Electro-Commun. Univ.,Neyagawa 572,Japan vestigated in our growth conditions. Those results are compared and discussed with the results reported in other literature. As the first stage of our investigations, the crystalline quality of $In_xGa_{1-x}As$ ($0\leq x\leq1$) was examined by measuring the full-width at half maximum(FWHM) of the x-ray rocking curve. The dependence of the quality on the lattice-mismatched degree was investigated. The $In_xGa_{1-x}As$ layers were grown on various substrates such as GaAs, InAs, and InP, respectively. The layer thickness was 2μm far above the critical thickness.

In the past understanding, the crystalline quality degrades monotonously with increasing the lattice mismatching. To the contrary, it was shown in our experiments that the quality degrades at first with increasing the lattice-mismatched degree, and then recovers and improves beyond the certain degree in the mismatching[17,18].

Since then, we focused on the investigation of the initial growth layers and the critical thickness. The surface morphology of the InAs layers with various thicknesses on GaAs substrates was observed with the atomic force microscope(AFM) in order to see the transition from two-dimmensional growth to three-dimenstional growth, i.e., the island formation. The electroluminescence(EL) and the photocurrent spectroscopy (PCS) were measured to see whether the first two-dimensional flat layer remains during the InAs island formation, i.e., the growth mode of InAs/GaAs is the Stranski-Kastanov.

The critical thickness was calculated with the theory using valence-force field(VFF) method[19] which can be applied to even a few monolayer thickness. The plan-view images of transmission electron microscope(TEM) were observed to see the critical thickness. The results were compared with those calculated with the theory. The InAs layers were grown on GaAs vicinal substrates to make the critical thickness thicker. It was confirmed with photoluminescence(PL) measurement.

Finally, the heteroepitaxial process and the growth modes are discussed.

2. HETEROEPITAXIAL GROWTH OF InAs LAYERS

The InAs layers were grown on semi-insulating and (001)-oriented GaAs substrates by molecular beam epitaxy equipped with the 30kV reflection high-energy electron diffraction(RHEED) system. The substrates were etched by $H_2SO_4 : H_2O_2$: $H_2O = 5 : 1 : 1$ solution for 1min. at $60°$C.

The native oxide on the substrate was thermally dissociated at $630°$C under As pressure in the growth chamber. The GaAs buffer layer of 1000Å -thickness was grown at $590°$C, and then the InAs layer at $480°$C. Finally, the GaAs capped layer of 100Å-thickness was grown to passivate the InAs layer. The layer structure of samples is shown in Figure 1. The cap layer was not grown on the InAs layer for the AFM observation. An amorphous cap layer was deposited below $150°$C to see clearly the plan-view image of the TEM observation. A crystalline cap layer was grown at $480°$C to avoid the absorption of an excited light for the PL measurement. The differences in the cap layers did not cause contradictory results, but coincident results from the observations and the measurements.

The growth conditions are summarized in Table 1. The growth temperature was chosen as $480°$C through the experiments [17,18,20-24]. The growth rate is lower

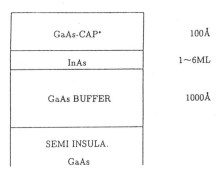

* NO CAP LAYER FOR *AFM* OBSERVATION
AMORPHOUS CAP LAYER FOR *TEM* OBSERVATION
CRYSTALLINE CAP LAYER FOR *PL* OBSERVATION

Figure 1. Layer structure of samples.

Table 1
Growth conditions for samples.

Growth method	MBE
Growth temp.	480°C
Substrates	GaAs, (001) orientation nominal angle off-angle to [110], 3.5°, 5.0°, off-angle to [1$\bar{1}$0], 3.5°, 5.0°
V/Ⅲ flux ratio	40 for 1-2ML (for AFM), 28 for 1-5ML
Growth rate	0.10ML/s with V/Ⅲ=40, 0.14ML/s with V/Ⅲ=28

at a higher value of the V/Ⅲ ratio, and then the thickness control of the growth layer becomes easy. The V/Ⅲ ratio of 40 was used to grow the layer for the AFM observations. Unless stated, all the layers were grown at the V/Ⅲ ratio of 28. In this study, the expression "we grow the X ML of InAs" means that "we supply the amount of In and As atoms for the X ML growth if the growth would last two-dimensionally".

3. OBSERVATIONS AND MEASUREMENTS OF INITIAL GROWTH LAYERS

3.1. RHEED observations

The RHEED patterns by electron beams incident along the [110] and [1$\bar{1}$0] directions were respectively observed during the growth of the InAs layers [21,22]. In both cases, the streak patterns lasted until about 2ML, and then the patterns abruptly became spotty. It indicates that the InAs layer grows two-dimensionally until about 2ML and then abruptly three-dimendionally. The facet of the InAs islands formed with the three-dimensional growth was deduced as the {115} or

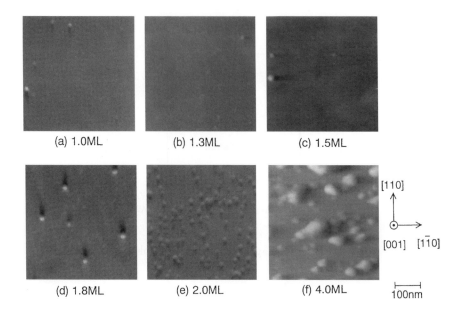

(a) 1.0ML   (b) 1.3ML   (c) 1.5ML

(d) 1.8ML   (e) 2.0ML   (f) 4.0ML

Figure 2. AFM images of the InAs layers grown on the just-angled (100)-oriented GaAs substrates.

{113}, surfaces from the electron beam incident along the $[1\bar{1}0]$ direction. However, a definite facet was not seen from the RHEED pattern by the electron beam along the [110] direction. The spotty pattern continued to about the 10ML growth, and then the pattern returned to the streak, i.e., back to the two-dimensional growth.

3.2. AFM observations

In order to see the transition thickness, the surface morphology of the InAs layer was observed by AFM [23,24]. The transition thickness here is defined as the maximum thickness at which the growth changes from the two-dimensional to the threee-dimensional mode. Normal 1.0, 1.3, 1.5, 1.8, 2.0, and 4.0ML-thick InAs layers were prepared without the GaAs capping layers for AFM observations. The AFM images observed are shown in Figure 2 [23,24]. The layer surfaces are considerably flat until 1.5ML. Although the numbers of the InAs islands are gradually increased with the thickness of the grown layer, the InAs islands could find few until 1.5ML. During the growth from 1.8ML to 2.0ML, many InAs islands are abruptly formed. Two different sizes of the InAs islands appear as seen in Figure 2(e), i.e., the small (about 250Å) and the large (greater than 400Å) sized islands.

Table 2
Density of InAs islands.

	InAs LAYER THICKNESS(ML)					
	1.0	1.3	1.5	1.8	2.0	4.0
ISLAND DENSITY	2.8	2.5	3.5	3.9	1.7(LARGE)	5.6(LARGE)
($\times 10^9$ cm^{-2})					46(SMALL)	12(SMALL)

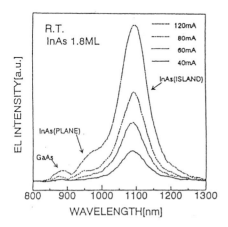

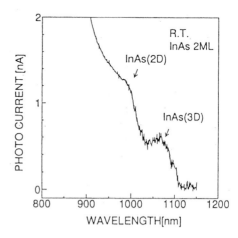

Figure 3. Electroluminescence from the double-heterojunction diode of p-GaAs/non-doped 1.8ML InAs/n-GaAs layers.

Figure 4. Photocurrent spectrum by the double-heterojunction diode of p-GaAs/non-doped 2.0ML InAs/n-GaAs layers.

The shape of the small island is rather hemispherical with some facets, and that of the large island is not similar to each other but various. The densities were slightly varied with the layer thickness as summarized in Table 2. The density of the small islands at 4.0ML became somewhat less than that at 2.0ML, while the density of the large islands increases remarkably.

3.3. Electroluminescence and photocurrent measurements

It is uncertain whether the first flat layer of InAs is disrupted during the formation of the InAs islands or remains even after the transition thickness. The following two samples were prepared: p-type confinement layer(1μm) of GaAs/non-doped active layer (1.8 and 2.0ML, respectively) of InAs/n-type confinement layer (0.42μm) of GaAs/n-type GaAs substrate. They form a double heterojunction. The EL and PCS were measured to see whether luminescence and absorption spectra by the flat layer appear in the characterisitcs.

The EL measurements for the 1.8ML-InAs active layer were implemented at room temperature. The results are shown in Figure 3. At the excitation current of 40 and 60mA, only the luminescence from the InAs islands was observed about 1090nm. However, other two luminescence, about 880 and 950nm, peaks appear at the higher excitation current, 80 and 120mA. They are considered as the luminescence from the GaAs layer and the InAs flat layer, respectively.

The PCS measured at room temperature for the 2.0ML-InAs active layer is given in Figure 4[22]. Two absorption peaks can be seen in the wavelength range longer than the absorption edge 860nm of the GaAs layer. It is considered that the peak about 980nm is caused by the InAs flat layer and that the peak about 1070nm by the InAs islands. Thus, the experimental results of EL and PCS indicate that the flat layer of InAs certainly remains.

4. CRITICAL THICKNESS

4.1. Theory

The theoretical results on the critical thickness of InGaAs on GaAs substrate have been derived by Matthews and Blakeslee [25] and People and Bean [26]. Their results show that the critical thickness of InAs on GaAs is about 5 and 8ML, respectively. They calculated the strain energy of the growth layer as a continuous elastic body. The question arises whether a few monolayers of the growth layer can be treated as a continuous elastic body. Further, the strain energy by misfit dislocations generated in a few monolayers is questionable to calculate by the theory for a continuous elastic body.

We have applied the VFF [19] method to calculate the strain energy of a heteroepitaxial growth layer. The number of lattice points on the base line was taken in the followings. Lattice-mismatched phenomena would be repeated in the distance of the least common multiple of InAs and GaAs lattice constants. There are approximately 25 lattice points of InAs and 27 lattice points of GaAs in the distance of the least common multiple. The 25 lattice points were taken along the base line as shown in Figure 5. The {115} facets determined by the RHEED observation was used to form a block of the lattice points for the strain calculation. Although the block forms an island composed of the 3ML height, total lattice points in the block equal to those of 2ML, when they deposit two-dimensionally. The white circle represents As atom, and the black circle Ga atom in the substrate and In atom in the epitaxial layers. The bond network is the projected figure of the zincblende structure to the $[1\bar{1}0]$ direction. The size along the $[1\bar{1}0]$ direction is assumed semi-infinite.

In the VFF method, the strain energies caused by the deviations in bond length and bond angle from a thermal equilibrium values are calculated. Thus, the strain energies of every bonds in a growth layer can be calculated by the VFF method. This method may be considered as a microscopic approach in contrast to the method based on a continuous elastic body, i.e., a macroscopic approach.

A total strain energy is increased with increasing the thickness of a lattice-mismatched layer with no misfit dislocation. The dislocation releases the strain

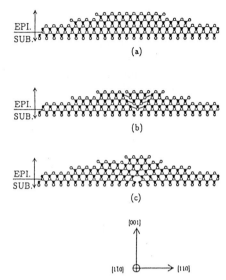

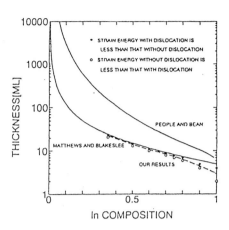

Figure 5. InAs islands formed with the 2ML-InAs growth on GaAs substrate. They have {115} facets. (a)with no misfit dislocation, (b)with a misfit dislocation(before calculation), and (c)with a misfit dislocation(after calculation).

Figure 6. Critical thickness of In-GaAs grown on GaAs substrate. It is calculated by microscopic approach, i.e., the theory based on VFF method to calculate energy of every strained bonds, and is compared with the results by macroscopic approach, i.e., the theory based on a continuous elastic body.

energy since the bonds in a lattice-mismatched layer are relaxed and approach to the bonds in its own bulk material. The strain energy in a monolayer away from the misfit dislocation becomes less, but an extra stress is induced near the misfit dislocation. Thus, a total strain energy would be less in a strained few monolayers with no misfit dislocation as compared with that in the layer with misfit dislocatin. However, the total strain becomes greater in the layer with no dislocation than that in the layer with dislocation, when the layer thickness extends beyond a certain thickness, that is, a critical thickness. It is resulted from the greater strain of the bonds in the layer with no dislocation due to no relaxation of the bonds. The critical thickness can be found as the maximum thickness below which a total strain energy of the layer with no dislocation is less than that of the layer with dislocation.

The strain energy was calculated in the layers shown in Figure 5 [20]. It shows the cases of the lattice points for the 2ML-growth with no dislocation (a) and with dislocation (b). At first, the bond of InAs was assumed equal to that of GaAs before the calculation. The position of individual atoms at which the strain energy becomes nimimum with varing the bond length and angle of In and As. Then the calculations are repeated and repeated until a total strain energy is no more reduced. The bond lengths in InAs islands are compressed along the substrate surface and tensiled along the direction perpendicular to the surface. The bond network with dislocation becomes the network shown in Figure 5(c) in which the total strain energy becomes minimum. The critical thickness is determined as the thickness at which the strain energy for the layer with the misfit dislocation becomes less as compared with the strain energy for the layer with no misfit dislocation. The results are given in Figure 6 [20].

The theoretical results based on the microscopic approach become close to those calculated by the macroscopic and continuum approach, when the critical thickness is thicker than about 10ML. However, the microscopic approach gives a thinner critical thickness near the InAs layer where it becomes 3ML. The theoretical result of 3ML agree with the experimental results of the plan-view TEM observations.

4.2. Observations of critical thickness and its extension

The critical thickness was experimentally investigated by the plan-view TEM observations and examined by the PL measurements. There are a couple of ways to extend the critical thickness in the growth technology such as a surfactant growth and a vicinal-substrate growth. In our study, the InAs layers were grown on the GaAs vicinal substrate. They were grown on five different GaAs substrates with the (001)-orientation of the just-angle, the off-angle to the [110]-direction by 3.5° and 5.0°, and the off-angle to the [1$\bar{1}$0]-direction by 3.5° and 5.0°, respectively.

4.2.1. Plan-view TEM observations

It has not been shown how many ML is the critical thickness of the InAs layer on GaAs. The critical thickness here is defined as the maximum thickness beyond which misfit dislocations are generated in the growth layer and/or the heterointerface. The InAs small islands become larger with the coalescence of them with increasing the growth layers. The dislocation is considered to generate in the large-sized island in which the strain energy is stored.

For plan-view TEM observations, the samples were grown with 100Å-thick amorphous GaAs capping layer rather than crystalline layer. The observation images viewed from [001] direction are shown in Figure 7. Very flat surface can be seen at 1ML because of the two-dimensional growth, but many dotted images indicating the InAs small islands appear at 2ML as seen in Figure 7(b). At 4ML, there are large-sized islands with fringes which are considered to be moire patterns caused by misfit dislocations. It suggests that the critical thickness would be 3ML. This agrees with the theoretical results calculated by the microscopic approach. Further observations of the TEM plan-view images showed that the critical thickness can be extended to 5ML grown on the GaAs substrate off-angled to [1$\bar{1}$0] direction [27].

4.2.2. PL measurements

The PL spectra of the five different InAs layers were measured at 77K. The samples were excited with the 5145Å Ar-ion laser and the results are shown in Figure 8. All PL peaks in Figure 8(a) are at 850nm and the FWHM is as narrow as 8meV. The results indicate the luminescence from the flat 1ML of InAs. In the PL spectra for the 2ML of InAs, the peaks shift to the lower energy and the

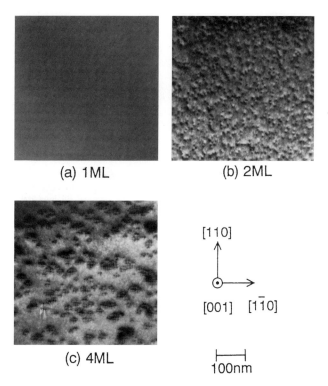

Figure 7.   Plan-view TEM images of the InAs layers grown on the just-angled (001)-oriented GaAs substrates.

FWHMs become as wider as 80meV as shown in Figure 8(b). The results suggest the luminescence excited in the InAs islands. The wider FWHMs are resulted from the fluctuated sizes of the islands. The PLs were observed only from the samples grown on the substrates misoriented toward [1$\bar{1}$0]-direction as shown in Figure 8(c). No luminescence was observed from other InAs layers. It is understood due to nonradiative centers nucleated by the misfit dislocations beyond the critical thickness. The results by the PL measurements of the InAs layers also suggest that the critical thickness can be made thicker on the GaAs substrate off-angled to [1$\bar{1}$0] direction.

5. DISCUSSION

5.1. Growth temperature

The growth temperature was chosen as 480°C through the experiments [17, 18, 20-24]. The growth temperatures of the InAs layer on GaAs substrate have spreaded from 420 to 540°C. The temperature 420°C was adopted by O.Brandt et al.[4] as the temperature same as the homoepitaxial growth of the InAs layer.

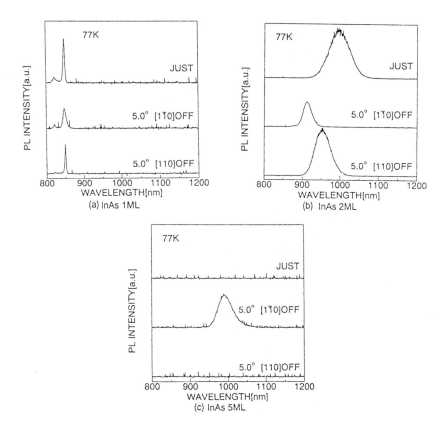

Figure 8.   Photoluminescence spectra of the InAs layers grown on just- and off-angled GaAs substrates.

The $In_{0.33}Ga_{0.67}As$ layer grown at 470°C was reported to be metastable in the sense that the lattice constant relaxes to the value of the layer grown at the higher temperature when the 470°C-layer was heated during the growth[28]. Thus, the low temperature growth would be better to avoid. In general, the homoepitaxial growth temperature of GaAs has been about 600°C [29]. It was experimentally found by S.S.Dosanjh et al.[10] that there is a temperature range of 530 to 560°C in which a single ML of InAs can be grown pseudomorphically. At a higher growth temperature such as 580°C, the recovery of RHEED oscillation following the InAs growth becomes very poor, because of the evaporation and/or segregation of the In atoms. The optimum growth temperature was suggested to be 540°C [10]. At the same time, the temperature 480°C also was adopted by X.Zhang et al.[9] to grow the InAs layer with a thickness range of 0∼300Å under As over pressure.

The $In_{0.43}Ga_{0.57}As$ layer was grown at $510°C$ under As-rich conditions, since this temperature prevented In evaporation from the surface [28,30]. The temperature $480°C$ which has been adopted in our study would not be too high nor too low for the InAs/GaAs heteroepitaxy.

5.2. Transition Thickness

It is uncertain whether the three-dimensional growth following to the two-dimensional growth occurs before the generation of the misfit dislocation, when the critical thickness is so thick as 40~50ML which is the case of $In_{0.2}Ga_{0.8}As$/GaAs. Even the three-dimensional growth occurs and the islands are formed, the island height would be neglected as compared with a thick critical thickness. It has not been necessary to distinguish between the thicknesses at which three-dimensional growth occurs and at which dislocations generate. However, they should be distinguished when the critical thickness becomes so thin as a few monolayers which are the case of InAs/GaAs. We define here a transition thickness for the maximum thickness at which a three-dimensional growth occurs. The transition thickness becomes important to obtain the heterojunction with a sharp and abrupt interface and the critical thickness to form quantum dot structure with no dislocations.

The transition thickness has been investigated many researchers and reported as (1) less than 2ML at the substrate temperature $420°C$ by F.Houzay et al. [1], H.Munekata et al. [2], (2) 2ML at $420°C$ by O.Brandt et al. [5], at $540°C$ by S.S.Dosanjh et al.[10]. They mostly estimated the thickness by the RHEED patterns. Very recently, J.M. Moison et al. [16] and we [23,24] have seen the transition thickness by AFM which reveals directly the surface morphology. The results are 1.75ML at $500°C$ [16] and 1.8ML at $480°C$ [23,24]. It is the first time to see the transition thickness by AFM observation. It seems the transition from the two-dimensional to the three-dimensional growth occurs before the 2ML-growth. Atom migration on the layer surface is enhanced at a higher growth temperature. However,it seems that the transition thickness is insensitive in the substrate temperature, since it has been reported about2ML at the temperature range from 420 to $540°C$.

The 8ML of the transition thickness has been realized by A.Trampert et al.[31] with the MBE growth under In-stable surface conditions, acting as a "virtual surfactant". It was interpreted as that the surfactant acts to reduce the misfit dislocation density at the heterointerface and to obtain the two-dimensional growth with a low defect density. The layer-by-layer growth is continued even after the dislocation generation.

The transition thickness can be defined for the Stranski-Krastanov growth mode. What mode we see in the growth has been illustrated in terms of the binding energy with the nearest neighbors and the adsorption energy [32]. However, it has been commonly understood that islands are formed when the strain energy and the surface energy of islands are less than those of a flat layer. The bonds in the islands would be less constrained to the substrate bonds than those in the two-dinmensional layer.

5.3. InAs islands

The InAs islands have been interested from two standpoints : the growth process in initial layer and the application to quantum dots. Their shape, size, and the density have been investigated. F.Houzay et al.[1] observed them grown at 420°C with TEM, and their results are (a) square shape with ⟨100⟩-orientation sides, (b) mostly 200Å wide but variation from 120 to2000 Å, and (c) the dislocation generation beyond 200Å width. J. M. Moison et al.[16] observed them grown at 500°C with AFM, and their results are (a) rounded pyramid, (b) 30∼40Å height and 120∼150Å width at half maximum,(c) (410) facets, and (d) interdot distance 600Å. We [23,24] observed them grown at 480°C with RHEED and AFM, and their results are (a) hemisphere, (b) about 250Å, (c) {113} or {115} facets [21], (d) the density summarized in Table 2, and (e) the 400Å-sized islands with dislocations. The island formation is considered much dependent on growth conditions.It is required further investigations to state definitely the islands. However, the followings are common among different studys.

The islands at the first stage form with the size of about 200Å and they are very similar shape and size, i.e., less variations. The islands sizes are increased with the coalescence with increasing the growth layer, and then dislocations are generated in them.

A strong PL emission was observed from the InAs islands grown on GaAs substrates [22]. Then, the islands which form at the initial stage of the heteroepitaxial growth of a lattice-mismatched system have been proposed to be utilized for the realization of a mesoscopic structure [21].

5.4. Critical Thickness

The maximum thickness beyond which misfit dislocations are generated has not been much investigated. The moire pattern was observed on the surface of the 7ML grown at 420°C by F.Houzay et al.[1]. The layer was grown with annealing for several minutes after each 0.5 or 1ML deposition under arsenic deposition. The presence of 60° dislocations at the island edges on the 4.8ML grown at 420°C was revealed with high resolution electron microscope by O.Brandt et al. [6]. The InAs layer was grown in increments of 0.6ML, each followed by a 2 min. annealing cycle at the growth temperature under low As_4 flux. The thickness of the embedded InAs film becomes 2.4ML for six deposition steps, i.e., 3.6ML InAs. It was considered that the difference between the deposited and actually incorporated amount of In directly corresponds to the fraction of In which is lost by the heating cycle. Thus, the embedded thickness would be less than the 4.8ML. It was reported that the dislocations were seen in the 4.8ML-growth of InAs, but not in the 3.6ML-growth of InAs.

The critical thicknesses of InAs grown on GaAs were calculated with the theory for a continuous elastic body by Matthews and Blakeslee [25] and People and Bean [26]. They are about 5ML and 7ML, respectively. The critical thickness derived in our study with the VFF method was 3ML. In the region of a less lattice-mismatched degree and thus a greater critical thickness, the results by the VFF method approach to those by the theory for a continuous elastic body. However, in the thinner critical thickness, both results become gradually apart each other.

It was reported by O. Brandt et al.[11] that a continuum elastic theory can be applied to the 3ML, but not to a thinner layer. The stress caused by the dislocation itself is concentrated in a couple of monolayers. It would be recommended to use the theory for a microscopic approach in order to calculate the strain energy of several monolayers involving the dislocations. The plan-view images of TEM observations revealed no moire pattern on the surface of 2ML InAs, but on the surface of 4ML InAs. The lattice constants of the 2ML of InAs calculated by the VFF method are 1.082 and 1.022 with and without dislocations, respectively [20]. They are normalized by the lattice constant of GaAs, and the normalized value of the lattice constant of an InAs bulk becomes 1.072. The normalized value derived from the RHEED pattern becomes 1.032 which is closer to the value without dislocation. The oretical values for the 4ML of InAs are 1.063 and 1.025 with and without dislocations, respectively. The experimental value is 1.076 which is closer to the value with dislocation. The experimental results obtained from the TEM and the RHEED observations agree with the results derived by the theory for a microscopic approach rather than for a macroscopic approach. If the assumption that the embedded thickness is reduced with the same rate as the case of the 3.6ML, 2.4/3.6 = 0.67, can be applied to the 4.8ML growth, it becomes 3.2ML. Although the critical thickness is considered dependent on growth conditions, it would be 3 to 4ML with a conventional growth process in which special process such as surfactant action is not utilized.

5.5. Growth mode

The different results concerning the growth mode of InAs on GaAs have been reported, and thus a definite conclusion has not been come yet. It is a very interesting question arising whether the initial flat layer of InAs remains at and after the island formation, that is, whether the growth is the Stranski-Krastanov mode or it transits from that mode to the Volmer-Weber mode. The three-dimensional growth at 420°C following to the two-dimensional growth was observed by the RHEED and the TEM, and then the Stranski-Krastanov mode was reported by F.Houzay et al. [1] at the early stage of this sort of the study. However, the remaining of the initial flat layer has not been confirmed. The 4.39Å of the flat InAs layer was shown by TEM and the Stranski-Krastanov mode was reported by O.Brandt et al. [5]. It is uncertain that the initial flat layer still remains after the island formation. The PL excitation measurement by S.S.Dosanjh et al. [10] of the 2MLs InAs grown at 540°C shows that there is no absorption relating to the 1 ML InAs. The first ML is disrupted on the onset of three-dimensional nucleation, and thus it suggests that the growth of InAs on GaAs does not proceed via a simple Stranski-Krastanov mode. [10]. The wedge-shaped InAs layer was grown at 480°C and its heterointerface was observed with TEM by X.Zhang et al. [9]. The observations say that the InAs island growth was accompanied by breaking up of the initial strained layer and dissociation of the exposed part of GaAs. However, it was stated whether the breaking-up of the 1 or 2ML at the nucleation of islands cannot be determined from this observation. Then, it was suspected from the PL of the $(InAs)_2(GaAs)_5$ strained layer superlattice that simultaneous breaking-up of the pseudomorphic layer occurs with three-dimensional nucleation. But, it must be noted from experimental results by Lin et al. [33] that the growth of GaAs on the InAs islands causes the intermixing at the heterointerface.

It was found by J.M. Moison et al. [16] that the initial flat layer of InAs decreases with increasing the InAs layer grown at 500°C beyond the transition thickness and to vanish around the thickness on the coalesence of the InAs islands. This was observed with RHEED and AES in-situ and AFM. The growth mode was stated as the depart from the classical Stranski-Kranstanov model.

We have observed (1) EL from the InAs flat layer of the 1.8ML sample grown at 480°C and (2) photocurrent spectrum of the flat layer of the 2.0ML sample grown at 480°C, as described in Section 3.3. It can be claimed that the growth mode is the Stranski-Krastanov until and just after the transition thickness. This growth process was tried to be interpreted in terms of the surface energy and the strain energy [23]. It should be confirmed in further study whether the flat layer covers with a uniform thickness all the GaAs substrate. It could be stated at this stage of the investigation that the growth mode of the InAs growth on GaAs is the Stranski-Krastanov before and just after the three-dimensional growth. It was observed by X.Zhang et al. [9] that not only the InAs layer, but also the GaAs layer near the interface are seriously disrupted after the critical thickness. It could be said that the growth transits to the Volmer-Weber mode. The growth processes cannot be interpreted with growth paramenters such as the binding and the adsorption energies with which the growth modes are classified [32].

The critical thickness becomes thicker on the off-angled substrate where the steps appear on the substrate. It is now in progress to prove it theoretically by calculating the nucleation energy of the misfit dislocation crossing over the step and by considering the anisotropic generation of misfit dislocation [27].

6. CONCLUSION

The initial growth layer, island formation, and the critical thickness of InAs grown at 480°C on GaAs have been experimentally and theoretically investigated. The RHEED, AFM, EL, and PCS have been used to investigate the initial growth layers. The VFF method has been applied to calculate the critical thickness, and the plan-view of TEM and PL have been used to observe the critical thickness. The growth mode of the InAs/GaAs heteroepitaxy has been discussed. They have been compared and discussed with the results reported in other literatures.

The results of the InAs layer grown at 480°C on the (001) GaAs substrate with As-stabilization are summarized in the following.

1. The maximum thicknesses at which the growth transits from the two-dimensional mode to the three-dimensional mode and at which misfit dislocations are generated have been suggested not to use together. The former has been defined as a transition thickness and the latter as a critical thickness. Some three-dimensional morphology appears on the surface of layer thinner than the critical thickness.

2. The transition thickness is about 2ML. The InAs islands begin to form very slightly less than 2ML and the growth transit completely at 2ML.

3. The small-sized islands are very similar each other in size and shape. The size is about 250Å. The large-sized islands (roughly 400Å) show various shapes because they are considered to form with the coalescence of the small-sized islands. They involve the misfit dislocation.

4. The critical thickness has been calculated using the VFF method. The results are almost same with those derived by the theory for a continuous elastic body in the region of a thick critical thickness, but they become thinner in the region of a few monolayers of the critical thickness.

5. The critical thickness has been experimentally observed as 3ML which agrees with the theoretical value.

6. The critical thickness can be make thicker by the growth on a vicinal substrate. It becomes 5ML on the (001)-oriented substrate inclined to $[1\bar{1}0]$ direction by 3.5° and 5°.

7. It is stated at least that the growth mode is the Stranski-Krastanov at and just after the transition thickness. Far beyond the critical thickness, the InAs flat layer is disrupted and the GaAs surface is exposed. The growth transits to the Volmer-Weber mode. The growth mode far beyond the critical thickness cannot be interpreted in terms of only the binding and adsorption energies which have been used for the classification of classical growth modes.

Some of conclusions above would be modified with other growth conditions, although they are examined with the results reported in other literature. Particularly, the results become different from the layer grown in a metastable state. For further study, (1) the transition thickness is required to study theoretically in a microscopic approach and (2) it must be revealed that at what monolayer of InAs the growth transits to the Volmer-Weber mode [34].

ACKNOWLEDGEMENTS

The author would like to express his gratitude Drs. Susumu Noda and Akihiro Wakahara and Mrs. Masao Tabuchi and Yohichi Nabetani for their discussions, sample preparations, and measurements. This work was supported in part by Grant-in-Aid #03243107 and #04227107 for Scientific Research on Priority Areas "Crystal Growth Mechanism in Atomic Scale" from the Ministry of Education, Science and Calture of Japan.

REFERENCES

1. F. Houzay, C. Guille, J.M. Moison, P. Henoc, and F. Barthe, J. Crystal Grouth 81 (1987) 67.

2. H. Munekata, L.L. Chang, S.C. Woronick, and Y.H. Kao, J. Crystal Growth 81 (1987) 237.

3. D.J. Eaglesham and M. Cerullo, Phys. Rev. Lett. 64 (1990) 1943.

4. O. Brandt, L. Tapfer, R. Cingolami, K. Ploog, M. Hohenstein, and F. Phillipp, Phys. Rev. B 41(1990) 12599.

5. O. Brandt, L. Tapfer, K. Ploog, M. Hohenstein, and F. Phillipp, J. Crystal Growth 111 (1991) 383.

6. O. Brandt, K. Ploog, L. Tapfer, M. Hohenstein, and F. Phillipp, J. Crystal Growth 115 (1991) 99.

7. S. Ohkouchi and I. Tanaka, Jpn. J. Appl. Phys. 30 (1991) L1820.

8. P.N. Fawcett, B.A. Joyce, X. Zhang, and D.W. Pashley, J. Crystal Growth 116 (1992) 81.

9. X. Zhang, D.W. Pashley, J.H. Neave, J. Zhang, and B.A. Joyce, J. Crystal Growth 121 (1992) 381.

10. S.S. Dosanjh, P. Dawson, M.R. Fahy, B.A. Joyce, R. Murray, H. Toyoshima, X.M. Zhang, and R.A. Stradling, J. Appl. Phys. 71 (1992) 1242.

11. O. Brandt, K. Ploog, R. Bierwolf, and M. Hohenstein, Phys. Rev. Lett. 68 (1992) 1339.

12. S. Ohkouchi and I. Tanaka, Ultramicroscopy 42-44 (1992) 771.

13. D. Loenald, M. Krishnamurthy, C.M. Reeves, S.P. Denbaars, and P.M. Petroff, Appl. Phys. Lett. 63 (1993) 3203.

14. J. Ahopelto, A.A. Yamaguchi, K. Nishi, A. Usui, and H. Sasaki, Jpn. J. Appl. Phys. 32 (1993) L32.

15. M. Aindow, T.T. Cheng, N.J. Mason, T.Y. Seong, and P.J. Walker, J. Crystal Growth 133 (1993) 168.

16. J.M. Mosion, F. Houzay, F. Barth, L. Leprince, E. Andre, and O. Vatel, Appl. Phys. Lett. 64 (1994) 196.

17. Sz. Fujita, Y. Nakaoka, T. Umemura, M. Tabuchi, S. Noda, Y. Takeda, and A. Sasaki, J. Crystal Growth, 95 (1989) 224.

18. M. Tabuchi, S. Noda, and A. Sasaki, J. Cryst. Growth 99 (1990) 315.

19. R.M. Martin, Phys. Rev. B 1 (1970) 4005.

20. M. Tabuchi, S. Noda, and A. Sasaki, J. Crystal Growth 115 (1991) 169.

21. M. Tabuchi, S. Noda, and A. Sasaki, Sci. & Tech. of Mesoscopic Structures, S. Namba, C. Hamaguchi, and T. Ando (eds.), Springer-Verlag, Tokyo, 1992, 379.

22. Y. Nabetani, T. Ishikawa, S. Noda, and A. Sasaki, J. Appl. Phys. 76 (1994) 347.

23. Y. Nabetani, N. Yamamoto, T. Tokuda, and A. Sasaki, Proceedings of the seventh Topical Meating on Crystal Growth Mechanism, Atagawa, Japan, January 1994, 303.

24. Y. Nabetani, N. Yamamoto, T. Tokuda, and A. Sasaki, 8th Int'l Conf. Vapour Growth and Epitaxy, Freiburg, Germany, July 1994, Paper #MB29, and J. Crystal Growth 146 (1995) 363.

25. J.W. Matthews and A.E. Blakeslee, J. Crystal Growth 27 (1974) 118.

26. R. People and J.C. Bean, Appl. Phys. Lett. 47 (1985) 322
27. Y. Nabetani, A. Wakahara, and A. Sasaki, J. Appl. Phys. 78 (1995) 6461.
28. E.G. Scott, D.A. Andrews, and G.J. Davies, J. Vac. Sci. & Tech. B 4 (1986) 534.
29. M. Heiblum, E.E. Mendez, and L. Osterling, J. Appl. Phys. 54 (1983) 6982.
30. G.J. Whaley and P.I. Cohen, Appl. Phys. Lett. 57 (1990) 144.
31. A. Trampert, E. Tournie, and K.H. Ploog, 8th Int'l Conf. Vapour Growth and Epitaxy, Freiburg, Germany, July 1994, Paper #MB26.
32. A.A. Chernov, Modern Crystallography III, Springer-Verlag, Berlin, 1984, 88.
33. X.W. Lin, J. Washburn, Z. Liliental-Weber, E.R. Weber, A. Sasaki, A. Wakahara, and Y. Nabetani, Appl. Phys. Lett. 65 (1994) 1677.
34. The two-dimensional InAs layer was observed at the heterointerface in between InAs islands, either small or large (private communication with X.W. Lin).

Advances in the Understanding of Crystal Growth Mechanisms
T. Nishinaga, K. Nishioka, J. Harada, A. Sasaki and H. Takei (Editors)

Effects of buffer layers in heteroepitaxy of gallium nitride

K.Hiramatsu[a], T. Detchprohm[a] H.Amano[b] and I.Akasaki[b]

[a]Department of Electronics, Nagoya University
Furo-cho, Chikusa-ku, Nagoya 464-01, Japan

[b]Department of Electrical and Electronic Engineering, Meijo University
1-501, Shiogamaguchi, Tempaku-ku, Nagoya 468, Japan

Heteroepitaxial growth of GaN has been carried out by using buffer layers on sapphire (0001) and Si (111) substrates by metalorganic vapor phase epitaxy (MOVPE) and hydride vapor phase epitaxy (HVPE). The buffer layers are very important to obtain high-quality GaN single crystals. Buffer layers we have employed were as follows: (1) MOVPE on sapphire (0001); a low-temperature-deposited AlN buffer layer, (2) MOVPE on Si (111); a high-temperature-deposited AlN or SiC buffer layer, (3) HVPE on sapphire (0001); a sputtered ZnO buffer layer. Thick GaN bulk single crystals (several hundreds of micrometers) without thermal strains can be obtained by HVPE using the ZnO buffer layer. The roles of these buffer layers were found to be (a) high-density nucleation of GaN on the buffer layers, (b) arrangement of crystalline directions of GaN islands and (c) quasi two-dimensional growth of GaN.

1. INTRODUCTION

GaN is one of the most promising semiconducting materials for optical devices in the region from blue to ultraviolet light, because it has a direct energy band gap of 3.39 eV at room temperature. It is extremely difficult to grow a large-scale bulk single crystal of GaN because of the high equilibrium pressure of nitrogen at the growth temperature over 1000°C. Therefore, vapor phase epitaxial methods such as metalorganic vapor phase epitaxy (MOVPE) and hydride vapor phase epitaxy (HVPE) have been conducted using dissimilar substrates like sapphire or Si for growing GaN single crystal films. However, it had been fairly difficult to grow high quality epitaxial films, in particular, with a smooth surface free from cracks, because of the large lattice mismatch and the large difference in the thermal expansion coefficient between GaN and those substrates as shown in Table 1.

To solve the problem the deposition of a thin buffer layer before GaN growth was proposed [1-3]. In the case of sapphire substrates, surface morphology as well as electrical and optical properties of GaN films has been improved remarkably by preceding deposition of AlN [1] or GaN [2] buffer layer before MOVPE growth of GaN films or by preceding deposition of ZnO [3] buffer layer before HVPE growth of GaN films.

Table 1
Lattice mismatch $\Delta a/a$ and differences in thermal expansion coefficients $\Delta\alpha/\alpha$.

GaN on substrate	$\Delta a/a$ (%)	$\Delta\alpha/\alpha$ (%)
GaN(0001)/Sapphire (0001)	+13.9	-34.2
GaN(0001)/Si(111)	-20.4	+55.3
GaN on buffer layer	$\Delta a/a$ (%)	$\Delta\alpha/\alpha$ (%)
GaN(0001)/AlN(0001)	+2.5	+5.5
GaN(0001)/ZnO(0001)	-1.9	+1.6
GaN(0001)/3C-SiC(111)	+3.4	+48.1

In this chapter, we show how crystalline properties of GaN grown films are improved in MOVPE growth and HVPE growth by using buffer layers deposited on sapphire and Si substrates.

2. HETEROEPITAXY ON SAPPHIRE SUBSTRATE BY MOVPE

A horizontal type MOVPE reactor operated at an atmospheric pressure was used for the growth. An optical grade polished sapphire with the (0001) (C-face) was used as the substrate. The misorientation was less than 0.5 degree. The substrate was placed on a graphite susceptor which was heated by r.f.. Prior to the growth, it was heat-treated at 1150°C for 10 min in a stream of H_2 to remove the surface damage and impurity. Trimethylgallium (TMG -15°C 15 sccm H_2), trimethylaluminum (TMA +15°C 50sccm H_2) and ammonia (NH_3 1.5 slm) were used as source materials for GaN or AlN growth. The carrier gas was hydrogen (H_2 1.5slm). In order to reduce the parasitic reactions of organometallics diluted with H_2 carrier gas, the NH_3 and the organometallics were mixed just before the reactor, and the mixture was fed to a slanted substrate through a delivery tube with a high velocity (110 cm/sec). Before the growth of the GaN film, the thin AlN buffer layer of about 50 nm thick was deposited at about 600°C. Then, the substrate temperature was raised to a growth temperature of about 1040°C, and a GaN film was grown.

The surface morphology as well as the electrical and optical properties of GaN film have been remarkably improved by the preceding deposition of a thin AlN layer as a buffer layer before the GaN growth. [1,4] In this chapter, effects of the AlN buffer layer on the crystallographic structure of the GaN film will be discussed. [5]

From electron diffraction spots the crystallographic relations between GaN, AlN and α-Al_2O_3 were found to be [0001]GaN || [0001]AlN || [0001]Al_2O_3 and [1$\bar{1}$00]GaN || [1$\bar{1}$00]AlN || [11$\bar{2}$0]Al_2O_3. The sharp spots of AlN indicate that the AlN is crystallized epitaxially on the sapphire substrate during the raising the temperature and/or the growth of GaN. The above relations agree with those occurring in GaN grown directly on (0001) sapphire without AlN.

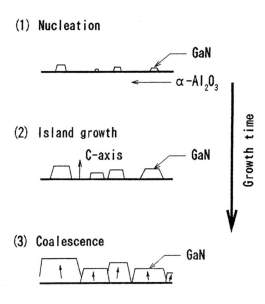

(1) Nucleation

GaN

α-Al₂O₃

(2) Island growth

C-axis   GaN

Growth time

(3) Coalescence

GaN

Figure 1. Schematic diagrams showing the growth process of GaN on a sapphire substrate as the cross sectional views without an AlN buffer layer.

Observation of initial growth stage and cross sectional TEM images of GaN films revealed the growth process of GaN without and with AlN buffer layer, which are shown in Figures 1 and 2, respectively.

In the case of GaN film without AlN buffer layer nucleation density of GaN on sapphire substrate is law as shown in Figure 1 (1). Many hexagonal GaN columns with different sizes and heights are formed and they grow three-dimensionally (Figure 1 (2)), resulting in rough surface and many pits at their boundaries as shown in Figure 1 (3) Furthermore, many crystalline defects generate near the boundaries between GaN grains, which is caused by misorientaion of each island.

The growth process of GaN film with AlN buffer layer is shown in Figure 2. The AlN buffer layer has amorphous-like-structure at the deposition temperature of 600°C, but when the temperature is raised to the growth temperature of GaN (1040°C), AlN is crystallized by solid phase epitaxy and then it exhibits the columnar structure. Since the AlN films were single-crystal-like from electron diffraction spots, orientations of AlN columnar crystals were found to be arranged each other.

Each GaN column is grown from a GaN nuclei which has been generated on the top of each columnar fine AlN crystallite Therefore, it is thought that high-density nucleation of GaN occurs owing to the high-density of the AlN columns, as shown

402

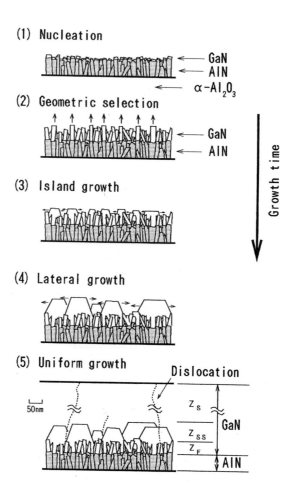

Figure 2. Schematic diagrams showing the growth process of GaN on an AlN buffer layer as the cross sectional views.

in Figure 2 (1), compared with the nucleation density of GaN grown directly on the sapphire substrate.

The columnar fine GaN crystals increase accordingly in size during the growth and the crystalline quality of GaN is improved in this stage. It is thought that the geometric selection of the GaN fine crystals occurs, as shown in Figure 2 (2). Each first fine crystal of GaN begins to grow along the c-axis, forming columnar structure. Each column has a various random orientation and do not keep growing uniformly. The number of columns emerging at the front gradually decreases with the front area of each column increasing accordingly. Because only columns survive that grow along the fastest growth directions (i.e. c-axis of each column is normal to the substrate), then all columns are arranged in the direction normal to the substrate, indicated in the arrows in Figure 2 (2).

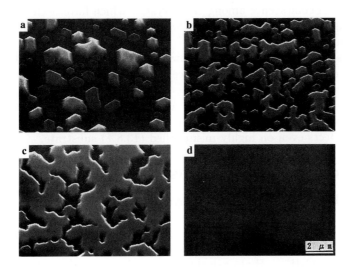

Figure 3. SEM images of various growth stages of GaN on AlN. The growth times are (a) 3 min, (b) 5 min, (c) 10 min and (d) 60 min.

In the next stage the trapezoid crystals are formed on the columnar crystals. As the front area of the column increases by the geometric selection, c-face appears in the front of each column and trapezoid islands with c-face are formed, as shown in Figure 2 (3). These islands preferentially grow up to become larger trapezoid crystals, which cover the minor islands nearby. Figure 3 shows the changes in surface morphology during the early growth stage of GaN. The stage of Figure 3a corresponds to the generation of the trapezoid island crystals after the geometric selection of the columnar crystals.

Subsequently, lateral growth and coalescence of the islands occur in the stages from Figure 3b to Figure 3c. The pyramidal trapezoid crystals grow at a higher rate in a transverse direction, as shown in Figure 2 (4), because the growth velocity of c-face is much slower. After the lateral growth the islands repeat coalescence each other very smoothly.

Finally, since crystallographic directions of all islands agree well with each other, one can obtain a smooth GaN layer with a small amount of defects as a result of the uniform coalescence, as seen in Figure 2 (5) and Figure 3d. Thus, the uniform growth due to layer-by-layer occurs creating high-quality GaN with low defect density and smooth surface.

From the above results, the roles of the AlN buffer layer on the GaN MOVPE growth are summarized as follows:

(1) the high-density nucleation of GaN occurs on the AlN columnar crystals,

(2) the geometric selection occurs among the GaN fine crystals which are able to arrange the crystallographic direction of GaN columnar crystals,

(3) because of coalescence among GaN crystals which have been much arranged in the crystallographic direction, crystalline defects near the interface between the grains are much reduced and

(4) because of the higher lateral growth velocity of the pyramidal trapezoid islands with c-face on the top, the surface is covered at an early growth stage and smooth surface is easily obtained.

Thus, non-uniform growth of GaN on AlN mentioned above plays an important role for realization of uniform growth and for obtaining a high-quality GaN film with a few defects even on highly-mismatched (13.8%) substrate.

3. HETEROEPITAXY ON Si SUBSTRATE BY MOVPE

3.1. Growth using 3C-SiC buffer layer [6]

3C-SiC was grown in a vertical-type, low-pressure chemical vapor deposition (CVD) reactor . The substrate was a Si(111) wafer misoriented 4 degree off toward ¡110¿ direction. The Si and carbon (C) source gases were dichlorosilane (SiH_2Cl_2) and isobutane (i-C_4H_{10}). Hydrogen chloride (HCl) was added to these source gases during the 3C-SiC growth in order to control the stoichiometry of the SiC layer. Typical flow rates for SiH_2Cl_2, i-C_4H_{10}, HCl and hydrogen (H_2) were 6, 10, 90

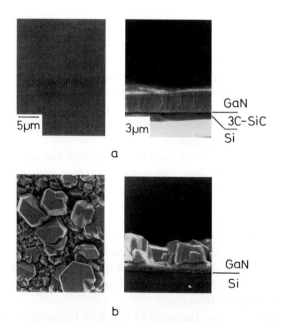

Figure 4. SEM photographs of the surface and cross section of GaN layers grown on (a) Si (111) substrates covered with 3C-SiC layer and (b) Si (111) substrates.

and 6000 sccm, respectively. The 3C-SiC layer was grown at 925°C. Under these conditions, a highly (111)-oriented 3C-SiC layer about 200 nm thick was grown on the Si(111) substrate. Growth of GaN film on these highly (111)-oriented 3C-SiC layers, which cover the whole surface of the Si substrates, were carried out in a horizontal-type MOVPE reactor operated at atmospheric pressure.

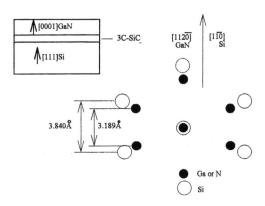

Figure 5. Schematic illustration of the relation in crystallographic orientation between the (0001) GaN layer and the (111) Si substrate using the 3C-SiC buffer layer.

Figure 4 shows SEM photographs of the surface and cross section of a GaN film grown on Si (111) covered with a 3C-SiC layer (Figure 4a) and grown directly on Si (111) (Figure 4b). As shown in Figure 4a, a GaN film with a smooth surface is obtained on Si covered with a 3C-SiC layer, which means that layer growth of GaN is achieved on the Si substrate using 3C-SiC as the intermediate layer. On the contrary, a GaN film grown directly on Si (111) has a very poor surface morphology. An island-like structure similar to that of Figure 1b is observed. It is thought that GaN is grown three-dimensionally on the Si substrate without the 3C-SiC intermediate layer. Cracks are observed in Figure 6a, which is caused by the large difference in thermal expansion coefficient between GaN and Si, as shown in Table 1.

RHEED patterns for GaN using 3C-SiC buffer layer show spotty, which indicates that the GaN film is single crystalline. On the other hand, the RHEED patterns of the GaN without 3C-SiC intermediate layer were very week and hallow-like. No evidence was obtained that the GaN film grown directly on Si was a single crystal. These results indicate that the quality of the GaN film grown on Si is improved by using 3C-SiC as the intermediate layer.

The epitaxial relationship between the GaN film and the Si substrate using the

3C-SiC intermediate layer, as revealed by X-ray Laue patterns, is shown in Figure 5. It is found that the (0001) plane of GaN is parallel to the (111) plane of Si, and the [$11\bar{2}0$] direction of GaN is parallel to the [110] direction of Si.

The large lattice mismatch between GaN and Si is reduced by using 3C-SiC buffer layer and hence growth of GaN is changed from three-dimensional mode to two-dimensional mode. The 3C-SiC buffer layer was not a single crystal but highly (111)-oriented crystals. We have no experiment of GaN growth on a single crystalline 3C-SiC buffer layer. Further studies on growth process of GaN on the SiC buffer layer and on effects of the 3C-SiC buffer layer will be required.

3.2. Growth using AlN buffer layer [7]

AlN buffer layers were deposited on Si (111) substrates at layer thickness of 100 - 200 nm. As long as the deposition temperature is between 800 and 1250°C, the surfaces of AlN layers were rather smooth with good coverage on Si surfaces. When the deposition temperature is higher than 1100°C, the RHEED patterns had spots which indicated that the AlN layers are single crystalline. On the other hand, when the temperature is lower than 800°C, the RHEED patterns had rings which indicated that the layer is polycrystalline. At the temperature of 900°C, polycrystalline AlN crystals are included in single crystal of AlN layer and with increasing temperature the AlN film becomes polycrystalline. Formation of Si_3N_4 due to the reaction between NH_3 and Si substrates might be possible; however, there is no indication from the RHEED patterns.

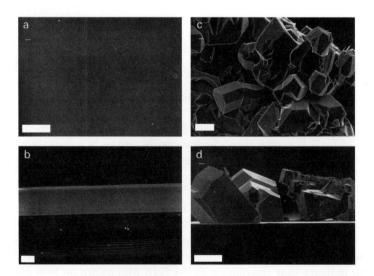

Figure 6. SEM photographs of (a) the surface and (b) cross section of GaN layers grown on Si (111) substrates covered with AlN buffer layer. SEM photographs of (c) the surface and (d) cross section of GaN layers grown directly on Si (111) substrates. The bars represent 5μm; (a), (c), (d) and 1μm; (b).

Figure 6 shows SEM photographs of the surface and cross section of the GaN films grown on Si(1 11) substrates with and without AlN intermediate layers which has been deposited at 1150°C. As shown in Figure 6a and 6b, the GaN film with a smooth surface is obtained on Si(111) covered with the AlN layer although cracks are observed on the surface. While, an island-like structure is observed on the GaN film grown directly on Si(111) (Figure 6c and 6d). The film thickness of GaN was about 2μm. When the deposition temperature of the AlN layer is higher than 900°C, strong streaks and Kikuchi lines are clearly observed in the RHEED pattern. This indicates that high-quality single crystalline GaN films were grown. On the other hand, when the temperature is 800°C, a spotty pattern is observed. This indicates that the surface of GaN becomes rough.

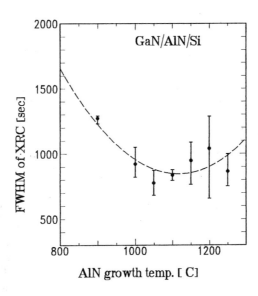

Figure 7. Dependence of the FWHM of the double-crystal x-ray rocking curves of the GaN layers on the AlN buffer layer deposited at different temperatures.

Figure 7 shows the variation of the FWHM of double-crystal X-ray rocking curves (XRCs) for the GaN films grown on Si with the AlN intermediate layers, which were predeposited at different temperatures. When the AlN deposition temperatures are higher than 1050°C, the FWHM of XRC is about 10 - 20 min, which also indicates good crystallinity of the GaN films. On the other hand, the FWHM of XRC increases abruptly when the deposition temperatures are lower than 900°C. These results are consistent with those of the RHEED analyses.

The epitaxial relationship between the GaN and Si substrate was revealed by X-ray diffraction measurements. It is found that the (0001) plane of GaN is almost parallel to the (111) plane of Si, and the [11$\bar{2}$0] direction of GaN is parallel to the [110] direction of Si. The cracking plane of GaN is mainly the 1010 plane. This result is same as GaN on Si(111) using 3C-SiC buffer layer.

As mentioned above we can obtain extremely smooth surface of GaN by using the AlN buffer layer. This is mainly because coverage of AlN is increased more than that of GaN. It is thought that the Si surface is easily covered with AlN because chemical reaction between Al species and Si substrate occurs more easily than that between Ga species and Si substrate. Further important reason is that the lattice mismatch of 16.9% between GaN and Si is much reduce to 2.5% in the case of GaN and AlN, which results in realizing two-dimensional growth of GaN instead of three-dimensional growth.

The crystalline quality as well as the surface morphology of GaN films is much improved with AlN buffer layers deposited over 900 °C. Since the AlN buffer layers fabricated over 900 °C are single crystal, single crystal of AlN is more important as buffer layer on Si substrates to obtain high-quality GaN films. Recently, high-quality GaN films can be grown by MOVPE on 6H-SiC using high temperature buffer layer of AlN. In this case, it is confirmed that the AlN buffer layer is single crystal and coalescence among GaN islands occurs and the surface becomes smooth at the early stage shorter than only one minutes. That is to say, GaN is grown on the buffer layer epitaxially from the early growth stage. Therefore, this AlN buffer layer is though to be a kind of single crystalline AlN substrate. In the case of Si substrate the AlN buffer layer also have single crystalline substrate.

In contrast, the AlN buffer layer on sapphire substrate mentioned the previous section is not single crystalline. The initial growth stage of GaN on the AlN buffer layer have several nonuniform growth stages as shown in Figure 2. Consequently, we find that crystalline structure of AlN buffer layer is different for sapphire and Si substrate. In order to understand the reason why the difference in the crystalline structure of the buffer layer we should need further studies on initial growth process and observation of fine crystalline structure near the interface between the GaN films and the buffer layers.

4. HETEROEPITAXY ON SAPPHIRE SUBSTRATE BY HVPE

4.1. Thermal strain in GaN grown layer [8,9]

Heteroepitaxial growth on a sapphire substrate is usually employed for the preparation of single crystalline GaN. Because of the large difference in the thermal expansion coefficient between GaN and sapphire, the sample bends if a thick film is grown and cracking often occurs near the heterointerface due to the thermal stress during cooling after the growth. The cracks occurring in the GaN(0001) / α-Al$_2$O$_3$(0001) heterostructure fabricated by hydride vapor phase epitaxy (HVPE), and found that the cracking is caused by the tensile stress applied in the sapphire substrate due to the large difference in the thermal expansion coefficient $(\alpha_{GaN} < \alpha_{sapphire})$. Lattice constants and strains of GaN are measured for film thicknesses in the range from 0.6 to 1200μm to investigate the process of the strain relaxation. GaN (0001) films were grown on sapphire (0001) substrates by the

MOVPE and HVPE methods. The thickness of the sapphire substrate was 250μm. The detailed growth conditions of MOVPE were reported in the previous chapter. The growth rate was about 3μm/h. A conventional HVPE system was also used for the growth of GaN. HCl was reacted with Ga at 850 °C, then Ga + HCl gases were introduced to the growth zone with NH_3 and a purified N_2 carrier gas at 1030 °C. The growth rate was high at 30 - 70μm/h and the film thickness was varied from 11 to 1200μm.

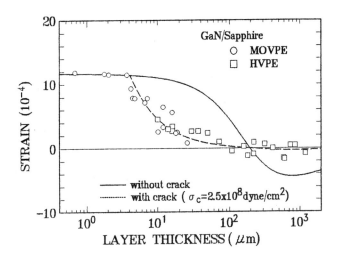

Figure 8. Comparison of the strain along the c-axis between the theory and experiment. The solid and dotted lines represent the stresses calculated without and with consideration of the relaxation due to cracking, respectively. The circles and squares correspond to the MOVPE-grown and HVPE-grown GaN, respectively.

Figure 8 shows the lattice constant as a function of the film thickness of GaN from 0.6 to 1200μm. The lattice constant for thin GaN films of a few microns thickness is relatively large, which originates from lattice deformation due to the thermal stress after cooling from the growth temperature because of the difference in the thermal expansion coefficients between GaN and sapphire. Then, the value of c decreases gradually until the film thickness reaches about 100μm. Beyond about 100μm, the value of c becomes constant. Thus the strain in the GaN film is almost completely relaxed at thickness greater than 100μm. From this value the intrinsic lattice constant c_0 is determined to be 5.1850Å.

The relaxation mechanism is studied by comparison of the experimental strain with the calculated strain considering the relaxation due to cracking of the sapphire substrate. Comparison of the strain at GaN surface along the c-axis between the theory and the experiment is shown in Figure 8. The strain measured experimentally is constant until the thickness reaches 4μm. The strain decreases abruptly and then is almost completely relaxed at a thickness greater than 100μm. This experimental data does not agree with the theoretical curve (solid line) calculated under the condition of no cracking. However, the data is in good agreement with the curve (dotted line) calculated considering the relaxation due to cracking at the critical stress of 2.5×10^8dyne/cm^2. Thus, the sudden relaxation of the strain is attributed to the cracking, not the bending of the heterostructure.

When the film thickness of GaN is 4 to 20μm, since "macrocracks" are not observed. Therefore, it is suggested that the interface defects such "microcracks" and/or dislocations result in the sudden relaxation of the strain. When the film thickness of GaN is greater than 20μm, "macrocracks" occur near the interface especially on the sapphire side and play an important role of the relaxation of the thermal strain.

4.2. Growth using ZnO buffer layer [3,10]

Growth of thick GaN substrates is not only desirable for making high performance optoelectronic devices, but also required for measuring the intrinsic properties of GaN, which are still vague. Because of the high equilibrium pressure of nitrogen at the growth temperature, it is extremely difficult to grow a large-scale "bulk" single crystal of GaN. Preparation of thick GaN crystalline films (bulk) by MOVPE methods is very difficult owing to low growth rate. With high growth rate, HVPE has been employed to prepare thick GaN single crystalline films. However, the reproducibility of growing GaN single crystal by this HVPE method was unpleasantly poor. ZnO buffer layer sputtered on a basal plane sapphire substrate has been used for the preparation of GaN film by the HVPE method. ZnO is expected to be an excellent buffer because: (1) its physical properties,

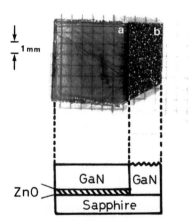

Figure 9. The photograph from the top view of a GaN layers grown on a sapphire substrate (a) with and (b) without ZnO buffer layers including the corresponding side view of the layers illustrated below.

given in Table. 1, are almost similar to those of GaN and (2) it can be etched by any acid, e.g., aqua regia, so it is possible to separate a GaN film from the sapphire substrate by etching the ZnO buffer layer away.

Grown films with the ZnO buffer layer was found to be single crystalline, while most of the films grown directly on the sapphire substrate by HVPE were poly-crystalline (please see Figure 9). This result implies that the sputtered ZnO layer forms an excellent buffer layer and also greatly improves the reproducibility of the growth of GaN by HVPE. In addition, we have observed that the reproducibility of growth using substrates with ZnO buffer layers.

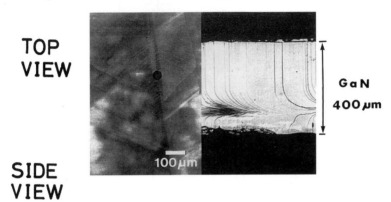

Figure 10. The differential interference micrographs of (a) the surface and (b) cross section of the single crystalline GaN prepared by using a ZnO buffer layer.

The single crystalline films of GaN alone were obtained by etching the sputtered ZnO layer away using aqua regia with an ultrasonic cleaner. After etching, the GaN films did not peel off easily, so that the sapphire substrates had to be pulled away by forceps. Figure 10 shows the thick single crystalline film of GaN after etching the ZnO buffer layer away. It confirms that the single-crystal films of GaN can be prepared by this method. Recently, using the thick GaN film as the substrate the homoepitaxial growth of GaN by MOVPE and application to a blue light emitting device have been realized. [11,12]

Next, change in the crystalline structure of the ZnO buffer layer due to the thermal annealing of the growth temperature of GaN is investigated by using X-ray diffraction method. Figure 11 shows the X-ray diffraction intensity of ZnO (0002) and sapphire (0006) for ZnO buffer layers as-deposited and annealed at $1090°C$ for 5 to 20min. The intensity of the as-deposited sample is weak and the half width is broad, indicating the crystalline quality is not so good. However, after the annealing time of 5min, the intensity increases and the half width becomes sharp, indicating c-axis orientation is arranged by recrystallization due to the thermal annealing. After the annealing time of 20min, the intensity is becoming weak. This suggests occurrence of the thermal decomposition of ZnO. Thus, it is found that the ZnO layer which is improved by the thermal annealing for a short time is very useful to obtain the high-quality GaN. On the other hand, the thermal annealing for a long time causes degradation of the GaN crystal according to degradation of the ZnO crystal.

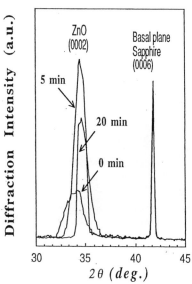

Figure 11. X-ray diffraction (0002) of ZnO lyaerers as-deposited and annealed at 1090°C for 5min and 20min as well as that (0006) of sapphire substrates.

5. SUMMARY

We described heteroepitaxial growth of GaN films by MOVPE and HVPE on sapphire and Si substrate on which buffer layers are deposited.

The buffer layers which were effective to realize high-quality GaN growth were (1) low temperature AlN buffer on sapphire substrate by MOVPE, (2) 3C-SiC buffer or high temperature AlN buffer on Si substrate by MOVPE, and (3) sputtered ZnO buffer on sapphire substrate by HVPE. The "bulk" single crystalline GaN of which film thickness is several hundred microns can be obtained by the HVPE method.

When buffer layers were not used in the heteroepitaxy of GaN, the crystalline quality of the GaN films were very poor and the surface morphology was very rough. The reason of the poorness in crystalline quality is attributed to the growth process which is summarized as (1) low-density nuceation of GaN on the substrate, (2) formation of many crystalline defects because of inhomogeneous coalescence among large GaN island of which crystalline direction is not different each other and (3) three-dimensional growth mode after the inhomogeneous coalescence.

In contrast, roles of the buffer layer are summarized as (1) high-density nucleation of GaN on the buffer layer, (2) homogeneous coalescence among the GaN islands of which crystalline direction is much arranged, (3) low crystalline defects near the interface between the GaN grains due to the homogeneous coalescence, (3) two-dimensional growth mode after the homogeneous coalescence. Thus, the change in the growth process which occurs by using the buffer layer plays an important role to grow high-quality GaN epitaxial films with low density of crystalline defects.

However, the optimum crystalline structure of AlN buffer layer is single crystalline for Si substrate and not single crystalline for sapphire substrate. In order to understand the reason of the difference in the crystalline structure further studies are required.

REFERENCES

1. H.Amano, N.Sawaki, I.Akasaki and Y.Toyoda: Appl. Phys. Lett. 48 (1986) 35.
2. S. Nakamura: Japan J. Appl. Phys., 30 (1991) L1705.
3. T.Detchprohm, K.Hiramatsu, H.Amano, and I.Akasaki: Appl. Phys. Lett., 61 (1992) 2688.
4. I.Akasaki, H.Amano, Y.Koide, K. Hiramatsu and N. Sawaki; J. Crystal Growth, 98 (1989) 209.
5. K.Hiramatsu, S.Ito, H.Amano, I. Akasaki, N. Kuwano, T. Shiraishi and K.Oki: J.Crystal Grorwth, 115 (1991) 628.
6. T. Takeuchi, H. Amano, K. Hiramatsu, N. Sawaki and I. Akasaki: J. Cryst. Growth , 115(1991)634.
7. A.Watanabe, T.Takeuchi, K.Hirosawa, H. Amano, K. Hiramatsu and I. Akasaki: J.Crystal Growth,
   128 (1993) 391.
8. T.Detchprohm, K. Hiramatsu, K. Itoh and I. Akasaki: Japan J. Appl. Phys., 31 (1992) L1454.
9. K.Hiramatsu, T.Detchprohm and I.Akasaki: Japan J. Appl. Phys., 32 (1993) 1528.
10. T.Detchprohm, H.Amano, K.Hiramatsu and I.Akasaki: J. Crystal Growth, 128 (1993) 384.
11. T.Detchprohm, K.Hiramatsu N.Sawaki and I.Akasaki: J. Crystal Growth, 137 (1994) 170.
12. T.Detchprohm, K.Hiramatsu, N.Sawaki and I.Akasaki: J. Crystal Growth, 145 (1994) 192.

Advances in the Understanding of Crystal Growth Mechanisms
T. Nishinaga, K. Nishioka, J. Harada, A. Sasaki and H. Takei (Editors)
© 1997 Elsevier Science B.V. All rights reserved.

Growth processes in the heteroepitaxy of Ge and $Si_{1-x}Ge_x$ on Si substrates using gas-source molecular beam epitaxy

Y. Yasuda, H. Ikeda and S. Zaima

Department of Crystalline Materials Science,
School of Engineering, Nagoya University
Furo-cho, Chikusa-ku, Nagoya 464-01, Japan

Growth processes of Ge and $Si_{1-x}Ge_x$ films on Si substrate surfaces by gas-source molecular beam epitaxy using Si_2H_6 and GeH_4 have been studied by *in situ* observations of reflection high-energy electron diffraction and replica electron microscopy. In $Si_{1-x}Ge_x$ ($0.25 \leq x \leq 1.0$) on (100)Si, three sequential growth steps are observed in the initial growth stages, which is an (8×2) surface structure, an {811}-faceted islands with a doubly-ordered ⟨111⟩ stacking structure and a {311}-faceted island. The formation of {811}- and {311}-faceted islands is caused by the relaxation of the film strain. In the case of Ge/(111)Si heteroepitaxy, at the first step of growth, δ-(7×7) and (1×1) structures are observed at substrate temperatures below 500°C and, on the other hand, they are not observed above 550°C, at a GeH_4 pressure of 6×10^{-4} Torr.

1. INTRODUCTION

There has been much interest in the study of Ge/Si and $Si_{1-x}Ge_x$/Si strained layer systems because of potential applications of these heterojunction structures to optoelectronic and microelectronic devices [1-3]. The lattice mismatch between Ge films or $Si_{1-x}Ge_x$ alloy films and Si substrates provides an extra degree of flexibility in tailoring electronic and optical properties of fabricated heterostructures. However, strain from this lattice mismatch also imposes a constraint on the critical thickness at which the alloy films can be pseudomorphically grown on the substrate at a given composition. The pseudomorphic growth of strained crystalline films is only possible for thicknesses less than the critical thickness. As the film thickness increases, the homogeneous strain energy of two-dimensional (2D) films becomes sufficiently large to render the induction of misfit dislocations energetically favorable, and three-dimensional (3D) clustering occurs.

The transition process during film growth from 2D to 3D structures must be elucidated in order to understand the growth mechanism. Moreover, in Ge/Si and $Si_{1-x}Ge_x$ heteroepitaxial growth systems, problems of atomic mixing of Ge and Si atoms and surface segregation of Ge atoms should be solved to form well-designed superlattices. Therefore, a complete understanding of heteroepitaxial growth mechanisms from an atomic viewpoint is indispensable for the realization of Ge/Si and $Si_{1-x}Ge_x$/Si superlattice devices.

2. INITIAL STAGE OF Ge FILM GROWTH ON Si SUBSTRATE

2.1. Ge films on (100)Si substrate

Figure 1 shows the *in situ* RHEED patterns taken from a (100)Si substrate surface and Ge films grown on a (100)Si surface at 300°C by GeH_4-source molecular beam epitaxy (MBE) [4]. The average thickness of the Ge film in Figure 1(d) was 13Å. The (100)Si surface was cleaned thermally at 1200°C. In Figure 1(a), the $\frac{1}{2}$-order Laue zone ($L_{\frac{1}{2}}$) and $\frac{1}{2}$-order rods are clearly observed, which indicates that a double-domain (2×1) surface structure is formed on the (100)Si clean surface. Streaks extending along the direction normal to the substrate surface are observed in Figure 1(b). On the other hand, diffraction spots appear in Figures. 1(c) and 1(d). These findings indicate that the growth mode of Ge films on (100)Si surfaces changes from a layer-by-layer growth to an island growth, which is called Stranski-Krastanov (SK)-type growth. Moreover, it is found from Figure 1 that there are three kinds of growth step in the initial growth stage of (100)Ge films on (100)Si surfaces. These phenomena were observed in the substrate temperature range of 300 to 600°C.

At the first step in the initial stage of growth (Figure 1(b)), the RHEED pattern exhibits a streaky pattern, and every streak extends in the $[100]_{Ge} \parallel [100]_{Si}$ direction. Streaks of $\frac{1}{8}$ order (indicated by arrows) are observed near the (01) and (0$\bar{1}$) rods. These findings mean that Ge films grow layer by layer with an (8×2) structure on the surface at this stage. The (8×2) structure consists of a dimer structure with missing a dimer row at every eight units, which was observed by scanning tunneling microscopy after our previous observation by RHEED [5].

At the second step in the initial stage of growth (Figure 1(c)), the RHEED pattern exhibits a spotty pattern, and, at the same time, streaks extending in two ⟨811⟩ directions are observed. It should be noted that streaks of fundamental spots extend in both [$\bar{8}1\bar{1}$] and [$\bar{8}11$] directions, and that fractional-order streaks (denoted by arrows in Figure 1(c)) extend alternately in the [$\bar{8}1\bar{1}$] and [$\bar{8}11$] directions in a doglegged shape. The streaks in the RHEED patterns indicate that the surface has crystal facets corresponding to the extending directions of the streaks. Therefore, it is concluded that the growing islands of Ge have predominantly {811} facets. On the other hand, the doglegged fractional-order streaks are interpreted as follows.

Figure 2 shows schematic diagrams explaining the RHEED pattern in Figure 1(c) [4]. In Figure 2(a), spots marked by ∘ and □ are due to the double ordering in the [111] and [1$\bar{1}\bar{1}$] directions, respectively. The fractional-order streaks in the [811] and [$8\bar{1}\bar{1}$] directions extend from the positions of $(h+\frac{1}{2}, k+\frac{1}{2}, l+\frac{1}{2})$ and $(h+\frac{1}{2}, k-\frac{1}{2}, l-\frac{1}{2})$ in the reciprocal lattice of the diamond structure, respectively. This diffraction pattern is interpreted as being due to the existence of the double periodicity of the (111) plane in the (811)-faceted region of Ge islands and of the (1$\bar{1}\bar{1}$) plane in the (8$\bar{1}\bar{1}$)- faceted region (Figure 2(b)). The (811) surface is considered to have a regularly stepped structure, which is composed of (100) terraces with the single-

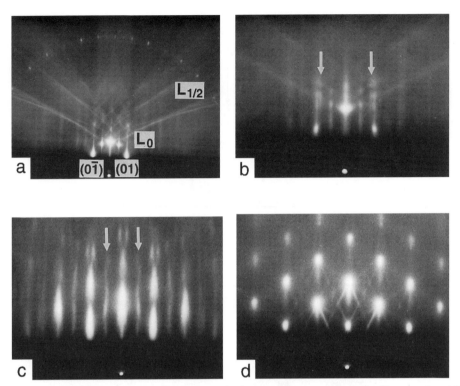

Figure 1. RHEED patterns showing typical growth processes of Ge films on (100)Si substrate surfaces. (a) Double-domain (2×1) clean Si surface; (b), (c) and (d) the first, second and final steps in the initial stage of growth, respectively. The substrate temperature was 300°C. The incident electron beam is parallel to the $[0\bar{1}1]_{Si}$ direction and the average thickness of the Ge film (d) is 13Å.

domain (2×1) structure and $[0\bar{1}1]$ steps with bilayer height [6]. Consequently, it is noted that the ordered structure is formed on only the {111} plane parallel to the side plane of ⟨011⟩ steps on {811} facets of growing Ge islands. The probable atomic models of the ordered structure are shown in Figures 2(c) and 2(d). The formation of this ⟨111⟩-ordered structure was observed in the substrate temperature range of 300 to 700°C.

At the final step in the initial stage of growth (Figure 1(d)), sharp $[\bar{3}1\bar{1}]$ and $[\bar{3}11]$ streaks from the fundamental spots are observed, and ⟨811⟩ streaks disappear. This result means that the preferential facet of Ge islands changes from {811} to {311} planes. In this stage, the growth temperature markedly influences the growth of Ge islands. Figure 3 shows replica electron micrographs taken from Ge films,

418

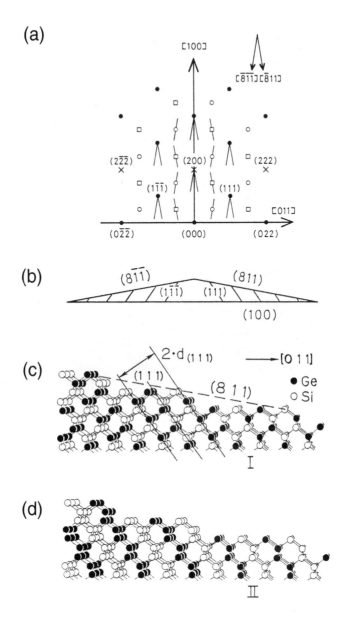

Figure 2. Schematic diagrams explaining the RHEED pattern in Figure 1(c). Spots marked by o and □ are due to the double ordering in the [111] and [1$\bar{1}\bar{1}$] directions, respectively. Spots marked by × appear due to the double diffraction. The probable atomic models of the ordered structure are shown in (c) and (d).

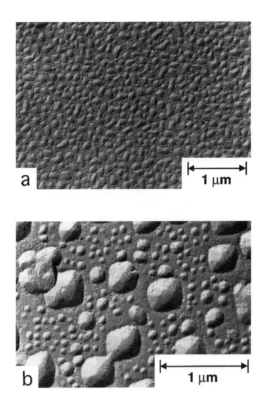

Figure 3. Replica electron micrographs taken from Ge films, during the island growth step, grown at substrate temperatures of (a) 300°C and (b) 600°C. The average thickness is about 9 nm.

during the island growth step, grown at substrate temperatures of (a) 300°C and (b) 600°C [7]. The average thickness is about 9 nm. At 300°C (Figure 3(a)), the islands have small size and irregular shapes. In contrast, rectangular Ge islands of various dimensions grew at 600°C (Figure 3(b)). The transition temperature between the two island morphologies was about 500°C. Considering that H coverage on Ge island surfaces is very small at substrate temperatures above 500°C [8], the difference in the island morphology is thought to be influenced by the adsorption of H atoms from GeH_4. That is, the morphology of growing islands is determined by the minimization of the total energy in the system, which consists of the interaction energy between the islands and the substrate, the volume energy and the surface energy of the islands. Consequently, energetically stable structures depend strongly

420

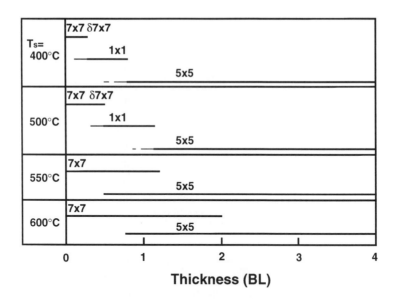

Figure 4. Summary of the growth processes of Ge films on (111)Si-(7×7) surfaces at different substrate temperatures and a GeH$_4$ pressure of 6×10^{-4} Torr.

on the existence of H atoms on the surface because the adsorbed H atoms change the surface energy and the interaction energy.

In the case of growth of Ge films on (100)Si by solid-source MBE (SSMBE), {501}-faceted islands initially appear after a layer-by-layer growth, and then {311}-faceted islands are formed [9]. The formation of {501}-faceted islands is thought to correspond to the absence of adsorbed H atoms on the surface [10].

2.2. Ge films on (111)Si substrate

Figure 4 summarizes the growth processes of (111)Ge films on (111)Si-(7×7) surfaces at different substrate temperatures and a GeH$_4$ pressure of 6×10^{-4} Torr [11]. There are marked differences in the growth processes between the substrate temperature ranges below 500°C and above 550°C. At 400 and 500°C, the (7×7) structure changes first to the δ-(7×7) structure, then to the (1×1) structure, and finally to the (5×5) structure. On the other hand, at 550 and 600°C, the (7×7) structure changes first to a mixed state of (7×7) and (5×5) structures, and finally to the (5×5) structure. Furthermore, the critical thickness at which each growth step starts increases at higher substrate temperatures. The formation of δ-(7×7) and (1×1) structures is a characteristic feature of Ge film growth on (111)Si-(7×7) using a GeH$_4$ gas source. The maximum substrate temperature at which δ-(7×7) and (1×1) structures are observed depends on the GeH$_4$ gas pressure and decreases from 500°C to 450°C when the GeH$_4$ pressure decreases from 6×10^{-4} to 1×10^{-4} Torr. This phenomenon indicates that the formation of δ-(7×7) and (1×1) structures is attributed to the adsorption of H atoms on the surface. In the case of SSMBE, the (5×5) structure is also observed, but the δ-(7×7) structure is not [12,13]. It has been reported that the δ-(7×7) structure appears on (111)Si surfaces when hydrogen atoms or alkaline metals adsorb on (111)Si surfaces [14,15]. These reports indicated that the (5×5) structure is a stable state of (111)Ge films on (111)Si substrates and support the above-mentioned conclusion that H adsorption is essential for the formation of δ-(7×7) and (1×1) structures.

Figure 5 shows size changes of domains having the (7×7), δ-(7×7) and (5×5) structures with growth time, t, at substrate temperatures of (a) 450 and (b) 600°C [11]. The GeH$_4$ pressure was 6×10^{-4} Torr. The domain sizes of the (7×7), δ-(7×7) and (5×5) regions were estimated by the dimensions of $(\frac{3}{7}, \frac{4}{7})$, $(\frac{3}{7}, \frac{4}{7})$ and $(\frac{2}{5}, \frac{3}{5})$ rod intensity in the diffraction patterns, respectively. As seen in Figure 5, the (7×7) and δ-(7×7) domains become smaller with increasing growth time. On the other hand, the (5×5) domains become larger in proportion to t at a temperature of 450°C (Figure 5(a)), and to $t^{\frac{1}{2}}$ at a temperature of 600°C (Figure 5(b)). In both cases, the domains finally reach a saturated size of 1000Å, which indicates the existence of saturated density. These two types of growth time dependence are always observed during growth in the GeH$_4$ pressure range of 1×10^{-4} to 6×10^{-4} Torr. The results shown in Figures 4 and 5 suggest that there exist two kinds of growth mechanism. That is, the growth process on the H-adsorbed surface at 450°C is characterized by the formation of δ-(7×7) and (1×1) structures, and that on the H-desorbed surface above 500°C is characterized by the formation of mixed regions of (7×7) and (5×5) structures.

Corresponding to the above changes in surface structure, the growth rate of the domains can be considered as follows. At temperatures below 450°C, the H coverage on the step edge of domains is large and the growth rate of the domains is thought to be limited by the linear density of H-desorbed sites in the step edge, which might act as adsorbed sites of Ge. In this case, the step growth rate is constant and independent of the domain size, which leads to a linear relationship of the growth time dependence of increasing domain size, r. On the other hand, at temperatures above 500°C, the growth rate is limited by the mass-transfer process of Ge from the surface to the domain edges because the edge sites of domains possess few adsorbed H atoms. In this case, the increment rate of the domain area, πr^2, is constant and independent of the domain area, because the number of Ge atoms supplied to a domain is determined by the saturated density. That is, $d(\pi r^2)/dt = const.$ and $r \propto t^{\frac{1}{2}}$. From these results, it is concluded that the main factor governing the marked differences in film growth is the adsorption on and

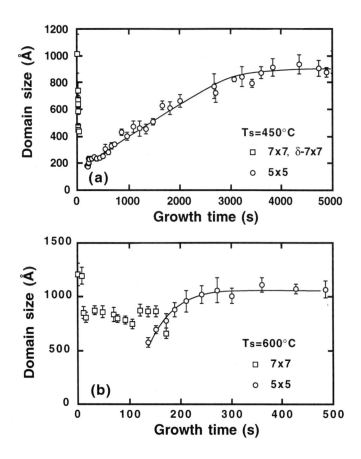

Figure 5. Size changes of domains having the (7×7), δ-(7×7) and (5×5) structures in Ge film grown on (111)Si-(7×7) surfaces with growth time at substrate temperatures of (a) 450°C and (b) 600°C. The GeH$_4$ pressure was 6×10^{-4} Torr.

desorption from the domain edge steps of H atoms supplied from GeH$_4$ molecules.

3. Si$_{1-x}$Ge$_x$ FILMS ON (100)Si SUBSTRATE

Figure 6 shows the relationships between Ge atomic fraction, x_p, in the gas phase and Ge composition, x, in Si$_{1-x}$Ge$_x$ epitaxial films grown on (100)Si substrate surfaces by Si$_2$H$_6$- and GeH$_4$-source MBE at several substrate temperatures of 300 to 600°C [16]. Here, x_p is defined by $P_{GeH_4}/(2P_{Si_2H_6} + P_{GeH_4})$, where $P_{Si_2H_6}$ and P_{GeH_4} are the partial pressures of Si$_2$H$_6$ and GeH$_4$, respectively. x is determined from measurements by Rutherford backscattering spectroscopy and X-ray photoelectron spectroscopy. Figure 6 also includes previously reported data for

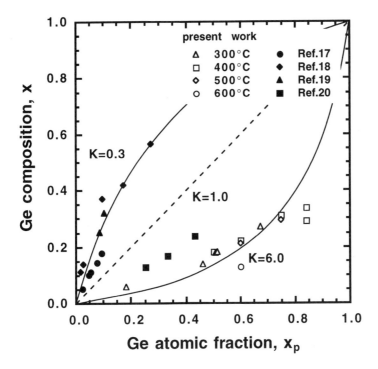

Figure 6. Relationships between Ge atomic fraction, x_p, in the gas phase and Ge composition, x, in $Si_{1-x}Ge_x$ epitaxial films grown on $(100)Si$-(2×1) substrate surfaces by Si_2H_6- and GeH_4-source MBE at several substrate temperatures of 300 to 600°C. This figure also shows data for combinations of SiH_4 and GeH_4 (Refs. 17, 18), SiH_2Cl_2 and GeH_4 (Ref. 19), and Si_2H_6 and GeH_4 (Ref. 20).

combinations of SiH_4 and GeH_4 [17,18], SiH_2Cl_2 and GeH_4 [19], and Si_2H_6 and GeH_4 [20]. It is found that the present curve of x_p vs. x is concave and deviates from the linear dependence (broken line), which means that the incorporation rate of Si atoms from Si_2H_6 into the film is larger than that of Ge atoms from GeH_4. This tendency contrasts with that for the growth from SiH_4 and GeH_4. Moreover, this nonlinear dependence of x on x_p is found to weakly depend on the substrate temperature in the examined range of 300 to 600°C.

Next, the relationship between x_p and x is discussed from the viewpoint of growth kinetics. Based on the kinetics of Si homoepitaxial growth from SiH_4 reported by Gates and Kulkarni [21] and from Si_2H_6 by Lin et al. [22], the following reaction steps can be assumed, which describe the $Si_{1-x}Ge_x$ film growth from Si_2H_6 and

GeH$_4$ by gas-source MBE (GSMBE):

$$Si_2H_6(g) + 2_- \rightarrow 2\underline{Si}H_3, \tag{1}$$

$$2\underline{Si}H_3 \rightarrow 2Si(s) + 3H_2(g) + 2_-, \tag{2}$$

$$GeH_4(g) + 2_- \rightarrow \underline{Ge}H_3 + \underline{H}, \tag{3}$$

$$\underline{Ge}H_3 + \underline{H} \rightarrow Ge(s) + 2H_2(g) + 2_-, \tag{4}$$

where the symbol $_-$ denotes a surface site corresponding to a dangling bond, and $\underline{Si}H_3$, $\underline{Ge}H_3$, and $\underline{H}$ show the species adsorbed on the sites. Reaction steps (1) and (3) represent dissociative adsorptions of Si$_2$H$_6$ and GeH$_4$ molecules, respectively. Reaction steps (2) and (4) are surface reactions that involve the decomposition of adsorbed hydrides and the desorption of adsorbed H atoms. The surface sites created by H desorption are consumed by the successive dissociative adsorption of Si$_2$H$_6$ and GeH$_4$. Under a steady-state condition, the growth rate of Si$_{1-x}$Ge$_x$ films, $R(x)$, is obtained as

$$\begin{aligned} R(x) &= R_{Si}(x) + R_{Ge}(x) \\ &= 2k_{Si}N_{DB}^2 P_{Si_2H_6} + k_{Ge}N_{DB}^2 P_{GeH_4}, \end{aligned} \tag{5}$$

where k_{Si} and k_{Ge} are the rate constants for adsorption reaction steps (1) and (3), respectively, and N_{DB} is the density of the dangling-bond surface sites. x is derived from Eq. (5) as

$$x = \frac{R_{Ge}(x)}{R_{Si}(x) + R_{Ge}(x)} = \frac{P_{GeH_4}}{2KP_{Si_2H_6} + P_{GeH_4}}, \tag{6}$$

where K is k_{Si}/k_{Ge}. The relationship between x and x_p is, therefore, obtained as

$$\frac{1}{x} = 1 - K + \frac{K}{x_p}. \tag{7}$$

Equation (7) is equivalent to the Markham-Benton formula [23] for the equilibrium adsorption isotherm in the presence of two kinds of adsorbate. Equation (7) means that the x vs. x_p relation is determined by the dissociative adsorption of Si$_2$H$_6$ and GeH$_4$. In Figure 6, the line obtained using Eq. (7) with $K = 6.0$ fits the present data well, and K is determined to be 6.0±0.5 in the substrate temperature range from 300 to 600°C, which means that the dissociative adsorption rate of Si$_2$H$_6$ is six times as large as that of GeH$_4$.

Applying the present model to the combinations of SiH$_4$ and GeH$_4$, and SiH$_2$Cl$_2$ and GeH$_4$, K=0.30±0.05 is obtained, as shown in Figure 6. This result indicates that the dissociative adsorption rate of SiH$_4$ is 0.3 times as large as that of GeH$_4$. From these results, it is concluded that the dissociative adsorption rate of hydride compounds on the surface determines the incorporation ratio of Si and Ge atoms into the films for these kinds of gas system.

As mentioned in §2.1, there exist three steps in the initial stage of the growth process of Ge films on (100)Si substrates by GeH$_4$-source MBE. In the case of

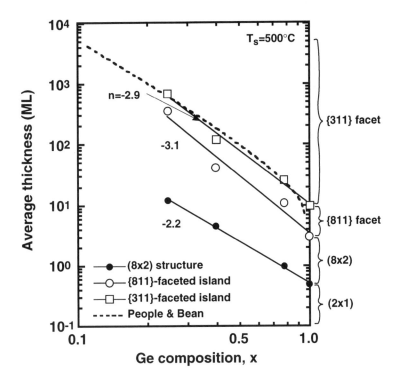

Figure 7. Ge composition (x) dependences of the average critical film thicknesses for the start of formation of (8×2) surfaces (•), $\{811\}$-faceted islands (o), and $\{311\}$-faceted islands (□) for $Si_{1-x}Ge_x$ ($0.25 \leq x \leq 1.0$) films grown on (100)Si-(2×1) surfaces at a substrate temperature of 500°C. The broken line is a dependence calculated using a commensurate model (see Ref. 24).

$Si_{1-x}Ge_x$ film growth on (100)Si substrates, the same three steps are also observed, that is, an (8×2) surface structure is initially formed on top of 2D $Si_{1-x}Ge_x$ layers, followed by $\{811\}$-faceted islands having a doubly ordered $\{111\}$ stacking structure composed of Si and Ge atoms, and finally by $\{311\}$-faceted islands [16]. Figure 7 shows film composition (x) dependences of the growth processes of $Si_{1-x}Ge_x$ ($0.25 \leq x \leq 1.0$) films on (100)Si-(2×1) surfaces at a substrate temperature of 500°C [16]. In this figure, the average critical film thicknesses, h_c, are shown, at which the (8×2) surface structure, $\{811\}$-faceted islands, and $\{311\}$-faceted islands start to be formed. The average film thicknesses were estimated from the film growth rate determined by the intensity oscillation of the specular reflection spots in RHEED patterns in the 2D layer growth step. It should be noted that there exist marked composition dependences of the growth process. The critical film thicknesses at which each growth step starts increase drastically with decreasing x, obeying the power law of $h_c \propto x^n$. The exponents, n, obtained in the x range from 0.2 to 1.0 are -2.2, -3.1, and -2.9 for the formation of the (8×2) structure, $\{811\}$-faceted islands, and $\{311\}$-faceted islands, respectively. It should be noted that h_c for 3D island formation exhibits x^{-3} dependence for both $\{811\}$- and $\{311\}$-faceted islands; on the other hand, h_c for the 2D (8×2) structure exhibits x^{-2} dependence. Figure 7 also shows the x dependence of the critical film thickness changing to the incommensurate region, calculated by People and Bean [24]. This model is based on the relaxation of the commensurate strain energy due to the introduction of screw dislocations. From Figure 7, the calculated result of People and Bean shows almost the same dependence of x^{-3} as the present experimental ones. Consequently, it is concluded that the formation of $\{811\}$- and $\{311\}$-faceted islands is directly related to the strain release of the films due to the introduction of crystalline defects and that the release takes place successively in two steps accompanying the formation of $\{811\}$- and $\{311\}$-faceted islands.

4. CONCLUSIONS

In the initial stage of growth of Ge and $Si_{1-x}Ge_x$ films on Si substrate surfaces by GSMBE using Si_2H_6 and GeH_4, the growth processes and H effects on the processes have been investigated by *in situ* observations of RHEED and replica electron microscopy. The main conclusions are summarized below.

In the initial growth stage of Ge films on (100)Si surfaces, there exist three kinds of growth step in the substrate temperature range from 300 to 600°C. At the first step in the initial stage of growth, Ge films exhibit layer-by-layer growth with an (8×2) structure on the surface. This is followed by the formation of $\{811\}$-faceted islands with $\langle 111 \rangle$-ordered structure on the surface. Finally, the preferential facet of Ge islands changes from $\{811\}$ to $\{311\}$ planes. The morphology of $\{311\}$-faceted islands is influenced by the existence of adsorbed H atoms.

In Ge film growth on (111)Si-(7×7) surfaces, δ-(7×7) and (1×1) structures are formed, which is a characteristic feature in the use of GeH_4 as a gas source. The formation of these structures is attributed to the adsorption of H atoms on the surface. In the next growth stage, the (5×5) structure is formed. The growth rate of (5×5) domains is limited by the H desorption process at substrate temperatures below 450°C and by the mass-transfer process of Ge on the surfaces without adsorbed H atoms above 500°C.

In the case of $Si_{1-x}Ge_x$ ($x = 0.2 \sim 1.0$) film growth on (100)Si substrates, the dissociative adsorption rate of Si_2H_6 is six times as large as that of GeH_4. There exist the same three steps in $Si_{1-x}Ge_x$ growth as in Ge growth on (100)Si. The average critical film thicknesses at which each growth step starts increase according to power laws with decreasing Ge composition, and it is concluded that the formation of $\{811\}$- and $\{311\}$-faceted islands is directly related to the strain release of the films.

REFERENCES

1. T. P. Pearsall and J. C. Bean, IEEE Trans. Electron. Device Lett. EDL-7 (1986) 308.
2. G. L. Patton, S. S. Iyer, S. L. Dlage, S. Tiwari and J. M. C. Stork, IEEE Trans. Electron. Device Lett. EDL-9 (1988) 165.
3. T. P. Pearsall, H. Temkin, J. C. Bean and S. Luryi, IEEE Trans. Electron. Device Lett. EDL-7 (1986) 330.
4. N. Ohshima, Y. Koide, S. Zaima and Y. Yasuda, Appl. Surf. Sci. 48/49 (1991) 69.
5. W. Watanabe, F. Iwasaki, M. Tomitori and O. Nishikawa, *Extended Abstracts of the 39th Spring Meeting, 1992* (The Japan Society of Applied Physics), p. 316.
6. Y. Koide, S. Zaima, K. Itoh, N. Ohshima and Y. Yasuda, J. Appl. Phys. 68 (1990) 2164.
7. Y. Koide, S. Zaima, N. Ohshima and Y. Yasuda, J. Crystal Growth 99 (1990) 254.
8. R. Tsu, D. Lubben, T. R. Bramblett, J. E. Greene, D. -S. Lin and T. -C. Chiang, Surf. Sci. 280 (1993) 265.
9. Y. W. Mo, D. E. Savage, B. S. Swartzentruber and M. G. Lagally, Phys. Rev. Lett. 65 (1990) 1020.
10. M. Tomitori, K. Watanabe, M. Kobayashi, F. Iwasaki and O. Nishikawa, Surf. Sci. 301 (1994) 214.
11. N. Ohshima, S. Zaima, Y. Koide, S. Tomioka and Y. Yasuda, Appl. Surf. Sci. 60/61 (1992) 120.
12. U. Köhler, O. Jusko, G. Pietsch, B. Müller and M. Henzler, Surf. Sci. 248 (1991) 321.
13. M. Aono, R. Souda, C. Oshima and Y. Ishizawa, Phys. Rev. Lett. 51 (1983) 801.
14. A. Ichimiya and S. Mizuno, Surf. Sci. 191 (1987) L765.
15. H. Daimon and S. Ino, Surf. Sci. 164 (1985) 320.
16. Y. Yasuda, Y. Koide, A. Furukawa, N. Ohshima and S. Zaima, J. Appl. Phys. 73 (1993) 2288.
17. B. S. Mayerson, K. J. Uram and F. K. LeGous, Appl. Phys. Lett. 53 (1988) 2555.
18. M. Kato, H. Murota and N. Ono, J. Crystal Growth 115 (1991) 117.
19. P. M. Garone, J. C. Sturm, P. V. Dchwartz, S. A. Schwartz and B. J. Wilkens, Appl. Phys. Lett. 56 (1990) 1275.
20. T. Tatsumi, K. Aketagawa, M. Hiroi and J. Sakai, *Extended Abstracts of the 52nd Autumn Meeting, 1991* (The Japan Society of Applied Physics), p. 315.
21. S. M. Gates and S. K. Kulkarni, Appl. Phys. Lett. 58 (1991) 2963.
22. D. S. Lin, E. S. Hirschorn, T. C. Chiang, R. Tsu, D. Lubben and J. E. Greene, Phys. Rev. B4 (1992) 3494.
23. E. C. Markham and A. F. Benton, J. Am. Chem. Soc. 53 (1931) 497.
24. R. People and J. C. Bean, Appl. Phys. Lett. 49 (1986) 229.

Advances in the Understanding of Crystal Growth Mechanisms
T. Nishinaga, K. Nishioka, J. Harada, A. Sasaki and H. Takei (Editors)

Crystal growth mechanisms in III-V/Si heteroepitaxy

T. Soga[a] and M. Umeno[b]

[a]Instrument and Analysis Center, Nagoya Institute of Technology, Gokiso-cho, Showa-ku, Nagoya 466, Japan

[b]Department of Electrical and Computer Engineering, Nagoya Institute of Technology, Gokiso-cho, Showa-ku, Nagoya 466, Japan

The epitaxial growth mechanisms in III-V compound semiconductors on Si substrate using metalorganic chemical vapor deposition method are reviewed. The initial growth mode of GaP on Si changes from island-type to layer-type with increasing the V/III ratio or the gas pressure. The stress, the lattice deformation and the defect structure of GaP on Si substrate grown under high V/III ratio are characterized with varying the GaP thickness. The structure of antiphase domain in GaP on Si with various misorientation angle is described. The growth conditions to obtain a low dislocation density GaP-on-Si with smooth surface morphology are presented. The growth details of GaAsP layers with low dislocation density grown on GaP/Si with compositionally-step-graded buffer layer are addressed.

1. INTRODUCTION

The research on the crystal growth of III-V compounds on Si substrate has been performed actively in a decade since high quality GaAs layers were successfully grown on Si substrates in 1984 [1–5]. Although various devices such as lasers [6] and solar cells [7] have been fabricated on Si substrate, the device characteristics are not satisfactory due to the existence of a high density of the threading dislocations in the epitaxial layer. The reduction of dislocation density is an important issue to obtain a high-performance compound semiconductor device on a Si substrate. The dislocations are generated due to the lattice mismatch, the thermal expansion mismatch, the crystal structure difference, the generation of an antiphase domain (APD), the surface contamination, etc.

The threading dislocations in the epitaxial layer on Si substrate are classified into (i) dislocations which originate from the dislocations in the Si substrate, (ii) dislocations generated by the coalescence of the islands at the initial stage, (iii) dislocations generated by the lattice mismatch and (iv) dislocations generated by the thermal stress during the cooling stage from the growth temperature.

Among these four types of dislocations, item (i) need not be taken into account because the dislocation density of Si is very low. In order to reduce the dislocation density according to items (ii), (iii) and (iv), many efforts have been done. The efforts to change the initial growth mode from three-dimensional to two-dimensional have been made to avoid the dislocations generated by the coalescence of the islands at the initial stage of the growth [8]. Although many methods using strained layer superlattice buffer layer [1,9,10], rapid thermal annealing [11], thermal cycle annealing [12,13], etc. have been adopted to reduce the dislocation density of item (iii), the dislocation density is still in the order of 10^6 cm^{-2}. The low-temperature growth has been investigated to decrease the generation of the dislocation by the thermal stress [14,15]. It is expected that the number of dislocations generated by the thermal stress is reduced when the growth temperature is low. Although the dislocation density of GaAs-on-Si to the order of 10^4 cm^{-2} has been obtained at the growth temperature of 350°C, the thermal stability is still a problem.

This chapter reviews the growth process of III-V compound semiconductors on Si substrates grown by metalorganic chemical vapor deposition (MOCVD). The nucleation of GaP on Si, the dislocation generation mechanism, antiphase domain (APD) structure of GaP on Si substrate and the dislocation-filtering of compositionally step graded buffer layer (CSGBL) for the highly lattice mismatched material are presented. Although highly mismatched systems such as GaAs-on-Si, GaAsP-on-Si and GaN/Si are interesting for the device application, the study on the crystal growth of GaP on Si is interesting for the fundamental understanding of the III-V compounds on Si substrates. It is expected that the understanding of the growth mechanism will lead us to achieve the growth of low-dislocation-density of III-V compounds on Si substrates.

2. CRYSTAL GROWTH AND CHARACTERIZATION

The epitaxial growth was performed using low pressure MOCVD. It consists of lamp-heated horizontal reactor with load lock chamber. The substrate is put on the SiC-coated carbon susceptor and the temperature was controlled by the thermocouple inserted into the susceptor. Source gases for Ga, As and P were trimethylgallium (TMG), AsH$_3$ and PH$_3$, respectively. The Si substrate orientation is (001) 0 - 6° tilted toward [110] direction. The misorientation angle is 4° off unless otherwise mentioned. Si substrates were degreased with organic solvents, followed by the repetition of the oxidation by H$_2$SO$_4$: H$_2$O = 4 : 1 and the removal of the oxides by 25 % HF solution. After loading the substrate into the reactor, the substrate was heated at 1000 °C for 10 min in a hydrogen ambient. The V/III ratio was varied from 100 to 6400 by changing the PH$_3$ flow rate, keeping the TMG flow rate as constant. The gas pressure has been changed from 76 to 380 torr. The epitaxial layer thickness has been varied from 20 nm to 3.7 μm. The growth temperature was kept constant at 900 °C. Three-dimensional growth was not observed under these growth conditions for GaP growth on GaP substrate. The samples were examined using Nomarski optical microscopy, cross-sectional transmission electron microscopy (TEM) and double crystal X-ray diffraction. The TEM operation voltage was 200 kV and the incident beam was made perpendicular to the misorientation direction.

3. NUCLEATION OF GaP ON Si SUBSTRATE

In general, the growth mode is largely divided into three categories, i.e., two-dimensional (2D) type, Volmer-Weber (3D) type and Stranski-Krastanov (2D + 3D) type [16,17]. In the case of the 2D mode, the dislocations are generated when the layer thickness exceeds the critical thickness [18]. On the other hand, the dislocations and stacking faults are generated at the coalescence of the islands formed at the initial stage of the growth in the case of the 3D mode[19]. Therefore, the defect density will be significantly reduced if the two-dimensional growth can be realized from the beginning of the growth. The three-dimensional growth mode has origins not only in the basic material property differences between the epitaxial layer and the substrate (e.g. lattice constant mismatch, polar/nonpolar effect, etc.) but also in the growth conditions.

Table 1
Important material parameters of III-V compound semiconductors and Si.

	GaN	GaP	GaAs	InP	Si
Lattice constant (Å)	3.19(a) 5.19(c)	5.45	5.65	5.87	5.43
Linear thermal expansion coefficient ($\times 10^{-6}$ K^{-1})	5.6(a) 3.2(c)	5.9	6.8	4.6	2.6
Crystal structure	WZ	ZB	ZB	ZB	Diamond

WZ: Wurzite, ZB: Zincblend

The material properties of some of the III-V compounds and Si are shown in Table 1. As shown in this table, the lattice constant of GaP is closest to that of Si. Therefore, the effect of lattice mismatch on the growth mode is expected to be minimum and other effects such as the surface migration effect, surface contamination, polar/nonpolar structure are emphasized.

Figure 1 shows the growth mode of GaP-on-Si for various gas pressure and the V/III ratio[20]. The GaP nominal thickness is 40 nm. All the surface morphologies of the samples were classified into three types, i.e., island-type, mixture-type (mixed-island/layer-type) and layer-type. The typical surface morphologies are also shown in Figure 1. It is indicated that the growth mode of GaP changes from island-type to layer-type with increasing V/III ratio. The V/III ratio at which the growth mode changes from island-type to layer-type decreases with increasing the gas pressure. Very high V/III ratio of 3200 is necessary to obtain a GaP layer without island-type crystal at the gas pressure of 76 torr [8]. High resolution TEM micrograph of GaP island grown under the V/III ratio of 800 and the gas pressure of 76 torr is shown in Figure 2 [8]. Island-type growth is clearly demonstrated. At the edge of the island, it is observed that the islands are not connected by a GaP

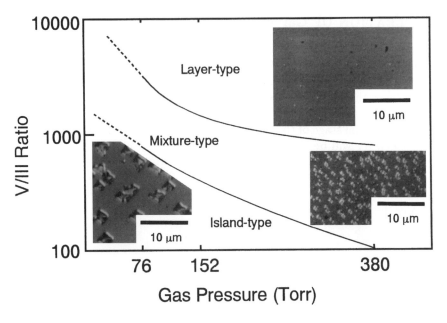

Figure 1. Growth mode and surface morphologies of GaP on Si substrate for various V/III ratio and gas pressure.

layer between themselves. This means that the growth mode of GaP on Si substrate is not Stranski-Krastanov type but Volmer-Weber type under these growth conditions. Faceting was observed at the boundary either on (111) or (211) type planes.

The island formation is interesting because these islands are not formed in the case of homoepitaxy; it is unique for the heteroepitaxy. In the case of the homoepitaxial growth, the source gases are usually incorporated into the step edge or terrace of the misoriented substrate. However, in the heteroepitaxy the island spacing is several orders of magnitude larger than that of the average step distance of the misoriented substrate (Figure 1). Therefore, the nucleation site is not governed by the substrate steps. Furthermore, the residual oxide or impurity is not the nucleation site because the GaP island density changes drastically with the growth conditions [21].

In the nucleation process of GaP on Si substrate, the diffusion of the growth species through the boundary layer, the surface migration and the nucleation at the nucleation sites should be taken into account. Comparing the heteroepitaxy of GaP-on-Si and the homoepitaxy, there should not be any difference in the diffusion process if the growth conditions are the same. Therefore, it is deduced that the only difference between the homoepitaxy and the heteroepitaxy is the difference in the migration of growth species on the Si substrate. The migration length of the migrating species on Si is considered to be longer than that of homoepitaxy. This is due to the weak interaction between Si and Ga or P atoms. An example of strong atomic interaction is the growth of a III-V compound semiconductor containing Al on Si substrates. It has been reported that AlGaP [22], AlGaAs [23] and AlAs [24] layers grown on Si substrate are flat from the beginning of the growth.

Before discussing the details of the migrating species for GaP on Si, the surface migration of Ga is discussed. Usually, the migration species for the deposition of Ga films on Si substrate is the cluster of Ga atoms. TMG is probably perfectly decomposed to Ga and a metal-radical at 900 °C [25]. Therefore, in the case of the deposition of Ga on Si, the migrating species are supposed to be Ga_x-type clusters. It is expected that the molecular mass of a cluster increases during the surface migration. Because the island density increases gradually with increasing V/III ratio, the migration species would be a $Ga_x P_y$-type cluster [21]. When the PH_3 flow rate is increased, the number of decomposed P atoms is increased, and a high density of $Ga_x P_y$-type cluster with large mass is easily formed. Clusters with large molecular mass are expected to migrate more slowly than those with smaller masses. When the cluster size exceeds the critical size, the clusters are deposited on the Si surface and the islands are formed. Moreover, when the concentration of P atoms on the Si surface is high, P atoms absorbed on the Si surface are increased. It results in the formation of the flat layer because the P atoms absorbed on Si capture the migrating species.

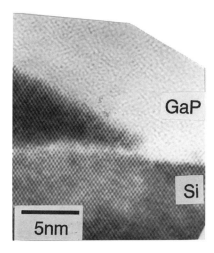
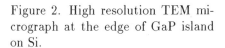

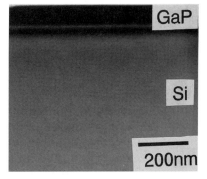

Figure 2. High resolution TEM micrograph at the edge of GaP island on Si.

Figure 3. Cross-sectional TEM micrograph of 70 nm-thick GaP layer on Si grown under high V/III ratio.

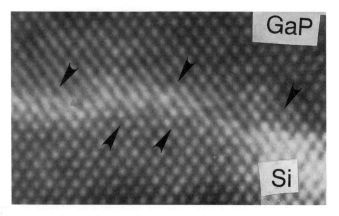

Figure 4. High resolution TEM micrograph of 3.7 μm-thick GaP layer on Si grown under high V/III ratio.

In order to explain the gas pressure dependence of the growth mode, other factors in the reactor must be considered. A possible factor affected by the gas pressure is the flow velocity in the reactor. TMG is almost completely decomposed at the growth temperature [25]. In contrast, the decomposition of PH_3 varies with the flow velocity because the decomposition rate for PH_3 is not so fast as TMG [26]. These results are qualitatively explained as follows. It is evident that the flow velocity is increased when the pressure is decreased. The pyrolysis of PH_3 takes place in the heated region in the reactor. Since the resident time for the source gases in the heated region becomes short, the decrease of pressure makes the PH_3 decomposition difficult. So the higher V/III ratio is necessary for lower growth pressure to produce a layer-type growth.

4. GENERATION OF MISFIT DISLOCATION AND STRESS RELAXATION

In the heteroepitaxial growth, the generation of misfit dislocation and the stress relaxation are very much related each other. The misfit dislocation generation and the stress relaxation of GaP layer on Si substrate grown under high V/III ratio (layer-type growth mode) are described. Because the defects associated with the coalescence of the islands are not generated, the observed dislocations are generated after the layer thickness exceeds the critical thickness.

Figure 3 shows the cross-sectional TEM micrograph of GaP on Si with the thickness of 70 nm taken by (220) bright field conditions [27]. The V/III ratio is 1600 and the gas pressure is 380 torr. It can be seen from the picture that the GaP surface is very flat and defects such as dislocations or structural defects are not at all observed. This means that GaP grows on Si coherently with compressive

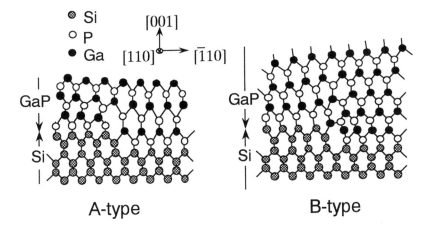

Figure 5. Atomic arrangement for GaP on Si with A-type dislocation and B-type dislocation.

stress at the initial stage. The TEM measurement for GaP on Si with various layer thickness shows that the dislocations at the interface are observed when the GaP layer thickness exceeds 90 nm.

Figure 4 shows the high resolution TEM image of the GaP/Si interface with the thickness of 3.7 μm [28]. All the dislocations have only one extra-half plane along with (111) plane. The Burgers' vector is inclined from the (001) plane. This means that all the observed dislocations are of the 60 °-type [29]. No Lomer misfit dislocation has been observed in GaP on Si. On the other hand, Lomer misfit dislocations have been observed in GaAs grown on Si substrate[30]. This contradiction might be due to the difference in the growth mode and the lattice mismatch [31]. In the case of the growth of GaAs on Si, GaAs grows three-dimensionally at the initial stage of the growth and the misfit dislocations are formed when the island size becomes larger to relax the lattice mismatch. Lomer misfit dislocations are preferably generated since the strain relaxed by the Lomer misfit dislocation is twice as that of 60 ° dislocation. In the case of GaP on Si, dislocations are not introduced at the initial stage of the growth. Hence, all the dislocations are generated above the critical thickness and in the cooling down process.

It should be worth noting that two kinds of dislocations with an extra-half plane in the Si substrate (A-type) and the GaP layer (B-type) are observed. The schematic models of atomic arrangements for these structures are shown in Figure 5[28]. In general, the dislocation generation should take place at random in the isotropic crystal [32], i.e., four kinds of A-type dislocations with Burgers' vector of $\frac{1}{2}[1\ 0\ 1]$, $\frac{1}{2}[0\ \bar{1}\ 1]$, $\frac{1}{2}[0\ 1\ 1]$ and $\frac{1}{2}[\bar{1}\ 0\ 1]$ can be generated. However, one silent feature came to be seen; instead of these four directional A-type dislocation, only

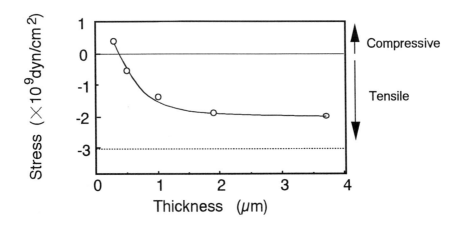

Figure 6. Stress of GaP layer on Si substrate as a function of the thickness.

two directional A-type dislocations with Burgers vector of $\frac{1}{2}[1\ 0\ 1]$ or $\frac{1}{2}[0\ \bar{1}\ 1]$ are observed. A possible interpretation for this generation of A-type dislocation with only two-kinds of Burgers' vectors in stead of the four previously reported types of dislocations is the difference of the situation of the site, i.e., dislocations are created at the step edge of the misoriented Si substrate rather than randomly isotropic generation. The direction of Burgers' vector for the B-type dislocations is the reverse of that of the A-type dislocation.

Assuming that the strain is completely accommodated at the growth temperature by $60°$ dislocations, the spacing between dislocations corresponds to 110 nm. However, the spacing between A-type dislocations is much smaller than the calculated. This difference might be due to the uniformity of the Si substrate steps. Furthermore, it is reasonable because no dislocation is observed in some micrograph, which has been taken at step-free site.

If the dislocations are generated by the lattice mismatch, extra-half plane should be in the Si substrate, which has a small lattice constant[30]. Because both the room temperature lattice constant and the thermal expansion coefficient of GaP are larger than those of Si, GaP should have the larger lattice constant than Si at the growth temperature. Hence, the lattice mismatch relaxation at the growth temperature is responsible for the generation of the A-type dislocation. The B-type dislocations cannot be explained merely on the ground of lattice mismatch.

Because dislocations are not generated at the initial stage of the growth, the B-type dislocations should have generated during the growth or cooling down process. If the lattice strain of GaP is relaxed completely at the growth temperature by introducing A-type misfit dislocations, the tensile stress is produced in the GaP layer during the cooling process, due to the difference of the thermal expansion

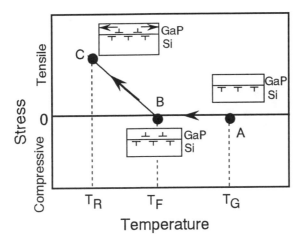

Figure 7. Schematic illustration of the stress and the type of misfit dislocation for GaP/Si heterostructure.

coefficients of GaP and Si. The thermal expansion coefficient of GaP is about 2.5 times larger than that of Si. In order to relax the tensile stress in the GaP layer, the dislocations with the extra-half plane in the GaP layer should be introduced. Accordingly, it is proved that the dislocations with the extra-half plane in the Si substrate are formed by the lattice mismatch and that those in the GaP layer are formed during the cooling process to relax the thermal stress. This experiment supports the earlier report that proved the generation of the threading dislocations in GaP on Si during the cooling down process [30].

The stress applied to the GaP layer as a function of the thickness measured by X-ray diffraction is shown in Figure 6 [34]. The dotted line shows the stress value for the thermal stress calculated by the bimetal model. A thin GaP layer has a compressive stress and the stress changes to tensile with increasing the thickness. Considering the thermal stress between the growth temperature and the room temperature, GaP layer must have compressive stress at the growth temperature when the layer thickness is thinner than the critical thickness. Since the strain energy which is caused by the misfit increases with the thickness, the misfit dislocations are generated with increasing the thickness. The change of stress in GaP on Si from compressive to tensile with the thickness is due to the increase of the misfit dislocation density at the growth temperature. The strain is smaller than the calculated thermal stress because the thermal stress is partly relaxed by the generation of misfit dislocations with the extra half plane in the GaP layer.

From the above experimental results, the dislocation and stress generation mechanisms become clear. Figure 7 shows the schematic illustration of the stress and the misfit dislocation as a function of the temperature. Although the dislocation is

438

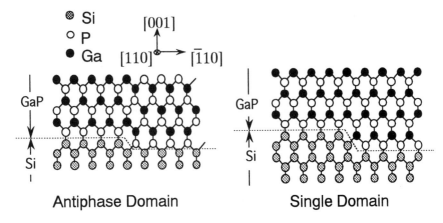

Figure 8. Antiphase domain structure and single domain structure at different step height.

not generated at the initial stage of the growth, the dislocations with the extra-half plane in the Si substrate are generated at the growth temperature (T_G) to relax the lattice mismatch (A). During the cooling down process, the thermal stress is relaxed by generating the misfit dislocation with the extra-half plane in the GaP layer (B). Below the dislocation frozen temperature (T_F: 350 - 400 °C), the thermal expansion mismatch produces the tensile stress in the epitaxial layer without generating new dislocations, resulting in the large tensile stress at room temperature (T_R, C).

5. GENERATION AND ANNIHILATION OF ANTIPHASE DOMAINS

The problem of the antiphase domain occurs at the Si surface step due to the polar/nonpolar structure. As shown in Figure 8, the antiphase domains are generated when the Si has a single (or odd) atomic step, whereas the single domains are formed when the Si surface is double (or even) atomic steps.

Two kinds of APD's are observed in GaP grown on exactly (001) Si substrate. One is the APD which is annihilated during the growth and the other is one which propagates to the surface. Figure 9 shows the typical TEM micrograph of APD which is annihilated during the growth [35]. The pictures were taken under dark field conditions using (002) (a) and (00$\bar{2}$) (b) reflections. In these figures, the contrasts of the domain region and matrix region are inverted by changing the reflection vector g from 002 to 00$\bar{2}$. In Figure 9(a), the domain contrast is bright and that of the matrix is dark. On the other hand, the opposite contrast is obtained in Figure 9(b). The amplitudes of 002 and 00$\bar{2}$ reflections have been calculated to unequal for most thicknesses in the case of zincblende structure [36]. It is concluded that these regions are APD's, based on the above-mentioned results and the same reasons described in ref.37. Therefore, it is proved that APD is annihilated during

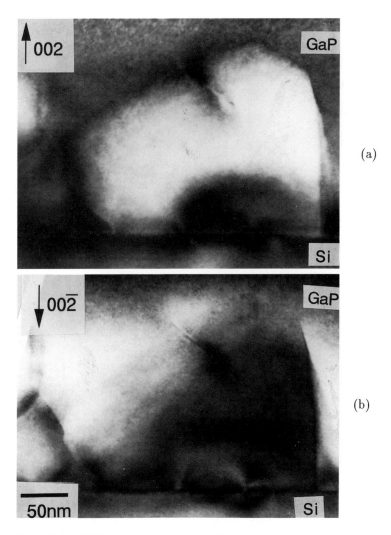

Figure 9. Dark field TEM micrographs of APD annihilated during the growth. The reflection vector is (002) (a) and (00$\bar{2}$) (b).

growth. This kind of structure is frequently observed in this sample. From the analysis of several micrographs, the size of APD is estimated to be about 0.2 μm. The antiphase boundary is normal to the (001) plane near the Si substrate. It is obvious that the APD annihilates upon changing the orientation of the boundary from the (001) normal to higher index planes so as to minimize the total energy.

Figure 10. Surface morphologies of 1 μm-thick GaP grown on 2° off, 4° off and 6° off (001) Si.

Similar domain structures have been observed for GaP grown on the 1° off Si substrate, i.e., not only the APD which propagates to the surface, but also thatwhich is self-annihilated during growth, has been observed. In the case of the 2° off (001) Si substrate, the small domain with the contrast similar to Figure 9 is observed. Therefore it is also the APD which is self-annihilated during growth. The size and the density of APD are smaller than those grown on the exactly oriented (001) Si substrate. In this sample, all the APD's are annihilated during the early stage of growth. No APD which propagates to the surface is observed. On the other hand, APD is not detected by TEM in GaP grown on 4° off and 6° off (001) Si substrates.

The mechanisms for generation and annihilation of APD's are discussed. The Si surface steps are usually composed of single atomic steps and double atomic steps. The initial growth mode of GaP on Si is two-dimensional under a high V/III ratio. In the initial stage of growth, the Si substrate is covered with P under a high PH$_3$ rate. Therefore, APD is introduced at the single atomic step position. In the case of the growth on (001) just Si, the spacing between the steps is assumed to be large compared with the misoriented substrate. Therefore, the size of the APD is large. On the other hand, the size of the APD is small in the case of the misoriented substrate.

The calculation shows that the (211) and (110) antiphase boundaries are enegetically more favorable to form than those for the (111) or (100) planes [38]. Therefore the appearance of the (110) antiphase boundary is in good agreement with the calculations. Also the total energy is increased with increasing thickness. In order to reduce the total energy, the antiphase boundary changes its orientation to the low-energy index-plane. In Figure 9, the higher-index plane of the antiphase do-

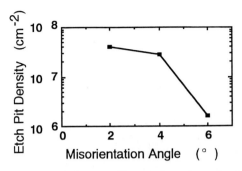

Figure 11. Etch pit density of GaP on Si as a function of misorientation angle.

main is estimated to be the (211) plane from the angle. This result also supports the calculation that the energy of the antiphase boundary for (211) is smaller than that of (110).

If the size of APD introduced in the initial stage is small, the APD annihilates in the early stage of growth. On the other hand, if the APD is large, a thick layer is necessary for all the APD to annihilate. Therefore APD remains at the surface. The TEM micrographic observations shows that the APD in GaP on 2 ° off (001) Si indicates the existence of single atomic steps. However, the density of single steps is lower than that of the exactly oriented (001) Si substrate. On the other hand, all the steps change to double atomic steps in the cases of 4° off and 6°off substrates after the annealing process. This is inferred from the fact that the APD is not detected in these samples.

Figure 10 shows the surface morphologies of 1 μm-thick GaP grown on 2° off, 4° off and 6° off (001) Si [35]. All the samples appear mirrorlike to the naked eye. However, the surface morphologies are completely different when observed under Nomarski optical microscope. The surface morphology of GaP on 2 ° off Si is wavy. On the other hand, very smooth surfaces with the cross-hatched pattern are observed for 4 ° off and 6°off Si substrates. The cross hatching is emphasized for 6°off Si compared with that for 4°off Si. The difference in the surface morphologies of 2° off, 4° off and 6°off Si substrates is related to the existence of APD at the interface. Once GaP grows on Si two-dimensionally in the first stage of growth, two-dimensional growth continues and a very smooth surface is obtained, if APD is not generated at the initial stage of the growth. The cross-hatched pattern is generated at the GaP surface due to the lattice strain of GaP on Si when the thickness increases. On the other hand, when the APD generates at the initial stage of the growth, the surface morphology becomes wavy due to the formation of a high-index plane at the GaP surface during the annihilation process of APD.

The crystal quality of GaP grown on Si substrate was evaluated by the etch pit density to see the effects of APD generated at the initial stage on the overgrown GaP layer. Figure 11 shows the etch pit density of 1 μm-thick GaP on Si as a

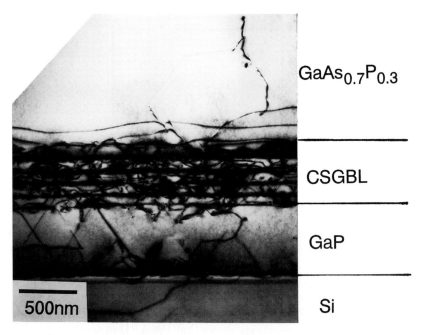

Figure 12. Cross-sectional TEM micrograph of $GaAs_{0.7}P_{0.3}$ on Si with CSGBL.

function of misorientation angle [35]. The etch pit density decreases with increasing misorientation angle, and etch pit density as low as 1.6×10^6 cm^{-2} has been obtained for GaP in the case of the 6° off Si substrate. This result indicates that a low density of dislocation is induced if the antiphase domain is not generated at the interface. In other words, the elimination of APD generation in the initial stage of growth is essential in order to obtain an atomically flat and low-dislocation-density GaP layer on Si.

6. CRYSTAL GROWTH OF HIGHLY MISMATCHED III-V COMPOUNDS ON Si SUBSTRATES

Since the low dislocation density GaP can be grown on Si substrate using the advantage of the small lattice mismatch, the next target is to grow a highly mismatched III-V semiconductors on GaP/Si with buffer layers. One method is to insert a lattice mismatched buffer layer so that the threading dislocations generated by the lattice mismatch are bent parallel to the interface by the misfit strain.

In order to reduce the dislocation density of the $GaAs_{0.7}P_{0.3}$ layers grown on Si substrate, compositionally step graded buffer layer is grown between the GaP and $GaAs_{0.7}P_{0.3}$. CSGBL is made of 5 layers ($GaAs_{0.13}P_{0.87}$, $GaAs_{0.27}P_{0.73}$, $GaAs_{0.41}P_{0.59}$, $GaAs_{0.54}P_{0.46}$ and $GaAs_{0.67}P_{0.33}$). Figure 12 shows the cross-sectional TEM micrograph of $GaAs_{0.7}P_{0.3}$ on GaP/Si with CSGBL with the individual layer thickness of 100 nm. The dislocation density of the GaP layer is low because of the small lattice mismatch. The effective bending of dislocations at CSGBL is observed although the each layer thickness is thicker than the critical thickness [18]. It indicates that the dislocations generated by the lattice mismatch are bent parallel to the interface and the dislocation density of the $GaAs_{0.7}P_{0.3}$ layer is lower than that of GaP layer. On the other hand, most of the dislocations propagate towards the surface when the individual layer thickness is 10 nm. Some dislocations propagate perpendicular to the surface and some dislocations are lying on the (111) planes. When the high strain energy is enough to effectively bend the dislocations, the dislocation-filtering effect is high, resulting in the low dislocation density.

7. SUMMARY

The mechanisms for the nucleation, the dislocation generation, the stress relaxation and the annihilation of APD in the GaP/Si heteroepitaxial growth have been reviewed. The growth mode appears to be island-type for the low values of V/III ratio and the low gas pressure. When the V/III ratio or the gas pressure was increased, the growth mode changes from island-type to layer-type for thin GaP layer thickness. The misfit dislocations were observed only when the layer thickness exceeds 90 nm. Two types of misfit dislocations which were generated by the lattice mismatch and the thermal expansion mismatch were observed. A high density of APD's which propagate to the surface and are annihilated during growth has been observed in the case of exactly (001) and 1° off Si substrates. All the APD's were self-annihilated during growth in the case of the 2° off Si substrate. No APD has been detected at the interface for 4° off and 6° off Si. The etch pit density decreases with increasing misorientation angle. GaP on Si substrate with the dislocation density as low as 1.6×10^6 cm^{-2} has been obtained without using intermediate layer. GaP and compositionally step graded buffer layers were effective to reduce the dislocation density of the highly lattice mismatched epitaxial layer on Si substrate.

The crystal growth of III-V compounds on Si substrate is important not only for the fundamental research of crystal growth but also for the device application combining the III-V technology and Si technology. The authors believe that the crystal growth technology of the heteroepitaxial growth of III-V compounds on Si substrate will proceed and new devices will be realized in the near future once the fundamental aspects of crystal growth mechanisms are understood.

ACKNOWLEDGEMENT

This work was partly supported by a Grant-in-Aid for Scientific Research on Priority Areas "Crystal Growth Mechanism in Atomic Scale" No. 03243220, 04227218, 05211212 from the Ministry of Education, Science and Culture of Japan.

REFERENCES

1. T. Soga, S. Hattori, S. Sakai, M. Takeyasu and M. Umeno, Electron. Lett. 20 (1984) 916.
2. M. Akiyama, Y. Kawarada and K. Kaminishi, Jpn. J. Appl. Phys. 23 (1984) L843.
3. W. I. Wang, Appl. Phys. Lett., 44 (1984) 1149.
4. B-Y. Tsaur and G. M. Metze, Appl. Phys. Lett., 45 (1984) 535.
5. W. T. Masselink, T. Henderson, J. Klem, R. Fisher, P. Pearah, H. Morkoc, M. Hafish, P. D. Wang and G. Y. Robinson, Appl. Phys. Lett., 45 (1984) 1309.
6. T. Egawa, H. Tada, Y. Kobayashi, T. Soga, T. Jimbo and M. Umeno, Appl. Phys. Lett. 57 (1990) 1179.
7. T. Soga, T. Kato, M. Yang, M. Umeno and T. Jimbo, J. Appl. Phys. 78 (1995) 4196.
8. T. Soga, T. George, T. Suzuki, T. Jimbo, M. Umeno and E. R.Weber, Appl. Phys. Lett. 58 (1991) 2108.
9. R. Fisher, D. Neuman, H. Zabel, H. Morkoc, C. Choi and N. Otsuka, Appl. Phys. Lett. 48 (1986) 1223.
10. Y. Watanabe, Y. Kadota, H. Okamoto, M. Seki and Y. Ohmachi, J. Crystal Growth 93 (1988) 459.
11. N. Chand, R. People, F. A. Baicocchi, K. W. Wecht and A. Y.Cho, Appl. Phys. Lett. 49 (1986) 815.
12. M. Yamaguchi, A. Yamamoto, M. Tachikawa, Y. Itoh and M. Sugo, Appl. Phys. Lett. 53 (1988) 2293.
13. M. Yamaguchi, J. Mater. Res. 6 (1991) 376.
14. K. Nozawa and Y. Horikoshi, Jpn. J. Appl. Phys. 30 (1991) L668.
15. H. Shimomura, Y. Okada and M. Kawabe, Jpn. J. Appl. Phys. 31 (1992) L628.
16. M. Volmer and A. Weber, Z. Physik. Chem. 119 (1926) 277.
17. J. A. Venable, G. D. T. Spiller and M. Hanbucken, Rep. Prog. Phys. 47 (1984) 399.
18. J. W. Matthews, Dislocations in Solids, F. R. N. Nabarro (ed.), North-Holland, Amsterdam, 1979, p. 461.
19. K. Tamamura, K. Akimoto and Y. Mori, J. Crystal Growth 94 (1988) 821.
20. T. Soga, T. Suzuki, M. Mori, T. Jimbo and M. Umeno, J. Crystal Growth 132 (1993) 134.
21. T. Soga, T. George, T. Jimbo and M. Umeno, Jpn. J. Appl. Phys. 30 (1991) 3471.
22. N. Noto, S. Nozaki, T. Egawa, T. Soga, T. Jimbo and M. Umeno, Mat. Res. Soc. Symp. Proc. 48 (1989) 247.
23. T. Soga, T. George, T. Jimbo and M. Umeno, Appl. Phys. Lett. 58 (1991) 1170.
24. O. Ueda, K. Kitahara, N. Ohtsuka, A. Hobbs and M. Ozeki, Mat. Res. Soc. Symp. Proc. 221 (1991) 393.
25. J. Nishizawa and T. Kurabayashi, J. Crystal Growth 93 (1988) 98.
26. G. B. Stringfellow, J. Crystal Growth 115 (1991) 418.
27. T. Soga, T. Jimbo and M. Umeno, Jpn. J. Appl. Phys. 32 (1993) L767.

28. T. Soga, T. Jimbo and M. Umeno, Appl. Phys. Lett. 63 (1993) 2543.

29. N. Otsuka, C. Choi, Y. Nakamura, S. Nagakura, R. Fisher, C. K. Peng and H. Morkoc, Mater. Res. Soc. Symp. Proc. 67 (1986) 85.

30. D. Gerthsen, D. K. Biegelsen, F. A. Ponce and J. C. Tramontana, J. Crystal Growth 106 (1990) 157.

31. T. Soga, T. Jimbo and M. Umeno, J. Crystal Growth 145 (1994) 358.

32. J. W. Matthews, Epitaxial Growth Part B, Academic, New York, 1975.

33. M. Tachikawa and H. Mori, Appl. Phys. Lett. 56 (1990) 2225.

34. T. Suzuki, M. Mori, Z. K. Jiang, T. Soga, T. Jimbo and M. Umeno, Jpn. J. Appl. Phys. 31 (1992) 2079.

35. T. Soga, H. Nishikawa, T. Jimbo and M. Umeno, Jpn. J. Appl. Phys. 32 (1993) 4912.

36. O. Ueda, T. Soga, T. Jimbo and M. Umeno, Defect Control in Semiconductor, K. Sumino (ed.), Elsevier Science, Amsterdam, 1990, p.1141.

37. O. Ueda, T. Soga, T. Jimbo and M. Umeno, Appl. Phys. Lett. 55 (1989) 445.

38. P. M. Petroff, J. Vac. Sci. Technol, B4 (1986) 874.

Advances in the Understanding of Crystal Growth Mechanisms
T. Nishinaga, K. Nishioka, J. Harada, A. Sasaki and H. Takei (Editors)
© 1997 Elsevier Science B.V. All rights reserved.

Atomic layer epitaxy of GaAs, AlAs and GaN using $GaCl_3$ and $AlCl_3$

F. Hasegawa

University of Tsukuba, Institute of Materials Science, Tsukuba, 305 Japan

In order to investigate possibility of carbon free atomic layer epitaxy (ALE) of III-V compound semiconductors, $GaCl_3$ and $AlCl_3$ were used for ALE growth of GaAs, AlAs and GaN. GaAs ALE growth was performed fairly well, but AlAs ALE growth was not obtained due to surface oxidation of AlAs during flow of the AsH_3 and purging hydrogen. The ALE window of GaN was very narrow probably due to slow reaction between $GaCl_3$ and NH_3, and single crystal was obtained only on the GaAs (111) surface. Surface photo-absorption analysis revealed that the surface was covered by the gallium chlorides rather than the gallium metal for this ALE growth using the trichloride sources.

1. INTRODUCTION

Atomic layer epitaxy (ALE) is an ultimate growth method which can control the growth rate by exactly one atomic layer. The feature of ALE growth is self limitation, i.e., the layer thickness per cycle is independent of the exposure time to the source gas. Although most of ALE growth of III-V compound semiconductors have been successfully performed using metalorganic sources such as trimethylgallium and dimethylaluminum-hydride [1-5], one of drawbacks of the metalorganic ALE is carbon contamination. In usual metalorganic vapor phase epitaxy (MOVPE), where the metalorganic group III source and the hydride group V source are simultaneously supplied, the carbon contamination from hydrocarbon radicals in the metalorganic can be suppressed by increasing the V/III ratio, i.e., partial pressure of the hydride such as AsH_3. The V/III ratio in the ALE growth, however, is effectively zero during the supply of the metalorganic group III sources, so the carbon contamination is inevitable. Of course, the methyl radicals can be removed in some extent by increasing the exposure time to the hydride group V source for GaAs[6]. However, since the Al-C bond is much stronger than the Ga-C and In-C bonds, the carbon contamination is much more serious for ALE growth of AlAs and AlGaAs than for GaAs and InP etc. Actually, the lattice constant of the ALE grown AlAs was slightly smaller than that of the normal AlAs [7].

ALE growth of GaAs and InP can be beautifully performed by using chloride sources such as GaCl and InC[8,9], and their ALE window such as the temperature range where the ALE growth can be obtained is much wider than that for the ALE growth using metalorganic sources [10]. Of course, there is no problem of the carbon contamination for the chloride ALE. The biggest drawback of the chloride

448

carbon contamination for the chloride ALE. The biggest drawback of the chloride ALE is that Al compounds such as AlAs and AlGaAs could not be grown as for the usual chloride VPE, because the Al metal reacts with the quartz reactor at above ∼600°C. Other problem is that it needs a high temperature source zone.

GaCl$_3$ and AlCl$_3$ are stable and have a reasonable vapor pressure at 70∼100°C. Therefore, if we can use these materials as the source of ALE growth, both problems of the chloride ALE would be solved. Actually, AlGaAs could be grown by direct reaction between GaCl$_3$, AlCl$_3$ and AsH$_3$ [11,12]. That is the reason why possibility of ALE growth using GaCl$_3$ and AlCl$_3$ were investigated in this work. GaAs and GaN could be grown reasonably well in ALE mode, but AlAs could not be grown in the ALE mode, though it was grown by simultaneous supply of AlCl$_3$ and AsH$_3$. This is probably due to oxidation of the AlAs surface during the AsH$_3$ supply and hydrogen purging. The results presented here indicates that ALE growth of AlAs and AlGaAs will be also possible using GaCl$_3$, AlCl$_3$ and AsH$_3$, if a purer AsH$_3$ and more sophisticated apparatus whose dew point is lower than ours are used.

2. EXPERIMENTAL

Figure 1 shows a schematic diagram of the growth reactor and surface photo-absorption (SPA) monitoring system. Alternate supply of GaCl$_3$ or AlCl$_3$, H$_2$ or N$_2$ purging gas and AsH$_3$ or NH$_3$ was controlled by computer controlled airvalves. The GaCl$_3$ and AlCl$_3$ were contained in stainless steel evaporators, and were conveyed to the reactor by H$_2$ or N$_2$ carrier gas. Temperatures of the evaporators were usually maintained at 60∼70°C for GaCl$_3$ and at 80∼100°C for AlCl$_3$, respectively. Gas lines including the valves between the evaporator and the reactor were heated up about 20°C higher than those of the evaporators, to prevent condensation of the GaCl$_3$ or AlCl$_3$. Partial pressures of those reactants were 1∼2×10^{-4} atm and

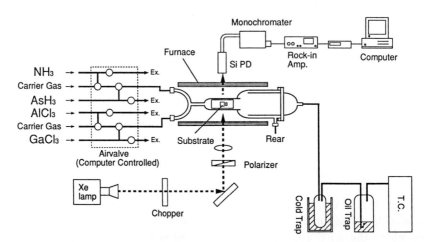

Figure 1. Schematic diagrams of the growth system and the surface photo-absorption monitoring system.

those of AsH$_3$ and NH$_3$ were $2\sim3\times10^{-3}$ atm and 1.5×10^{-1} atm, respectively. The reactor pressure was 1 atm, and the total gas flow rate was about 500 ccm. The gas velocity at the substrate position was estimated to be about 10 cm/sec.

For the SPA measurements, a Xe lamp (300W) was used as the light source. A chopped p-polarized light was irradiated to the substrate surface with an about 70° incident angle and the reflected light was detected by a Si p-n photodiode through a monochromator, as shown in Fig.1. A 1 cm^2 GaAs substrate, usually with (100) plane, was used for the SPA measurements. Small GaAs (100) chips partly covered with SiO$_2$ were put on the up-stream and down stream of the SPA measurement wafer to monitor the grown thickness. The GaAs substrate was thermally cleaned at 550°C for 10 min. under AsH$_3$ + H$_2$ ambient before the ALE growth and SPA measurements.

3. RESULTS and DISCUSSION

3.1. ALE growth of GaAs and its analysis by the SPA method [13]

The feature of ALE growth is independence of the growth rate per cycle on the exposure time to the source gas. The result for GaAs growth is shown in Fig.2. The substrate temperature was 400°C. The growth rate saturates at nearly one monolayer(ML) per cycle in a few seconds exposure to the GaCl$_3$. Reason why the growth rate is slightly less than 1 ML/cycle seem to be size effect of the GaCl$_3$, since the growth rate decreased slightly with increase of the carrier gas velocity. As will be discussed later in the SPA analysis, the GaCl$_3$ is inferred to adsorb on the GaAs surface in a chloride form, and when the gas velocity is too high, some of these chlorides might be swept away.

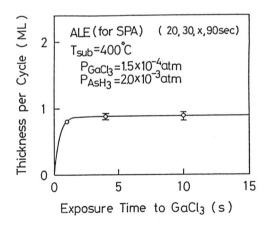

Figure 2. Growth rate of GaAs at 400°C as a function of GaCl$_3$ exposure time.

Crystal quality of the grown layer was confirmed by the Raman shift; only the LO (longitudinal optical) phonon shift was observed. The crystal quality degrades with decrease of the growth temperature as shown in Fig.3, where ratio of the

Raman shift signals, LO/(LO+TO) , here TO represents peak of the transverse optical phonon shift, is plotted against the growth temperature. ALE growth of a high quality GaAs was obtained only between 350°C and 400°C. Gas etching of the substrate occurred at higher substrate temperatures. This phenomenon was always observed when the GaCl$_3$ was used as a source [11].

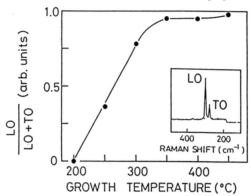

Figure 3. Dependence of the crystal quality of ALE grown GaAs estimated by the LO Raman shift peak intensity.

In order to investigate growth mechanism of the ALE using the GaCl$_3$, the SPA signal was observed. Figure 4 show typical SPA signals at 400°C with incidence azimuths of [110] (curve a) and [$\bar{1}$10] (curve b), respectively. The wavelength of monochromatized light in this experiment was 488 nm. We define the change in reflection intensity as $\Delta R/R_{As} = (R_{GaCl_3} - R_{As})/R_{As}$, where R_{GaCl_3} and R_{As} denote the reflection intensities before and after the AsH$_3$ supply, respectively. During an alternate supply of GaCl$_3$ and AsH$_3$, absolute values of $\Delta R/R_{As}$ were approximately 5.5% and 1.0% for incidence azimuths of [110] and [$\bar{1}$10], respectively. However, the reflection intensity was $R_{As} > R_{GaCl_3}$ for both incidence azimuths.

For the [110] incidence azimuth, the reflection intensity decreases rapidly when GaCl$_3$ was supplied, and keeps constant during the H$_2$ purge. On the other hand, an increase and saturation of reflection intensity can be seen during the H$_2$ purge for the [$\bar{1}$10] incidence azimuth after rapid decrease by exposure to GaCl$_3$.

The spectral dependencies of SPA signal for the incidence azimuths of [110] and [$\bar{1}$10] are shown in Fig.5(a) and 5(b), respectively. The growth temperature was 400°C. The results for triethylgallium (TEGa)-AsH$_3$ system reported by Kobayashi and Horikoshi [14] are also shown by broken lines for comparison. Our result for a [110] incidence azimuth is similar to the one for TEGa-AsH$_3$ system, but the spectrum for a [$\bar{1}$10] incidence azimuth is quite different from theirs. The GaAs surface after TEGa supply is assumed to be covered by Ga atoms since the growth temperature is 560°C, which is high enough to decompose TEGa to Ga atom completely. Since our SPA spectrum for a [$\bar{1}$10] incidence azimuth is different from theirs, the GaAs surface after GaCl$_3$ supply and H$_2$ purge must be covered by some different species from Ga atoms, such as gallium chloride or its complexes.

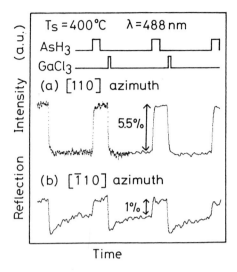

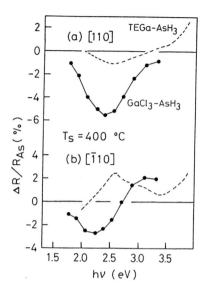

Figure 4. Typical SPA signals at 400°C with (a) [110] and (b) [$\bar{1}$10] incidence azimuths. The partial pressures of GaCl$_3$ and AsH$_3$ were 5.9×10^{-4} atm and 2.1×10^{-3} atm, respectively. A gas flow sequence is AsH$_3$ for 30s, H$_2$ for 30s, GaCl$_3$ for 10s and H$_2$ for 160s.

Figure 5. Spectral dependence of the SPA signal with (a)[110] and (b) [$\bar{1}$10] incidence azimuths at 400°C. A gas flow sequence for measurement is AsH$_3$ for 350s, H$_2$ for 30s, GaCl$_3$ for 20s and H$_2$ for 160s. The results for TEGa-AsH$_3$ system at 560°C are also shown by broken lines for comparison.

To clarify the above described result furthermore, the reflection intensity after GaCl$_3$ supply and H$_2$ purge was compared to that of Ga atom surface. The Ga atom surface (R_{Ga}) was formed by the desorption of arsenic (As) atoms from GaAs surface, which is the same method reported by N.Kobayashi and Y.Kobayashi [15]. The results with [110] and [$\bar{1}$10] incidence azimuths are shown in Fig.6 curve(a) and curve(b), respectively. The substrate temperature was 440°C and the gas flow sequence is shown in the figure. The reflection intensities for [110] and [$\bar{1}$10] azimuths change due to the desorption of As atoms. After the formation of As-stabilized surface (R_{As}) again by AsH$_3$ supply and H$_2$ purge, GaCl$_3$ was supplied. The reflection intensity decreased for both incidence azimuths and reached the saturation level during the H$_2$ purge. This level is different from that for Ga atom surface (R_{Ga}). Therefore, the surface after the GaCl$_3$ supply and H$_2$ purge is not considered to be covered by Ga atoms. A possible adsorbed species are gallium chlorides. In other words, the surface after the GaCl$_3$ supply and H$_2$ purge might be a Cl-terminated Ga surface as far as the ALE growth using GaCl$_3$ is concerned. The desorption of Cl atoms probably occurs when AsH$_3$ is supplied.

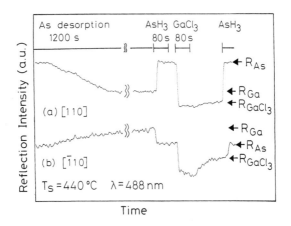

Figure 6. A change in the reflection intensity with (a) [110] and (b) [$\bar{1}$10] incidence azimuths during an arsenic desorption from GaAs surface and GaCl$_3$ supply on the As-stabilized surface. A gas flow sequence is H$_2$ purge for 1200s, AsH$_3$ for 80s, H$_2$ for 30s, GaCl$_3$ for 80s and H$_2$ for 160s.

3.2. Optical investigation of AlAs growth with alternate or simultaneous supply of AlCl$_3$ and AsH$_3$ [16]

As the first step, AlCl$_3$ and AsH$_3$ was alternately supplied in the same way as the GaAs ALE growth and the SPA signal was detected. The substrate temperature was 400°C and wave length of the detected light was 488 nm. A similar SPA signal to the case of GaAs ALE growth using GaCl$_3$ and AsH$_3$ was observed at the beginning of the growth but the signal was attenuated quickly and was extinguished in about 10 periods of times.

In order to study whether this result is due to a too low substrate temperature or other phenomenon, AlCl$_3$ and AsH$_3$ were supplied simultaneously by changing the substrate temperature or sequence of the gas supply. Taking advantage of the SPA system, reflected light was monitored during the growth. The reflected light changed greatly and oscillated during the supply of AlCl$_3$ and AsH$_3$, probably due to interference of the incident light. According to a simple analysis, inverse of the period of oscillation should be proportional to the growth rate. Therefore, temperature dependence of the growth rate can be estimated from the change of the period in one run by changing the substrate temperature.

This optical monitoring could be also used to observe change of the growth situation for different reactant gas supplies or growth conditions. Figure 7 shows change of the optical interference when the AlCl$_3$ supply was interrupted for 5 min.. The oscillation could not be obtained when the AlCl$_3$ supply was started again, indicating that a normal growth could not be obtained after the interruption of AlCl$_3$ supply. On the other hand, when supply of AsH$_3$ was interrupted and the substrate temperature was lowered to 540°C, the oscillation indicating the normal growth was observed when the AsH$_3$ supply was begun again at the substrate temperature of 580°C as shown in Fig.8. These facts suggest that when the AlCl$_3$ supply is stopped and the AlAs surface is exposed to AsH$_3$/H$_2$, the AlAs surface is oxidized and further normal growth of AlAs becomes difficult. On the other hand, when the AsH$_3$ supply is stopped, the AlAs surface is not oxidized and is kept clean as far as the AlCl$_3$/H$_2$ supply is continued. It probably means that the AlCl$_3$ has elimination effect of residual H$_2$O or/and oxygen in the carrier gas and the AlAs surface is kept clean without oxidation.

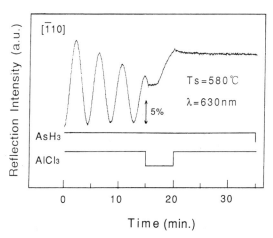

Figure 7. Change of the reflection intensity oscillation due to the optical interference by interruption of the AlCl$_3$ supply.

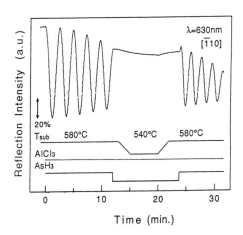

Figure 8. Reflection intensity change due to the optical interference by interruption of the AsH$_3$ supply with the AlCl$_3$ supply continued and by lowering the substrate temperature to 540°C

Since AlAs layers could be grown by simultaneous supply and the surface was kept clean by flowing AlCl$_3$, GaAs ALE growth on the AlAs layer was tried at growth temperature of 350°C to see exchange of AlCl$_3$ and GaCl$_3$ on the surface. Change of the surface was monitored by the reflected light of 488 nm. The results are shown in Fig.9. After observing the SPA signal of GaAs ALE growth, an AlAs layer of about 1000Å(1.5 cycle of interference oscillation) was grown at 580°C. Maintaining the AlCl$_3$ supply, the substrate temperature was lowered to 350°C and GaCl$_3$ was added. The reflected light intensity was increased, indicating that species covering the AlAs surface was changed. The signal did not change very much when the AlCl$_3$ supply was stopped. The signal changed a little when the GaCl$_3$ was replaced by AlCl$_3$ and vice versa, but its magnitude was much less than the one observed when the GaCl$_3$ was added to AlCl$_3$, suggesting that the surface was covered by GaCl or GaCl and GaCl$_3$ complexes. Actually, a normal SPA signal of GaAs ALE growth was observed by alternate supply of GaCl$_3$ and AsH$_3$ as seen in Fig.9, though it became slightly noisy on the AlAs layer.

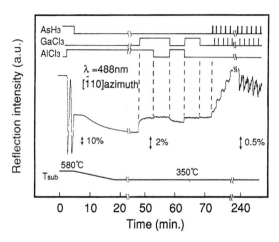

Figure 9. Reflection intensity change probably due to an exchange between AlCl$_3$ and gallium chloride on the AlAs surface.

3.3. ALE growth of GaN on GaAs substrates [17]

When the GaN layers were grown on (001) and (111) GaAs surfaces at 500°C, typical SPA signals indicating layer-by-layer growth were observed. When GaN layers were grown on (001) and (111) substrates, the absolute values of $\Delta R/R_N$ were approximately 0.7% for incidence azimuths [$\bar{1}$10] and [11$\bar{2}$], respectively. The reflection intensity was $R_N > R_{GaCl_3}$ in the same way as the case of GaAs ALE and remained constant during the N$_2$ purge. These results indicate that the change in the reflection intensity was caused by a change in the type of adsorbed molecules. Layer-by-layer growth was observed at substrate temperatures between 450°C and 500°C. Neither the growth nor the SPA signal could be observed at 400°C, probably because NH$_3$ of group V source did not react with Ga well due to too low temperature. Gas etching, rather than the growth, was observed at 550°C. This phenomenon was also observed in our previous studies on GaAs growth from GaCl$_3$

and AsH$_3$.[11]

Figure 10 shows the dependence of the growth rate of GaN on the exposure time to GaCl$_3$, in comparison with that for GaAs ALE growth. When GaAs was grown by ALE at 400°C, the growth rate saturated at the GaCl$_3$ supply of a few seconds as shown in Fig.2, but the growth rate of GaN did not necessarily show a distinct saturation point. Mixing of GaCl$_3$ and NH$_3$ was suspected due to incomplete purging by N$_2$. Several layers were grown with different purging times to investigate this possibility. It was confirmed that there was no dependence of the growth rate on the N$_2$ purging time. These results indicate that the growth rate is limited by the surface reaction of GaCl$_3$ with the N surface of GaN or NH$_3$ adsorbed on the GaN surface.

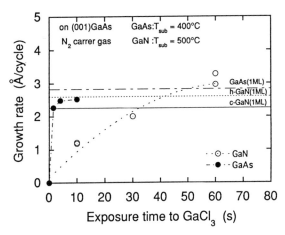

Figure 10. Dependence of the growth rate of GaN on the GaCl$_3$ exposure time compared with that of GaAs, (o) GaN, (•) GaAs.

In order to clarify the difference between the growth of GaN and that of GaAs, the SPA signals for GaN and GaAs ALE growth were compared. The results are shown in Fig.2. The incident azimuth is [110] in this case. When GaAs was grown at 350°C, the reflection intensity changed instantaneously when the GaCl$_3$ was supplied, and remained constant during the GaCl$_3$ supply. On the other hand, when GaN was grown at 500°C, the change in the reflection intensity was very slow and did not saturate during 60 s of GaCl$_3$ supply. This is probably due to the much slower reaction between GaCl$_3$ and N or NH$_3$ than that between GaCl$_3$ and As or AsH$_3$.

Figures 12 (a, a') and 12(b, b') show the XRD and RHEED patterns of GaN layers grown on (001) and (111) GaAs substrates, respectively. When a GaN layer was grown on (001) GaAs, the XRD peak intensity of GaN was very weak. In this case, the RHEED pattern is a halo, suggesting that the grown layer was almost amorphous. On the other hand, when a GaN layer was grown on (111) GaAs, the (0002) diffraction peak of hexagonal GaN was observed at around 34°. In this case the RHEED pattern is that of hexagonal GaN for the [01$\bar{1}$0] azimuth.

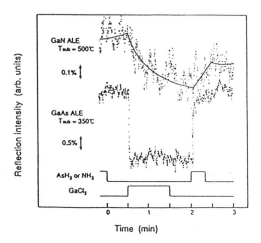

Figure 11. Comparison of the SPA signals for GaAs and GaN growth.

Furthermore, the symmetry of the pattern indicates that the GaN layer is a single crystal and has wurtzite structure. These results indicate that layer-by-layer growth of hexagonal GaN can be performed on (111) GaAs surfaces by the chloride growth method.

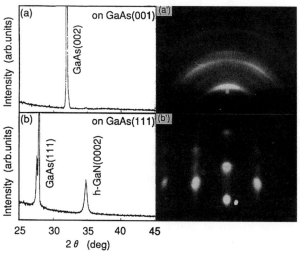

Figure 12. XRD and RHEED patterns of GaN grown (a, a') on (001) GaAs substrate and (b, b') on (111) GaAs substrate.

4. CONCLUSIONS

Possibility and mechanism of atomic layer epitaxy (ALE) of III-V compound semiconductors using $GaCl_3$ and $AlCl_3$ were investigated, since these sources are inherently carbon free. GaAs ALE growth was performed fairly well, but AlAs ALE growth was not obtained due to surface oxidation of AlAs during flow of the AsH_3

and purging hydrogen. A layer by layer growth of GaN was observed but its ALE window was very narrow probably due to slow reaction between $GaCl_3$ and NH_3. Single crystalline GaN could be obtained only on GaAs (111) substrates. Surface photo-absorption analysis revealed that the surface was covered by the gallium chlorides rather than the gallium metal for this ALE growth using the trichloride sources.

Acknowledgments

This work was performed by many graduate students, most of them are now working at industries. The author would like to express his sincere thanks to Dr. R. Kobayashi (now at NEC), Messrs. K. Ishikawa (Hitachi), S. Narahara (Kobe Seiko), M. Akamatsu (Fuji Xerox) and H. Tsuchiya.

REFERENCES

1. J.Nishizawa, H.Abe and T.Kurabayashi, J. Electrochem. Soc. 132 (1985) 51.
2. S.M.BeDair, M.A.Tishler, T.Katsuyama and N.A.El-Masry, Appl. Phys. Lett. 47 (1985) 51.
3. A.Doi, Y.Aoyagi and S.Namba, Appl. Phys. Lett. 48 (1986) 1787.
4. M.Ozeki, K.Mochizuki,N.Ohtsuka and K.Kodama, J. Vac. Sci. & Technol. B5 (1986) 1184.
5. M.Ishizaki, N.Kano, J.Yoshino and H.Kukumoto, Jpn. J. Appl. Phys. 30 (1991) L428.
6. K.Mochuzuki, M.Ozeki, K.Kodama and N.Ohtsuka, J. Crystal Growth, 93 (1988) 557.
7. N.Kano, M.Ishizaki, M.Deura, J.Yoshino and H.Kukumoto, The 10th Technical Meeting of Atomic Order Processing, No.151, Group of Japan Society for Promotion of Science (1992) p.50.
8. A.Usui and H.Sunagawa, Jpn. J. Appl. Phys. 25 (1986) L212.
9. A.Koukitu, H.Nakai, A.Saegusa, T.Suzuki, O.Nomura and H.Seki, Jpn. J. Appl. Phys. 27 (1988) L744.
10. H.Watanabe and A.Usui, Proc. Int. Symp. Gallium Arsenide and Related Comounds, Las Vegas, 1986, Inst. Phys. Conf. Ser. No.83, (1987) p.1.
11. F.Hasegawa, H.Yamaguchi and K.Katayama, Jpn. J. Appl. Phys. 27 (1988) L1546.
12. H.Yamaguchi, R.Kobayashi, Y.Jin and F.Hasegawa, Jpn. J. Appl. Phys. 28 (1989) L4.
13. R.Kobayashi, S.Narahara, K.Ishikawa and F.Hasegawa, Jpn. J. Appl. Phys. 32 (1993) L164.
14. N.Kobayashi and Y.Horikoshi, Jpn. J. Appl. Phys. 29 (1990)L702.
15. N.Kobayashi and Y. Kobayashi, Jpn. J. Appl. Phys. 30 (1991)L1699.
16. M.Akamatsu, S.Narahara, T.Kobayashi and F.Hasegawa, Appl. Surface Science 82/83 (1994) 228.
17. H.Tuchiya, M.Akamatsu, M.Ishida and F.Hasegawa, Jpn. J. Appl. Phys. 35 (1996) L748.

Surface diffusion processes of Ga and Al in MBE
- formation of 10-nm scale GaAs ridge structures -

T. Noda[a] [b], S. Koshiba[a] [c], Y. Nakamura[a,c] , Y. Nagamune[a,] *, and H. Sakaki[a,b,c]

[a]Research Center for Advanced Science and Technology, University of Tokyo,
4-6-1 Komaba, Meguro-ku, Tokyo 153, Japan

[b]Institute of Industrial Science, University of Tokyo,
7-22-1 Roppongi, Minato-ku, Tokyo 106, Japan

[c]Quantum Transition Project, JRDC,
Park Bldg. 4F 4-7-6 Komaba, Meguro-ku, Tokyo 153, Japan

In this study we investigate diffusion processes of Ga and Al on (001)-(111)B facet structures on patterned substrates in molecular beam epitaxy (MBE) and the optimization of growth for the 10-nm scale GaAs ridge structure. The shape of the facet structure is analyzed to quantitatively evaluate the surface diffusion of atoms between two planes. We have found that the growth rates on the (111)B and (001) planes depend strongly on the As flux and are mainly determined by the inter-plane migration of Ga. As the growth proceeds, the facet structure develops into a ridge structure. The shrinkage of the ridge width is found to continue until the ridge width reaches a critical value, which depends strongly on substrate temperature, T_s, As flux, and alloy composition. This width becomes as small as 10 nm for AlAs ridges for a wide range of growth conditions. It also holds true for GaAs ridges, when grown at temperatures below 580°C with high As flux. The formation of ridge structures by the selective MBE in the presence of atomic hydrogen is also described.

1. INTRODUCTION

Epitaxy of III-V compound semiconductors on patterned substrates by molecular beam epitaxy (MBE) has attracted considerable interest due to potential application to the fabrication of quantum wires and other novel structures [1-7], also provides information on the microscopic growth mechanism [8-11].

Many papers on the MBE growth of GaAs on a patterned (001) substrate with mesa stripes along the $< 110 >$ orientation have described the formation of a (001)-(111)B facet structure, as shown in Figure 1(a). This facet structure results from the difference between the local growth rates on (001) and (111)B planes,

---

*Permanent address: Electrotechnical Laboratory,
1-1-4 Umezono, Tsukuba-shi, Ibaraki 305, Japan

which are related with the intra- and inter-plane diffusion of Ga and Al and their incorporation processes. As the growth proceeds, the width of the top (001) plane is reduced and finally a ridge structure is formed. This shrinkage of the ridge width is expected to stop at some critical value, W_{crit}, once the incorporation rate of Ga on the very narrow (001) plane is reduced.

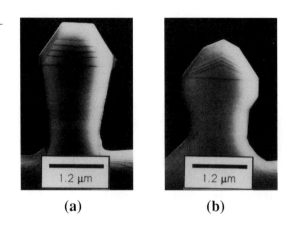

(a)   **(b)**

Figure 1. Cross-sectional SEM images of samples simultaneously formed on the mesa stripes along the $< 110 >$ (a) and $< 1\bar{1}0 >$ (b) directions. Note (001)-(111)B facet planes in (a) and (001)-(11n) (n=1 or 3) facet planes in (b). The average fluxes of Ga and As on the (001) plane were 1.6 and 3.1×10^{14} molecules/cm²s, respectively.

In contrast, the structure grown on the mesa stripe along the $< 1\bar{1}0 >$ direction consists of (111)A and (113)A facets, as shown in Figure 1(b). This suggests that the mechanisms of facet formation are quite different for the two cases. In this work, we describe our study on the inter-plane diffusion of atoms on the (001)-(111)B facet structure and also the mechanism which determines the ridge width, W_r.

2. SAMPLE PREPARATION

For our experiment a (001) GaAs substrate was first patterned by photolithography and reactive-ion etching using $SiCl_4$ [4]. The mesa stripes were prepared along the $< 110 >$ direction with a width of $1 \sim 10~\mu m$ and a depth of about $2 \sim 3~\mu m$. The growth of GaAs was performed at substrate temperatures, T_s, between 520°C and 620°C to form the (001)-(111)B facet structure. The growth rate of GaAs on a flat (001) surface was 0.25 $\mu m/hr$ and that of AlAs was 0.10 $\mu m/hr$. In most experiments, the As_4/Ga flux ratio was fixed to be 2 and the As flux, F_{As}, was 3.1 $\times 10^{14}$ molecules/cm²s when measured on the (001) plane.

In order to follow the time evolution of the growth front of GaAs, a thin AlAs layer was inserted as a marker after the deposition of a GaAs layer of the specified thickness. The substrate holder was rotated during this sequence so as to supply the fluxes uniformly over the sample. This lead to the formation of a symmetric (001)-(111)B facet structure on the mesa stripe.

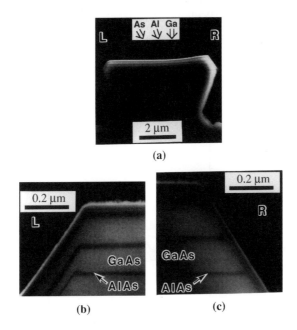

Figure 2. Cross-sectional SEM images of a (001)-(111)B facet structure. Samples were grown on the 6 μm-wide mesa with the beam fluxes coming in as shown in (a). The magnified SEM images of the top left (marked by **L**) and right (**R**) corners are shown in (b) and (c), respectively. In these images, bright regions are GaAs and dark regions are AlAs.

We grew additional structures on top of this facet structure to measure the growth rate on each plane and to evaluate the diffusion length of atoms. In this case, the holder was fixed to a particular angle to control the incident fluxes of Ga, Al and As on the (111)B and (001) planes independently. The typical geometry of materials is shown in Figure 2. High resolution scanning electron microscope (SEM) was employed to investigate cross sections of samples. All images were taken without stain etching after cleavage.

3. SURFACE DIFFUSION OF Ga AND GROWTH MECHANISMS OF (001)-(111)B FACET STRUCTURES

3.1. Growth rates of GaAs on (001) and (111)B facet planes

GaAs/AlAs facet structures were grown on $6\mu m$-wide mesa stripes at 580°C with the fluxes specified in section 2. Figure 2(a) shows a cross-sectional SEM image of the sample and also the direction of incoming fluxes. Magnified images of two top corners are shown in Figures 2(b) and 2(c). One can clearly see that the growth rate of GaAs on the (111)B plane, $R_{(111)B}$, is much lower than that on the (001) plane. Note also that the growth rate on the (001) plane, $R_{(001)}$, is enhanced near both corners. This tendency is particularly strong on the right corner of the facet, where the incoming As flux is low. We studied how the growth rate ratio, $R^* = R_{(111)B}/R_{(001)}$ changes with incoming fluxes. Here, we varied the fluxes on the (111)B plane while fluxes on (001) planes were kept constant.

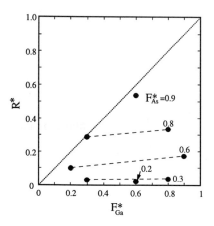

Figure 3. The ratio R^* of the GaAs growth rate on the (111)B plane to that on the (100) plane as functions of normalized Ga flux, F_{Ga}^*. The dotted line indicates the growth rate without any inter-plane migration.

Figure 3 shows the relation of R^* with the normalized Ga flux, F_{Ga}^* which is defined as $F_{Ga(111)B}/F_{Ga(001)}$, where $F_{Ga(111)B}$ and $F_{Ga(001)}$ are the Ga fluxes on the (111)B and (001) planes, respectively. If R^* is simply determined by the ratio of their incident fluxes, the rate should follow the dotted line in Figure 3. However, all experimental data lie below the dotted line, irrespective of the normalized As flux, F_{As}^*, ranging from 0.2 ~ 0.9. This result indicates that the growth rate on the (111)B plane is always lower than that on the (001) plane. Note that R^* depends less on F_{Ga}^*, but sensitively on F_{As}^*. This finding demonstrates that the growth rate on the (111)B plane can be easily reduced if $F_{As(111)B}$ is set to below 1/5 of $F_{As(001)}$.

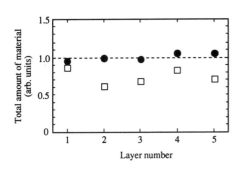

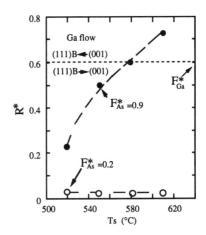

Figure 4. The total amount of GaAs incorporated during a fixed period in the facet structure growth of Figure 1. Closed circles and open squares are for those grown on mesa along the < 110 > and < 1$\bar{1}$0 > directions, respectively. When no Ga is lost from the (001)-(111)B facets, the data should fall on the dotted line.

Figure 5. The growth rate ratio R^* as functions of substrate temperature, T_s, for the two As fluxes. Here the normalized Ga flux F_{Ga}^* is 0.6 for both cases. The ratio determined only by the Ga flux ratio with negligible flow of Ga between the (001)-(111)B planes is indicated by the dotted line.

3.2. Origins of reduced growth rate of GaAs on the (111)B plane

The following three processes could be responsible for the reduction of the growth rate on the (111)B plane: (i) the re-evaporation of Ga from the (111)B plane, (ii) the downward flow of Ga from the (111)B plane to the underlying (110) plane, and (iii) the upward flow of Ga from the (111)B plane to the top (001) plane. To estimate the contribution of processes (i) and (ii), we measure the cross-sectional areas of the GaAs/AlAs multi-layered structure of Figure 1(a) and evaluate how many Ga atoms are lost via processes (i) and (ii) during the growth. Closed circles in Figure 4 shows the total amount of GaAs incorporated during each period of the growth. It is clear that the total amount of GaAs incorporated on the (111)B and (001) planes per unit deposition time is conserved and that this amount is equal to that estimated from the growth rate on a flat (001) surface. Open squares in Figure 4 show the corresponding data for the facet structure of Figure 1(b) grown on the mesa running along the < 1$\bar{1}$0 > direction. Note that an appreciable amount of Ga deposited is lost. The loss is even more noticeable when the (113)A facet appears on the mesa. Hence, one can conclude that the reduction in the growth rate on the (111)B surface is predominantly caused by process (iii), the diffusion of Ga from the (111)B to the (001) plane. It is noteworthy that Ga atoms diffusing downward from the (111)B to the side (110) plane are negligible, i.e., Ga atoms are effectively reflected.

3.3. Flow of Ga atoms between (001) and (111)B planes

As pointed out in Figure 3, the net flow of Ga atoms between the (001) and (111)B planes depends on the normalized As flux F^*_{As} [4]. Similarly, it may depend on the substrate temperature T_s. We have studied how R^* varies with T_s, when F^*_{Ga} is kept constant at 0.6. The solid circles in Figure 5 display the data for F^*_{As} = 0.9, while the open circles for F^*_{As} = 0.2. One can see in the case of low F^*_{As} that R^* is very low and thus Ga flows from the (111)B to the (001) plane, irrespective of T_s. In the case of high F^*_{As}, however, R^* increases rapidly with increase of T_s and eventually exceeds the dotted line above which Ga flows from the top (001) to the (111)B plane. This indicates that Ga flows from the (111)B to the (001) plane at low T_s but its direction is reversed at $T_s > 580°C$. This suggests that the incorporation rate of Ga decreases with T_s much faster on the (001) plane than on the (111)B plane, probably due to the T_s-dependent population of As atoms on the growth front. This point will be discussed later.

The flow of Ga is also affected by the absolute value of the As flux. Next, we studied the net flow of Ga for two different As fluxes while the ratio of the As fluxes on two facet planes was kept constant at 0.9. We have found from cross-sectional SEM images that Ga flows from the (111)B to the (001) plane at higher As flux (2.8~3.1 x 10^{14} molecules/cm^{2}s on the (001) plane), whereas Ga flows from the (001) plane to the (111)B plane at lower As flux(1.8~2.0 x 10^{14} molecules/cm^{2}s on the (001) plane) [15]. Unless the boundary barrier between the (001) and the (111)B planes is affected, this change of the net flow is caused by the difference in the growth condition dependences of the incorporation rate of the two planes. That is, as the As flux is reduced, the incorporation rate on the (001) plane is reduced more rapidly than that on the (111)B plane. Hayakawa et al. [12] and Shen et al. [13] also observed this reversal of the Ga flow. The former pointed out that the formation of an As-trimer on the (111)B plane enhances the migration of Ga under high As flux [12].

4. SURFACE DIFFUSION LENGTH OF GaAs AND AlAs ON A (001)-(111)B FACET STRUCTURE

4.1. Diffusion length on the (001) plane

As shown in Figure 2, a cliff-like structure at the edge of the (001) plane is formed due to the flow of Ga atoms from the (111)B to the (001) plane. As the position on the (001) plane moves away from the edge, the thickness of GaAs and AlAs decreases exponentially. Hence, one can determine the diffusion length, λ, of Ga on the (001) plane along the $< 1\bar{1}0 >$ direction. The T_s dependence of λ is plotted by closed circles in Figure 6. Note that λ ($\lambda_{(001)}$) of Ga on the (001) GaAs plane is about 1 μm when T_s = 580°C and increases exponentially with $-1/T_s$, which gives an activation energy of 0.8 eV.

The same cliff-like structure was formed for the AlAs. It is found that λ of Al on the (001) AlAs plane is about 20 nm at 580°C, which is about 1/40 of that of Ga, indicating that the flow of Al from the (111)B to the (001) plane is quite small. Similarly, an activation energy of about 1.0 eV is obtained for Al, as plotted by open circles in Figure 6. Hata et al. studied the flow of Ga atoms between the (111)B and (001) planes from the micro-RHEED observation [11] and found that $\lambda_{(001)}$ of Ga is $1 \sim 8 \ \mu m$. This diffusion length is somewhat longer than ours, probably because of the low growth rate used in their experiment ($= 0.1 \ \mu m/hr$).

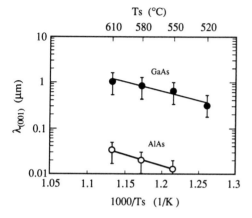

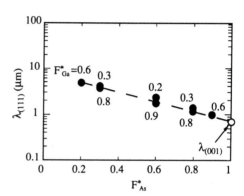

Figure 6. The diffusion lengths of Ga and Al on a (001) surface as functions of reciprocal substrate temperature. The activation energies are 0.8 eV for Ga and 1.0 eV for Al.

Figure 7. The diffusion length $\lambda_{(111)B}$ of Ga (closed circles) on the (111)B plane at the substrate temperature of 580°C vs normalized As flux.

4.2. Diffusion length on the (111)B plane

The thickness of GaAs on the (111)B plane is almost uniform over the whole plane, as shown in Figure 2, suggesting that λ ($\lambda_{(111)B}$) of Ga on the (111)B plane is either much longer or much shorter than the length of the (111)B plane. In the former case, the thickness variation on the (111)B plane is efficiently averaged out over the (111)B plane by the diffusion process. In the latter case, most Ga atoms deposited are incorporated at the initial site and the thickness of the grown layer is expected to be determined by the incident Ga flux on the (111)B plane, which is not consistent with our experimental results. Therefore, we conclude that $\lambda_{(111)B}$ of Ga is considerably larger than the length of the (111)B facet ($\sim 1 \ \mu m$). In order to further examine this point, we apply a simple model of the diffusion through the (001)-(111)B interface. As described in ref[15], one can show that R^* at the boundary between two planes is expressed by the following relation.

$$R^* = R_{(111)B}/R_{(001)} = (a_{(111)B}/a_{(001)}) \cdot (T_{(001)}/T_{(111)B})(\lambda_{(001)}/\lambda_{(111)B})^2 \quad (1)$$

466

Here, a is interatomic distance and T the transmission probability of adatoms through the boundary for each plane. Equation (1) indicates that R^* at the boundary depends neither on F_{Ga}^* nor on the width of the facet.

By assuming that $T_{(001)} = T_{(111)B}$ and using $\lambda_{(001)}$ ($= 0.84\ \mu m$) of Ga at $580°C$, $\lambda_{(111)B}$ of Ga is evaluated. The values of $\lambda_{(111)B}$ as a function of the normalized As flux are plotted by closed circles in Figure 7, which shows that the diffusion length decreases with an increase of the As flux on the (111)B plane but is larger than the width of the (111)B facet.

5. FORMATION OF 10-nm SCALE RIDGE STRUCTURES

As mentioned before, the formation of a sharp ridge structure results from the migration of Ga atoms from the (111)B to (001) plane, as illustrated in Figure 8. However, it is found that the sharpening of the ridge or the shrinkage of the ridge width, W_r, stops at a critical value, which depends on T_s.

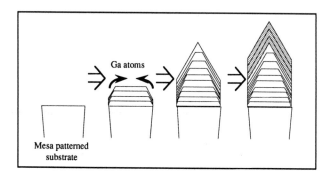

Figure 8. Schematic illustration of the ridge formation.

To understand this mechanism, the stability of the ridge structure was studied by examining the cross section of a stack of six GaAs/AlAs layers grown at various temperatures between 537°C and 617°C. The result is shown in Figure 9. The growth conditions are the same as described in section 2. To identify the GaAs ridge shape, 10-nm AlAs layers were inserted. The GaAs buffer (marked by #1) underlying the GaAs/AlAs stack structure was grown at 557°C and has a slightly rounded shape. When an additional 100-nm GaAs layer was grown at the same temperature, W_r is reduced further, becoming as narrow as 20 nm as shown by layer #2 in Figure 9.

The ridge structure grown at higher temperatures tends to widen and W_r exceeds 80 nm at $T_s = 617°C$, as shown by layers #4 and #5. By further depositing GaAs at 537°C (#6), the ridge shape becomes quite sharp again. These experimental findings indicate that the ridge shape can be well controlled by T_s. It is also found that the use of high As pressure is effective in reducing W_r.

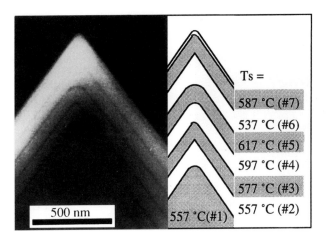

Figure 9. SEM micrograph of the ridge cross section consisting of a stack of GaAs/AlAs layers grown at various temperatures.

The cross-sectional transmission electron microscope (TEM) image of the ridge consisting of a stack of GaAs/AlAs/GaAs layers grown at 570°C is shown in Figure 10. The bright regions in the figure are AlAs layers. One can see from this micrograph that the width of the GaAs ridge is 14 ± 1 nm and that of the AlAs ridge is 10 ± 1 nm. This result suggests that the AlAs ridge becomes sharper than the GaAs ridge probably due to the quicker incorporation rate of Al which gives a shorter diffusion length and a smaller interval of steps [15].

The T_s dependence of W_{crit} is plotted in Figure 11. It is found that W_{crit} increases with the increase of T_s, exceeding 30 nm if $T_s > 590$°C. This rapid increase of W_{crit} above 600°C may be caused by different temperature dependences of growth processes on two planes, which suppress the preferential incorporation of Ga on the (100) plane. Thus, to fabricate the ridge QWI structures, T_s should be lowered below 580°C during the growth of the GaAs quantum well. We speculate that W_{crit} is likely to be correlated with the size of islands on a flat (001) surface at a given T_s. We have also found that the critical width of GaAs ridges decreases at high As pressure and becomes less than 10 nm.

The mechanism preventing the ridge width from shrinking below W_{crit} may be related to the thermodynamical stability of the nanostructure. The migration of Ga from the (111)B to the top (001) plane and preferential incorporation on the (001) plane force the ridge structure to be sharper. On the other hand, the dissociation of Ga atoms from the ridge region and their interfacet migration or desorption lead to a thermodynamical instability of the ridge and result in the widening of ridge structures. Hence the ridge width is determined by the balance existing between these two processes.

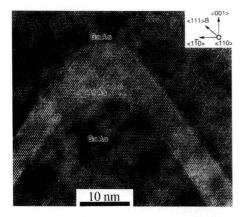

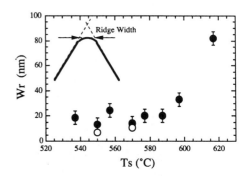

Figure 10. TEM micrograph of GaAs/AlAs/GaAs heterostructures formed at the top of the epitaxially grown ridge structure.

Figure 11. The ridge width, W_r, measured as a function of the substrate temperature. Note that the width of the AlAs ridge shown by open circles could be even smaller, if the growth continued.

6. SURFACE DIFFUSION IN SELECTIVE MBE GROWTH USING ATOMIC HYDROGEN AND FINITE PATTERN EFFECT

MBE growth on a substrate with SiN_x mask patterns in the presence of atomic hydrogen enables one to grow GaAs selectively in the window regions [15–17]. This is because atomic hydrogen enables Ga atoms impinging on SiN_x masks to re-evaporate, leading to selective area deposition in uncovered regions. This method has attracted broad interest because of possible application of forming quantum wires and other laterally-defined structures. The application of this method to form quantum wire structures was reported by Sugaya, et al. [16] and authors [17]. Kawabe et al. has shown that GaAs deposited in the line windows eventually grows in the form of ridge [15]. This indicates that the flow of Ga atoms from (111)B to top (001) plane is unaffected by hydrogen ambient. We have applied this method to form very sharp ridges and have studied the diffusion processes particularly for the ridges of finite-length.

Figure 12 shows a cross-sectional SEM image of the top part of the ridge structure consisting of a stack of GaAs/AlAs multi-layers formed via hydrogen-assisted growth at different substrate temperatures T_s. The dark regions of the image are 7-nm thick AlAs layers which serve as markers and the light regions are GaAs layers. For this growth, 1.6-μm wide line windows were first patterned by e-beam lithography into a SiN_x film on a (001) GaAs substrate so that the line runs along the < 110 > direction. During the MBE growth, thermally cracked atomic hydrogen of 0.90 ccm was supplied. Growth rates of GaAs and AlAs are the same as those given in section 2 and the flux ratio of As_4 to Ga is 5 ~ 6. One can see that the 10 nm scale ridge structure is successfully formed at $T_s < 580°C$, indicating that the atomic hydrogen neither disturb the sharp ridge formation nor widen the critical ridge width.

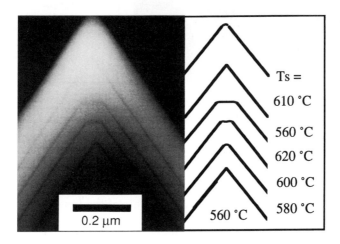

Figure 12. Cross-sectional SEM image and the illustration of a ridge structure grown on a SiN$_x$ patterned (001) GaAs substrate. Dark regions are 7-nm thick AlAs layers which serve as markers and light regions are GaAs layers.

When the length of stripe patterns is finite, the net flow of material along the length direction becomes important. We studied such processes by examining the ridge shape grown on short stripes (less than 6 μm) connected with wide window regions for T_s in the range of 600 ~ 620°C.

Figure 13 shows an SEM image seen from the top in the vicinity of a short stripe window, which is 1.6 μm in width and 6 μm in length. One can see that the width of the top (001) plane exceeds 300 nm, while it is as narrow as 30 nm in the case of long stripes. This indicates that the shrinkage process of the top (001) plane is partially suppressed in the case of short stripes [17]. This can be explained by the surface diffusion of Ga along the length direction, which is negligible for long stripes.

From SEM observations, it is found that the top (001) width at the central portion of the stripe increases exponentially as the stripe length is reduced. This suggests that, in addition to the inter-plane diffusion between the (001) and (111)B planes, the diffusion along the length direction, longitudinal diffusion, plays an important role in determining the ridge shape [17]. It is also found that the amount of material incorporated on a ridge is more than locally supplied, suggesting that Ga flows into the short stripe region from adjacent wide window regions. One can

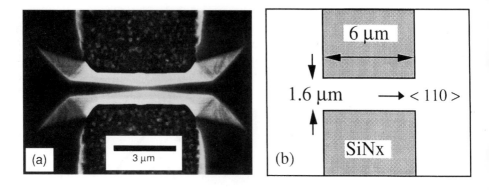

Figure 13. Top SEM image of a short stripe (a) and an illustration of its pattern (b).

expect that Ga migrates via the (111)B plane towards the stripe center from the end regions of stripes, since the shrinkage rate of the top (001) width with respect to the stripe length is 3 μm, which is close to λ on the (111)B plane, when the stripe width is fiexed at 1.6 μm. This situation is illustrated in Figure 14.

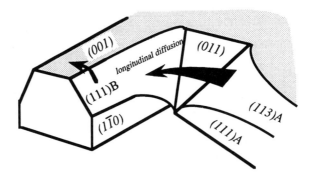

Figure 14. Schematic illustration of dominant diffusion processes in the case of short stripe patterns.

7. CONCLUSIONS

We have obtained the following results from cross-sectional SEM observations on the (001)-(111)B facet structures grown by MBE. (1) The growth rates on the (111) and (001) planes are mainly determined by the interplanar flow of adatoms between two planes, which strongly depends on the growth temperature as well as the As flux. (2) The flow of Ga from the (111)B plane to the underlying (1$\bar{1}$0) plane is negligible. (3) The diffusion length, λ, of Ga on a (001) surface is about 1 μm at 580°C, while λ of Al is about 0.02 μm. The activation energy is 0.8 eV for Ga diffusion and 1.0 eV for Al, respectively. (4) λ of Ga on (111)B planes is 1 $\sim$ 10 μm at 580°C, depending strongly on As flux. (5) The width of the ridges decreases, untill it reaches a critical value, which depends on the substrate temperature and As flux. (6) The shrinkage process of the top (001) plane is suppressed by the presence of both ends of stripe patterns when a stripe pattern is shorter than $\sim$ 6μm. Based on these findings, we have successfully fabricated a ridge structure with a lateral dimension of less than 20 nm. These microscopic observations elucidating the growth mechanism of MBE will serve as a basis for the fabrication of well-defined nanostructures, e.g., quantum wires and quantum boxes.

REFERENCES

1. H. Sakaki, Jpn. J. Appl. Phys. **19** (1980) L735.
2. Y. Arakawa and H. Sakaki, Appl. Phys. Lett. **40** (1982) 939.
3. L.Pfeiffer, K.W. West, H.L. Stormer, J.P. Eisenstein, K.W. Baldwin, D. Gershoni, and J. Spector, Appl. Phys. Lett. **56** (1990) 1697.
4. Y. Nakamura, S. Koshiba, M. Tsuchiya, H. Kano, and H. Sakaki, Appl. Phys. Lett. **59** (1991) 700.
5. S. Koshiba, H. Noge, H. Akiyama, Y. Nakamura, A. Shimizu, and H. Sakaki, Institute of Physic Conference Series 129 (IOP, Bristol, 1993), p.931.
6. S. Koshiba, H. Noge, H. Akiyama, T. Inoshita, Y. Nakamura, A. Shimizu, Y. Nagamune, M. Tsuchiya, H. Sakaki, and K. Wada, Appl. Phys. Lett. **64** (1994) 363
7. S. Koshiba, H. Noge, H. Ichinose, H. Akiyama, Y. Nakamura, T. Inoshita, T. Someya, K. Wada, A. Shimizu, and H. Sakaki, Solid State Electron. **37** (1994) 729.
8. M. Kawabe, and T. Sugaya, Jpn. J. Appl. Phys. **28** (1989) L1077.
9. Y. Horikoshi, M. Kawashima, and H. Yamaguchi, Jpn. J. Appl. Phys. **25** (1986) L868.
10. S. Nagata and T. Tanaka, J. Appl. Phys. **48** (1977) 940.
11. S. Koshiba, Y. Nakamura, M. Tsuchiya, H. Noge, H. Kano,Y. Nagamune, T. Noda and H. Sakaki, J. Appl. Phys. **76** (1994) 4138.
12. M. Hata, T. Isu, A. Watanabe and Y. Katayama, J. Vac. Sci. Technol. B8 (1990) 692.
13. T. Hayakawa, M. Morishima, and S. Chen, Appl. Phys. Lett. **59** (1991) 3321.
14. X. Q. Shen and T. Nishinaga, Jpn. J. Appl. Phys. **32** (1993) L1117.
15. T. Sugaya, Y. Okada, and M. Kawabe, Jpn. J. Appl. Phys. **31** (1992) L713.
16. T. Sugaya, M. Kaneko, Y. Okada, and M. Kawabe, Jpn. J. Appl. Phys. **32** (1993) L1834.
17. T. Noda, Y. Nagamune, S. Koshiba, and H. Sakaki, 9th Int. Conf. on MBE, Malibu, USA (1996).

Advances in the Understanding of Crystal Growth Mechanisms
T. Nishinaga, K. Nishioka, J. Harada, A. Sasaki and H. Takei (Editors)

Heteroepitaxy of cadmium telluride on sapphire

M. Kasuga, D. Kodama, H. Hagiwara, K. Kagami, K. Yano and A. Shimizu

Department of Electrical Engineering and Computer Science, Yamanashi
University, 4 Takeda, Kofu 400, Japan

The strain relaxation mechanism in heteroepitaxial layers of CdTe on sapphire
was investigated for layers grown by vapor phase transport (VPT) and by hot wall
epitaxy (HWE). X-ray diffraction study showed that in VPT layers, pseudomorphic relaxation prevails where tilt in epilayers was observed and the dispersion
of lattice constants was small. In HWE, however, non-pseudomorphic relaxation
prevails where no tilt was observed and the dispersion was somewhat large. These
advantages of VPT in crystal quality are due to *in situ* treatment and the chemical
reactions involved. The change in the spacing of lattice planes in HWE layers is
explained by the difference in expansion coefficients, the effect of defects and the
bimetal effect.

1. INTRODUCTION

Cadmium telluride (CdTe) has a wide variety of applications such as for substrates for IR detecting HgCdTe, X-ray and γ-ray detectors, and photovoltaic cells,
due to the possibility of fabricating both p- and n-type crystals. Since the processing of wafers from CdTe bulk crystals is difficult due to the high melting point,
high vapor pressure and complicated procedures, epitaxial growth of CdTe films
has been desired.

In some cases, single-crystal films of CdTe have been reported to grow epitaxially
on sapphire substrates in spite of the different crystal systems, and mismatch in
lattice constants and in thermal expansion coefficients. Regarding the mechanism
of lattice matching there are many points to be clarified. One of them is related
to the our finding that there is a tilt of 0.1 to 0.6° between the film plane of CdTe
(111) and sapphire (0001) when CdTe is deposited by the vapor phase transport
(VPT) method on sapphire [1]. Ebe et al. also found tilt of 0.3 to 1.3° for the
same combination of epitaxy in a MOCVD system [2]. By hot wall epitaxy (HWE),
however, CdTe (111) grew on sapphire (0001) with no tilt in our study as well as
in that of Tatsuoka et al. [3].

In this work, CdTe films were grown on sapphire substrates by VPT and HWE
and the tilt angle, α, of the film and the off-angle, β, of the substrate surface were
measured. Then the correlation between the tilt and off-angle was discussed, as
well as their effects on crystal quality. The variation of the lattice constant of CdTe
perpendicular to the interface was also observed and discussed.

2. HETEROEPITAXY OF CADMIUM TELLURIDE ON SAPPHIRE BY VAPOR PHASE TRANSPORT

2.1. Growth method and results

The CdTe layer was grown by the open tube method where CdTe powder of 5N purity was used as the source material [1]. The source CdTe in the high-temperature (about 700°C) zone is transported by atmospheric hydrogen carrier gas and deposited on sapphire substrates in the low-temperature (about 350°C) zone with a growth rate of 0.02-0.3 μm/h. The *in situ* heat treatment in hydrogen at 1000°C before growth proved to be indispensable for obtaining single-crystal films. After the heat treatment, the substrate zone was heated to a given point and the source container made of quartz was moved to this point to start the transport reaction.

The substrate used was the sapphire c face. Although we did not intentionally use any off-angle for the substrate surface, we found that the alignment where sapphire (0001) diffraction is maximum differs from that where CdTe (111) diffraction is maximum. This means that there exists a tilt, which we call α, between the substrate and the grown layer. This tilt angle, α, and the half-width at half-maximum (HWHM: $\Delta 2\theta / 2$), which is a measure of crystal quality of the grown layer, are plotted in Figure 1 as a function of layer thickness of each sample. Although α fluctuates it remains independent of thickness. On the other hand, HWHM has a tendency to decrease as the layer becomes thicker. This suggests that the lattice constant recovers to some uniform state.

Ebe et al. chose dimethylcadmium and dimethyltellurium as source materials for growing CdTe with a IV/II ratio of 4.5-5.5 at 400°C. In their experiment, a sapphire (0001) substrate with an offset angle of 1-5 degrees toward ($1\bar{1}02$) was used with hydrogen treatment at 1100°C just before the deposition. Figure 2 shows the relationship between offset angle β and tilt α taken from their results. Although α varies widely, β increases with α.

2.2. Tilt angle α vs offset angle β of substrate

In the VPT growth of CdTe on a sapphire substrate, a single-crystal film could be obtained in spite of large mismatch of the lattice constants and the thermal expansion coefficients. In this case, tilt angle of about half a degree was observed between the substrate basal plane and CdTe (111) plane. We can demonstrate the cause of this tilt α using the model depicted schematically in Figure 3. Here we assume the crystal grown to be an ensemble of elastic cubic cells whose heights are stretched from h to c for periodic distance L along the interface where h and c are lattice spacings of the sapphire substrate and the CdTe layer, respectively. When we define the tilt between lattice planes of the layer grown and sapphire substrate to be α, and the tilt between the substrate surface, i.e., the interface, and substrate lattice plane to be β, we get

$$\tan \alpha = \frac{c - h}{L} \tag{1}$$

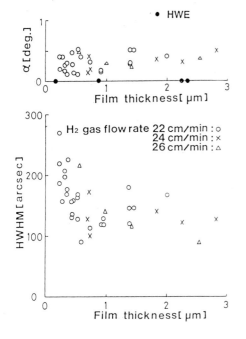

Figure 1. CdTe(111) layers grown by VPT: (a) tilt and (b) HWHM measured by X-ray diffraction

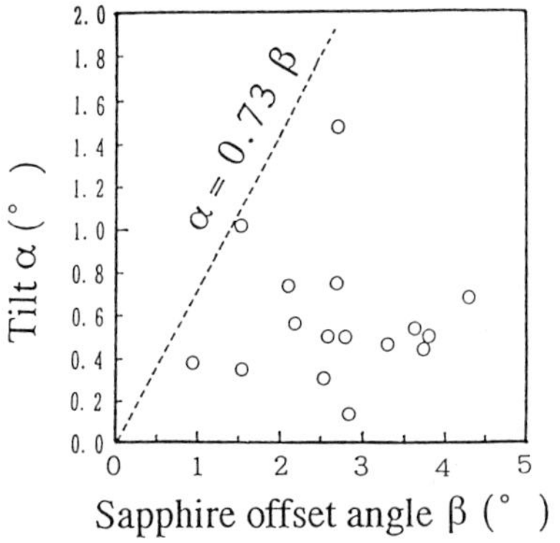

Figure 2. Tilt of CdTe(111) plane in the CdTe layers grown by MOVPE[2]

$$\tan \beta = \frac{h}{L}. \tag{2}$$

We can obtain, by small angle approximation,

$$\alpha = \frac{c - h}{h} \beta. \tag{3}$$

Substituting $h = 2.16$ and $c = 3.74$ for the sapphire basal plane and the CdTe (111) plane, respectively, we get

$$\alpha = 0.73\beta. \tag{4}$$

This equation shows that tilt angle is proportional to the off-angle and the difference in lattice spacing $(h - c)$, and suggests that no tilt results if the off-angle of the substrate is zero.

This model was originally proposed by Nagai for the GaAs/Ga$_x$In$_{1-x}$As system, where he assumed that the "horizontal" mismatch is completely adjusted and the stress is released only "vertically" [4]. Therefore, strictly speaking, we should take

476

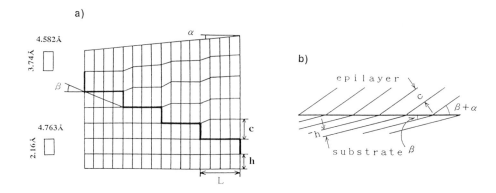

Figure 3. (a) Nagai model for the origin of the tilt and (b) the interfacial coalescence model

into account the horizontal deletion or compression of the two-dimensional square lattice. Eq.(3) is therefore a rough estimation. This type of elastic deformation is referred to as pseudomorphic [5].

There is another expression of the model, as shown in Figure 3(b), which we call the "interfacial coalescence model" [3,5]. This model assumes the coalescence of both lattice planes at the interface in epitaxial growth. As is easily shown, Figure 3(b) is geometrically identical to (a) if we incline the interface by angle β.

Although we have not intentionally employed an off-set-angled substrate, our CdTe single-crystal layers seem to obey this model because they have finite tilt which is substantially independent of film thickness and have HWHM which recovers to small value with increasing thickness. Ebe et al. also explained their tilt data (Figure 2) on the basis of this model. The solution to Eq. (4) is shown in Figure 2 as the broken line.

The above discussion is based on an extreme assumption that all the interfacial stress caused by lattice mismatch can be absorbed by the elastic deformation of cells. In general, however, this assumption does not hold strictly due to the introduction of point defects and/or dislocations, and leads to $\alpha \leq 0.73\beta$, as is shown in Figure 2. In other words, α/β can be regarded as a quality factor which reflects the perfection of the crystals grown.

2.3. Tilt and orientation of layers determined by X ray diffraction

To systematically investgate the effect of the off-angled substrate on the quality of epitaxial layers, it is necessary to measure precisely not only α and β, but also corresponding azimuthal angles. For consistency with the model in Figure 3 (a)

Table 1

Tilt and azimuthal angle of CdTe/sapphire determined by X-ray diffraction

Sample No.	$\alpha(°)$	$\beta(°)$	$\xi_c(°)$	$\xi_s(°)$	intensity(kcps)	2HMHW(°)	α/β
♯ 1	0.109	0.232	98	-121	4.0	0.06	0.47
♯ 2	0.133	0.202	67	-124	2.5	0.05	0.658
♯ 3	0.088	0.417	134	-71	0.2	0.156	0.212
♯ 4	0.092	0.493	-136	49	0.5	0.108	0.195

we take the basal plane of the sapphire substrate as reference plane, the tilt and azimuth of the CdTe layer as α and ξ_c, and those of the sapphire surface as β and ξ_s, respectively.

We have devised a method of measuring these tilts and azimuths by X-ray diffraction, details of which are shown in the appendix. Examples of measured data obtained by this procedure are shown in Table 1. Although the α values do not scatter very much from sample to sample, we can distinguish a group with large α/β ratio, ♯1 and ♯2 and a group with small α/β ratio, ♯3 and ♯4. The former shows high diffraction intensity with small FWHM suggesting higher grade of crystallinity. The latter shows, in contrast, lower crystallinity and thus the ratio α/β is a quality factor of the layer. The reference line for ξ_c and ξ_s was taken to be the longer edge of the rectangular specimen. The angular differences of azimuths for maximum inclination, ξ_c - ξ_s, prove to be roughly 180 degrees, which indicates that net planes of the grown layer inclines to the opposite side of the tilt of the substrate surface, as schematically shown in Figure 3 (a).

Furthermore, the model shown above suggests that the substrate surface steps, especially "one sided steps", plays an essential role for the pseudomorphic growth in heteroepitaxy.

3. HETEROEPITAXY OF CADMIUM TELLURIDE ON SAPPHIRE BY HOT WALL METHOD

3.1. Growth procedure and results

Hot wall epitaxy of CdTe film was carried out in the reactor shown in Figure 4 where CdTe powder of 5N purity was used as the source material and (0001) sapphire was used as the substrate. During the growth of CdTe, the substrate, source and hot wall temperatures were maintained at 300°C, 360 − 410°C and 370 − 420°C, respectively. The sapphire substrate was chemically etched in the mixture of H_3PO_4 + H_2SO_4 at 300°C, but hydrogen *in situ* treatment was not possible in this apparatus. The whole reactor was evacuated by a cryopump system to the base pressure of 2×10^{-6} Torr. In twelve hours, CdTe films 0.1 to 50μm thick were obtained.

Grown layers were examined by X-ray diffraction and proved to have CdTe (111) // sapphire (0001) orientation. In HWE, unlike in VPT, no tilt was observed, as is shown in Figure 1. The X-ray back-Laue pattern of the films exhibited spots of

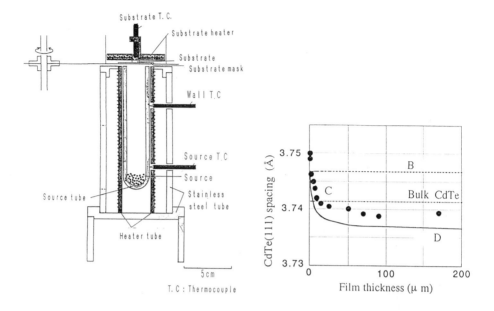

Figure 4. HWE system for CdTe/sapphire epitaxy

Figure 5. Lattice spacing of CdTe(111) planes vs film thickness

single, or sometimes twinned, crystals. Considering the high electrical resistivity of about 6×10^6 Ωcm and low electron mobility of about 10 cm^2/V.s these films seem to be composed of columnar crystals with many grain boundaries.

Tatsuoka et al. also reported no tilt in HWE where they employed a substrate with off-angle of 2.7° from (0001) sapphire.

X-ray diffraction of CdTe films with various thicknesses showed that the spacing of the (111) plane changes with film thickness, as is shown in Figure 5. Lattice spacing of films less than 5μm thick is larger than that of bulk crystal which is shown in the figure by the dotted line. The spacing decreases as film thickness increases and becomes less than the bulk value above the thickness of 10μm.

3.2. Characterization of strain

In order to understand the complicated dependence of the lattice spacing of CdTe (111) on layer thickness, it is necessary to consider the following four effects which are also schematically shown in Figure 6.

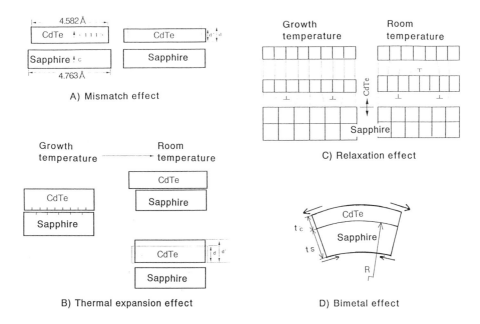

Figure 6. Four effects of strain

(A) Mismatch in lattice constants

The lattice constant in the interface plane of CdTe ($a_c = 4.582$Å at room temperature) is less than that of sapphire ($a_s = 4.763$Å at room temperature). Therefore, if coherent film growth took place, the CdTe lattice would stretch in the direction parallel to the interface, resulting in CdTe (111) spacing smaller than the bulk spacing (see Figure 6A). Assuming that the volume of the CdTe layer remains constant under stretching, the lattice constant after deformation, d', would be given by

$$d' a_s^2 = d a_c^2, \tag{5}$$

where d denotes the lattice spacing of bulk CdTe. Eq.(5) gives $d = 3.466$Å, which is far smaller than the measured value even if the Poisson ratio of CdTe, 0.31, and the high growth temperature are taken into account. Thus we conclude that coherent growth does not occur, i.e. the growth mode of the CdTe layer is non-pseudomorphic.

We can speculate why no tilt is observed in CdTe/ sapphire HWE on the basis of non-pseudomorphism. The growth mode would not be explained by the elastic theory due to the large number of defects and poor crystallinity. Comparison of HWE with VPT suggests that the inferior crystal quality is due to the lack of hydrogen-related reactions before and/or during the growth process, that is, in the surface cleaning procedure and/or any chemical reactions involved.

(B) Difference in thermal expansion coefficient

The thermal expansion coefficient of CdTe, $\alpha_c = 4.5 \times 10^{-6} \text{deg}^{-1}$ is smaller than that of the sapphire basal plane, $\alpha_s = 6.6 \times 10^{-6} \text{deg}^{-1}$. Therefore, the CdTe layer expanded perpendicular to the substrate surface due to lateral compressive stress during cooling from growth temperature (see Figure 6B). In this case, corresponding to Eq.(5), following relation holds:

$$a' = a_c(1 + \alpha_c \Delta T) \times (1 - \alpha_s \Delta T) \tag{6}$$

$$da_c^2 = d'a'^2, \tag{7}$$

where ΔT is the difference between growth temperature and room temperature. Here we also assumed invariance in the volume of the CdTe layer and non-pseudomorphism; in other words, interfacial defects which disrupt coherent growth are introduced. This effect results in increased CdTe(111) spacing, as is shown by the dashed line B in Figure 5.

(C) Relaxation effect

As the CdTe layer grows thicker, the part of CdTe far from the interface tends to relax due to the introduction of dislocations, thus avoiding excess compression and recovering the proper (bulk CdTe) lattice constant in the cooling process. This effect is shown in Figure 6(C) and around C in Figure 5. However, this mechanism is not sufficient to explain the reduction of the lattice spacing to less than bulk value.

(D) Bimetal effect

After the CdTe film grows even thicker, the CdTe layer tends to shrink, due not to the thermal expansion coefficient of sapphire, but to its own thermal expansion coefficient in its cooling process. Thus the CdTe layer and sapphire bend, just as does a bimetal system. The curvature of the bending pulls the outermost CdTe layer laterally reducing the spacing of CdTe (111) to less than the bulk value, as shown in Figure 6(D). The one-dimensional beam model gives the radius of curvature, R, and the lateral deformation, ε, on the basis of balancing the force and moment at the interface [6].

$$R = \frac{t_s + t_c}{2\Delta\alpha\Delta T} + \frac{E_s t_s^3 + E_c t_c^3}{6\Delta\alpha\Delta T(t_s + t_c)} \frac{E_s t_s + E_c t_c}{E_s t_s E_c t_c} \tag{8}$$

$$\varepsilon = \frac{t_s + t_c}{2R} \tag{9}$$

$$da_c^2 = d'\{a_c(1 + \varepsilon)\}^2 \tag{10}$$

Here E denotes Young's modulus and t denotes thickness. Suffixes c and s stand for CdTe and sapphire, respectively. $\Delta\alpha$ denotes the difference between α_s and α_c.

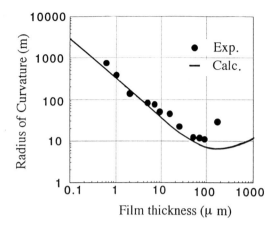

Figure 7. Radius of curvature vs film thickness of CdTe

When the film becomes thicker than 10μm the measured (111) spacing approaches the value specified by Eqs. (8), (9) and (10) as is shown in region D in Figure 5.

In order to determine the radius of curvature of the crystals grown, deviation of the X-ray diffraction angle was measured as a function of displacement of the specimen. The radius of curvature thus estimated is shown in Figure 7 as a function of film thickness, and good agreement of the measured and calculated values shows that the bimetal effect actually occurs.

4. CONCLUSION

In the heteroepitaxy of CdTe/sapphire, where a large difference in lattice constants exists, pseudomorphic strain relaxation takes place in VPT and non-pseudomorphic strain relaxation takes place in HWE. In the former case (VPT), the lattice plane of the grown layer tilts relative to that of the substrate and the crystal quality is good. In the latter case (HWE), the grown layer has no tilt and the crystal quality is poor. This difference is assumed to be due to *in situ* pretreatment of the substrate and/or hydrogen-related chemical reactions in the former case.

The observed variation in lattice spacing with layer thickness in HWE can be understood quantitatively in terms of three effects, i.e., difference in thermal expansion coefficients, relaxation with defects, and the bimetal effect. Coherent growth does not occur.

Heteroepitaxy with an off-angled substrate has been reported for several other combinations such as ZnSe/Ge and GaAs/Si [5]. To investigate the heteroepitaxy

482

mechanism for these cases, careful experiments using the measurement procedure introduced in the appendix should be useful.

This study was supported by a Grant-in-Aid for Scientific Research on Priority Areas, "Crystal Growth Mechanism in Atomic Scale", No.03243102, from the Ministry of Education, Science and Culture, Japan.

REFERENCES

1. M.Kasuga, H.Futami and Y.Iba, J.Cryst.Growth 115 (1991) 711
2. H.Ebe, A. Sawada, K.Maruyama, Y.Nishijima, K. Shinohara and H.Takigawa, J.Cryst. Growth 115 (1991) 718
3. H.Tatsuoka, H.Kuwabara, Y.Nakanishi and H.Fujiyasu, Thin Solid Films 213 (1992) 1
4. H.Nagai, J.Appl. Phys. 45 (1974) 3789
5. J.E.Ayers, S.K.Ghandi and L.J.Schowalter, J.Crystal Growth 113 (1991) 430
6. Z.Feng and H.Liu, J. Appl. Phys. 54 (1983) 83

APPENDIX: MEASUREMENT METHOD OF TILT BY X-RAY DIFFRACTION

The lattice plane of the substrate is defined as the x-y plane. The lattice plane of the grown layer is assumed to incline by α toward the direction of azimuth, ξ (see Figure A1). The unit vector of the intersection of the inclined plane and the x-z plane is defined as $\mathbf{t}_x$, and that of the intersection of the inclined plane and the x-y plane is defined as $\mathbf{t}_y$. θ_x and θ_y are angles between $\mathbf{t}_x$ and the x axis, and between $\mathbf{t}_y$ and the y axis, respectively. The unit vector normal to the inclined plane is defined as

$$\mathbf{n} = (-\sin\alpha\cos\xi, -\sin\alpha\sin\xi, \cos\alpha) \tag{11}$$

$$\mathbf{t}_x = (\cos\theta_x, 0, \sin\theta_x) \tag{12}$$

$$\mathbf{t}_y = (0, \cos\theta_y, \sin\theta_y) \tag{13}$$

. $\mathbf{n}$ is perpendicular to $\mathbf{t}_x$ and $\mathbf{t}_y$:

$$\mathbf{n} \cdot \mathbf{t}_x = -\sin\alpha\cos\xi\cos\theta_x + \cos\alpha\sin\theta_x = 0 \tag{14}$$

$$\mathbf{n} \cdot \mathbf{t}_y = -\sin\alpha\sin\xi\cos\theta_y + \cos\alpha\sin\theta_y = 0. \tag{15}$$

From these two equations, we obtain

$$\tan^2\alpha = \tan^2\theta_x + \tan^2\theta_y \tag{16}$$

$$\tan\xi = \frac{\tan\theta_y}{\tan\theta_x}. \tag{17}$$

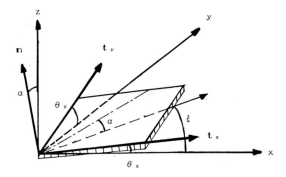

Figure A1. Tilt and azimuth of the plane inclined against xy plane

Therefore, α and ξ are calculated directly from θ_x and θ_y.

The measurements of θ_x and θ_y are described as follows (see Figure A2).

a) Fix the x and y axes on the surface of sapphire (i.e., these coordinate axes always move with the surface).

b) Set the positive direction of the x axis, i.e., $\theta = 0$, along the incident X-ray.

c) Move the Geiger counter in the direction, $2\theta_B$, of the X-ray reflected by the sapphire lattice plane.

d) Rotate the sample by θ around the y axis to make the surface coincide with the direction of reflection, θ_B, for the sapphire lattice plane. The x and y axes in this case are set in space and are called x_o and y_o axes, respectively(Figure A2-a).

e) Rotate θ around the y axis to obtain maximum reflection.

f) Rotate the sample around the x axis to obtain maximum reflection. This angle can be ignored.

g) When the surface and the sapphire lattice plane are inclined, the x-axis does not coincide with the x_o axis. Therefore θ is usually different from θ_B and we get $\theta_x = \theta_B - \theta$ (Figure A2-b).

h) Rotate the sample around z-axis by 90°. This is an in-plane rotation.

i) Return to b), where the negative y direction coincides with the incident X-ray. Here, $\theta = 0$ (Figure A2-c).

j) Repeat c) to f) with x axis and y axis exchanged until $\theta_x = \theta_B - \theta$ is obtained.

k) Here we have angles θ_x and θ_y of the sapphire lattice plane against the surface.

l) Fix x' and y' on the sapphire lattice plane as in step a). x' and y' axes move with the lattice plane in this case.

m) Move Geiger counter to the position of X-ray reflection by CdTe lattice plane $2\theta_B$.

n) Repeating d) through j), the crossing angles between CdTe lattice plane and sapphire lattice plane, θ_x and θ_y, are obtained.

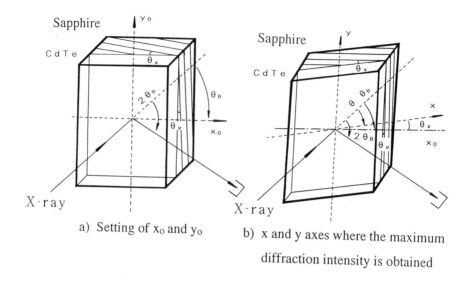

a) Setting of x_o and y_o

b) x and y axes where the maximum diffraction intensity is obtained

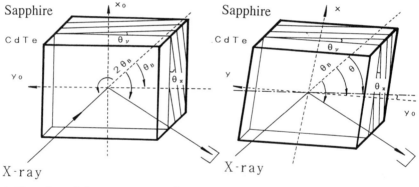

c) Rotation of the sample by 90° in $x_o y_o$ plane

d) x and y axes where the maximum diffraction intensity is obtained again

Figure A2. Alignment for the measurement of θ_x and θ_y by X-ray diffraction

Advances in the Understanding of Crystal Growth Mechanisms
T. Nishinaga, K. Nishioka, J. Harada, A. Sasaki and H. Takei (Editors)
485

Formation of strain-free heteroepitaxial structures by annealing under ultrahigh pressure

H. Ishiwara and T. Hoshino

Precision and Intelligence Laboratory, Tokyo Institute of Technology,
4259 Nagatsuta, Midoriku, Yokohama 226, Japan

Theoretical and experimental discussions concerning a novel method to solve the thermal mismatch problem in heteroepitaxial growth have been reviewed. It has theoretically been predicted for structures such as Ge/Si and GaAs/Si that the effect of the different thermal expansion coefficients can be compensated by that of the elastic strain generated by hydrostatic pressure. The theoretical prediction has experimentally been confirmed using two kinds of samples; one sample is an amorphous Ge film deposited on a Si(100) substrate in which the Ge film is expected to grow epitaxially on the Si in solid phase upon annealing under ultrahigh pressure (UHP) up to 2.2 GPa, and the other sample is a GaAs film epitaxially grown on a Si substrate in which rearrangement of the interfacial atomic bonding is expected to be induced by annealing under UHP.

It has been found that the residual strains in Ge and GaAs films linearly decrease with increase of pressure and the strain in GaAs films becomes zero at a pressure around 1.9 GPa. It has also been found that the strain depends weakly on the annealing temperature ranging from 300°C to 500°C. When characterized using Rutherford backscattering spectrometry, no degradation of the crystalline quality of the UHP-annealed films is apparent.

1. INTRODUCTION

Recently, heteroepitaxial growth using dissimilar films and substrates has become important for the fabrication of novel electronic and photonic devices. In these heterostructures, however, the crystalline quality of the films is often imperfect, because the film and substrate have different physical properties, such as lattice constants, thermal expansion coefficients, and surface energy. In particular, mismatch of the thermal expansion coefficient has a strong influence upon fabrication of minority carrier devices, since it results in residual defects such as dislocations, even near the film surface. It is known that the lifetime of laser diodes fabricated using a GaAs-on-Si structure is as short as several hours, due to thermal mismatch defects present in the GaAs film.

In this paper, we first review the novel annealing method as a solution to the thermal mismatch problem in heteroepitaxy, which is based on the knowledge that soft materials have a relatively large thermal expansion coefficient. Then, we discuss experimental results which demonstrate that the residual strains in Ge [1, 2] and GaAs films on Si substrates can be linearly decreased by increasing the pressure during annealing and that the strain in GaAs films becomes zero at a pressure around 1.9 GPa [3-5].

2. THEORETICAL CONSIDERATION

The basic concept underlying this method is that soft materials have a relatively large thermal expansion coefficient, α. That is, it is expected in a heteroepitaxial structure that the effect of the different thermal expansion coefficients of the film and substrate can be compensated by that of elastic strain ε, which is generated by applying hydrostatic pressure to the sample. This method can be illustrated using a bimetal model as shown in Figure 1 [2]. In this model, a film is adhered to a substrate at an elevated temperature either at atmospheric pressure (AP) or at ultrahigh pressure (UHP). Since α of the film is assumed to be larger than that of the substrate, the sample bends downwards with decreasing temperature, as shown in Figure 1(a). However, since ε in the film is also assumed to be larger than that of the substrate, the sample bends upwards when the pressure is decreased from UHP. Therefore, we anticipate realization of a strain-free condition upon decreasing both temperature and pressure simultaneously, as shown in Figure 1(b).

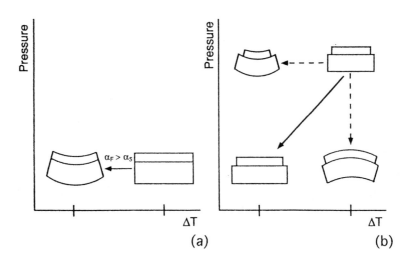

(a) (b)

Figure 1. Bimetal model used to illustrate the annealing method at UHP. The thermal expansion coefficient, α_F, of the film is assumed to be larger than the thermal expansion coefficient, α_S, of the substrate.

In order to realize the strain-free condition, it is necessary that the net strain in the film, which corresponds to the difference between the thermal strain and elastic strain, is equal to that in the substrate. In other words, the effective thermal expansion coefficient, α', of the film, which is defined by $\alpha - \varepsilon/\delta T$, must be equal to that of the substrate, where δT is the difference between the growth temperature and room temperature (RT). Using the approximation that α and ε do not depend on temperature, this condition is expressed by a simple linear relation between δT and pressure p, as follows. When hydrostatic pressure is applied to a cubic crystal, ε in the crystal is expressed by [2]

$$\varepsilon = p/(C_{11} + 2C_{12}), \tag{1}$$

where C_{11} and C_{12} are the diagonal and off-diagonal components of the elastic stiffness constants. Thus, α' is expressed by

$$\alpha' = \alpha - p/(C_{11} + 2C_{12})\delta T. \tag{2}$$

Equating the α' values of the film and substrate materials, we obtain the following relationships for the Ge/Si, GaAs/Si and ZnSe/GaAs systems. It is interesting to note that these relationships do not change even if the materials used for the film and the substrate are exchanged.

$$
\begin{aligned}
p &= 3.0\delta T \quad (p : \text{MPa}, \delta T :^\circ \text{C}) \quad \text{for} \quad \text{Ge/Si} \tag{3}\\
p &= 4.2\delta T \quad (p : \text{MPa}, \delta T :^\circ \text{C}) \quad \text{for} \quad \text{GaAs/Si} \tag{4}\\
p &= 1.3\delta T \quad (p : \text{MPa}, \delta T :^\circ \text{C}) \quad \text{for} \quad \text{ZnSe/GaAs} \tag{5}
\end{aligned}
$$

These data show that the pressures of 1.14 GPa (11.4 kbar) and 1.60 GPa (16.0 kbar) are necessary in order to obtain strain-free Ge and GaAs films on Si substrates by annealing at 400°C. However, the pressure (0.49 GPa) is much lower in the case of ZnSe film on the GaAs substrate, which has recently become important for the fabrication of blue-green laser diodes.

3. EXPERIMENTAL PROCEDURE

3.1. Pressure Apparatus

Ultrahigh pressure up to 2.2 GPa was generated using an oil-pressed piston-cylinder apparatus, as shown in Figure 2(a) [6]. This apparatus is advantageous in that samples are minimally damaged during the pressure experiment. In this experiment, Ar gas was used as a pressure transmission medium. The diameters of the piston and cylinder were 20 mm and 200 mm, respectively, as shown in Figure 2(b). Figure 2(c) shows the details of the arrangement around the samples. As can be seen in the figure, three rings of rubber, bronze and stainless steal were used for sealing the Ar gas. An electric furnace with a maximum temperature of 600°C was placed in the cylinder and, in order to keep the temperature profile uniform, the samples were enclosed in a Cu cap. Typical samples were about 3×5 mm^2

488

in size and their temperatures were monitored using attached thermocouples. In order to maintain high heating efficiency, the furnace was surrounded by thermally insulating parts composed of BN.

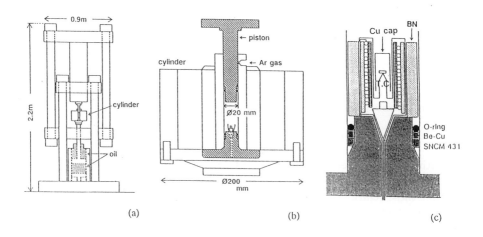

Figure 2. Schematics of (a) a piston-cylinder ultrahigh-pressure apparatus, (b) a magnified view of the piston and cylinder, and (c) details of the arrangement around the sample.

In order to decrease the amount of residual oxygen in the cylinder, the filling and releasing processes of Ar gas at 15 MPa (150 bar) were repeated 5 times, with the samples in the cylinder. Then, the insulating parts were baked at 200°C for 20 min in Ar gas of 15 MPa so that the absorbed air was ejected from the pores in the insulator. Finally, the filling and releasing processes were again repeated 5 times. During annealing, the temperature of the samples was controlled within ±2°C using a thyristor-controlled power supply. The Ar gas pressure in the cylinder was calculated from oil pressure using the area ratio (1 : 100) between the piston and the press head, and the calculated values were sometimes calibrated by the change in resistance of a manganin wire in the cylinder. Typical error in the calculated values was about 10 %.

3.2. Sample Preparation

In the case of Ge-on-Si structures, amorphous Ge films 160 to 850 nm thick were deposited on a single-crystal Si(100) substrate using a vacuum evaporation method, and they were crystallized in the solid phase by annealing under AP or UHP [1,2]. The annealing temperature and time ranged from 375°C to 460°C, and from 30 min to 100 min, respectively. Since the solid phase epitaxial growth rate of amorphous Ge has been reported to be about 60 nm/min at 375°C [7], these annealing conditions were considered to be sufficient for growing the films up to the surfaces. In the UHP annealing, the pressure was first increased to a necessary value, then the sample temperature was increased. After the annealing, both temperature and pressure were decreased every 3 minutes, typically in 10 steps, so that a proportional relationship approximately holds down to the vicinity of AP.

In the case of GaAs-on-Si structures, it is generally difficult to grow GaAs films by solid phase epitaxy from an amorphous phase. Therefore, the films were first grown on Si substrates by metalorganic chemical vapor deposition (MOCVD) and were subsequently annealed under UHP. If the atomic bonding at the GaAs/Si interface is fully rearranged during the UHP annealing, the residual strain in the GaAs film is expected to be reduced.

In MOCVD using trimethyl gallium (TMG) and AsH$_3$, the substrate temperature was initially kept at 500°C and then increased to 750°C, for the purpose of improving the crystalline quality of the films. The thickness of GaAs films was about 3 μm and the etch pit density of the films was on the order of 10^6 cm^{-2}. In some samples, GaAs films were etched in a 10-μm-wide line-and-space pattern along the $\langle 011 \rangle$ direction using a solution with NH$_4$OH : H$_2$O$_2$: H$_2$O = 1 : 2 : 5. The rearrangement of atomic bonding at the interface and the subsequent shrinkage of the GaAs film is anticipated to be completed in a shorter annealing time for these patterned samples than for broad-area samples. The samples were annealed at temperatures ranging from 300°C to 500°C either at AP or at UHP up to 2.1 GPa. The annealing time was varied with in the range of 20 min to 160 min.

The residual strain in the films was measured from the peak shift in X-ray diffraction analysis as well as by Raman spectroscopy using an Ar laser of 514.5 nm. The crystalline quality was characterized by Rutherford backscattering spectrometry (RBS) with 1.5 MeV He ions and through observation of the etch pit density on the surface of the GaAs films. The etch pits were formed by dipping the samples for 30 sec in molten KOH kept at 350°C.

4. EXPERIMENTAL RESULTS

4.1. Ge/Si(100) Structures

Figure 3 shows the laser Raman spectra for Ge/Si(100) samples annealed at 400°C for 30 min at AP, 1.14 GPa, and 2.10 GPa [2]. The spectrum for a bulk Ge sample is also shown for comparison. We can see from the peak shift in the figure that tensile strain is generated in all Ge films on Si substrates and it is decreased with increase of pressure. The strain values are 2.93×10^{-3}, 1.56×10^{-3}, and 0.55×10^{-3} for the samples annealed at AP, 1.14 GPa, and 2.10 GPa, respectively.

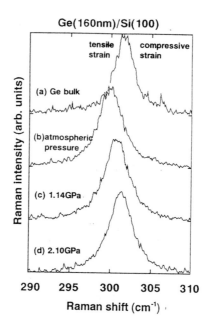

Figure 3. Laser Raman spectra for a bulk Ge crystal and Ge films on Si substrates annealed at AP, 1.14 GPa, and 2.10 GPa.

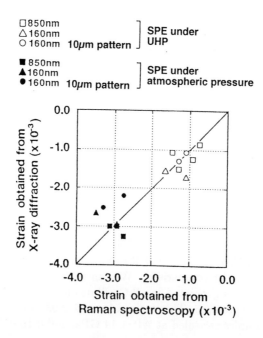

Figure 4. Comparison of the residual strain in various samples.

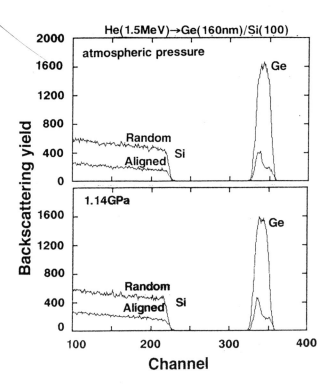

Figure 5. RBS random and aligned spectra for Ge films annealed at AP and 1.14 GPa.

The residual strain in various Ge samples was further investigated using X-ray diffraction analysis, and the results were compared with those obtained by Raman spectroscopy. In this experiment, the Ge samples were annealed at 400°C for 30 min under AP or UHP of 1.14 GPa and the film thickness was either 160 nm or 850 nm. The results are summarized in Figure 4 [1], in which the horizontal axis shows the strain values derived from the Raman spectroscopy, while the vertical axis shows the strain values calculated from X-ray diffraction analysis. Open marks show the results for the UHP-annealed samples, while closed marks show those for the AP-annealed samples. The results for the patterned samples are also included in this figure. We can see from this figure that the strain values obtained by the two methods agree fairly well and that the strain in the UHP-annealed samples is roughly 1/3 of that in the AP-annealed samples.

Next, the crystalline quality of the Ge films was characterized using RBS. Figure 5 shows RBS spectra for the 160-nm-thick Ge samples annealed at AP and 1.14 GPa. We can see from this figure that the channeling minimum yield χ_{min} (ratio of the aligned yield to the random one) of the UHP-annealed sample is almost the same as that of the AP-annealed one. In another experiment in which an 850-nm-thick Ge film was deposited on Si, degradation of the channeling minimum yield was observed in the UHP-annealed sample. Therefore, we can say that pronounced degradation of the film quality can be prevented in the UHP-annealed samples by thinning the Ge films and that the quality in the UHP-annealed sample is almost the same as that in the AP-annealed one, as long as it is characterized by RBS. Cross-sectional TEM micrographs also revealed that many defects existed near the interfacial regions of the Ge films, and their densities were found to be almost equal between the AP-annealed and UHP-annealed samples. These defects are considered to be generated due to lattice mismatch between Si and Ge.

Variations of the residual strain and χ_{min} are plotted in Figure 6 against the annealing pressure. We can see from the figure that the residual strain in the films decreases linearly with increase of the pressure, while χ_{min} of the films remains constant. We propose that a strain-free Ge-on-Si structure can be grown at 2.5 GPa, on the basis of Figure 6. This value is about twice as large as the value predicted using the linear approximation. The origin of this discrepancy is not well understood at present. The following factors may explain this discrepancy: (1) temperature dependences of α and C_{ij}, and (2) shear stress generated by different film and substrate strains.

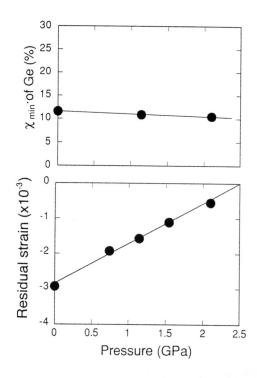

Figure 6. Variation of the residual strain and the channeling minimum yield of he Ge films with pressure during annealing. The strain values were derived from the Raman data.

4.2. GaAs/Si(100) Structures

Figure 7 shows the annealing time dependence of the residual strain in GaAs films on Si substrates, in which the annealing temperature and pressure were kept at 300°C and 1.22 GPa, respectively. In this figure, the negative values of the vertical axis correspond to the tensile strain along the in-plane direction and the broken line shows the residual strain in the as-grown sample. We can see from this figure that the strain in the UHP-annealed samples is less than 1/3 of that in the as-grown sample, and this value is independent of annealing times longer than 20 min. This result suggests that the rearrangement of atomic bonding at the film/substrate interface is completed within 20 min under these conditions.

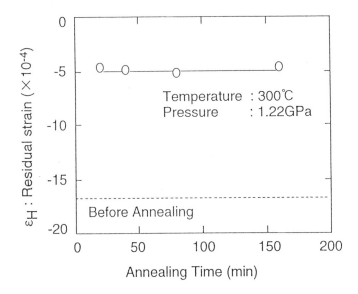

Figure 7. Variation of residual strain in GaAs films with annealing time.

Figure 8 shows the variation of the residual strain plotted against the annealing pressure. We can see from this figure that the residual strain in the films decreases linearly with increase of the pressure and that the strain-free GaAs film can be obtained at a pressure around 1.9 GPa. The linear pressure dependence is similar to that in the case of Ge-on-Si structures [2], and this result agrees fairly well with the simple theoretical calculation based on the bimetal model, in which the strain-free condition in the GaAs-on-Si structure is satisfied at 1.6 GPa for annealing at 400°C. Furthermore, we assert that an actual strain-free condition has been achieved by us for the first time in the case of GaAs-on-Si structures. We can also see from this figure that the strain depends weakly on the annealing temperature. This weak temperature dependence cannot be explained by the bimetal model shown in Figure 1. In order to explain this phenomenon, a more realistic model is necessary , which may include the hypothesis that the movement of dislocations stops at around 300°C and that the atomic bonding at the interface is fixed at that temperature.

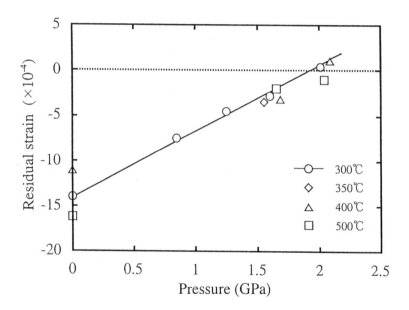

Figure 8. Variation of residual strain in GaAs films with pressure at various annealing temperatures. The positive and negative values on the vertical axis correspond to compressive and tensile strains along the in-plane direction, respectively.

Figure 9 shows RBS random and aligned spectra for the samples with broad-area GaAs films, which were annealed at 400°C for 20 min at AP and at 2.09 GPa. The probe beam was 1.5 MeV He^+ ions. Since the GaAs film thickness is about 3 μm, no Si spectrum appears in this figure. We can see from comparison of the aligned spectra that the channeling minimum yield in the UHP-annealed sample is slightly higher than that of the AP-annealed one, particularly in the low-energy region in the spectrum. The difference between the two spectra is small. However, a similar finite difference was repeatedly observed in other samples sets which were annealed under different temperature and pressure conditions. There was no large difference between etch pit densities of the AP- and UHP-annealed samples, and these values were on the same order as the value for the as-grown film (7.4×10^6 cm^{-2}). We conclude from results of both RBS and etch pit observation that the crystalline quality of GaAs films worsens slightly upon UHP annealing.

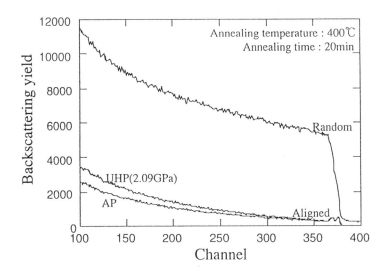

Figure 9. RBS random and aligned spectra for AP- and UHP-annealed samples with broad-area GaAs films.

Finally, the same annealing experiment was performed for the samples with patterned GaAs films. The pattern structure was a 10-μm-wide line-and-space formed along the ⟨011⟩ direction. Figure 10 shows the RBS spectra for the samples annealed under the same conditions (at the same time) as those of Figure 9. Since the GaAs films in these samples cover about 1/2 of the sample surface, the random backscattering yields by Ga and As atoms are about 1/2 of those in Figure 9. The step in the random yield around the channel number of 250 corresponds to the scattering from Si atoms in the substrate, which are exposed on the sample surface in the gap regions in the pattern.

We can see from this figure that the aligned spectrum for the UHP-annealed sample completely overlaps with the spectrum for the AP-annealed sample, which means that the crystalline quality of the former GaAs film is as good as that of the latter GaAs film, as determined by RBS analysis. Decrease of the residual strain in the UHP-annealed sample was also observed from X-ray diffraction analysis. In this particular sample, the residual strain was about $+3 \times 10^{-4}$, which suggests that the pressure during annealing was too high for the patterned sample. We conclude from these results that strain-free epitaxial growth of GaAs films on Si substrates can be realized without degrading the crystalline quality of the films, at least under the condition in which the GaAs films are etched in a line-and-space pattern.

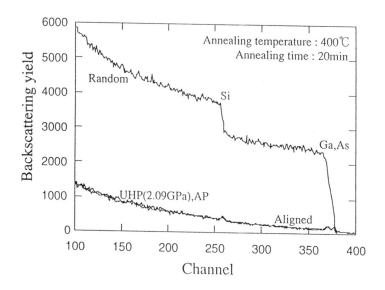

Figure 10. RBS random and aligned spectra for AP- and UHP-annealed samples with patterned GaAs films.

5. SUMMARY

In summary, a novel method as a solution to the thermal mismatch problem in heteroepitaxial growth was proposed and its validity was experimentally verified using Ge-on-Si and GaAs-on-Si structures. It was found from Raman spectroscopy and X-ray diffraction analysis that the residual strain in Ge and GaAs films on Si substrates decreased linearly with increase of pressure during annealing and that strain-free GaAs films were obtained at around 1.9 GPa. Their crystalline quality was found to be slightly worse than that of as-grown or AP-annealed films when the films were thick and they were not etched in a line-and-space pattern. It was also found, however, that the film quality was not degraded when GaAs films were etched in a line-and-space pattern. It is concluded from these results that the UHP annealing method is useful for decreasing the residual strain in heteroepitaxial structures.

REFERENCES

1. H. Ishiwara, T. Sato and A. Sawaoka ; Mat. Res. Soc. Sympo. Proc. 239, 467 (1992)
2. H. Ishiwara, T. Sato and A. Sawaoka ; Appl. Phys. Lett. 61, 1951 (1992)
3. H. Ishiwara, T. Hoshino, M. Usui and H. Katahama ; Ext. Abst. of 1993 Intern. Conf. on Solid State Device and Materials, Chiba, D-2-2
4. H. Ishiwara, T. Hoshino and H. Katahama ; Appl. Phys. Lett. 66, 2373 (1995)
5. H. Ishiwara, T. Hoshino and H. Katahama ; Materials Chemistry and Physics 40, 225 (1995)
6. H. Ishiwara, H. Wakabayashi, K. Miyazaki, K. Fukao and A. Sawaoka ; Jpn. J. Appl. Phys. 32, 308 (1993)
7. J.M. Poate, K.N. Tu and J.W. Mayer ; "Thin Films, Interdiffusion and Reactions", (Wiley & Sons, New York, 1978) p.473

PART V

Crystal Growth and Mechanisms in Complex Systems

Advances in the Understanding of Crystal Growth Mechanisms
T. Nishinaga, K. Nishioka, J. Harada, A. Sasaki and H. Takei (Editors)
© 1997 Elsevier Science B.V. All rights reserved.

Growth mechanism of oxide superconductor crystals

H. Takei*

Institute for Solid State Physics, the University of Tokyo
Roppongi, Minato-ku, Tokyo 106, Japan

The recent studies on growth of the oxide superconductor crystals $YBa_2Cu_3O_x$ (YBCO) and $(TlO)_2Ba_2Ca_{n-1}Cu_nO_{2n+2}$ (n=1-3) (TlBCCO) from the mixed state of solid with melt are reviewed. It has been clarified that recrystallization process in the peritectic mixtures formed between the solid and the melt is essential for the growth. The in-situ analyses in the mixed states have revealed that the growth rate is determined by diffusion of raw materials in the liquid. Single crystals of YBCO and TlBCCO with large aspect ratios have been successfully grown from the peritectic states. Such a crystallization technique makes a road to obtain metastable or high-melting crystals with high chemical reactivity in an easier way, because the growing temperatures are able to decrease much lower than those of melting. Another advantageous technique for obtaining large, thick crystals of YBCO has been developed using the modified TSSG method where the solute feeding and the solution transportation processes are separated under the peritectic condition.

1. INTRODUCTION

The discovery of the superconductive oxides with high transition temperatures has motivated research on the growth of cuprate crystals. Numerous investigations have clarified that the compounds are chemically unstable at high temperatures above 1100°C and tend to decompose into non-superconductive phases [1]. Another important point is the high chemical reactivity of molten cuprates which dissolve almost all kinds of crucible materials. Such a dissolved impurity contaminates the growing crystals. These points restrict the research within a use of solution growth technique to be operated at temperatures as low as possible.

Recently, there have been developed several new methods for obtaining single crystals using solid-melt mixtures, which are called "half-melt" or "partial melt", in nearly isothermal conditions [2]. Such a crystallization technique is advantageous for growing such metastable or high-melting point crystals with high chemical re-activity in an easier way, because the growing temperatures are able to decrease much lower than those of melting. One example is the advanced abnormal grain growth technique which has been applied for producing single crystals of Ni- or

---

*Present address: Graduate School of Science, Osaka University, 1-16 Machikaneyama, Toyonaka-shi, Osaka 560, Japan

(Mn,Zn)-ferrites [3], YIG [4], pure tungsten metal [5] and $Cd_x Hg_{1-x}Te$ [6], where liquid phases formed at grain boundaries play an essential role.

Another is the method using a chemical reaction between a solid and coexisting melt by the following peritectic reaction:

$$S_1 + L \rightarrow S_2 \qquad (T = T_p) \qquad (1)$$

at the peritectic temperature T_p, where S_1 and S_2 mean the high and low temperature phases, respectively, and L is the coexisting liquid phase. These are represented by the typical binary phase relation of S_1 and S_2, as shown in Fig. 1, where the decomposition reaction :

$$S_2 + L \rightarrow S_1 \qquad (T = T_d) \qquad (2)$$

proceeds at the decomposition temperature $T_d (= T_p)$ with elevating temperature. The peritectic reaction (1) occurs as an inverse reaction of the decomposition (2) when the molten system is in a cooling stage and passes through T_p.

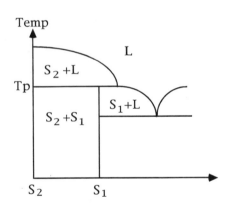

Figure 1. A schematic phase diagram for an incongruent melting system.

Figure 2. A typical YBCO single crystal grown under a "half melt" condition [8,9].

Popovich et al. [7] first reported that single crystal boules of superconductive $TlBiTe_2$ as S_2 were obtained under the peritectic condition where the solid $Bi_2 Te_3$ of S_1 and another melt phase (L) coexisted. However, they did not explain how the S_2 crystallization proceeded. One of the most successful results has been obtained on the growth of oxide superconductor crystal, namely, $YBa_2 Cu_3 O_x$ (YBCO) [8, 9] which has a superconductive transition above 90K, by applying the peritectic reaction technique between the solid $Y_2 BaCuO_5$ and the surrounding liquid.

The present article is a brief review for the crystallization mechanism of oxide superconductors from solid-liquid mixtures accompanied with a peritectic reaction between them. As a typical example, the ternary system Y_2O_3–BaO–CuO was mainly adopted for this purpose. The growth process of metastable, superconductive crystals of Tl–Ba–Ca–Cu–O system , namely, $Tl_2Ba_2CaCuO_8$ (Tl-2212) and $Tl_2Ba_2Ca_2Cu_3O_{10}$ (Tl-2223) were also discussed.

2. GROWTH OF $YBa_2Cu_3O_x$ (YBCO) CRYSTALS

In the 90K superconductor $YBa_2Cu_3O_x$ (YBCO), the pseudo-binary system of Y_2BaCuO_5 (Y-211) and 3BaO-5CuO is important for considering the crystal growth mechanism, because the stoichiometric YBCO phase lies on the tie line between the two terminal compositions. As YBCO melts incongruently and decomposes into Y-211 and liquid, the phases YBCO and Y-211 correspond to S_2 and S_1 in the above formula (2), respectively. That is, the reaction formula in the cooling stage around T_p is expressed as:

$$Y_2BaCuO_5 + L \rightarrow 2YBa_2Cu_3O_x \qquad (T = T_p). \qquad (3)$$

Many studies have revealed that the phase relation of the system is not so simple as that in Figure 1, but the general tendency of this system can be briefly explained by such a simplified relation [10,11]. The peritectic temperature T_p is around 1000°C.

It has been found that the growth features of YBCO crystals strongly depend on the content of liquid phase BaO+CuO. An abnormal grain growth occurred When the composition of the mixtures was close to the stoichiometry of YBCO, and the maximum heating temperature was kept around 1000°C. Cm-sized YBCO crystals with irregular shape were often obtained [12,13]. It is considered that a trace amount of liquid generated at grain boundaries may assist the grain growth. Such a crystal is called "block crystal" and in spite of its poor crystallinity, the superconductivity is excellent: the transition temperature $T_c > 90K$ and the transition width $\Delta T_c < 1K$[14]. On the other hand, normal solution growth progresses only when the content of solvent BaO+CuO is much higher. The temperature of the system must be kept above 1000°C so as to make the system in a complete solution. The grown YBCO crystals were, however, very small thin plates with mm-size, and are named as "plate crystals" [15,16]. The most disadvantageous point in "plate crystals" is unexpected contaminations by crucible materials [1,17]. The solvent BaO+CuO in the molten state above 1000°C becomes very reactive and dissolves any kinds of crucible materials such as noble metals or ceramics. Such an impurity considerably degrades the physical properties of obtained crystals to a lower and looser superconductive transition.

In the middle concentration of BaO+CuO, the "half melt" (in other words, "partial melt" or "pseudo-melt") state appears under relatively low soaking temperatures below 1000°C. By cooling the half melt state very slowly, large, well-habited YBCO bulk crystals of about $7 \times 7 \times 7mm^3$ in maximum size were obtained, as shown in Figure 2. From the developed polyhedral shape, they were called them "polyhedral crystals" [8,9]. The results that they behave very sharp ($\Delta T_c < 1K$)

and high transition ($T_c > 91K$) indicate a low contamination of impurities when we used sintered yttria (Y_2O_3) as a crucible [8,9,18]. In the following section will be described the crystal growth mechanism in such a heterogeneous condition.

3. GROWTH MECHANISM OF YBCO CRYSTALS IN THE SOLID-MELT MIXTURE

The peritectic reaction (1) which usually occurs at the interface of solid S_2 and the surrounding melt L below T_p, brings about a formation of resultant solid layers S_1 covering the mother solid S_2. This covering delays further reaction because the diffusion of the reactant molecules S_2 through the solid layer of S_1, which is formed at the first stage of the reaction, is commonly very slow in comparison with that in the liquid, and the reaction (1) is actually stopped at the final stage. This is a reason why the peritectic mixture of S_2 covered with S_1 is obtained, and in such a situation, no crystal growth occurs.

In the YBCO system, many studies have showed apparent growth of single crystals from the peritectic mixtures. To analyze the growth mechanism, two kinds of investigations have been conducted: one is by a quenching method and another is by an in-situ measurement.

3.1. Quenching method

Asaoka et al. have studied the growth feature of YBCO crystals from the mixed phase by inspecting heated-quenched specimens using an X-ray diffraction technique, a polarized microscope and an X-ray microanalyzer (XMA) [19,20]. Changes in the X-ray diffraction patterns at different quenching temperatures are shown in Figure 3. At 1010°C, the main peaks only correspond to the solid Y-211phase, indicating that the condition lies in the two-phase region of the solid Y-211 and the melt. When the mixtures reached 980°C, the diffraction patterns of the YBCO phase appeared, which is thought to be formed by the peritectic reaction (1). As the temperature was lowered to 970°C, the peak intensity due to the YBCO phase increased, whereas that of Y-211 peaks considerably decreased. These data support the thought that the growth of YBCO crystals is controlled by the peritectic reaction (1). Similar results have been obtained in the in-situ analysis described in 3.2.

Figure 4a is a typical example of the microscope observations where the smooth (on the upper left) and rough (on the right) interfaces coexist in the liquid phase. The same position was analyzed by XMA, as shown in Figure 4b, where two-dimensional yttrium mapping was conducted. It is apparent that the content of yttrium in the liquid phase is heterogeneous. The mapping pattern suggests an occurrence of yttrium diffusion from the rough to the smooth interfaces. The most typical photograph is shown in Figure 5, where the diffusion field of yttrium spreads in the liquid region from the rough or the sharply curved interface. These results lead to the consideration that a recrystallization process by dissolution and precipitation plays an important role in the peritectic mixture of the YBCO system. That is, small YBCO particles with a high curvature tend to dissolve, and large

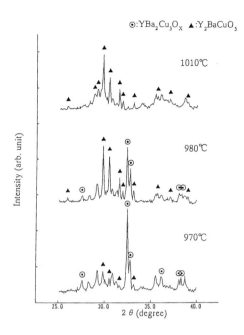

Figure 3. Changes in the X-ray patterns of YBCO samples quenched from various temperatures [19,20].

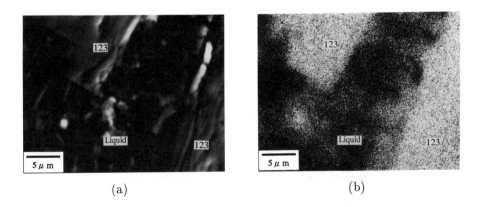

(a)　　　　　　　　　　　　　　(b)

Figure 4. Microscope (a, upper) and SEM-XMA (b, lower) observation of the quenched specimens from the "half melt" condition [19,20].

ones with a low curvature or especially with a flat interface gathers the solute, following the Gibbs-Thomson effect, named "Ostwald ripening". As a result, the well-habited crystals having the infinite curvature becomes large, as illustrated in Figure 6.

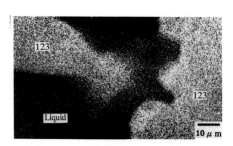

Figure 5. A SEM-XMA pattern of the quenched YBCO representing a diffusion field of yttrium from the rough interface [19,20].

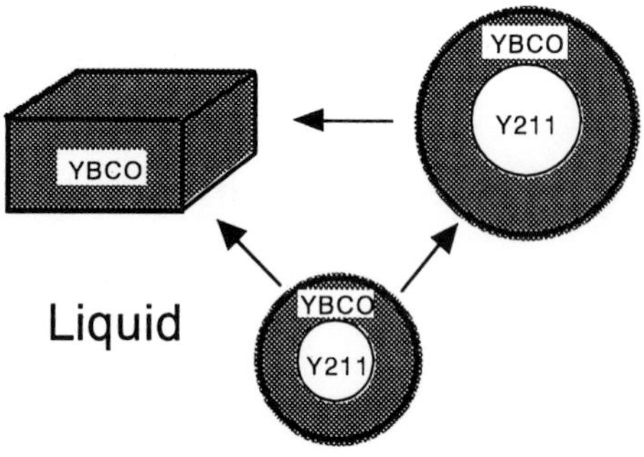

Figure 6. An illustration of the "Ostwald ripening" process under a peritectic condition.

3.2. In-situ analysis

To clarify the growth process of YBCO crystals more precisely, in-situ measurements were conducted by a use of a time-resolved X-ray powder diffraction apparatus [21–23]. The essential layout was designed so as to fulfil the following conditions: 1) the highest temperature is 1100°C, 2) the sample chamber can be evacuated, or filled with air, oxygen or inert gas, 3) the specimen kept in a horizontal position can be rotated around a vertical axis, 4) the total counting time of scattered X-rays in the allowed diffraction angle $\Delta(2\theta) < 40°$ is less than 1 min. A cross-section of the developed system is schematically shown in Figure 7. A position-sensitive proportional counter (PSPC) for rapid counting of diffracted X-rays was vertically installed at the right-hand arm and a radiation source was set at the left-hand arm. The specimens packed into a bored plate of single crystal MgO were heated with Pt heaters. Recently, an optical microscope with a CCD camera was added vertically at the upper part of the furnace so as to observe the surface of specimens in a heating stage [24]. The observed picture was recorded with a video-recorder.

The results of the measurements for the YBCO system are summarized as follows.

1. The peritectic reaction (3) proceeded between 962 and 955°C in a cooling rate of 1°C/min, and the peritectic mixture of YBCO, Y-211 and liquid appeared at this temperature range.

2. By keeping the mixture at 960°C for 71min, a complete chemical change from Y-211+L to YBCO by the reaction (3) was observed. This means that the peritectic state is unstable and the recrystallization progressed while the temperature was high.

3. The X-ray diffraction intensities of the YBCO phase increased nonlinearly and tended to saturate, briefly obeying a simple parabolic law, as shown in Figure 8. The data were treated for calculation by using the Einstein's diffusion law:

$$(I/I_{max})^2 = k(t - t_0), \tag{4}$$

where I is the diffraction intensities of the reaction time t, I_{max} is its maximum value, k is the velocity constant and t_0 is the threshold time for starting the reaction. The results were reasonably explained by this equation. It suggests that the growth mechanism is controlled by some kind of diffusion process.

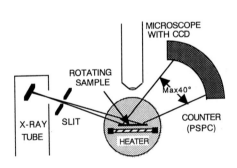

Figure 7. A schematic drawing of the in-situ measurement system using both a PSPC for diffracted X-rays and an optical microscope [21, 22].

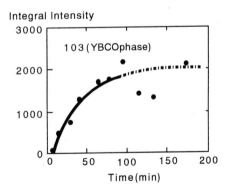

Figure 8. The change in the X-ray intensities of YBCO with time for the mixture of Y211 and liquid on cooling from 1010°C to 970°C [23].

4. The temperature dependence of the I–t relations obtained between 940 and 960°C was analyzed by the traditional Arrhenius equation:

$$k = k_0 exp(-E/RT),\qquad(5)$$

where k_0 is the velocity constant at $t = t_o$, E is the activation energy, R is the gas constant and T is the absolute temperature. The value of E was calculated to be 28 kJ/mol, which should be corresponded to the activation energy of liquid phase diffusion.

4. PULLING OF YBCO CRYSTALS BY THE MODIFIED TOP-SEEDED SOLUTION GROWTH (TSSG) TECHNIQUE

One of the most sophisticated application of the peritectic reaction (3) for the YBCO crystal growth is a modified top-seeded solution growth (TSSG) technique developed by Yamada et al. [25–27] The schematic configuration of the method is illustrated in Figure 9. The Y-211 powder is placed as a nutrient at the bottom of an yttria crucible and the solvent of BaO+CuO is poured on Y-211. The temperature of the bottom part is set about 15°C higher than that of the surface and a textured crystal rod with the c-axis orientation of $SmBa_2Cu_3O_x$ (SmBCO) is immersed at the top of the solution. The use of SmBCO is advantageous because it is isostructural to YBCO and has a higher decomposition temperature than YBCO. Cm-sized bulk crystals of YBCO having the same orientation of Sm-BCO were successfully obtained by continuous pulling of the seed, as illustrated in Figure 9. The O_2-annealed crystal behaved the superconductive transition at 89K with the transition width less than 2K.

It became clear from the experiments that the temperature of the melt surface was very critical for YBCO growth. When the temperature was higher than T_p of about 1000°C, only the Y-211 phase was precipitated on the seed. This is a result of simple dissolution-recrystallization process of Y-211 by the provided negative temperature gradient from the bottom to the top, because the whole system was in the Y-211 stable region. On the other hand, YBCO crystals were obtained on the seed when the surface temperature was close to 1000°C at which the YBCO phase became stable. The reaction (3), which usually proceeds in a common isothermal condition at solid-melt interfaces of a peritectic mixture, is divided by such a temperature gradient into the following three steps: 1) Y-211 dissolves into the liquid at the bottom, 2) the generated solute of Y-211 is transported towards the surface and 3) through the reaction (3), YBCO crystallizes at the seed. In this condition, the Y-211 molecules formally exist in the solution phase.

5. CRYSTAL GROWTH OF Tl-BASED SUPERCONDUCTORS

Oxide superconductors of the Tl-Ba-Ca-Cu-O system are important in low temperature science and technology, because they involve several compounds with superconductive transitions T_c much higher than 100K. Especially, the compounds

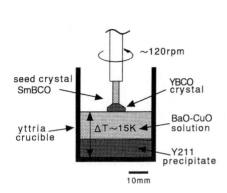

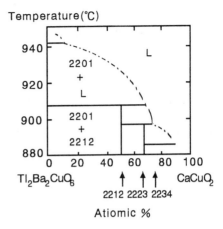

Figure 9. The YBCO pulling system using the modified TSSG method where the raw materials Y211 for the peritectic reaction lies under the liquid phase [26,27].

Figure 10. The pseudo-binary phase diagram of the Tl-Ba-Ca-Cu-O system reported by Kotani et al. [29].

$Tl_2Ba_2CaCu_2O_8$ (Tl-2212) and $Tl_2Ba_2Ca_2Cu_3O_{10}$ (Tl-2223) have very high T_c values of around 110K and 120 K, respectively. Few investigations, however, have been carried out on obtaining single crystals of Tl-2223 [28–33] because of difficulty in dealing with thallium oxide which has high chemical reactivity and volatility at high temperatures and very toxic vapors. These require that the growth experiment should be conducted under heating temperatures as low as possible.

Since the phase diagrams of the Tl-Ba-Ca-Cu-O system reported by several authors represent complex relationships [29,34], crystal growth experiments have faced some difficulty in the determination of the optimum conditions for the preparation of high-quality crystals. Kotani et al. have reported a tentative phase relation of the pseudo-binary system between $Tl_2Ba_2CuO_6$ (Tl-2201) and $CaCuO_2$, as shown in Figure 10, where Tl-2212 and Tl-2223 exist on the tie line [29]. As these compounds melt incongruently, the growth technique of using a peritectic condition is thought to be superior to the common solution growth method, because the growth temperature can be lowered with it. The reactions are described as follows:

$$Tl_2Ba_2CuO_6 + L \rightarrow Tl_2Ba_2CaCu_2O_8, \qquad (6)$$

and

$$Tl_2Ba_2CaCu_2O_8 + L \rightarrow Tl_2Ba_2Ca_2Cu_3O_{10}. \qquad (7)$$

By controlling the growth conditions of coexisting solid $Tl_2Ba_2CuO_6$ (Tl-2201) or Tl-2212 with a liquid phase, L, in accordance with Figure 11, single crystal

510

Figure 11. Tl-2212 crystal grown in a coexisting condition of Tl-2201 and liquid(L) [35–38].

Figure 12. Tl-2223 crystal grown in a coexisting condition of Tl-2212 and liquid (L) [35–38].

plates of Tl-2212 and Tl-2223 have been successfully obtained, as shown in Figure 12, where the maximum dimensions are $2 \times 2 \times 0.5$ and $0.8 \times 0.8 \times 0.4$ mm^3, respectively. X-ray analyses showed that the crystals were tetragonal with developed planes of (0 0 1), and the edge directions were parallel to the a- and a'-axes [35–38]. Microscope and SEM observations revealed that the (0 0 1) surfaces of the plates were covered with many thin layers having tiny curved steps, indicating that crystal plane developed through lateral advancement of the layers with steps. Compositional heterogeneity due to syntactic intergrowth was not observed by XMA or X-ray diffraction analyses.

The superconductive transitions of these crystals were very sharp at T_c, indicating the high crystallinity, as typically shown in Figure 13, where the results of the magnetic susceptibility measurements using a DC-SQUID on the as-grown Tl-2223 crystal was presented. The applied magnetic field was 3 Oe parallel to the 001 plane. The T_c value was 115 K and the transition width ΔT_c for 80% total demagnetization was within 3 K. In the Tl-system superconductor crystals, the interlayer structure of Tl-2201 and Tl-2212 or Tl-2212 and Tl-2223 was often observed [35,38,39]. Such a structure induces a broad superconductive transition at T_c [38,39], as shown in Figure 14. The interlayer structure usually appeared when the growth temperature was set at higher temperatures than 900°C. The small ΔT_c values of the present crystals are the results of using the low growth temperatures below 900°C. It should be noted that the differences between the field cooling (FC) and zero field cooling (ZFC) measurements are very close to each other in the 0 K-extrapolated diamagnetic susceptibilities , which suggest a weak pinning force of magnetic fluxes in paramagnetic domains. Studies on oriented thin films of Tl-2223 have revealed that the magnetic pinning force is strongly influenced by tilt angles between subgrains [40–42]. The weak pinning of the present crystal would

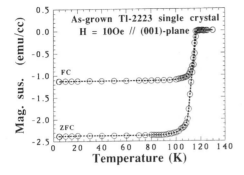

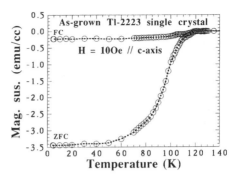

Figure 13. Magnetic susceptibility of Tl-2223 crystal grown between 900 and 870°C. The magnetic field was applied parallel to the {001} plane [38].

Figure 14. Temperature dependence of the magnetic susceptibility of Tl-system superconductor crystal grown above 900°C. The broad transition is due to the interlayer structure of Tl-2212 and Tl-2223 [38].

be induced by the absence of crystal imperfections, such as small grain boundaries, internal strains, point defects or substitutional impurities.

6. CONCLUSION

The present article has reviewed the recent studies on the crystal growth of YBCO and Tl-system oxide superconductors from the mixed state of solid with melt under the peritectic condition. It has been clarified that the recrystallization from the peritectic mixture formed between the solid and melt is an essential process for growing such oxide crystals. The in-situ analysis has revealed that the growth rate is determined by the diffusion of raw materials in the liquid phase. This result is quite similar to that for the usual solution growth process. Single crystals of YBCO and Tl-based cuprates, $(TlO)_2Ba_2Ca_{n-1}Cu_nO_{2n+2}$ (n=1-3), with large aspect ratios have been grown from the states of mixed solid with melt. Large, thick crystals of YBCO have been obtained by using the modified TSSG method where the solute feeding and the solution transportation processes are separated under the peritectic condition. Such a crystallization technique is advantageous for growing metastable or high-melting point crystals with high chemical reactivity in an easier way, because the growing temperatures are able to decrease much lower than those of melting.

512

ACKNOWLEDGMENTS

The author acknowledges to Dr. M. Hasegawa, Mrs. F. Sakai and Mr. M. Koike of the Institute for Solid State Physics, the University of Tokyo for their co-research, and to Dr. H. Takeya of National Research Institute for Metals (Tsukuba), Dr. H. Asaoka of the Japan Atomic Energy Research Institute, Dr. Y. Oyama of Optron Co. Ltd., Dr. Y. Matsushita of Michigan State University, Prof. M. Kikuchi and Dr. S. Nakajima of Tohoku University and Dr. T. Suzuki of Fukui University, for their assistance, suggestions and discussions. The main part of the present article is a summary of our study supported by the Special Scientific Research Project "Crystal Growth Mechanism in Atomic Scale" supplied from the Ministry of Education, Science, Culture and Sports, Japan.

REFERENCES

1. Y. Hidaka and M. Suzuki, Prog. Cryst. Growth and Charact. , 23 (1991) 179.
2. H. Takei, J. Jpn. Assoc. Crystal Growth , 18(1991)381.
3. S. Tanji, S. Matsuzawa, N. Wakatsuki and S. Soejima, IEEE Trans. Mag. , MAG-21 (1985) 1542.
4. V. A. Timofeeva, L. M. Belyaev, N. D. Ursulak, A. V. Belitsky, A. B. Bykov and V. M. Prilepo, J. Cryst. Growth, 52 (1981) 633.
5. M. Katoh, S. Iida, Y. Sugita and K. Okamoto, J. Cryst. Growth, 112 (1991) 368.
6. A. W. Vere, B. W. Straughan, D. J. Williams, N. Shaw, A. Royle, J. S. Gough and J. B. Mullin, J. Cryst. Growth,59(1982) 121.
7. N. S. Pipovich, V. K. Shura and D. W. Gitsu, J. Cryst. Growth, 61(1983) 406.
8. H. Takei, H. Asaoka, Y. Iye and H. Takeya, Jpn. J. Appl. Phys. , 30 (1990) L1102.
9. H. Asaoka, H. Takei, Y. Iye, M. Tamura, M. Kinoshita and H. Takeya, Jpn. J. Appl. Phys. , 32 (1992) 1091.
10. K. Dembinski, M. Gervais, P. Odier and J. P. Countures, J. Less-Comm. Met. , 164/165 (1990) 177.
11. K. Oka, K. Nakane, M. Ito, M. Saito and H. Unoki, Jpn. J. Appl. Phys. , 27 (1988) L1065.
12. H. Takei, H. Takeya, Y. Iye, T. Tamegai and F. Sakai, Jpn. J. Appl. Phys. , 26 (1987) L1425.
13. M. Murakami, M. Morita, K. Doi and K. Miyamoto, Jpn. J. Appl. Phys. , 28 (1989) 1189.
14. Y. Iye, T. Tamegai, H. Takeya and H. Takei, Jpn. J. Appl. Phys. , 26 (1987) L1425.
15. Y. Hidaka, Y. Enomoto, M. Suzuki, M. Oka, A. Katsui and T. Murakami, Jpn. J. Appl. Phys. , 26(1987) L726.
16. L. F. Schneemeyer, J. V. Waszczak, T. Siegrist, R. B. van Dover, L. W. Rupp, B. Batlogg, R. J. Cava and D. W. Murphy, Nature, 328 (1987) 601.
17. F. Licci, C. Frigeri and H. J. Scheel, J. Cryst. Growth, 112 (1991) 606.

18. S. N. Barilo, A. P. Ges, S. A. Guretskii, D. I. Zhigunov, A. V. Zubets, A. A. Ignatenko, A. N. Igmentsev, I. D. Lomako, A. M. Luginets, V. N. Yakimovich, L. A. Kurochkin, L. V. Markova and O. L. Krot, J. Cryst. Growth, 119(1992) 403.

19. H. Asaoka, H. Takei and K. Noda, Jpn. J. Appl. Phys. , 33 (1994) L932.

20. H. Asaoka, JAERI Review, #96-006 (1996) 37.

21. Y. Oyama, H. Yabashi, M. Hasegawa and H. Takei, Ferroelectrics, 152(1994) 361.

22. Y. Oyama, M. Hasegawa and H. Takei, Jpn. J. Appl. Phys. , 33(1994) 4779.

23. Y. Oyama, M. Hasegawa and H. Takei, J. Cryst. Growth, 143(1994) 200.

24. T. Suzuki, Ph. D. Thesis for the University of Tokyo, (1995) #3107.

25. Y. Yamada, M. Nakagawa, K. Ishige and Y. Shiohara, Adv. Superconductivity, IV (1993) 305.

26. Y. Yamada, M. Tagami, M. Nakamura, Y. Shiohara and S. Tanaka, Adv. Super conductivity, V (1993) 561.

27. Y. Yamada and Y. Shiohara, Physica C, 217(1993) 182.

28. D. S. Ginley, B. Morosin, R. J. Baughman, E. L. Venturini, J. E. Schirber and J. F. Kwak, J. Cryst. Growth, 91 (1988) 456.

29. T. Kotani, T. Kaneko and H. Takei, Jpn. J. Appl. Phys. 28 (1989) L1378.

30. . R. Zhifeng, M. J. Naughton, P. Lee and J. H. Wang, J. Cryst. Growth, 112 (1991) 587.

31. K. Kawaguchi and M. Nakao, J. Cryst. Growth, 99(1990) 942.

32. S. Matsuda, A. Soeta, T. Doi, K. Aihara and T. Kamo, Jpn. J. Appl. Phys. , 31 (1992)L1229.

33. G. Brandstäter, F. M. Sauerzoof, H. W. Weber, A. Aghei and E. Schwarzmann, Physica C, 235-240 (1994) 2797.

34. T. K. Jondo, C. Opagiste, J. L. Jorda, M. Th. Cohen-Adad, F. Sibieude, M. Couach and A. F. Khoder, J. Alloys and Compds. , 195 (1993) 53.

35. M. Hasegawa, Y. Matsushita and H. Takei, Adv. Superconductivity, VII (1995) 723.

36. M. Hasegawa, Y. Matsushita, Y. Iye and H. Takei, Physica C, 231 (1994) 161.

37. M. Hasegawa, Y. Matsushita, Y. Iye, F. Sakai and H. Takei, Physica C, 235-240 (1994) 3137.

38. Y. Matsushita, M. Hasegawa, F. Sakai and H. Takei, Jpn. J. Appl. Phys. , 34(1995) L1263.

39. H. Takei, F. Sakai, M. Hasegawa, S. Nakajima and M. Kikuchi, Jpn. J. Appl. Phys. , 32 (1993) L1403.

40. M. Oussena, S. Porter, A. V. Volkozub, P. A. J. de Groot, P. C. Lanchester, D. Ogborne, M. T. Weller, B. Balakrishnan and D. McK. Paul, Phys. Rev. B, 48 (1993) 10575.

41. T.Nakabatake, Y. Saito, K.Aihara, T. Kamo and S. Matsuda, Jpn. J. Appl. Phys. , 32 (1993) L484.

42. M. Kawasaki, E. Sarnelli, P. Chaudhari, A. Gupta, A. Kussmaul, J. Lacey and W. Lee, Appl. Phys. Lett. , 62 (1993) 417.

Advances in the Understanding of Crystal Growth Mechanisms
T. Nishinaga, K. Nishioka, J. Harada, A. Sasaki and H. Takei (Editors)

An atomic level analysis of crystal growth mechanism in complex systems by means of nano-optical microscopy and AFM.

H. Komatsu[a]*, S. Miyashita[a], T. Nakada[a], G. Sazaki[a] and A. A. Chernov [b]

[a]Institute for Materials Research, Tohoku University,
Katahira 2-1-1, Aoba-ku, Sendai 980-77, Japan

[b]Institute of Crystallography, Russian Academy of Science,
Leninsky Prospect 59, 117333 Moscow, Russia

The atomic topography of various crystals and interfaces in complex systems was studied using nano-optical microscopes and an atomic force microscope (AFM). The items we studied are as follows; (1) The accuracy of AFM in the z-direction was examined by combining phase-contrast microscopy with interferometry. (2) A profile of a hollow dislocation on the surface was colsely checked with the theory proposed by Frank. (3) The dissolution process on a calcite cleavage in water was observed by AFM. (4) Structural transformation in the liquid-solid interface of organic molecules and substrates was analysed by AFM and (5) a mechanism of electrodeposition of Ag on graphite was delt with by combining voltammetry and AFM.

1. INTRODUCTION

In electronic device technology, a scanning tunneling microscope (STM) and an atomic force microscope (AFM) are routinely used for observing the atomic configuration of crystal surfaces in a clean and high vacuum. But in the natural world, where solution growth predominates, we lack experience of applying STM or AFM to a crystal, which grows in a complex system and far from the clean environment.

When we examine a crystal surface growing in a solution, we need to observe a whole surface before selecting a specific area for STM or AFM observation. For that, nano-optical microscopes, which are capable of detecting molecular level differences on a crystal, are needed and must be combined with STM or AFM for analyzing the growth or dissolution mechanism of a crystal. Optical interferometry is also useful to measure the step height of a crystal in an unit cell order. We report here our experimental trials conducted on complex systems, combining various microscopic tools, which are found promising for the future applications.

---

*e-mail address; komatsu@komatsu.imr.tohoku.ac.jp

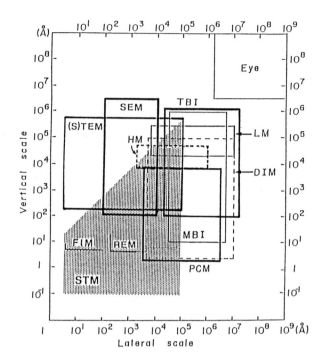

Figure 1. The detection limits of various types of microscopes[1]. TBI:two-beam interferometer, MBI:multiple-beam interferometer, LM:low-power optical microscope (NA 0.1), HM: high-power optical microscope (NA 1.4), DIM: differential interference microscope, PCM: phase-contrast microscope, SEM: scanning electron microscope, EM: electron microscope, TEM: transmission electron microscope, FIM: field ion microscope, REM: reflection electron microscope, STM: scanning tunneling microscope.

2. ACCURACY OF AFM MEASUREMENT APPLIED TO THE STEP HEIGHT[1]

At present there is no stringent check of the accuracy of atomic force microscopy when it is applied, in air or solution, to the step height measurement on a crystal surface. Since crystals grow or dissolve through the surfaces, it is vitally important to observe or measure what takes place at the interface in an atomic level. This is a summary of our work on the cross check of the detection limits and accuracy of AFM with nanoscale optical microscopy.

The detection limit of various kinds of microscopes has already been reported [1]. We cite here the diagram in Figure 1. As a standard for the step height (z-direction) measurement by AFM, we chose a Frank spiral growth pattern on an SiC crystal. According to the theory of spiral growth mechanism proposed by Frank [2], a single step pattern originating from a single screw dislocation retains the same step height, which is regulated by the strength of the Burgers vector of the screw dislocation. An elemental step height on a surface of a hexagonal silicon carbide is known to be 0.2518nm; thus the step height of any single Frank spirals on the basal face should be the integral multiple of 0.2518nm. This is an extremely good standard for a step height (vertical direction) since a silicon carbide crystal is mechanically hard, chemically stable to a high temperature and inactive to any materials at room temperature.

Single crystals of silicon carbide were prepared by the Lely method (sublimation and subsequent recrystallization of a raw material of silicon carbide polycrystals at about 2500°C).[3] The most common crystal structure is known as a hexagonal system (α form), which has more than 200 polytypes. In any polytypes of SiC the length of the c-axis is a multiple of 0.2518nm. We cleaned a hexagonal crystal using a ultrasonic vessel in a detergent solution and subsequently in water before observing the surface.

An incident phase-contrast microscope was used for observing a Frank spiral on the basal face, and the step height of the spiral was measured by means of two-beam interferometry (TBI) using a Michelson type interferometer. As shown in Figures 2 (a) and (b), the total height of the 18 spiral layers was measured to be about half a wavelength of a monochromatic light secured by an interference filter (λ=592nm, $\Delta\lambda/2$=6nm). The measurement error is assumed to be $\pm\lambda/(20 \times 2 \times 18)$ since it is possible to read the shift of a fringe within the accuracy of $\lambda/20$. Thus the individual step height was estimated to be about 16.5±1nm, which corresponds to 66 elemental layers of a silicon carbide crystal stacked parallel to the basal plane.

The same surface was measured by AFM as shown in Figure 3 (a), which shows step heights fluctuation considerably, in spite of the fact that every step should have the same height. No systematic errors occurred when the probe was scanned either from the higher side of the step or the lower one. In any case the step height measured by present AFM is always higher than the height measured by TBI.

A step height measurement by a combination of phase-contrast microscopy and TBI is more reliable than that of AFM. The possible errors in measuring a step height using AFM will be referred to at least two causes. The first is a geometrical relationship between the tip of the probe and the step edge. There is no proof that the sensing point of the atomic force on a tip is fixed at a certain point; on the contrary, the sensing point changes as the tip approaches a step edge.

Secondly, the linearity of the piezoelements is not assured when they are applied to a large step height of the order of 20-30nm. The last point can be examined by measuring several steps having different heights and repeating the measurement on the same step. It must be added that accumulated electricity on the tip, the surface or a step edge also can be a cause of errors in a step height measurement. It is desirable to calibrate the z direction using a reliable standard every time when we change a tip.

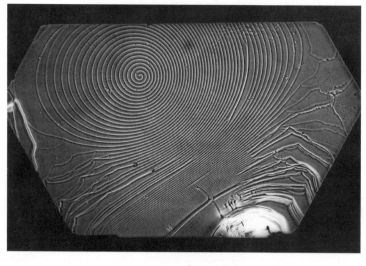

a

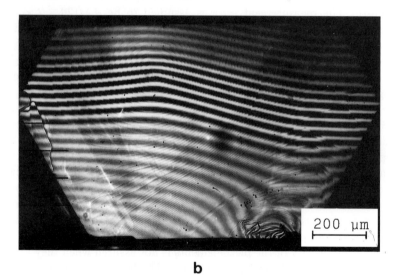

b

Figure 2. Basal face of an SiC crystal[1]. (a) An incident phase-contrast micrograph (positive-contrast). (b) A two-beam interferogram.

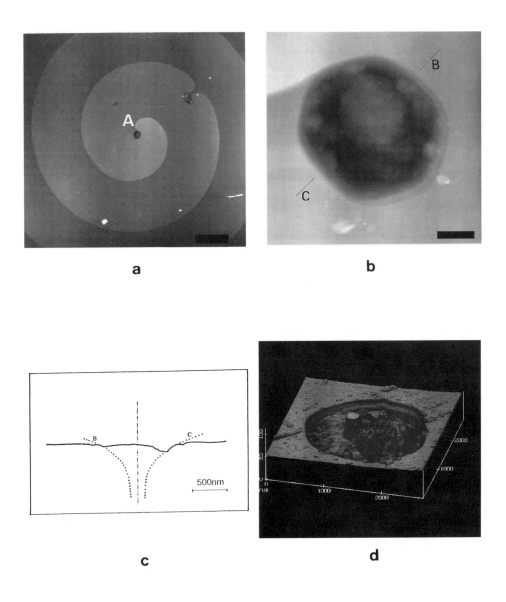

Figure 3. AFM images on an SiC single crystal[1,14]. (a): large scale image. (scale bar:10μm) (b): center of the spiral. (scale bar:500nm) (c): section profile along **BC** in (b). (d): a typical dimple.

3. A HOLLOW DISLOCATION AND MACROSTEPS ON AN SiC SINGLE CRYSTAL[4]

AFM easily enables us to observe crystal surfaces in an atomic scale and we can examine crystal growth theories which were difficult to demonstrate so far. As a case study we studied the structure of the hollow dislocations [5,6] on the silicon carbide crystal mentioned above.

Figure 2 (a) is a photograph of an SiC crystal taken by a phase-contrast microscope. We can see many interesting growth patterns on the basal face, such as a single spiral step which forms a vicinal surface of 0.13° and various stages of modifications of step fronts due to defects. Near the edge of the crystal, we see the step width fluctuation, which is probably due to the thermal effect caused by radiation reflected from the wall of the growth cell. Dimples or hollows which are the outcrops of the edge dislocations were seen. In this crystal, the step did not decompose into small steps within the distance of $300 \mu m$ from the spiral center.

Figure 3 (a) is an AFM image of the center of the spiral which has a hollow core (875nm in radius) generating a step. Within 4 μm from the spiral center, the step was linear and gradually changed to an Archimedian spiral far from the center. At the point **A**, the curvature of the step changed from negative to positive.

Figure 3 (b) is a magnification of Figure 3 (a). We could see the step wound itself around the core. Figure 3 (c) is a section profile across the core (along the line **BC** in Figure 3 (b)). Since the center of the core was covered with some other materials, we could not go further into a hollow tube nor a detailed profile of it.

Frank analyzed the profile of a hollow dislocation in terms of the strain field and surface free energy of the hollow core, when it crops up to the surface [5]. He expressed an equilibrium radius of the hollow core, r, to be

$$r = \mu b^2 / 8 \pi^2 \gamma, \tag{1}$$

where μ, γ and b are the rigidity modulus, the specific surface free energy and the strength of the Burgers vector, respectively. Since μ/γ depends on the direction, this explains the formation of the hexagonally shaped core. We neglected the reconstruction at the surface and estimated the binding energy per one Si-C bonding as 3.17 [7], the value of γ can be roughly estimated to be 6.2J/m^2. For the first approximation, we assume that the value of μ/γ is constant below the melting or sublimation temperature. Using the value of 1.7×10^{11}Pa for μ [8], the value of r was evaluated as 95nm.

Frank also predicted the profile of the crater of a dislocation. According to his theory, the profile of the crater was funnel-shaped. However, calculated profile cannot be applied to that obtained by AFM since in the predicted one, the height increased logarithmically with the distance from the center, while the measured one became constant far from the center. In order to fit the two profiles, a new boundary condition will be needed. This boundary condition will describe the extent to which the strain generated by the central dislocation will be neglected. Regardless their behavior far from the center, they coincided partially well, if we regard r as 125nm. It is expressed by dotted lines in Figure 3 (c).

Near the dislocation center, the step took a hexagonal facet. Probably, the step would tend to make facets to release the excess surface energy since the radius of curvature is small around the core.

The relation between the curvature of the step and the strain field created by the central screw dislocation was discussed in detail [9,10]. The change of the curvature shows that this spiral pattern was created when the crystal was growing. According to the simulation analysis [10], this spiral pattern corresponds to the case when $r/r_c = 0.6$, where r_c denotes the radius of the critical nucleus on a stress-free surface. The value of r_c could be estimated to be 200nm.

The length of the Burgers vector in this crystal was about one hundred times larger than that of the unit cell of an SiC. How such a large step was created? Komatsu [11] and Gotoh and Komatsu [12] have proposed the mechanism leading to the formation of dislocations having giant Burgers vectors.

Frank also predicted the form of a dimple or a crater which should appear when a dislocation meets the free surface of a crystal. Figure 3 (d) shows an AFM image of a typical dimple (370nm in radius), which is created at the intersection of an edge dislocation with the surface.

4. IN SITU OBSERVATION OF GROWTH AND DISSOLUTION OF CALCITE IN SOLUTION[4]

AFM can be applied to observing atomic images of a crystal in a liquid. The necessary practical conditions are; (1) the solution must be transparent, (2) the solubility and supersaturation should be small and (3) a flat crystal surface. Calcite ($CaCO_3$) is a good candidate for observing the surface by AFM in solution since we can easily obtain a flat cleavage and the solubility to water is small (solubility product at $25°C$ is 5×10^{-9}).

A cleaved calcite was put on a small Peltier element and they were set to the cell for observation by AFM and the sample was immersed in water. Temperature of the cell was controlled by the Peltier element. Calcite grows as temperature is increased since the solubility of calcite for water decreases as temperature is increased.

Figure 4 (a) shows the dissolution process of the cleaved calcite in water at $24°C$. As time passed, microsteps appeared and moved back (an arrow in the figure). The direction of these steps was parallel to the edge of a cleavage rhombohedron. The height of these steps was 0.34nm, which is one half of the spacing of (100) plane of the rhombohedron. This step height corresponds to that of one molecular layer formed by Ca and CO_3 ions arranged into a plane. Calcite dissolves as these steps go back to the [100] directions. Lozenge-shaped etch pits were also formed by these steps. Figure 4 (b) shows an atomic image of the surface of a calcite in solution. A single step is clearly demonstrated that the height is the same as that of one molecular plane. Figure 4 (c) is an atomic image of a calcite after dissolution was terminated. We could get more stable image in solution than in air.

We raised the temperature of the sample up to $57°C$. The crystal surface is sensitive to the temperature change and AFM images drastically changed with a

522

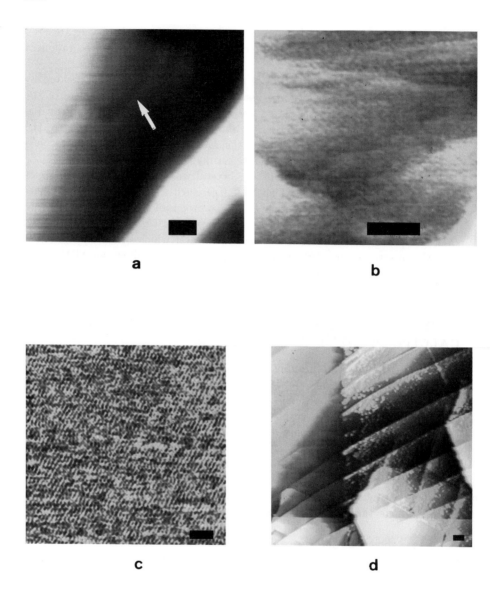

a

b

c

d

Figure 4. AFM micrographs of calcite[4]. They were taken *in situ*. (a): dissolution process. An arrow indicates a dissolution step of one molecular layer height. (scale bar:100nm) (b): an atomic image of steps. (scale bar:10nm) (c): an atomic image obtained after dissolution terminated. (scale bar:5nm) (d): surface morphology of a cleaved surface after the temperature was increased to 57°C . Small crystals were observed to grow on the surface. (scale bar:500nm)

small temperature variation and it is difficult to obtain a stable image. Figure 4 (d) shows an AFM image of the crystal surface after the system reaches equilibrium. Small island crystals were observed to nucleate epitaxially on the surface. Unfortunately, since the tip of the cantilever "contacts" the crystal surface, nucleation of the islands was suppressed in the area where the tip was scanned. We confirmed this by changing the observing area. This artifact must be taken into consideration when we observe the process of crystal growth by in situ AFM.

5. STRUCTURAL TRANSFORMATION IN LIQUID/SOLID INTERFACE

The understanding of the atomic structures of interfaces is a key to analyze the mechanisms of crystal growth. Recent development of STM and AFM made it possible to observe the interfacial structures. This is one of the applications of AFM to reveal the interfacial structure of n-alcohol on mica or graphite [13].

Immediately before the measurements, a substrate of mica or highly oriented pyrolytic graphite (HOPG) about 10mm square and of 0.5mm thickness was cleaved in air and mounted on the sample holder. Then the substrates and a cantilever were immersed in n-alcohol, which was purchased from Wako (octanol (98%) and undecanol (98%)) and Sigma (nonanol (98%), decanol (98%), and dodecanol (99%)). The n-alcohol had been purified by passing through a molecular sieve of type 4A (zeolite desiccant, mean pore diameter 0.4nm) to remove water and other contaminants before the experiments. All experiments were conducted in the substrate temperature range of 25 to 60°C , which was controlled within ±0.1°C using a Peltier element attached below the substrate crystal.

Figure 5 shows a typical force curve for the tip approaching on an HOPG surface immersed in n-dodecanol at 25°C . The tip-sample distance was calculated by subtracting the lever deflection from the sample displacement. This force curve shows the same tendency as that previously measured by O'Shea et al.[14]. A series of periodic repulsive 'walls' is also observed in this figure. The periodicity, which is the mean spacing between the successive walls, is 0.49±0.04nm, which is very close to the distance between adjacent C-C chains in higher alkanes (~0.49nm) and n-dodecanol crystals (0.476nm) [15].

Furthermore, the measured periodicities of the other n-alcohols are almost the same, independent of their carbon numbers (n=8-12). Therefore, these results strongly suggest that the molecules of all the n-alcohols investigated here are oriented parallel to the HOPG surfaces as shown in Figure 6. To validate these results, we further took AFM images, which showed clearly that the molecules lie on the substrate in parallel orientation. The effect of the temperature on this system is minimal, and between 25 to 60 °C , no drastic change was seen in either the force curves or the AFM images.

The force between the tip and the mica substrate immersed in n-dodecanol at 25 °C is shown in Figure 7 (a). The oscillation force curve is similar to that obtained in the experiment on HOPG[13]. The mean periodicity of the oscillation force (0.86±0.04nm) is, however, significantly different from that of HOPG. Surprisingly,

524

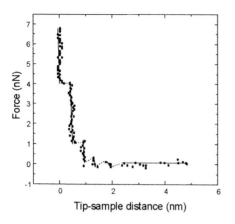

Figure 5. The force acting on the tip as a function of tip-sample distance for an HOPG substrate immersed in n-dodecanol[13]. The tip-sample distance is obtained by subtracting the lever deflection from the sample displacement. The speed of the tip as it approaches the sample is 50nm/s.

Figure 6. AFM image of the n-dodecanol/HOPG interface taken at a constant repulsive force of $\sim 2 \times 10^{-9}\text{N}$[13]. Scan area is approximately $13 \times 13\text{nm}^2$.

the periodicity is also different from the extended length ($\sim$1.7nm) of a dodecanol molecule. We assumed two possible structure models for this system as follows: (a) a double layer model, where double layers consisting of two molecules whose carbon chains are parallel to the mica surface, or (b) a tilted layer model, where layers are composed of molecules which are somewhat tilted from the surface normal. In order to examine these models, the periodicities of other n-alcohols on mica surfaces were measured because the periodicity should change with the number of carbon atoms in the latter model. As shown in Figure 8, the periodicities of the structure forces are the same for all n-alcohols except for n-octanol within an experimental error. Therefore, it is most likely that the n-alcohols form a double layer on a mica surface at 25°C.

In the dodecanol/mica system, the effect of temperature is significant in contrast to the dodecanol/HOPG system. Figure 7 (b) shows the force curve taken at 40°C. The steep walls with short periodicity in Figure 7 (a) are changed into walls with gentler slopes and longer periodicity. The structure change is reproducible and reversible when the substrate temperature is changed. The mean periodicity of the structure force (2.26±0.10nm) at 40°C is comparable to the extended length ($\sim$1.7nm) of the n-dodecanol molecule.

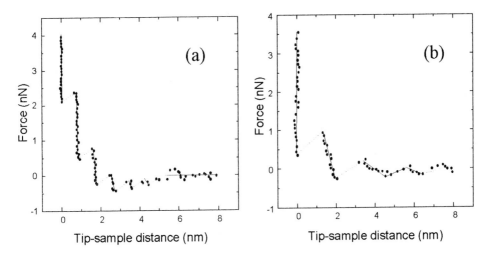

Figure 7. The force-distance curve for a mica substrate immersed in n-dodecanol at (a) 25°C and (b) 40°C [13].

We also measured the force curve for other n-alcohols and obtained that the mean periodicity increases as the number of carbon atoms increases. The calculated value of 0.12nm is very close to half the value of the projected zigzag C-C-C distance of n-alcohol crystals (0.127nm). This implies that the alcohol molecules are oriented perpendicular to the mica surface. It should be noted that the temperature dependence of the molecular arrangement resembles that of a stearic acid evaporated onto a mica substrate [16].

6. ELECTROCHEMICAL DEPOSITION OF Ag ON GRAPHITE

Electrodeposition is also an important crystal growth process for the production of layered materials, composite structures, and coatings. AFM and STM are the one of the most suitable techniques to investigate the electrochemical processes since atomic resolution is routinaly achieved on a solid liquid interface. In paticular, AFM is less susceptible to the electrical potential distribution on a substrate than STM. We will demonstrate here an atomic level study of an electrodeposition by in-situ AFM in a solution [4].

A silver wire was used as a reference electrode and a platinum wire was used as a counter electrode. All potentials are given with respect to the Ag-Ag$^+$ equilibrium potential of this system. The working electrodes, which were also AFM substrates, were an air-cleaved HOPG. Experiments were performed in 0.1M HNO$_3$ + 5×10^{-3}M AgNO$_3$ solution.

Prior to subjecting HOPG substrate to the potential ramp, the substrate was imaged to check the surface morphology at zero electric potential. Atomic steps

526

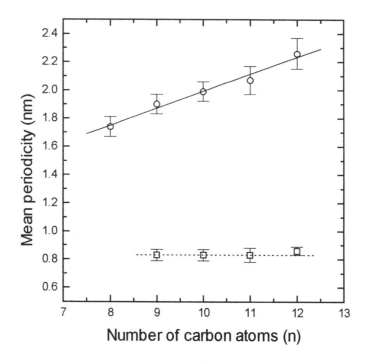

Figure 8. Mean periodicity *vs.* number of carbon atoms in a molecule of n-alcohol at high (○) and low (□) temperatures[13].

and atomic images were routinely observed on a fleshly cleaved HOPG even in an electrolyte. At this potential, any detectable Ag particles or impurities could not be observed in the AFM images for long periods of time (>30min.). Then, in order to remove residual impurities on HOPG and to examine the existence of underpotential (undersaturated) deposition (UPD)[17] in this system, we swept the potential of HOPG between 0.0V to 0.3V *vs.* Ag reference electrode. Recently, much attention has centered on the UPD process, in which a monolayer or a submonolayer of a foreign metal adatom is deposited at potentials positive to the reversible Nernst potential. Although UPD of Ag on some substrate was reported,[18] any detectable particles or superstructures were not found in the AFM images. Moreover, the voltammograms did not show any deposition or stripping peaks in the potential region. Therefore, we conclude that UPD does not occur in this system. This result suggests that a weak interaction between Ag and HOPG, because the UPD is regarded as a result of the formation of an adatom-substrate (Ag-HOPG) bond before the formation of a weaker adatom-adatom (Ag-Ag) bond.

Figures 9 (a)-(c) are the sequence of in-situ electrochemical AFM images showing the overpotential (supersaturated) deposition of the silver on HOPG. These figures

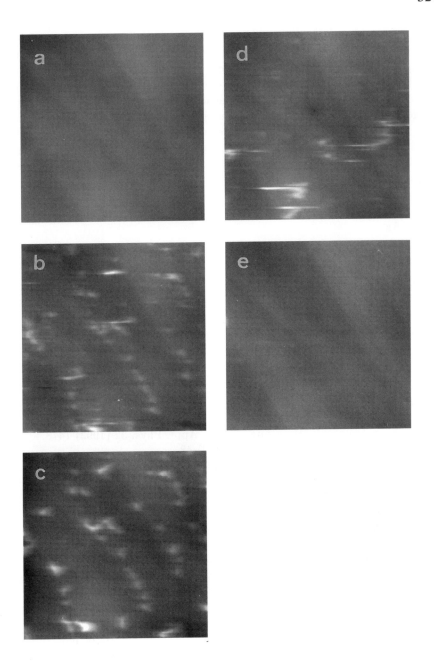

Figure 9. A sequence of AFM images of the Ag deposition (a - c) and dissolution (d and e) on a graphite electrode[4]. Scan size is approximately $32 \times 32 \mu m^2$.

are taken at a potential of -0.3V. As shown in these figures, numbers of particles are nucleated and grow along the step edges. In addition, there is no significant changes in atomic images on "flat" area, where no Ag particles exist. Hence, it is concluded that Ag is grown on graphite in the island growth mode (Volmer-Weber mode) at this potential. The growth aspects are almost identical in the potential range of 0.0 to -0.5V, except for the number of nuclei and growth rate. It is noted that some particles nucleated on the terrace have moved or disappeared while the particles at a step edge are becoming larger in size. This indicates that the silver can be mobile and Ag atoms are either formed or caught at a step edge where further growth can occur.

The deposited silver particles are easily removed by reversing the current. Figures 9 (d) and (e) are the sequence of the AFM images showing the electrodissolution of the Ag.

ACKNOWLEDGEMENTS

This research was supported by a Grant-in-Aid for Scientific Research and Priority Areas "Crystal Growth Mechanism in Atomic Scale" from the Japanese Ministry of Education, Science and Culture. We are thankful to Dr. Y. Inomata, the director of NIRIM (Tsukuba), for providing SiC crystals.

REFERENCES

1. H. Komatsu and S. Miyashita, Jpn. J. Appl. Phys. 32 (1993) 1478.
2. F. C. Frank, Disc. Farad. Soc. 5 (1949) 48.
3. J. A. Lely, Ber. Deut. Keram. Ges. 8 (1955) 229.
4. H. Komatsu, S. Miyashita and T. Nakada, J. Jpn. Assoc. Crystal Growth 21 (1994) s469 (*in Japanese*).
5. F. C. Frank, Acta Cryst. 4 (1951) 497.
6. F. R. Nabaro, Theory of crystal dislocation (Oxford, Clarendon Press, 1967) p303.
7. W. A. Harriosn, Electronic Sturucture and the Properties of Solids, The Physics of the Chemical Bond (W. H. Freeman and Company, San Francisco, 1980) Ch. 7.
8. P. T. B. Shaffer, Plenum Press Handbook of High Temperature Materials (Plenum Press, N. Y., 1964) No. 1, Material Index.
9. I. Sunagawa and P. Bennema, J. Crystal Growth 53 (1981) 490.
10. B. van der Hoek, J. P. van der Eerden, P. Bennema and I. Sunagawa, J. Crystal Growth 58 (1982) 356.
11. H. Komatsu, J. Mineral. Soc. Jpn. 14 (1980) Special No. 2, 142.
12. Y. Gotoh and H. Komatsu, J. Crystal Growth 54 (1981) 163.
13. T. Nakada, S. Miyashita, G. Sazaki, H. Komatsu and A. A. Chernov, Jpn. J. Appl. Phys. 35 (1996) L52.

14. S. J. O'Shea and M. E. Welland, Appl. Phys. Lett. 60 (1992) 2356.
15. D. A. Wilson and E. Ott, J. Chem. Phys. 2 (1934) 239.
16. F. Matsuzaki, K. Inaoka, M. Okada, and K. Sato, J. Cryst. Growth, 69 (1984) 231.
17. W. J. Lorenz, H. D. Hermann, N. Wuthrich, and F. Hilbert, J. Electrochem. Soc. 121 (1974) 1167.
18. C. Chen, S. M. Vesechky and A. A. Gewirth, J. Am. Chem. Soc. 451 (1992) 451.

Advances in the Understanding of Crystal Growth Mechanisms
T. Nishinaga, K. Nishioka, J. Harada, A. Sasaki and H. Takei (Editors)
© 1997 Elsevier Science B.V. All rights reserved.

A comprehensive treatise on crystal growth from aqueous solutions

T. Ogawa

Department of Physics, Gakushuin University Mejiro, Tokyo, 171, Japan

A comprehensive treatise on crystal growth from aqueous solutions is phenomeno-logically proposed here on quantitative observations of the rates of growth through concentration and temperature gradients on growing interfaces using basic data such as solubility, density, thermal conductivity, refractive index and viscosity of the solutions and partial molar volume, molar optical polarizability, proton reso-nance under the magnetic field and diffusion coefficient of solute in the solutions, all of which have been measured here for this purpose.

1. INTRODUCTION

Crystal growth from aqueous solutions is a very important process from both a scientific and an industrial point of view, since a great many materials are purified and produced as various industrial goods by this technique. Crystal growers are thus in urgent need of a comprehensive treatise that incorporates only the constants and parameters which are measurable.

An approximation to analytically treat some saturated aqueous solutions is suc-cessfully proposed here using solubility, density, partial molar volume, molar op-tical polarizability and proton resonance of the solutions, all ofwhich have been obtained for this purpose. This treatment is very similar to the regular solution [1], the effectiveness of which has been confirmed by crystal growth from aqueous solutions.

2. BASIC DATA OF AQUEOUS SOLUTIONS

2.1. Solubility [3]

The term $R\ln(X_{BS,T})$ is plotted against $1/T$ in Figure 1, where R is the gas constant, T is temperature of the solution and $ln(X_{BS,T})$ is a natural logarithm of $X_{BS,T}$ which is the molar fraction of the solute B in its saturated solution at T and is given by

$$X_{BS,T} = N_{BS,T}/(N_{AS,T} + N_{BS,T}). \tag{1}$$

Here, $N_{AS,T}$ is the number of solvent A (water) molecules in mole to saturate $N_{BS,T}$ moles of the solute at T. According to the regular solution theory, $R\ln(X_{BS,T})$ is the mixing entropy of the solute B and given by

$$R\ln(X_{BS,T}) = (\mu_{BS,T} - \mu_{B,T})/T - \ln\gamma. \tag{2}$$

Here, $\mu_{BS,T}$ and γ are, respectively, the chemical potential and activity coefficient of the solute molecules in the saturated solution at T, and $\mu_{B,T}$ is the chemical potential in the ideal solution at T. Equation (2) is confirmed with many inorganic aqueous solutions as shown in Figure 1 [3], so that we can claim that saturated aqueous solutions are ideal mixtures of solute molecules and water molecules, where they play as individual and independent particles, respectively.

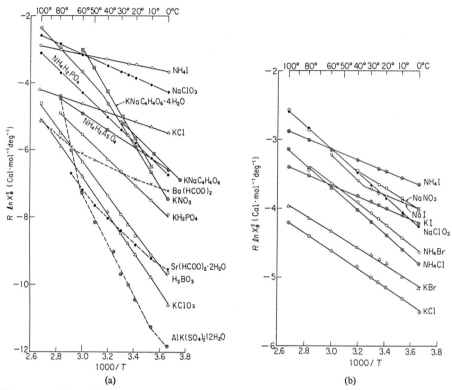

Figure 1. The mixing entropy of aqueous solutions against $1/T$

2.2. Density of Saturated Aqueous Solutions and Partial Molar Volume of the Solute in the Solutions [3,4]

When densities of saturated aqueous solutions were measured as a function of temperature and also calculated as ideal mixtures of solute and solvent molecules, the ratio of measured density to calculated one is nearly equal to unity as shown in Table 1. This suggests that the partial molar volume of the solute is nearly equal to the volume of its crystalline state.

Table 1
Ratio of the density calculated as an ideal mixture to that obtained by its saturated aqueous solution as a function of temperature [3]

Temp.(°C)	ADP	KDP	RS	NaClO$_3$	F.Sr	Ba(HCOO)$_2$
10	0.977		0.973			
20	0.979	0.97	0.992	1.01	0.980	
30	0.981	0.97	0.997	1.02	0.980	0.99
40	0.989	0.97	1.016	1.03	0.981	0.99
50	0.994	0.97	1.022	1.04	0.982	0.99
60	0.998	0.98	1.033			

[ADP: NH$_4$H$_2$PO$_4$, KDP : KH$_2$PO$_4$, F.Sr : Sr(HCOO)$_2$2H$_2$O,
RS: Rochelle Salt: NaKC$_4$H$_4$O$_6$4H$_2$O: sodium potassium tartrate tetra-hydrate]

Table 2
Refractive index of a RS aqueous solution at 25°C as a function of its concentration in molarity [4]

Molarity	index	Molarity	index	Molarity	index
0.1	1.33573	1.0	1.36196	1.9	1.38501
0.2	1.33847	1.1	1.36477	2.0	1.38776
0.3	1.34178	1.2	1.36750	2.1	1.39044
0.4	1.34450	1.3	1.37034	2.2	1.39282
0.5	1.34760	1.4	1.37244	2.3	1.39541
0.6	1.35032	1.5	1.37498	2.4	1.39721
0.7	1.35318	1.6	1.37760	2.5	1.39994
0.8	1.35592	1.7	1.37989	2.525	saturation
0.9	1.35868	1.8	1.38267		

2.3. Refractive Indices of Rochelle Salt Solutions and Molar Polarizability of the Solute and Solvent in the Solution [4]

The refractive index of a Rochelle Salt (RS) solution was measured within ±0.00003 by a differential refractometer using Hg: 546.1 nm radiation at 25.0± 0.1°C and shown in Table 2.

If we assume that the index n of the solution containing N_A moles of H$_2$O and N_B moles of crystalline RS molecules per cm^3 is given by the Lorentz-Lorenz relation

$$(n^2 - 1)/(n^2 + 2) = (4\pi/3)(N_A\alpha_A + N_B\alpha_B), \tag{3}$$

the molar optical polarizabilities of the solvent and solute, α_A and α_B, are obtained, respectively (Table 3).

Table 3
Optical polarizabilities of solvent α_A and solute α_B in cm^3/mol obtained from a RS aqueous solution as a function of the averaged concentration [4]

Molarity	α_A	α_B	Molarity	α_A	α_B
0.35	0.8854	11.178	1.45	0.8895	11.103
0.45	0.8845	11.251	1.55	0.8896	11.130
0.55	0.8858	11.150	1.65	0.8859	11.204
0.65	0.8850	11.196	1.75	0.8832	11.278
0.75	0.8848	11.281	1.85	0.8815	11.320
0.85	0.8835	11.337	1.95	0.8851	11.208
0.95	0.8832	11.341	2.05	0.8793	11.359
1.05	0.8818	11.397	2.15	0.8819	11.287
1.15	0.8835	11.280	2.25	0.8811	11.316
1.25	0.8875	11.161	mean value	0.884	11.24
1.35	0.8877	11.158		± 0.003	± 0.09

The polarizability of pure water is determined as 0.88617 cm^3/mol using $n = 1.33938$ at 25.0°C and 546.1 nm radiation, which is equal to α_A in Table 3 within experimental error. The polarizability of the RS crystal α_B is estimated as 11.2 cm^3/mol using its density 1.766g/cm^3 and the averaged refractive index

$$n = (n_\alpha + n_\beta + n_\gamma)/3 = 1.4947, \qquad (4)$$

because the RS crystal is optically biaxial but its anisotropy is very small. This polarizability also agrees with α_B in Table 3 [4].

From Table 3 the RS solutions look like mixtures of water molecules and crystalline RS molecules for all concentrations, which supports that the RS solution can be similarly treated as a "regular solution".

2.4. Proton Resonance of RS Solutions [3,5]

The proton signals from the RS solutions were measured as a function of concentration by a high resolution NMR spectrometer in a 60 Mhz range using a dual cell, one of which was for pure water to obtain a standard signal and the other for the solution. We observed two signals from the solution: one of them, signal A, due to the protons belonging to solvent water and the other, signal B, caused by the protons in RS molecules [5].

Since the integrated intensity (I.I.) is proportional to the number of resonating protons, the I.I. ratio of signal B to signal A is plotted as a function of concentration in Figure 2. This figure indicates that four water molecules per mole belong to a RS molecule as "water of crystallization" and thus do not act as a part of solvent water, which is shown by the solid line in the figure. The broken line is obtained by assuming that the four water molecules separate from the RS molecules and act as a part of solvent water [3].

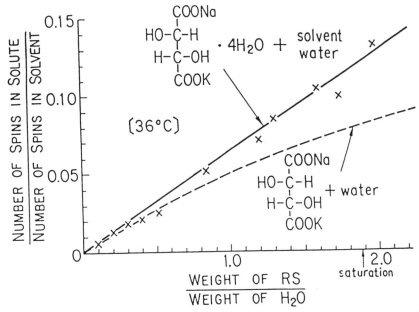

Figure 2. Ratio of the integrated intensity of signal B to the intensity of signal A against the concentration: (weight of RS)/(weight of water)

2.5. Diffusion Coefficient of the Solute in RS Solutions [5]

To obtain the intra-diffusion coefficient, tartaric acid with radio-active isotope, ^{14}C, was composed to obtain radioactive Rochelle salt. The coefficient of the solute in the solution saturated at $25.00 \pm 0.01°C$ was obtained as $2.3 \pm 0.05 \times 10^{-6} cm^2/s$.

2.6. Viscosity of the Rochelle Salt Solutions [4]

Viscosity η was measured by an Ostwald method at $T = 25.0 \pm 0.01°C$ as a function of concentration in molar fraction $X_{B,T} = N_{B,T}/(N_{A,T} + N_{B,T})$, and given by

$$ln\eta = -4.696 + 23.14X_{B,T}. \tag{5}$$

The viscosity of the 2.50 M solution is $(4.444 \pm 0.001) \times 10^{-2}$ poise at 25.0 °C, which is shown here for convenience of discussion.

The viscosity of the 1.0 and 2.5 M solutions was measured as a function of temperature from 20 to 40 °C, by which their activation energies were respectively obtained as 3.78 and 5.02 kcal/mol.

2.7. Thermal Conductivity of Rochelle Salt Solutions [6]

The thermal conductivity of RS solutions was measured as a function of concentration at 40 °C and obtained as $1.25 \times 10^{-3} cal^{-1}K^{-1}s^{-1}cm^{-1}$ in the saturated solution.

3. CRYSTAL GROWTH

3.1. Degree of Super-Saturation

When concentration of the saturated solution is given by eq.(1), the degree of super-saturation σ is defined by

$$\sigma = (X_{BSS,T} - X_{BS,T})/X_{BS,T} = X_{BSS,T}/X_{BS,T} - 1, \tag{6}$$

Here, $X_{BSS,T}$ is the mole fraction of the solute in a supersaturated solution and given by

$$X_{BSS,T} = N_{BSS,T}/(N_{AS,T} + N_{BSS,T}), \tag{7}$$

where $N_{BSS,T} > N_{BS,T}$ because of super-saturation. Usually, $0 < \sigma << 1$ and then eq.(6) is nearly equal to

$$\sigma = ln(X_{BSS,T}/X_{BS,T}) = ln(X_{BSS,T}) - ln(X_{BS,T}) = (\mu_{BSS,T} - \mu_{S,T})/(RT), \tag{8}$$

where $\mu_{S,T}$ and $\mu_{SS,T}$ are the chemical potential of the solute in the saturated and super-saturated solutions, respectively.

The degree of super-saturation is, therefore, proportional to the difference between chemical potential of the solute molecules in the saturated and supersaturated solution, that is, σ is equal to the motive force of crystallization [7].

3.2. Crystal Growth as a Mutual Interaction Between Material Deposition and Heat Generation at a Growing Interface

Growth rate of crystals is proportional to total amount of material flow onto a growing interface where the heat of crystallization is also proportionally generated by the amount of solute deposition. Since the deposition rate of the solute must be decreased with increment of temperature at the interface, the growth rate will be slow down if the heat of crystallization does not diffuse from there. The slowdown of the deposition rate induces less generation of heat and then the temperature will decrease. When the temperature reduces a low enough point, the material flow onto the interface again increases. Therefore, "relaxation oscillation" or "breathing" of the material flow as well as the temperature at the interface occurs during the crystal growth, even if the thermal convection current is eliminated in the cell. This oscillation is one of the origins of growth striation.

Since crystal growth is a typical irreversible process due to mutual interference between material flow J_B and heat flow q_B, Onsager's relations will be the most expected ones for crystal growth, which are given by

$$-J_B = (L_{11}/T)grad(\mu_{BSS,T}) + L_{12}grad(1/T) \tag{9}$$

and

$$q_B = (L_{21}/T)grad(\mu_{BSS,T}) + L_{22}grad(1/T). \tag{10}$$

However, the mutual interaction between material deposition and heat generation previously mentioned is completely different from the phenomenon given by the terms L_{12} and L_{21} in eqs.(9) and (10); these are usually very small and are thus neglected here.

Therefore, the material and heat flows are respectively given by

$$J_B = -L_{11}(grad\mu_{BSS,T})/T = -(RL_{11}/X_{BSS,T})grad(X_{BSS,T}) \tag{11}$$

and

$$q_B = L_{22}grad(1/T) = -(L_{22}/T^2)gradT. \tag{12}$$

where eq.(11) is equal to Fick's law of diffusion and eq.(12) is the fundamental relation of thermal conduction.

According to the principle of continuity, we have the same equation for the material flow and the heat flow as follow;

$$\partial\rho/\partial t + divJ = 0, \tag{13}$$

where ρ is number of deposited molecules for $J = J_B$ and the total amount of heat of the crystallization for $J = q_B$ in eq.(13), respectively.

If there is a stationary state, the following two Laplace equations will be effective:

$$\Delta X_{BSS,T} = 0 \tag{14}$$

and

$$\Delta T = 0. \tag{15}$$

Quasi-stationary states, where eqs.(14) and/or (15) are practically satisfied, will be realized only after a very long period or by a very slow growth rate.

3.3. Temperature Gradient around a Growing Rochelle Salt Crystal [10]

To eliminate a thermal convection current of the solution during crystal growth, a disc-like and two dimensional thin cell was made by transparent acrylic resin plate to measure growth rate and temperature profile around a growing RS crystal. The growth rate was obtained by optical observations with a constant time interval, while the temperature was continuously measured by six thermocouples of very fine copper and constantan wires (both wires 0.1 mm in diameter) set as shown in Figure 3.

The temperature of the solution was determined through the six 0.3 mm diameter copper posts; the solution side of the post was insulated by very thin resin film to prevent generation of electric potential between the solution and the post as an electric battery. The other end of the post was connected with the fine couple. Temperature of the cell holder was kept constant within 0.05 °C/day, and inside homogeneity and stability of the cell were kept within (0.005 °C/cm)/day. Temperature difference between the junction located at r and a terminal at r_o with a constant temperature of T_o was measured.

To measure the temperature profile, a RS solution to grow the seed was prepared so as to be saturated at 40 °C or to contain 3.46 moles of RS per liter, where the

538

growth cell was kept at $38.0 \pm 0.05°C$. Thus, the excess concentration $\Delta C = 3.46$ - 3.33 = 0.13 mol/l since the RS solution is saturated by 3.33 mol/l at 38.0 °C.

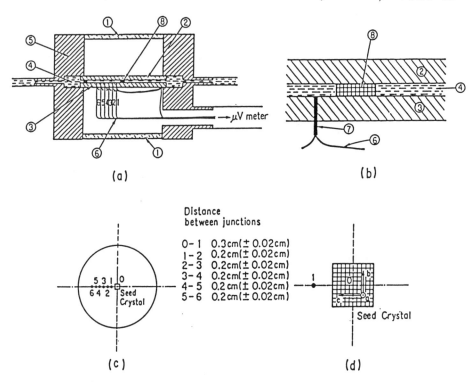

(a)   (b)

Distance
between junctions

0- 1   0.3cm(± 0.02cm)
1- 2   0.2cm(± 0.02cm)
2- 3   0.2cm(± 0.02cm)
3- 4   0.2cm(± 0.02cm)
4- 5   0.2 cm(± 0.02cm)
5- 6   0.2cm(± 0.02cm)

(c)   (d)

Figure 3(a) and (b): Vertical section of the cell; (1) window; (2) and (3) regin plates of the cell; (4) aqueous solution; (5) cell holder; (6) junctions (7) copper post with 0.3 mm in diameter to pick up temperature of the solution; (8) the seed crystal. (c) and (d): plan view to indicate the junctions and the seed crystal.

The vertical axis in Figure 4 is the temperature difference, $T(r) - T_o$, and the numerous on the horizontal axis indicate the positions of the six junctions.

Assuming a stationary state, the profiles of concentration C(r) and temperature T(r) around a growing crystal are, respectively, given by eqs.(14) and (15) or by cylindrical coordinates as

$$(1/r)\partial[r\partial C(r)/\partial r] = 0 \qquad (16)$$

and

$$(1/r)\partial[r\partial T(r)/\partial r] = 0. \qquad (17)$$

where r is the radial coordinate. Assuming the seed as a tiny circular disc located at the center of the cell, we have the following equations at the outside of the seed or in the range of r > R where R is the position of the interface:

$$C(r) = \{C(R) - C_o\}\{ln(r)/ln(R/r_o)\} + \{C_oln(R) - C(R)ln(r_o)\}/ln(R/r_o) \qquad (18)$$

and

$$T(r) = \{T(R) - T_o\}\{ln(r)/ln(R/r_o)\} + \{T_o ln(R) - T(R)ln(r_o)\}/ln(R/r_o). \quad (19)$$

Here, $C(R)$ and $T(R)$ are, respectively, the concentration and temperature at R, and C_o and T_o are those at r_o where r_o is radius of the cell periphery. C_o and T_o must be constant during this measurement because the periphery of the cell is in contact with the reservoir of the solution with the concentration C_o as shown in Figure 4a; it also has a constant temperature because the periphery is located far from the growing seed and is controlled by a precise instrument to maintain a constant temperature.

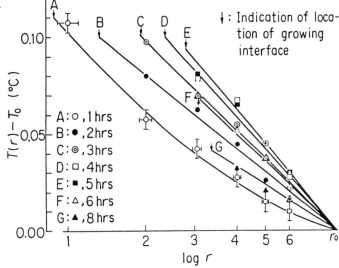

Figure 4. Temperature profiles around the seed crystal. Abscissa $ln(r)$: the distance from the 1st junction in log-scale. 1,2...6 indicate the positions of the junctions in Figure 3c. Ordinate $T(r)$: the temperature difference between T_o and each junction.

Equation (19) claims that the decrement of the temperature is proportional to $ln(r)$ if a stationary state can be realized. The linear relation between the decrement of temperature and $ln(r)$ is confirmed at 2 h or after a much longer time from injection of the solution as indicated in Figure 4. It thus appears that a stationary state has been realized. However, is this a true stationary state? We will continue analytical work under the assumption that a stationary state can be achieved.

The material flow is governed by the concentration gradient at the interface and then the growth rate V is given by

$$V = \Omega_B D_B \{\partial C(r)/\partial r\}_{r=R}, \quad (20)$$

where Ω_B and D_B are the molar volume and the diffusion coefficient of the solute, respectively.

540

The heat of crystallization caused by the solute deposition must diffuse from the interface through the surrounding solution because the temperature gradient inside the growing crystal should be zero at a stationary state. Thus we have

$$L_B D_B \{\partial C(r)/\partial r\}_{r=R} + k\{\partial T(r)/\partial r\}_{r=R} = 0, \tag{21}$$

where L_B is the heat of crystallization and k is thethermal conductivity of the solution.

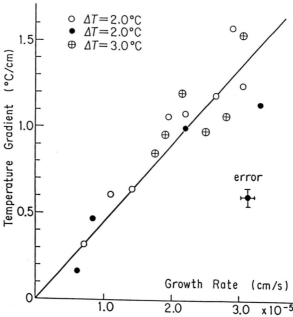

Figure 5 Relation between temperature gradient at the growing interface and growth rate.

By substituting eq.(21) into eq.(20), the growth rate is proportional to the concentration gradient at the interface as shown by eq.(22), and also proportional to the temperature gradient there given by eq.(23):

$$V = \Omega_B D_B \partial C(r)/\partial r_{r=R} \tag{22}$$

and

$$V = -(\Omega_B k/L_B)\partial T(r)/\partial r_{r=R}. \tag{23}$$

Here, eq.(23) was experimentally confirmed as shown in Figure 5. The value of L_B/k is estimated to be 6.84×10^6 cm s K mol^{-1} from the slope in Figure 5, assuming that Ω_B is equal to the molar volume of crystalline RS:

$$\Omega_B = 1.59 \times 10^2 cm^3 mol^{-1}, \tag{24}$$

In section 2-7 the thermal conductivity k was measured as

$$k = 1.25 \times 10^{-3} cal^{-1} K^{-1} s^{-1}. \tag{25}$$

Therefore, the heat of crystallization is estimated as

$$L_B = 8.6 \times 10^3 cal mol^{-1}. \tag{26}$$

which is very plausible because the heat of solution at infinite dilution is 12.3×10^3 cal mol^{-1}.

3.4. Concentration Gradients around Growing Crystals
3.4.1. Diffusion layer [10]
When eqs.(18) and (19) are substituted into eq.(21), we have

$$L_B D_B \{C(R) - C_o\} - k\{T(R) - T_o\} = 0. \tag{27}$$

Since the solute molecules are more quickly incorporated into a growing crystal through its interface compared to the time interval of the growth rate measurement, the concentration at the interface, C(R), should be equal to the saturated concentration there, $C_S(T(R))$, that is,

$$C(R) = C_S(T(R)) = C_S(T_o) + (\partial C_S/\partial T)_o\{T(R) - T_o\}, \tag{28}$$

Now, C(R) and T(R) are solved as the solutions of simultaneous equations (27) and (28), and then C(r) and T(r) are also obtained by substitution of C(R) and T(R) into eqs.(18) and (19). The growth rate is, therefore, determined as

$$V = \Omega_B \Delta C / \{F R ln(r_o/R)\}, \tag{29}$$

where

$$F = 1/D_B + (L_B/k)(\partial C_S/\partial T)_o. \tag{30}$$

The excess concentration ΔC was prepared as

$$\Delta C = C_o - C_s(T_o) = 1.3 \times 10^{-4} mol cm^{-3}, \tag{31}$$

in section 3-3. The temperature coefficient of the solubility at To was measured as

$$(\partial C_S/\partial T)_{40^\circ C} = 6.2 \times 10^{-5} mol cm^{-3} K^{-1}, \tag{32}$$

at 40 °C, and the diffusion constant was obtained in section 2-5 as

$$D_B = 2.5 \times 10^{-6} cm^2 s^{-1}. \tag{33}$$

By substitution of the numerical values given by eqs.(24), (25), (26), (31), (32) and (33) into eq.(29) using eq.(30), the growth rate is estimated as 1.1×10^{-7} cm s^{-1} at R = 0.38 cm, 2h after the injection of the solution. This rate is much too small compared with the 2.9×10^{-5} cm s^{-1} observed here. We thus numerically checked the value of $\{T(R) - T_{\mathrm{o}}\}$ which is analytically given by

$$T(R) - T_{\mathrm{o}} = L_{\mathrm{B}}\Delta T(\partial C_{\mathrm{S}}/\partial T)/(kF). \tag{34}$$

The observed value of $T(R) - T_{\mathrm{o}}$ is $0.10 \pm 0.01°$C, but the numerical value evaluated by eq.(34) is 0.003 °C using $\Delta T = 40.0 - 38.0 = 2.0°$C. This is also much smaller than the measured value.

Here, we only measured the temperature profiles around growing crystals but did not measure the concentration. The temperature profile will approach a stationary state as $\exp(-aK_{\mathrm{B}}t/r_{\mathrm{o}}^2)$ and the concentration distribution will vary as $\exp(-aD_{\mathrm{B}}t/r_{\mathrm{o}}^2)$ where K_{B} is the thermal diffusivity of the solution and "a" is a constant determined by the cell dimensions. Since the order of magnitude of K_{B} and D_{B} are, respectively, 10^{-3} and 10^{-6}cm^{2}s^{-1}, the temperature distribution becomes stationary much faster than the concentration does. We assume that the concentration distribution is still not stationary during our measurements, and then the diffusion field is limited around a growing crystal as schematically shown in Figure 6.

Since the concentration gradient is limited to within r_{e} but the temperature gradient is spread throughout the cell, r_{o} in eq.(18) is replaced by r_{e} but there is no change in eq.(19). Thus, the rate V_{e} and temperature at the interface $T_{\mathrm{e}}(R)$ are, respectively, given by

$$V_{\mathrm{e}} = \Omega_{\mathrm{B}}\Delta C/\{F_{\mathrm{e}}Rln(r_{\mathrm{e}}/R)\}, \tag{35}$$

and

$$T_{\mathrm{e}}(R) - T_{\mathrm{o}} = fL_{\mathrm{B}}\Delta T(\partial C_{\mathrm{S}}/\partial T)/(kF_{\mathrm{e}}), \tag{36}$$

where F_{e} and f are, respectively, given by

$$F_{\mathrm{e}} = 1/D_{\mathrm{B}} + f(L_{\mathrm{B}}/k)(\partial C_{\mathrm{S}}/\partial T)_{\mathrm{o}}. \tag{37}$$

and

$$f = ln(r_{\mathrm{o}}/R)/ln(r_{\mathrm{e}}/R). \tag{38}$$

To make a good agreement of the observed value with the value calculated by eqs.(35) and (36) using eqs.(24), (25), (26), (31), (32) and (33), the thickness of diffusion field

$$\delta = r_{\mathrm{e}} - R \tag{39}$$

should be a few hundred micrometers, which is a very proper thickness and usually called the "diffusion layer".

The factor F_{e} indicates the resistance against growth of crystals, in which $1/D_{\mathrm{B}}$ is the resistance due to the flow of the solute and $\{f(L_{\mathrm{B}}/k)(\partial C_{\mathrm{S}}/\partial T)_{\mathrm{o}}\}$ is the resistance due to the heat flow. The ratio of this heat resistance to the whole one is given by

$$\{f(L_{\mathrm{B}}/k)(\partial C_{\mathrm{S}}/\partial T)_{\mathrm{o}}\}/F_{\mathrm{e}} = \Delta T_{\mathrm{obs}}/\Delta T, \tag{40}$$

which is estimated as 0.05 at 2h after the injection.

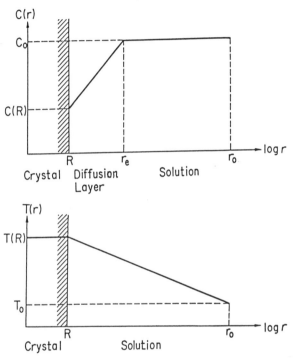

Figure 6. Schematic drawing of the concentration and temperature profile around a growing crystal. The temperature profile becomes stationary throughout the entire cell but the diffusion field of the solute is limited to within r_{e}.

3.4.2. Concentration gradient and growth rate

The concentration gradient is located within a thin layer with a few hundred micrometers in thickness on the interface of a crystal growing from aqueous solutions, which has been measured by Moire fringe method [6, 10] and by optical interferometry [12].

Figure 7 was obtained by growth of RS crystals at 12, 20, 30 and 40 °C, where optical index gradients at growing interfaces were measured against the growth rate. The optical index is translated into the concentration as discussed in section 2-3. Each solid line in Figure 7 was determined by the least-squares-fitting method using all the observed points but not using the origin of the coordinates. However, all the solid lines pass through the origins, which means that the growth rate is exactly proportional to the concentration gradient.

The diffusion coefficient of the solute is determined against temperature by substituting the slope obtained in Figure 7 and $\Omega_B = 1.59 \times 10^2 cm^3 mol^{-1}$ into eq.(20).

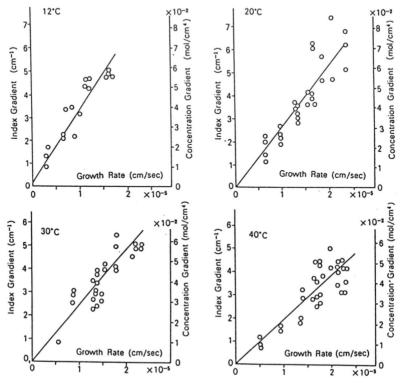

Figure 7. Concentration gradient and growth rate of RS crystals at 12. 20, 30, 40 °C.

The diffusion coefficients given in Table 4 are plotted against 1/T in Figure 8, from which the activation energy of the coefficient is given by

$$E = 0.136eV = 3.14kcal/mol \tag{41}$$

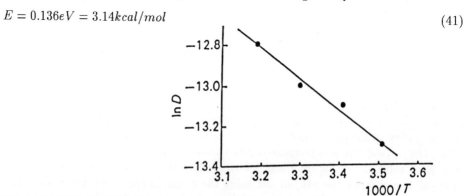

Figure 8. The coefficients of Table 4 against 1/T.

Table 4
Diffusion coefficients determined by the concentration gradient of Figure 7.

12 °C	20 °C	30 °C	40 °C	
1.70	1.95	2.19	2.85	$\times 10^{-6} \mathrm{cm}^2 \mathrm{s}^{-1}$

3.5. Growth Units

In the previous treatments in sections 2-2, 3-2 and 3-4-2, the growth unit is transcendentally assumed as Ω_B which is the molar volume of crystalline state of the solute.

Here, the Stokes-Einstein radius r_{SE} given by the next relation

$$r_{SE} = kT/(6\pi\eta D_B) \tag{42}$$

is numerically estimated as 0.2 nm, where k is the Boltzmann constant. On the other hand the equi-volume radius:

$$r_{ev} = (3\Omega_B/4\pi)^{1/3}, \tag{43}$$

is estimates as 0.4 nm [8, 13].

The growth units should be electrically neutral although a few ions and clusters will be present in the solution, because, if all the units charge up in the solution, a sort of clustering must be generated around the units to reach an electric stable state which will protest to be smoothly built into the seed crystal. This resistance to the growth will introduce another diffusion behavior and give us the other co-efficients when we obtain them by the slope of the gradient to the rate using Ω_B. However, there was no discrepancy between the diffusion coefficients obtained by an ordinary method mentioned in section 2-5 and the coefficient obtained from the crystal growth discussed in section 3-4-2.

To grow a perfect crystal every growth unit should be completely same from view points of its chemical structure and size, which must be the smallest one.

The growth units are, therefore, believed as the electrically neutral monomers composed by the same numbers and species of atoms included in their chemical formula.

4. CONCLUSIONS

A comprehensive treatment proposed here is that a saturated aqueous solution can be treated as an ideal mixture of solute monomers and solvent monomers, where all the monomers play individually and independently. Every monomer mentioned here is composed by the same numbers and species of atoms indicated by its chemical formula and then it is electrically neutral. Therefore, this mixture is successfully treated by the similar way of the "regular solution" as discussed here. Of course, a few ions and clusters will be present but they are minorities.

546

The rate of a crystal growing from an aqueous solution is clearly proportional to the temperature gradient at its interface and also proportional to the concentration gradient there, which is localized within the diffusion layer with a few hundred micrometers. From the physical aspects obtained here it is believed that the growth units must be the monomers of the solute.

ACKNOWLEDGEMENT

The author expresses his cordial thanks to his colleagues and students for supporting this research work, especially, Prof. T. Hashitani, Tokyo University of Agriculture and Technology, the late Prof. T. Kuroda, Hokkaido University, K. Satoh, K. Ohotsuka, E. Hirano, Y. Shiomi, T. Kurumizawa, N. Yamanouchi, Y. Tagawa and M. Naito. This work was begun from 1953 at Kobayashi Institute of Physics and continued by 1958, and started again in Gakushuin University at 1963 and continued to present. During these periods this research work was financially supported by Mombusho which was the most effective research fund for this basic research.

REFERENCES

1. J.H. Hildebrand, J. Am. Chem. Soc., 51 (1929) 66, J.H. Hildebrand and R.L. Scott, "Regular Solutions" Prentice-Hall Inc., (1962).
2. R. Fujishiro and T. Kuroiwa, "Youeki-no-Seishitsu 1", Tokyo-Kagaku-dojin, (1972). (In Japanese)
3. T. Ogawa, Jpn J. Appl. Phys., 16 (1977) 689-695.
4. T. Ogawa and K. Satoh, J. Chem. Eng. Data, 21 (1976) 33-35.
5. T. Ogawa, K. Satoh, T. Kurumizawa, N. Yamanouchi, and K. Ohotsuka, J. Crystal Growth, 56 (1982) 151-156.
6. T. Ogawa and K. Ohtsuka, J. Jpn Asso. Crystal Growth, 9 (1982) 161-174.
7. M. Ataka and T. Ogawa, J. Mater. Res., 8 (1993) 2889-2892.
8. T. Erdey-Gruz, "Transport Phenomena in Aqueous Solutions", Adam Hilger London, (1974)
9. D. D. Fitts, "Non equilibrium Thermodynamics" McGraw- Hill, (1962)
10. Y. Shiomi, T. Kuroda and T. Ogawa, J. Crystal Growth, 50 (1980) 397-403
11. E. Hirano and T. Ogawa, J. Crystal Growth 51 (1981) 113-118
12. M. Mantani, M. Sugiyama and T. Ogawa, J. Crystal Growth 114 (1991) 71-76
13. S. Glasstone, K.J.Laidller and H. Eyring, "Theory of Rate Process", McGraw-Hill, New York, 1941, p.516

Advances in the Understanding of Crystal Growth Mechanisms
T. Nishinaga, K. Nishioka, J. Harada, A. Sasaki and H. Takei (Editors)
© 1997 Elsevier Science B.V. All rights reserved.

Crystallization of sol-gel derived ferroelectric thin films with preferred orientation

S.Hirano, T.Yogo and W.Sakamoto[a*]

[a]Department of Applied Chemistry, School of Engineering, Nagoya University
Furo-cho, Chikusa-ku, Nagoya 464-01, Japan

Epitaxial lithium niobate ($LiNbO_3$, LN) and potassium tantalate-niobate
($KTa_xNb_{1-x}O_3$, KTN) thin films were synthesized through a reaction control of
metal alkoxide solutions. The structures of LN and KTN precursors in the solution
were analyzed by NMR spectroscopy. Starting metal alkoxides including metal-
oxygen-carbon bonds were found to undergo bond rearrangement yielding LN and
KTN precursors under controlled reaction conditions. An appropriate amount of
water for prehydrolysis of LN solution ranges from 1.0 to 2.0 equiv. $LiNbO_3$/Ti
films with an epitaxy were crystallized on sapphire C and Pt(111)/ sapphire C
substrates. Perovskite KTN films were crystallized on MgO(100) substrates using
H_2O/O_2 vapor treatment at 300°C followed by crystallization at 675°C. KTN films
on Pt(100)/MgO(100) of perovskite phase were also crystallized at 700°C. KTN
films were confirmed to have an epitaxy on Pt(100)/MgO(100) substrates by X-ray
pole figure measurement. $KTa_{0.65}Nb_{0.35}O_3$ films on Pt(100)/ MgO(100) substrates
showed P-E hysteresis at 225K.

1. INTRODUCTION

Lithium niobate ($LiNbO_3$, LN) has the illumenite structure, and has various
attractive properties, such as acoustic, piezoelectric, acoustooptic and electrooptic
properties [1]. LN thin films have been receiving much attention to their application
in optical waveguide devices [2]. Waveguides have been fabricated of single-crystal
LN. Usually, LN single crystal is nonstoichiometric, because the crystals have been
grown from a harmonic congruent melt corresponding to the composition of 48.45
mol% Li_2O. The presence of defects in the congruent LN crystal is responsible for
an uncontrolled diffusion of element during waveguide fabrication.

Potassium tantalate-niobate ($KTa_xNb_{1-x}O_3$, KTN) belongs to the perovskite
oxides, and changes the Curie temperature with the Ta/Nb ratio [3].
$KTa_{0.65}Nb_{0.35}O_3$ has the highest quadratic electrooptic coefficient at room temperature, which is associated with its remarkable photorefractive effect [4]. A modified
Kyropolous method has been utilized for the growth of KTN single crystals. How-

*This work was supported by a Grant-in-Aid for Scientific Reserach from the Ministry of
Education, Sicence and Culture of Japan, No. 04227102.

ever, the growth of perfect KTN single crystals is difficult because of the usual striation and compositional inhomogeneity.

Sol-gel process is one of the solution processes, and has several advantages, such as high purity, good homogeneity, ease of composition control and low temperature crystallization. Hirano et al. reported the synthesis of stoichiometric LN films with a preferred orientation on sapphire substrates using metal-organics [5–8]. The stoichiometric LN films were crystallized on sapphire substrates as low as 250°C. Calcination of gel film in water vapor/oxygen atmosphere is one of the key processing factors for the crystallization of LN films at low temperatures. Hirano et al. also demonstrated the prominent effect of water vapor treatment on the synthesis of $Pb(Zr,Ti)O_3$ [9] and $Pb(Mg,Nb)O_3$ films [10] of perovskite phase. They also reported the synthesis of oriented KTN films using metal-organics below 700°C [11]. Polycrystalline KTN films were also prepared by sol-gel method [12,13]. In these syntheses, however, the crystallization temperature of KTN perovskite phase is high in order to suppress the formation of pyrochlore phase, which degrades the ferroelectric properties of KTN.

This paper reviews the crystallization of sol-gel derived ferroelectric thin films with preferred orientation. Metal-organic precursors were designed through a reaction control of metal-oxygen-carbon bonds of component metal alkoxides. The structures of precursors in the solution were analyzed by NMR spectroscopy. The gel films on the substrates underwent the atomic rearrangement during calination and crystallization yielding epitaxial films. The calination and crystallization conditions were also investigated. Epitaxial LN and KTN thin films were successfully synthesized on Pt(111)/sapphire C and Pt(100)/MgO(100) substrates, respectively.

2. SYNTHESIS OF LiNbO$_3$ THIN FILMS

2.1. Syntheisis of LN precursor and partial hydrolysis

The processing scheme of the LN precursor solution is shown in Figure 1. Lithum ethoxide (LiOEt) was reacted with niobium ethoxide ($Nb(OEt)_5$) in absolute ethanol at a refluxing temperature for 24h, and then various amounts of water from Rw=1 ($Rw=H_2O/LiNb(OEt)_6=1.0$) to 4 euiv. were added to the LN solution. After reflux for 24h, the solution was condensed to 0.2 mol/l, which was used as a coating solution.

The authors have already reported that the LN precursor films crystallize on the sapphire substrates at 250°C using partial hydrolysis. The added water to the LN precursor influences significantly the crystallization temperature. An interaction between added water and the LN precursor in the solution was investigated by NMR spectroscopy.

The reaction product between LiOEt and $Nb(OEt)_5$ was analyzed by ^{13}C and ^{93}Nb NMR spectroscopy. Figure 2 shows ^{13}C NMR spectra of $Nb(OEt)_5$ and the LN precursor. $Nb(OEt)_5$ shows two kinds of ethoxy groups as shown in Figure 2a. However, LN precursor contains one kind of ethoxy group (Figure 2b), which suggests the presence of equivalent ethoxy group in the LN precursor. The LN precursor shows one signal at -1156 ppm in ^{93}Nb NMR spectra as shown in Figure 3.

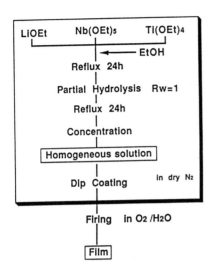

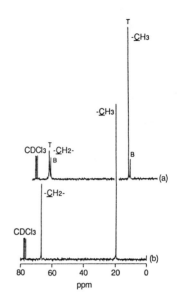

Figure 1. Processing scheme of LN thin films from LiOEt and Nb(OEt)$_5$.

Figure 2. ^{13}C NMR spectra of (a) Nb(OEt)$_5$ and (b) LN precursor.

This result suggests that the LN precursor has a Li[Nb(OEt)$_6$] structure including [Nb(OEt)$_6$] octahedron. Although Li[Nb(OEt)$_6$] crystal is reported to have a bent O-Nb-O bond [14], the LN precursor in the solution has a more symmetrical structure because the precursor is more free in stretching motions of chemical bonds in the solution.

Both the coordination of solvent molecules to partially positive Li cation and the ligand exchange reaction play important roles in the initial stage of the hydrolysis reaction. The effect of water on the LN precursor was analyzed by ^{7}Li and ^{93}Nb NMR spectroscopy. The ^{7}Li signal of Li[Nb(OEt)$_6$] increased in half-value width with increasing the water amount from 1.0 to 4.0 equiv. The change in the ^{93}Nb signal of the LN precursor at -1156 ppm with water is shown in Figure 3. The signal shows no change to the amount of water up to Rw=2.5. At Rw=2.7, the signal begins to decrease in intensity, and a new broad signal centered at -700 ppm appears. The signal at -1156 ppm disappears at Rw=3.0. The added water molecules coordinate mainly to Li of Li[Nb(OEt)$_6$] up to Rw=2.5. However, water begins to coordinate [Nb(OEt)$_6$] and substitute OEt for OH promoting the hydrolysis-condensation reaction at Rw=2.7. The broad signal at -700 ppm is considered to assign to the condensed (Nb-O)$_n$ species.

The precursor film formed from the coating solution of Rw=2.7 was heat-treated at 400°C. The film crystallized to LiNbO$_3$ with a small amount of Li$_3$NbO$_4$. The excess amount of water forms segregated (M-O)$_n$ of Li and/or Nb, which leads to the formation of a second phase. On the other hand, single-phase LN was formed

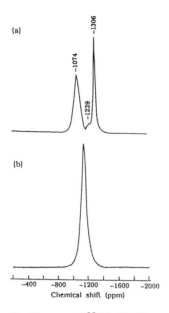

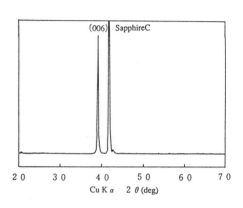

Figure 3. Change of ^{93}Nb NMR spectra with amount of added water (Rw).

Figure 4. XRD profile of LiNbO$_3$/Ti film crystallized at 350°C on a sapphire C substrate.

from the precursor treated with the added water of Rw=2.0

After LiOEt was reacted with Nb(OEt)$_5$ at room temperature for 3h, 2.0 euiv. water (Rw=2.0) was added. No refluxing was conducted on the LN precursor before partial hydrolysis. The ^{93}Nb NMR spectrum of the solution shows a main signal at -1156 ppm and a broad signal at -700 ppm. When the powder from the solution crystallized at 350°C, a mixture of LiNbO$_3$, LiNb$_3$O$_8$ and Li$_3$NbO$_4$ was formed. No single phase LiNbO$_3$ was obtained from the solution. Without reflux treatment, the formation of LN precursor is not completed. The unreacted Nb(OEt)$_5$ undergoes a hydrolysis affording segregated (Nb-O)$_n$. The segregation does not disappear by thermal diffusion of elements at 400°C. The refluxing is a key factor required to complete the reaction between LiOEt and Nb(OEt)$_5$.

2.2. Preparation of epitaxial LN films

The LN films crystallized on Si substrates were polycrystalline, while the films on sapphire substrates showed preferred orientations depending upon crystallographic planes of sapphire. Only the 110, 012 and 006 reflections of LN were observed on sapphire R, A and C substrates, respectively [6–8].

The Ti-doped waveguide of LN is fabricated by a Ti-diffusion method [2]. However, the interface between the waveguide and matrix is not well-controlled by the diffusion method. Ti is expected to be doped uniformly in the LN film by the sol-gel method. The LN precursor including 5mol % Ti was prepared from LiOEt,

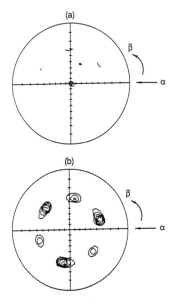

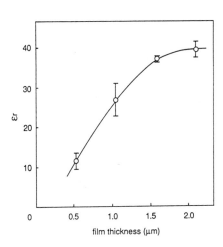

Figure 5. X-ray pole figures of LiNbO$_3$/Ti film on a sapphire C substrate (a) {006} pole, (b) {012} pole.

Figure 6. Change of dielectric constant of LN film with film thickness.

Nb(OEt)$_5$ and Ti(OEt)$_4$. Ti(OEt)$_4$ was reacted with lithium and niobium ethoxide yielding a complex alkoxide. A precursor film crystallized at 350°C on the sapphire C subatrates. The XRD profiles of the film is shown in Figure 4. The Ti-doped LN film shows only the 006 relection, which supports an excellent c-axis preferred orientation. Figure 5a shows the X-ray pole figure of the Ti-doped LN film for {006} pole. The storng spots appears at $\alpha=90°$, which shows a (006) plane orientation. Also, the pole figure for a {012} pole shows clear spots at $\alpha=36.6°$ (Figure 5b). This result corresponds not only to the preferred orientation but also to the three dimensional regularity of LN crystallites perpendicular to the substrate.

Similarly, the X-ray pole figure of the LN films crystallized on the Pt(111)/sapphire C substrates showed clear spots with a three-fold symmetry. The epitaxy of LN on the Pt(111)/sapphire C substrates reflects a crystallographic matching between Pt(111) and LN. The mismatch between Pt(111) and LN is 6.72 %, which is smaller than that between sapphire C and LN (8.20 %).

The refractive indices of LN and 5 mol% Ti-doped LN films were 2.28 and 2.34, respectively. The dielectric constant of the LN film (film thickness, 1.0μm) was 25 at 100 kHz at room temperature. The change of dielectric constant with film thickness is shown in Figure 6. The dielectric constant of the film above 1.5μm-thick showed a constant value of 38.

3. SYNTHESIS OF KTN THIN FILMS

3.1. Syntheisis of KTN precursor and partial hydrolysis

The synthesis procedure of the KTN precursor and thin films is shown in Figure 7. Tantalum ethoxide and niobium ethoxide (molar ratio, 65/35) were dissolved in absolute ethanol, and reacted for 24h. A stoichiometric amount of pottasium ethoxide in ethanol was added to the solution, and refluxed for 48h. The solution was condensed to 0.2 mol/l, and used for a coating solution.

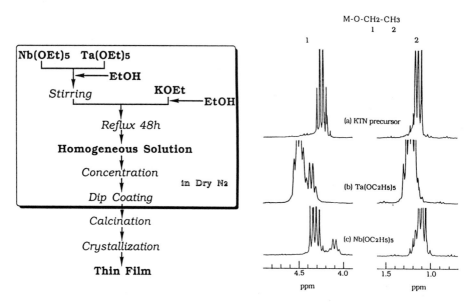

Figure 7. Processing scheme for the synthesis of KTN precursor and thin films.

Figure 8. ^{1}H NMR spectra of (a) KTN65 precursor, (b) Ta(OEt)$_5$, and (c) Nb(OEt)$_5$.

The structure of KTN65 (x=0.65) precursor was analyzed by NMR spectroscopy. Figure 8 shows ^{1}H NMR spectra of the KTN65 precursor, tantalum ethoxide and niobium ethoxide. Each ethoxide and the KTN precursor show methyl triplets and methylene qaurtets of ethoxy groups. The ethoxy group in the KTN65 precursor shows different chemical shifts from those of starting tantalum and niobium ethoxide. This fact reveals a formation of double alkoxide. Two kinds of methyl and methylene protons of tantalum and niobium ethoxide correspond to the formation of dimer in solution (Figure 9a).

Figure 10 shows ^{93}Nb NMR spectra of niobium ethoxide and the KTN65 precursor. Niobium ethoxide shows two signals at -1000 ppm and -1307 ppm, which indicates a formation of the equilibrium mixture of monomer and dimer. When the solution of niobium ethoxide was diluted with ethanol, the signal at -1307 ppm decreased in intensity, while the signal at -1000 ppm increased. This fact reveals that the former is due to dimeric niobium ethoxides, the latter to monomeric species. The KTN65 precursor shows a single signal at different chemical shift of -1145 ppm (Figure 10b), which shows the formation of one niobium nucleaus by the reaction of tantalum ethoxide, niobium ethoxide and pottasium ethoxide. From the NMR spectra, the KTN precursor consists of a $[Nb(OEt)_6]$ octahedral structure as shown in Figure 9b.

When the molar ratio of K, Ta and Nb deviates from the stoichiometry, the main signal was accompanied by other signal at -1000 ppm. The off-stoichiometric precursor results in the formation of pyrochlore phase as a by-product after crystalllization. The signal at -1145 ppm began to split by addition of 0.5 equiv. water. This suggests the hydrolysis of $[Nb(OEt)_6]$ species. Since the KTN precursor is less stable to water than the LN precursor, no partial hydrolysis was used for the KTN precursor.

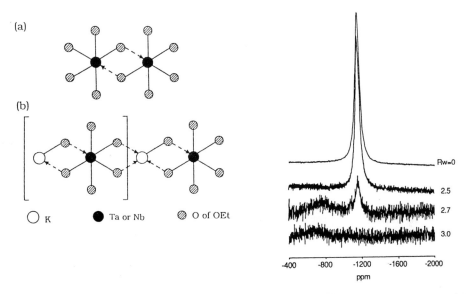

Figure 9. Structures of (a) M(OEt)$_5$ and (b) KTN precursor.

Figure 10. ^{93}Nb NMR spectra of (a) Nb(OEt)$_5$ and (b) KTN65 precursor.

The powders with compositions $KTa_xNb_{1-x}O_3$ ($x=1, 0.75, 0.65, 0.5, 0.25, 0$) were prepared from the corresponding precursor solutions by hydrolysis and evaporation. The powders were heat-treated at temperatures between 650°C and 850°C. The XRD profiles of KTaO$_3$ heat-treated at 650°C and 850°C are shown in Figure 11a and 11b, respectively. The pyrochlore and perovskite phase are formed at

650°C and 850°C, respectively. The powder of KNbO$_3$ composition crystallizes to the perovskite structure at 650°C (Figure 11c). Therefore, the crystallization and transformation depend upon the composition. The KTN powders were found to form a solid solution for $x=0.75\sim0.25$ by XRD analysis [13].

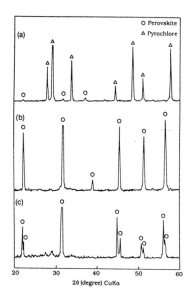

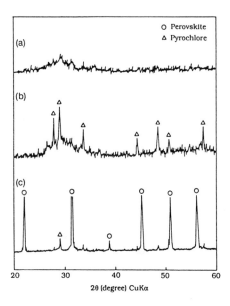

Figure 11. XRD profiles of powders of (a) KTaO$_3$ fired at 650°C, (b) KTaO$_3$ fired at 850°C and (c) KNbO$_3$ fired at 650°C.

Figure 12. XRD profiles of KTN65 powders heat-treated at(a) 300°C, (b) 650°C and (c) 750°C.

Figure 12 shows XRD profiles of the KTN65 powder ($x=0.65$) at (a) 300°C, (b) 650°C and (c) 750°C. The powder is amorphous at 300°C, which crystallizes to pyrochlore at 650°C, and then to perovskite at 750°C. The KTN65 powder easily crystallized in the pyrochlore phase at lower temperatures.

3.2. Synthesis of epitaxial KTN thin films

Various substrates were dipped into the coating solution, and withdrawn at a fixed speed producing the precursor films on the substrates. The precursor films were dried at room temperature for a few minutes under nitrogen, and were calcined and crystallized using two methods. For the type A method, the precursor films were calcined at 300°C and then crystallized at 675°C in an O$_2$ flow. The type B included the calcination of films in H$_2$O/O$_2$ gas. To increase the film thickness, the above procedures were repeated several times.

The precursor films was prepared from the coating solution of KTN65. MgO(100) was selected as a suitable substrate due to the good lattice matching to KTN. The

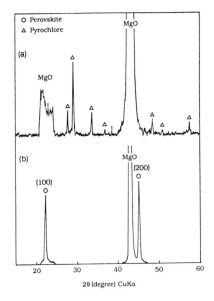

Figure 13. XRD profiles of KTN65 thin films on MgO(100) substrates after heat treatment through (a) type A and (b) type B.

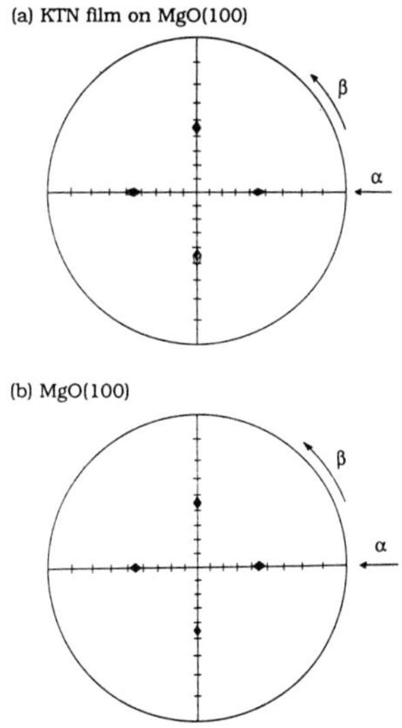

Figure 14. X-ray pole figures of (a) KTN65/MgO(100) and (b) MgO(100).

precursor films were tretaed by the two methods above mentioned. Figure 13 shows the XRD profiles of KTN65 films. The pyrochlore phase is observed using method A as shown in Figure 13a. On the other hand, the perovskite KTN65 with the (100) preferred orientation is obtained by method B (Figure 13b). The crystallinity of the film by method B is much superior to that by method A in terms of the direct formation of the perovskite phase. From these results, a mixture of H_2O and O_2 gases duirng calcination has a pronounced effect on the elimination of remaining organic components. The removal of organic groups promotes the subsequent crystallization of perovskite KTN. The hydrolysis of precursor films on MgO is considered to promote the crystallization of perovskite KTN rather than the pyrochlore phase. Similar effects were observed for the crystallization of perovskite $Pb(Zr,Ti)O_3$ and $Pb(Mg,Nb)O_3$ films.

Thin films on various substrates, such as Si(100), SiO_2(0110) and sapphire(R), by method B. However, only the pyrochlore phase was observed on these substrates even at 750°C by XRD analysis.

The oriented KTN65 film was analyzed by the X-ray pole figure measurement for {110} pole. The result is shown in Figure 14. The pole figure of MgO(100) substrate for {110} is also shown in Figure 14. Figure 14a shows clear spots at $\alpha=45°$ with a four-fold symmetry, which supported the epitaxial growth of KTN65 on the MgO(100) substrates.

556

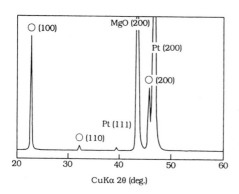

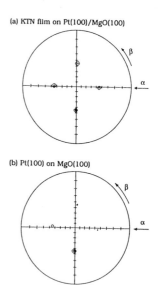

(a) KTN film on Pt(100)/MgO(100)

(b) Pt(100) on MgO(100)

Figure 15. XRD profile of KTN65 thin film on a Pt(100)/MgO(100) substrate crystallized at 700°C.

Figure 16. X-ray pole figures of (a) KTN65/Pt(100)/MgO(100) and (b) Pt(100)/MgO(100).

The KTN65 thin film was prepared on the Pt(100)/MgO(100) substrate in order to evaluate the electric properties of the films. In order to synthesize the oriented films, a buffer layer was precrystallized on the Pt(100)/MgO(100) substrates using a dilute KTN precursor solution with a concentration of 0.01 mol/l. The crystallization of KTN65 film was influenced by the crystallization of the preapplied film on the Pt(100)/MgO(100) substrate. The underlying KTN film developed by a first dip coating was crystallized at 700°C after calcination in flows of water vapor and oxygen. The KTN65 films on the underlying film were crystallized in perovskite at 700°C. Figure 15 shows the XRD profile of KTN65 films on the Pt(100)/MgO(100) substrates. The KTN65 films thus prepared shows only the 100 and 200 reflections, have a prominent (100) plane preferred orientation.

Figure 16 shows the X-ray pole figures of (a) KTN65/Pt(100)/MgO(100) and (b) Pt(100)/MgO(100). Pt(100) grows on MgO(100) epitaxially as shown in Figure 16b. When the X-ray pole figure of the KTN65 films on Pt(100)/MgO(100) was measured for {110} pole, clear spots with a four-fold symmetry were constructed at $\alpha=45°$. The KTN65 films were found to crystallize epitaxially on both the MgO(100) and Pt(100)/MgO(100) substrates.

The orientation of the films depend strongly upon the substrates. The cubic phase of KTN65 has a lattice parameter of 400.0 pm, and that of MgO is 421.3 pm. The lattice mismatch between KTN(100) and MgO(100) is 5.32 %. The deposited Pt layers on MgO(100) had a (100) orientation with a three-dimensional

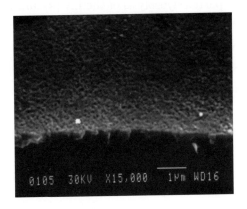

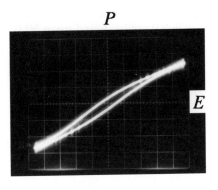

Figure 17. Edge-on profile of KTN65 thin film on a Pt(100)/MgO(100) substrate crystallized at 700°C.

Figure 18. P-E hysteresis loop of KTN65 film crystallized at 700°C on a Pt(100)/MgO(100).

alignment as shown in Figure 16b. Platinum has an fcc atomic packing with the lattice parameter of 392.3 pm. The mismatch between KTN(100) and Pt(100) is 1.96 %, whcih is much smaller than that between KTN and MgO. Thus, the crystallization of KTN with both in-plane and out-of-plane orientations results from the crystallographic matching of KTN(100) to Pt(100).

Figure 17 shows an edge-on photograph of the KTN film on the Pt(100)/MgO(100) substrate. The surface of the film is smooth, and has no pores and cracks.

Figure 18 shows a P-E hysteresis curve of the KTN65 films on the Pt(100)/MgO(100) substrate at 225K. The remanent polarization and coercive field were $1.5\mu C/cm^2$ and 8.7kV/cm (Film thickness, $0.7\mu m$), respectively. The dielectric constant was 1026 at 10kHz at room temperature.

4. CONCLUSIONS

Epitaxial LN and KTN films were successfully synthesized on platinum/metal oxide composite substrates from designed LN and KTN precursors. The precursors were synthesized by a reaction control of metal-organics. The crystallization conditions were controlled yielding the epitaxial LN and KTN films. The results are summarized as follows:

(1) Partial hydrolysis and water vapor treatment are required for syntheisis of the stoichiometric LN epitaxial thin films at low temperatures.

558

(2) The appropriate amount of water for partial hydrolysis of the LN precursor is 2.0 equiv. based upon NMR spectroscopy.

(3) The LN films were found to grow epitaxially on the sapphire and Pt(111)/ sapphire substrates.

(4) The KTN precursor comprises a double alkoxide including [Nb(OEt)$_6$] octahedron unit.

(5) The Provskite KTN thin films crystallized on the MgO(100) substrates using H$_2$O/O$_2$ calcination.

(6) The KTN thin films crystallize epitaxially on both the MgO(100) and Pt(100)/ MgO(100) substrates, and show a ferroelectric hysteresis.

The current procedure using the reaction control of metal-organics combined with crytallization under the controlled conditions is a novel method for synthesis of the pyrochlore-free perovskite film with excellent orientations at a relatively low temperature.

REFERENCES

1. R.S. Weis and T.K. Gaylord, Appl. Phys. A 37 (1985) 191.
2. M.M. Abouelleil and F.J. Leonberger, J. Amer. Ceram. Soc. 72 (1989) 1311.
3. S. Triebwasser, Phys. Rev. 114 (1959) 63.
4. R. Orlowski, L.A. Boatner and E. Kratig, Opt. Commun., 35 (1980) 45.
5. S. Hirano and K. Kato, Adv. Ceram. Mater. 2 (1987) 142.
6. S. Hirano and K. Kato, Adv. Ceram. Mater. 3 (1988) 503.
7. S. Hirano and K. Kato, J. Non-Cryst. Solid 100 (1988) 538.
8. S. Hirano and K. Kato, Solid State Ionics 32/33 (1989) 765.
9. S. Hirano, T. Yogo, K. Kikuta, Y. Arai, M. Saitoh and S. Ogasawara, J. Amer. Ceram. Soc. 75 (1992) 2785.
10. S. Hirano, T. Yogo, K. Kikuta and W.Sakamoto J. Sol-Gel Sci. Technol. 2 (1994) 329.
11. S. Hirano, T. Yogo, K. Kikuta, T. Morishita and Y. Ito, J. Amer. Ceram. Soc. 75 (1992) 1701.
12. J.V. Mantese, A.L. Micheli, N.W. Schubring, A.B. Catalan Y.L. Chen, R.A. Waldo and C.A. Wong, J. Appl. Phys. 72 (1992) 615.
13. D.D. Liu, J.G. Ho, R.C. Pastor and O.M. Stafsudd, Mater. Res. Bull. 27 (1992) 723.
14. D.J. Eichrost, D.A. Payne, S.R. Wilson and K.E Howard, Inorg. Chem. 29 (1990) 1458.

Advances in the Understanding of Crystal Growth Mechanisms
T. Nishinaga, K. Nishioka, J. Harada, A. Sasaki and H. Takei (Editors)
559

Anisotropy in microscopic structures of ice-water and ice-vapor interfaces and its relation to growth kinetics

Y. Furukawa[a] and H. Nada[b]

[a]Institute of Low Temperature Science, Hokkaido University, Sapporo 060, Japan

[b]National Institute for Resources and Environment, Tsukuba 305, Japan

Interface structures of ice crystals are discussed on the basis of the experimental results and the molecular dynamics simulations. Firstly, experimental results about the quasi-liquid layers (qll) originated by surface (interface) melting phenomena at the ice surfaces (interfaces) are summarized. It should be emphasized that the anisotropic surface melting occurs on the $\{0001\}$- and $\{10\bar{1}0\}$-faces of an ice crystal at temperatures below the melting point of bulk ice crystal. The melting layer at the interface between ice crystal and glass substrate was also detected at temperatures just below the melting point. Secondary, results of molecular dynamics simulations for the ice surface are mentioned. Anisotropic surface melting was first detected between the $\{0001\}$- and $\{10\bar{1}0\}$-faces, and the temperature dependencies of qll thickness were basically consistent with the experimental results. Molecular dynamics simulations for ice/water interfaces were also carried out and the anisotropic interface structures were clarified between the ice$\{0001\}$/water and ice$\{10\bar{1}0\}$/water interfaces. Finally, we emphasize that the pattern formation of ice crystal should be explained in relation to the anisotropic micro-structures of ice surface or interface.

1. INTRODUCTION

It is well known that the surfaces of an ice crystal and the interfaces between ice and other material are covered with the thin water layers (that is, quasi-liquid layer (qll)) at temperatures near the melting point ($0°C$). This is the so-called surface melting phenomenon, which is one of the phase transitions of surface structures. It goes without saying that such a melting transition at the surface or interface is a common phenomenon in any solid when it is in the temperature range near its melting point [1–3]. The physical properties of surface melting on such solids as the rare gas crystals and metals have considerably been elucidated by both the theoretical and experimental research in this decade. However, surface melting phenomenon on such a molecular crystal as ice still includes a lot of ambiguous properties. It is also very interesting to investigate how the melting transition relates to the equilibrium form of crystal, the growth kinetics, the growth forms of crystal and so on.

On the other hand, many natural phenomena observed in the cold regions of the earth should be originated by the existence of ice crystals. Specially, many interesting phenomena, which occur through the ice surfaces or interfaces, are well known. For instance, it has been expected that the drastic changes on the growth forms of natural snow crystal as a function of growth temperature should be explained in conjunction to the anisotropic surface melting on an ice crystal [4]. Such common phenomena as the regelation [5] and frost heave [6] strongly relate to the existence of qll at the interface between ice crystal and other material. Though, to understand the mechanisms of these phenomena, we should know the microscopic surface (interface) structures of ice and the dynamic properties of qll, there has not been enough systematic investigations about the ice surfaces and interfaces. One reason is in the difficulty to observe the ice surface or interface in molecular level, because the equilibrium vapor pressure of ice is so high at temperatures near the melting point that the new experimental techniques for surface analysis in the high vacuum condition can not be applied to the ice study.

Recently the authors have studied the ice surfaces and ice interfaces at the temperature range near the melting point using the methods of experiments and computer simulations. We would like to stress that the ice crystal is most important material in the sense of both the crystal growth relating to the melting transition, and the basic study of environmental problems as mentioned above. In this paper, we firstly summarize the results of optical experiments about the ice surface. Secondary, we mention about the results of molecular dynamics (MD) simulations for the ice surfaces and ice-water interfaces. Finally we briefly discuss the relation between the anisotropic surface (interface) structures and the pattern formation of ice crystals.

2. EXPERIMENTAL STUDIES ON MICROSCOPIC STRUCTURES AND DYNAMIC PROPERTIES OF ICE SURFACES

2.1. Microscopic structures of ice surfaces

Some ellipsometric measurements [7–9] were carried out to obtain the thickness d and the index of refraction of qll n_{qll} on both the $\{0001\}$- and $\{10\bar{1}0\}$-faces of an ice crystal at temperatures just below the melting point. Figure 1 shows the thickness of qll as a function of supercooling ΔT $(= T_m - T)$. Solid and broken lines show the results for the $\{0001\}$- and $\{10\bar{1}0\}$-faces, respectively, which were measured for ice samples put in the air by Furukawa $et\ al.$ [8].

The transition layers were detected at temperatures above -2°C and -4°C for the $\{0001\}$- and $\{10\bar{1}0\}$-faces, respectively, and d steeply increased with increasing temperature for both faces. n_{qll} was 1.330 (otherwise, its density is $0.991g/cm^3$) for both faces and for whole temperature ranges. Since this value is between $n_{\text{water}}=$ 1.333 and $n_{\text{ice}}=1.308$, and obviously smaller than n_{water} beyond the experimental precision, we conclude that the transition layer observed should be regarded as a quasi-liquid layer (qll). It is to be noted that the temperature dependence of the qll thickness extremely differ. The temperature dependence for the $\{10\bar{1}0\}$-face is proportional to $\Delta T^{-1/3}$ above -2°C, but the gradient for the $\{0001\}$-face is much

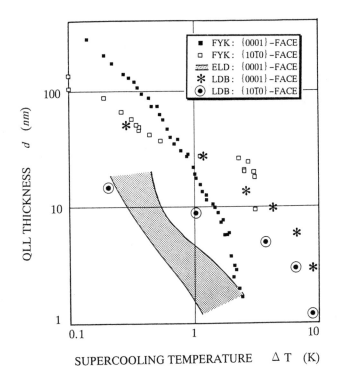

Figure 1. Summary of published data on the qll thickness at ice/air (FYK:from ref.[8]) and ice/vapor interfaces (ELD:from ref.[9] and LDB:from ref.[10]).

steeper. As a result, the relation between d_{0001} on $\{0001\}$ and $d_{10\bar{1}0}$ on $\{10\bar{1}0\}$ is reversed at about $-1°\mathrm{C}$. Consequently, we conclude that the anisotropic surface melting was firstly detected on the $\{0001\}$- and $\{10\bar{1}0\}$-faces of an ice crystal, and the structure of molecular arrangement in qll should be very close to that in bulk water but its physical property must be strongly influenced from the ice substrate underneath qll.

In addition, Elbaum $et\ al.$[9] measured a light reflectivity from ice surfaces and analyzed the qll thickness for the ice surfaces put both in the pure water vapor and in the air. They detected qll on the $\{0001\}$-face and found that d_{0001} for the ice sample put in water vapor is much larger than d_{0001} in the air.

On the other hand, Lied $et\ al.$[10] measured the X-ray scattering intensities from the ice surfaces in the water vapor. As a result, they detected the transition layers with the random arrangement of water molecules on both the $\{0001\}$- and $\{10\bar{1}0\}$-faces above $-10°\mathrm{C}$. This critical temperature is considerably lower than those obtained by the optical experiments.

Eventually, all experimental results give the direct evidence surface melting on ice crystals. However, it is to be recognized that the critical temperatures of surface melting are not consistent in the experimental results. We expect that the reasons of this discrepancy may be in the effect of environmental conditions around ice sample, the difference of ice surface preparations, and/or the difference of visual structures depending on the experimental methods. Furthermore, it is still difficult to gain a correct understanding about the relation between the temperature dependence of the qll thickness and the intermolecular force acting in an ice crystal. Though the dependence of $d_{qll} \propto \Delta T^{-1/3}$ observed in the optical experiments is usually expected from the van der Waals interaction between molecules [11], recent theoretical [12] and experimental results [13] indicate that such the simple interaction force is not applicable to the water molecules in an ice crystal. More precise and consistent studies should be put in practice on the basis of both the theoretical and experimental viewpoints in the future.

2.2. Physical properties of qll at interface between ice crystal and substrate

Similar melting transition at the interface between an ice crystal and other material. Since the molecules in the substrate interact with the water molecules in the ice crystal, physical properties of qll at the interface should be different from those of qll at the surface. Many reports have been published about the structures of interfaces. All of them, however, have not given any direct evidence showing the existence of qll at the interface.

Furukawa and Ishikawa [14] recently carried out an ellipsometric measurement for the interface between ice crystal and glass substrate. The result is summarized as follows. At temperatures below -1°C, a layer with the refractive index of 1.42 and the thickness of about $10 nm$, which may be originated from micro-roughness of the glass surface, was observed. When the temperature reaches above -1°C, a layer with the refractive index close to n_{water} appeared at the interface and its thickness rapidly increased as the temperature approaches to the melting point of ice. Consequently, we conclude that the direct evidence of the interfacial melting at the ice/glass interface is first given in this measurement and the threshold temperature of melting transition is at least lower than -1°C in this case.

On the contrary, the self-diffusion coefficient (D_{qll}) of water molecules in qll gives a measure of the physical property of interface. Recently, Ishizaki et al.[15] carried out NMR (Nuclear Magnetic Resonance) measurement for a sample which was prepared by freezing the mixture of water and soil particles, and determined D_{qll} for qll which is expected to exist between the ice crystal and the particle, using a Spin-Locking method. Namely, D_{qll} was in the order of $4 \times 10^{-10} cm^2/s$ at the temperature range between -1 and -10°C, which is shown in Figure 2. This value is about four to five orders in magnitude smaller than $D_{water} \approx 10^{-6} cm^2/s$ and about two orders larger than $D_{ice} \approx 10^{-14} cm^2/s$. This result indicates that the physical property of qll may be completely different from that of supercooled bulk water.

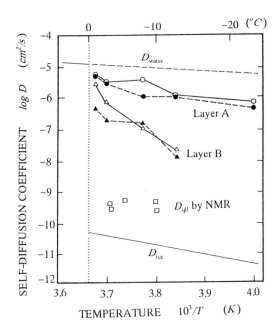

Figure 2. Summarized results for the self-diffusion coefficients of water molecules obtained by both the experiments and MD simulations. Results for each molecular layer obtained by MD are shown by the solid lines ({0001}-face) and the broken lines ({10$\bar{1}$0}-face). Experimental results for the quasi-liquid layer, measured by NMR method, are shown by the open squares. For reference, the self-diffusion coefficients for both bulk water and ice are also indicated by the thin lines.

3. MOLECULAR DYNAMICS SIMULATION OF ICE SURFACE

3.1. Simulation method

The TIP4P water molecular model[16–18], which is one of the rigid models, was used for the calculation of interaction among water molecules. In this model, the interaction force between water molecules is given as the sum of the Lennard-Jones force which works between the oxygen atoms and the Coulomb forces which work among the charged points on each molecule. Due to the difficulty of the application of Eward method for the system including the surfaces, the Coulomb interactions were cut off through a switching function at the distance of 8.95Å[19].

A simulation box was prepared as follows. First, 864 water molecules of TIP4P are arranged to make an ice crystal structure under the peridic boundary conditions imposed in all directions (x, y and z-directions). If only the periodic boundary condition in z-direction is eliminated, bilateral x-y faces are changed to the free surfaces being contact with the vapor phase. In this study, both the surface orientations of

564

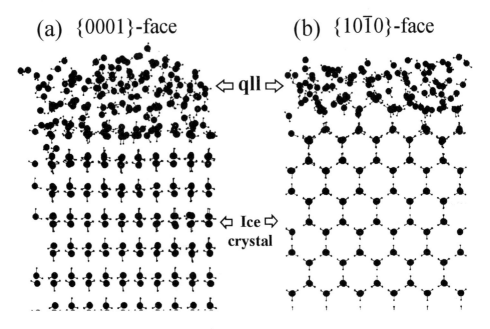

(a) {0001}-face   (b) {10$\bar{1}$0}-face

$\Leftarrow$ qll $\Rightarrow$

$\Leftarrow$ Ice $\Rightarrow$
crystal

Figure 3. Actual arrangements of water molecules around the surfaces in simulation boxes, projected along the x-y plane (namely, surface). Solid circles and small dots indicate the averaged positions of Oxygen atoms and Hydrogen atoms, respectively. (a) {0001}-face and (b) {10$\bar{1}$0}-face. Transition layers with random arrangements of water molecules (namely, quasi-liquid layers) are clearly observed on both the systems. Simulation temperature is 265K.

{0001} and {10$\bar{1}$0} were prepared to clarify the anisotropy of the surface structures.

Computations were carried out by the leap-frog algorithm. Initial translational velocities of each molecule in the ice crystal were assigned according to the Maxwell-Boltzmann distribution equation. To obtain the equilibrium state of the system, the time step of integration was initially set on $2.0 \times 10^{-15} s$ and the re-scaling of average molecular velocities was applied to the system in the period of $4.0 \times 10^{-14} s$.

In this simulation, the equilibrium state was established within the total simulation time of $1.0 \times 10^{-10} s$. After this time, the calculation was continued by the time step of $5.0 \times 10^{-15} s$ and finished at the total simulation time of $1.2 \times 10^{-10} s$. Simulations were carried out at the temperatures of 250, 260, 265, 270 and 272K, respectively.

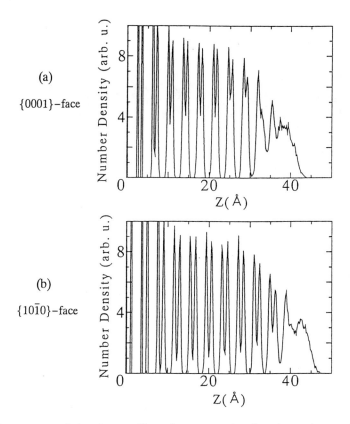

(a)

{0001}–face

(b)

{10$\bar{1}$0}–face

Figure 4. Time-averaged density profiles of water molecules along the z-axis (perpendicular to the ice surfaces) for the {0001}-surface (a) and the {10$\bar{1}$0}-surface(b), respectively. Double peaks originated from the ice crystal structure are observed at central regions of the simulation boxes. The peaks, however, disappear at the surface regions.

3.2. Anisotropic structures of ice surfaces obtained by MD simulations

Figure 3 shows actual arrangements of water molecules near the surfaces at the temperature of 265K in the simulation boxes, which are projected along the direction parallel to the surfaces. The solid circles and small dots indicate the positions of Oxygen atoms and Hydrogen atoms, respectively, which were averaged for the period of 10ps. The thin lines connected between the neighbouring atoms show the hydrogen bonds.

It is to be noted for both of the systems with different surface orientations that all water molecules included in several molecular layers at the surfaces move away from the ice lattice positions and are randomly arranged.

To indicate the detailed surface structures, the time-averaged density profiles

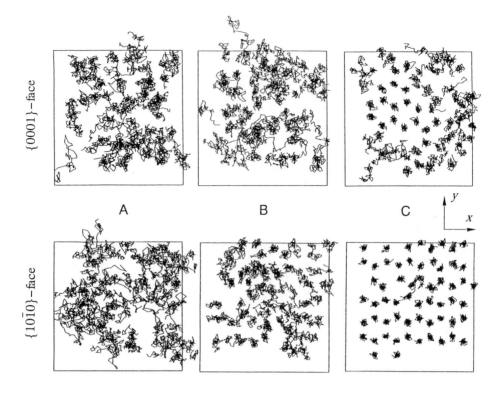

Figure 5. Illustrations of the trajectories of water molecules in layers A, B, and C. They mean 1st (outermost), 2nd and 3rd layers, respectively. The thickness of each molecular layer is 3.65Å and 3.90Å for the {0001}- and {10$\bar{1}$0}-faces, respectively. Note that the solid- and liquid-like regions coexist in one layer for layers B of both surfaces. Simulation temperature is 270K.

along the z-direction (normal to the surfaces) of each simulation box were calculated. Figure 4 shows illustrations of the density profiles for both the {0001}- and {10$\bar{1}$0}-faces at 272K. Periodic twin peaks observed in the central region of both profiles reflect the ice crystal structures. In contrast, the distinct twin peaks disappear at the surface regions, which correspond to the disordered surface layers as shown already in Figure 3. These results also indicate that the internal structures of transition layers are not uniform and gradually change from the disorder state to the crystalline order state. It is noted that both surface structures at the same temperature are obviously different from each other. Consequently, we can claim that the *anisotropic* surface melting on an ice crystal was first realized by this MD simulation [18,20].

Next, let us show the trajectories of the translational movements of all water molecules included in each layer with a mono-molecular thickness, which is labeled in the sequence of A, B and C from the outermost layer, (see Figure 4). Illustrations in Figure 5 show the trajectories of water molecules in each molecular layer for both systems at 270K. In layers A of both system, all molecules are equally moving out of the crystalline lattice points. Such translational motions strongly suggest that these layers should have the structures close to the bulk water. On the contrary, some of the water molecules included in layers B are still moving about, but almost all of the water molecules are captured around the crystal lattices. Furthermore, all molecules included in layers C are localized around each lattice site, except for several molecules in case of {0001}-face. It suggests that layers B and C might be in consequence of the mixture of both *solid-like* and *liquid-like* molecules, and the structures should be drastically changed in the layer by layer mode.

Here we introduce an order parameter S for the two-dimensional arrangement of water molecules in each layer to determine exactly the thickness of disordered layer from the MD simulation results. The order parameter is defined by

$$S = \sum \cos(\boldsymbol{k}_m \cdot \boldsymbol{r}_{ij})/N^2, \tag{1}$$

where $\boldsymbol{k}_m$ is the reciprocal lattice vector, $\boldsymbol{r}_{ij}$ the directional vector from the molecule i to the molecule j, and N the number of water molecules included in a molecular layer. It is known that $S = 1.0$ for the arrangement in the ideal ice crystal and $S = 0.03$ for that in the bulk water [19]. Assuming that only the layers with $S \leq 0.1$ are included in a disordered (namely, quasi-liquid) layer, we can exactly determine the qll thickness. Figure 6 indicates the temperature dependencies of qll thickness, obtained for both the {0001}- and {10$\bar{1}$0}-faces by the present simulations.

It is to be noted that the relation $d_{0001} < d_{10\bar{1}0}$ at lower temperature region is reversed at higher temperature region. This result is qualitatively consistent with the result of ellipsometric measurement as shown in Figure 1. As a result, it is to be emphasized again that the anisotropic properties of surface melting on an ice crystal were confirmed on the basis of both the experiments and MD simulations.

3.3. Self-diffusion coefficients of water molecules in qll

The self-diffusion coefficient D of water molecules gives a measure for the dynamic property of transition layer. Root mean square displacements of water molecules are usually related to D by the Einstein's relation of

$$\langle(r(t))^2\rangle \propto 6Dt, \tag{2}$$

where $\langle(r(t))^2\rangle$ is the root mean square of displacements in three-dimensions, and t the time. Consequently, it is possible to determine the self-diffusion coefficient only when the linear relation between $\langle(r(t))^2\rangle$ and t is clearly obtained from the simulation results.

The self-diffusion coefficients obtained for each molecular layer on both the {0001}- and {10$\bar{1}$0}-faces are plotted by the thick solid and broken lines in Figure

568

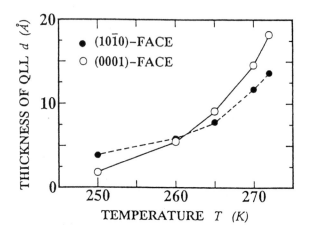

Figure 6. Temperature dependence of the qll thickness obtained by the MD simulations. It is to be noted that the thickness increases as the temperature approaches the melting point, and the anisotropy of thickness is reversed at about 260K.

2, respectively. Only the coefficients for layers A and B were able to be calculated in this way, but the calculations for further layers were very difficult because of the too short simulation time of MD. However, it is emphasized that the value of D in the layer A is fairly close to that in the bulk water and the value in the layer B drastically decreases by one to two orders in magnitude. This result indicates that the dynamic properties are drastically changed from liquid-like to solid-like as a function of depth from the qll surface.

Finally, it should be noted that the self-diffusion coefficients obtained by NMR are much smaller than those for layers A and B obtained by the present MD simulations. It means that the value obtained by NMR gives an average of self-diffusion coefficients for all molecular layers included in qll.

4. MOLECULAR DYNAMICS SIMULATION OF ICE/WATER INTERFACE

4.1. Simulation model

Recently importance for the investigation of solid/liquid interfaces quickly increases with relation to the exact understanding of melt growth kinetics. In addition, this interface may be taken for *the interface between crystal and qll*. Nevertheless the microscopic structures of liquid/solid interfaces in molecular level have hardly been made clear. Its main reason is in the difficulty observing directly solid/liquid interfaces using any sophisticated experimental techniques. In this section, we briefly mention the result of MD simulation carried out for the ice/water interfaces [21], which shows an applicability of the MD simulation method to the fascinating studies of solid/liquid interface structures, melt growth kinetics and so on.

The MD simulation of ice/water interface was carried out using a simulation box prepared as follows. Firstly two simulation boxes including 288 water molecules of TIP4P are prepared. All water molecules are arranged to make an ice crystal structure in one box and to make a bulk water in another box. Next, an ice/water interface is constructed by joining the two ends of these boxes. Periodic boundary conditions were imposed in the directions of x and y and computations were carried out by the same way as mentioned in the previous section. Time steps of the computation were $2.0 \times 10^{-15} s$ and the total simulation time was $1.0 \times 10^{-10} s$ in this case. To clarify anisotropic properties of ice/water interfaces, simulations were carried out for both the ice{0001}/water system and the ice{$10\bar{1}0$}/water system. The only temperature at the both ends of simulation box was fixed at 230K for these systems.

4.2. Molecular-level structures of ice/water interfaces

Figures 7(a) illustrate the trajectories of water molecules for the ice {0001}/water (left) and ice{$10\bar{1}0$}/water (right) systems, respectively. The initial positions of ice/water interfaces are shown by the arrows. On the other hand, Figures 7(b) show the time-averaged number density profiles of water molecules in the z-direction for both systems, respectively. Explicit twin peaks reflecting the double-layered structure of ice are observed in the left-hand region of each profile. It is to be noted that one can see some small peaks in the right-hand region of the bulk water. The existence of these small peaks suggests that these regions have a somewhat more highly ordered structure than the bulk water. Moreover, these profiles indicate that each ice/water interface has a diffuse structure throughout the thickness of several molecular layers. The difference between the interfacial structure of the ice{0001}/water and ice{$10\bar{1}0$}/water systems is also explicitly observed as a difference in the transitional pattern of oscillation peaks in the central region of profiles.

Recently, MD simulations under the non-equilibrium condition were also carried out to observe the growth process of ice crystal from the water [22]. As a result, the anisotropic growth kinetics were elucidated between the ice{0001}/water and ice{$10\bar{1}0$}/water systems. Namely, the former grew by the layer by layer mechanism but the latter did by the collective growth mechanism. This anisotropic growth kinetics strongly corresponds to the anisotropy of ice/water interface structures as mentioned above.

5. RELATION BETWEEN CRYSTAL MORPHOLOGY AND ANISOTROPIC STRUCTURE OF INTERFACE

It is very important to understand the pattern formation of snow and ice crystals to clarify the relationship between the anisotropic structures of surfaces or interfaces and the growth kinetics.

Kuroda and Lacmann [4] theoretically indicated that the anisotropic surface

(a)

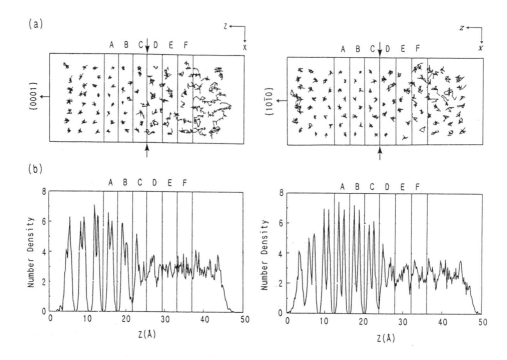

(b)

Figure 7. (a) Trajectories of water molecules in a layer perpendicular to the interface. The left-side figure is for the ice{0001}/water system and the right-side is for
the ice{10$\bar{1}$0}/water system. Layer thickness is 3.9Å and 3.65Å for each system,
respectively. (b) Density profiles in the z-directions for each system showed in the
upper figures, respectively. The initial position of interfaces are indicated by the
arrows.

melting on the {0001}- and {10$\bar{1}$0}-faces originates the complicated habit change of
snow crystal morphologies. Though their theoretical prediction for the anisotropic
surface melting on an ice crystal qualitatively coincides with the experimental results, we have to emphasize that a lot of ambiguities and mysteries are still remained
for the relation between the habit change of snow crystal and the anisotropic surface structures. For example, the temperature dependencies of qll thickness differ
from the theoretical expedition, and there is a wide contradiction in the physical
properties of transition layer between the theoretical and experimental results.

Furukawa and Kohata[23] observed the growth forms of negative crystal which
is a hole in an ice single crystal, and discovered that, though the {0001}-face is
remained as a smooth surface even at the melting point, the {10$\bar{1}$0}-face changes
from smooth to rough at the temperature of -2°C. This result indicates that the
roughening transition can not occur on the {0001}-face even at the melting point

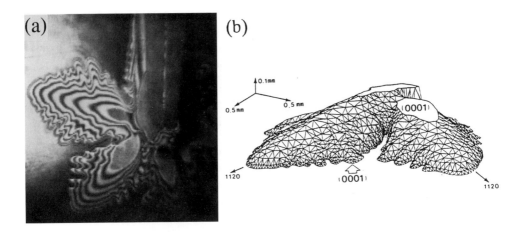

Figure 8. (a) An ice dendritic crystal growing in the supercooled water. The interference fringes originated the ice crystal thickness distribution are observed inside the crystal. (b) Three-dimensional pattern of ice crystal analyzed from the interference fringes of (a). Only {0001}-faces are observed as the smooth interfaces. For details, see ref.[24].

but on the {10$\bar{1}$0}-face at -2°C. This roughening transition on {10$\bar{1}$0}-face is expected to occur *at the interface between ice crystal and qll,* because the surface melting on the {10$\bar{1}$0}-face occurs at temperatures above -4°C. Consequently we conclude that the threshold temperature of roughening transition is *higher* than that of surface melting. It is also to be noted that the temperature dependencies of qll thickness on the {10$\bar{1}$0}-face drastically changes at -2°C (see Figure 1). The mutual relation between the surface roughening and the surface melting for the ice crystal surfaces must be made clear to discuss the pattern formation of snow crystal in the future.

On the other hand, the growth kinetics of ice crystals in the supercooled water are strongly influenced by the anisotropic ice/water interfaces. Recently, Furukawa and Shimada[24] carried out in-situ observations of the ice crystal growth in the supercooled water.

Figure 8(a) shows a picture of an ice crystal growing at the tip of thin glass capillary, which was observed through a Mach-Zehnder interferometer. Analyzing the interference fringes originated from the thickness distribution of ice crystal, they obtained the three-dimensional pattern of ice crystal, as shown in Figure 8(b). As a result, they clarified that the {0001}-interface is always observed as a smooth interface (facet) but the {10$\bar{1}$0}-interface never appears as such a smooth interface. Furthermore, the growth rate of {10$\bar{1}$0}-interface is two to three orders of magnitude larger than that of the {0001}-interface. Our MD simulation results also indicated that the structure of the ice{0001}/water interface is smoother compared with that of the ice{10$\bar{1}$0}/water interface. Consequently, we can claim that the simulation results for ice/water interfaces properly coincide with the experimental observations of ice pattern formation.

6. CONCLUSIONS

(1) The quasi-liquid layers at the ice surfaces and the ice/substrate interfaces, which are originated from the melting transition were experimentally detected at temperatures near the melting point. Anisotropic properties of surface melting on the {0001}- and {10$\bar{1}$0}-faces were clearly indicated and the self-diffusion coefficient in qll was in the order of $4 \times 10^{-10} cm^2/s$, which is five orders in magnitude smaller than the value of bulk water.

(2) The molecular dynamics simulation was carried out for the system including the ice crystal surfaces. The anisotropic surface melting, whose behaviors are qualitatively consistent to the experimental results, was observed. The MD simulation for the system including an ice/water interface was also carried out. As a result, the anisotropic interfacial structures, namely the smooth ice{0001}/water interface and the diffuse ice{10$\bar{1}$0}/water interface, was detected through the growth process.

(3) Growth forms of snow and ice crystals were discussed on the basis of the anisotropic surface and interfacial structures observed by the experiments and MD simulations.

The authors wish to thank Mr. T. Irisawa of the Computer Center, Gakushuin Univ., for his kind suggestions regarding the MD simulation. This work was supported by Grant-in-Aid for Scientific Research on Priority Areas "Crystal Growth Mechanism in Atomic Scale" No.05211203 from the Ministry of Education, Science, Sports and Culture.

REFERENCES

1. J. W. M. Frenken and J. F. van der Veen, Phys. Rev. Lett. 54(1985)134.
2. M. Hiroi, T. Mizusaki, T. Tuneto and A. Hirai, Phys. Rev. B40(1989)6581.
3. J. P. van der Eelden, in *Handbook of Crystal Growth* 1a, edited by D. T. J. Hurle, (North-Holland, Amsterdam, 1993)307.
4. T. Kuroda and R. Lacmann, J. Cryst. Growth 56(1982)189.
5. R. R. Gilpin, J. Colloid Interface Sci. 77(1980)435.
6. J. G. Dash, Contemporary Physics 30(1989)89.
7. D. Beaglehole and D. Nason, Surf. Sci. 96(1980)363.
8. Y. Furukawa, M. Yamamoto and T. Kuroda, J. Cryst. Growth 82(1987)655.
9. M. Elbaum, S. G. Lipson and J. G. Dash, J. Cryst. Growth 129(1993)491.
10. A. Lied, H. Dosch and J. H. Bilgram, Phys. Rev. Lett. 72(1994)3554.

573

11. J. G. Dash, in *Phase Transitions in Surface films* 2, edited by H. Taub *et al.*, (Plenum Press, New York, 1991)339.
12. M. Elbaum and M. Schick, Phys. Rev. Lett. 66(1991)11713.
13. L. A. Wilen and J. G. Dash, Phys. Rev. Lett. 74(1995)5076.
14. Y. Furukawa and I. Ishikawa, J. Cryst. Growth 128(1993)1137.
15. T. Ishizaki, M. Maruyama, Y. Furukawa and J. G. Dash, J. Cryst. Growth 163(1996)455.
16. O. A. Karim and A. D. J. Haymet, J. Chem. Phys. 89(1988)6889.
17. W. J. Jorgensen, J. Chandrasekhar, J. D. Madura, R. M. Impey and M. L. Klein, J. Chem. Phys. 79(1983)926.
18. H. Nada and Y. Furukawa, Trans. Mat. Res. Soc. Jpn. 16A(1994)453.
19. G. J. Kroes, Surf. Sci. 275(1992)365.
20. H. Nada and Y. Furukawa, To be submitted.
21. H. Nada and Y. Furukawa, Jpn. J. Appl. Phys. 34(1995)583.
22. H. Nada and Y. Furukawa, J. Cryst. Growth (1996)in press.
23. Y. Furukawa and S. Kohata, J. Cryst. Growth 129(1993)571.
24. Y. Furukawa and W. Shimada, J. Cryst. Growth 128(1993)234.

Growth of InGaSbBi Bulk Crystals on InSb Seeds and Rapid Permeation of Ga into InSb during Growth

M.Kumagawa and Y.Hayakawa

Research Institute of Electronics, Shizuoka University,
3-5-1 Johoku, Hamamatsu, Shizuoka 432, Japan

$In_{1-x}Ga_xSb_{1-y}Bi_y$ ($0 < x \leq 0.21$, $0 < y \leq 0.005$) quaternary bulk single crystals were grown on InSb seed crystals using a rotary-Bridgman method. All grown crystals were 10 mm in diameter and more than 10 mm in length. Owing to segregation, the compositional ratio of Bi (y) increased and that of Ga (x) decreased as crystals grew. During growth of InGaSbBi, both Ga and Bi permeated into the InSb seed and there appeared domains of InBi. For comparison, $InSb_{1-y}Bi_y$ ($0 < y \leq 0.05$) and $In_{1-x}Ga_xSb$ ($0 < x \leq 0.16$) were grown on InSb. It turned out that Bi did not permeate into InSb without Ga, but Ga permeated without Bi. The incorporation of Ga produced the excess In, and as a result InBi domains were formed. In order to investigate the permeation phenomenon of Ga into the InSb seed, the In-Ga-Sb solution was contacted with an InSb substrate for a constant period of time at a constant temperature. The permeation distance of Ga clearly increased with increasing both the temperature and the period. The apparent diffusion coefficient of Ga ranged between 10^{-8} and 10^{-7} cm^2/s when the diffusion coefficient was supposed to be a function of concentration. These values were extremely high in contrast to self-diffusion coefficients of In and Sb in InSb. Incorporation of Ga atoms into InSb produced the excess In atoms; as a result, precipitation of In occurred.

1. INTRODUCTION

III-V mixed semiconductors composed of three and/or four compositions are very useful for device fabrication, because band-gap energy and lattice constant can be controlled by adjusting their compositional ratio. Recently, the study on highly lattice-mismatched hetero epitaxial growth such as GaAs on Si has attracted much attention for preparing opto-electronic devices [1–3], however, the reduction of misfit dislocations still remains as a big problem to be solved. If large bulk single crystals of III-V mixed semiconductors can be used as substrates of devices, the problem of lattice-mismatch will be overcome.

Ternary semiconductors such as InGaAs [4,5], GaAsP [6], InSbBi [7], and InGaSb [8,9] have been grown at present; however, it is very difficult to grow large ternary and quaternary single crystals of high quality, because constitutional supercooling appears in front of the growth interface and brings about unstable growth [10,11]. In order to reduce its occurrence, Bachmann et al. [8] increased the rotation rate of a seed crystal to $100 \sim 200$ rpm in the Czochralski method and grew bulk crystals of $In_{1-x}Ga_xSb$ ($0 < x < 0.1$).

We have developed a new technique to stir a melt by introducing ultrasonic vibrations (10 kHz) into the melt from the crucible bottom in the Czochralski method. This vibration-introduced technique was applied to grow InSb crystals [12–17] of homogeneous impurity distribution and also InGaSb ternary single crystals [18–21] of large size. In the latter case, the growth of single crystals with higher mixed ratio of compositions was found to be possible by increasing the output power of ultrasonic vibrations. We also have modified the conventional Bridgman method so as to make a growth ampoule rotate at high speed [22–24], and have obtained successful results on the growth of InSb, InSbBi, InGaSb, and InGaAsSb crystals. We named this technique the rotary Bridgman method. This method has been, for the first time, applied to grow InGaSbBi quaternary bulk single crystals, which have not been grown up to the present as far as we know. This material is very attractive, because it is composed of semimetal (InBi) and semiconductors (InSb and GaSb). Band-gap and lattice constant can be separately varied in the wide range.

We have investigated the growth morphology, compositional profiles, and lattice constant of $In_{1-x}Ga_xSb_{1-y}Bi_y$ single crystals, and have found the incorporation of Ga and Bi and formation of domains in an InSb seed crystal [25] . In order to grow InGaSbBi single crystals of high quality, it is important to investigate these mechanism.

This chapter describes (1) the quality of $In_{1-x}Ga_xSb_{1-y}Bi_y$ single crystals grown by the rotary Bridgman method and (2) the permeation of Ga and Bi into an InSb seed crystal and the formation of InBi domains. For comparison, $InSb_{1-y}Bi_y$ and $In_{1-x}Ga_xSb$ crystals were also grown. (3) To make the phenomenon of rapid permeation of Ga in an InSb substrate clear, the Ga compositional profiles when the In-Ga-Sb solution was contacted with the InSb substrate for a constant period at a constant temperature are investigated.

2. EXPERIMENTAL PROCEDURE

A growth arrangement for growing InGaSbBi, InSbBi, and InGaSb is schematically shown in Figure 1. A semitransparent resistance furnace was connected on a seesaw plate, which was able to be inclined by driving motor I to adjust the level of the solution. This furnace was very useful to make the seeding procedure easy, because the inside of a growth ampoule was seen during growth.

An InSb seed crystal and source materials were charged in the growth ampoule. After sealed under the pressure of about 3×10^{-7} Torr, the ampoule was settled to a connecting rod of motor II. The seeding procedure was carefully achieved, at first by adjusting the solution temperature and then by covering about $60\% \sim 90\%$ of a growth surface of the seed with solution.This brought about the relative motion between the seed crystal and the solution, and as a result the stirring effect was enhanced. Crystal growth was performed by lowering the solution temperature at

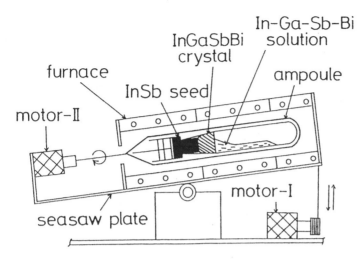

Figure 1. Schematic representation of the rotary Bridgman method.

a speed of 0.6°C/h with the ampoule rotation rate of 100 rpm by using the motor II. The compositional ratio of Ga in the In-Ga-Sb-Bi source was changed as 0, 2, and 4 at.%.

Using the conventional liquid phase epitaxy (LPE) apparatus, the permeation of Ga into an InSb substrate was carried out to make the permeation mechanism clear. Figure 2 shows the In-Ga-Sb ternary phase diagram [26] in which dotted and solid lines are liquidus isotherms and isoconcentration of x in the solid state $In_{1-x}Ga_xSb$ in mole fraction, respectively. The compositional ratios (at.%) in the solution at 480, 430, and 380°C are indicated at A, B, and C, respectively. The Ga solid composition in equilibrium to the solution is 0.2 molefraction. The compositions, Ga, In, and Sb, were six-nine grade. To restrict the contact area between the solution and the (111)-oriented InSb substrate, the substrate surface was covered with the carbon plate with a stripe of 2 mm in width extending in the $[1\bar{1}0]$ direction. The In-Ga-Sb solution was dropped down onto the InSb substrate and then held at a constant temperature of 380, 430, or 480°C for 15 or 30 min. After that, the solution was wiped off the substrate by sliding the boat; then the temperature was decreased rapidly at a rate of 22°C/min.

Grown crystals were cut parallel to the growth direction and were mounted in plastic. They were finally polished with an alumina abrasive of 0.05µm diameter, followed by etching for 1 min with a solution of $KMnO_4(0.05M)$: HF : $CH_3COOH = 1 : 3 : 3$ (vol%) at room temperature. Growth morphologies were observed under an optical microscope, an infrared microscope, and a scanning electron microscope. X-ray topographs were taken by an x-ray-diffraction apparatus. Compositional profiles on the sample surface were analyzed by an electron probe micro analysis (EPMA), energy dispersive spectroscopy (EDS), and secondary-ion-mass spectroscopy (SIMS). The lattice constant was measured by four-crystal x-ray diffractometry.

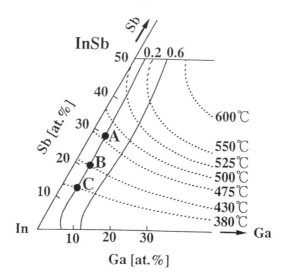

Figure 2. In-Ga-Sb ternary phase diagram. Dotted and solid lines are liquidus isotherms and isoconcentration of x in the solid state $In_{1-x}Ga_xSb$ in mole fraction, respectively. The compositional ratios (at.%) in the solution at 480, 430, and 380°C are shown at A, B, and C, respectively. The Ga solid composition in the $In_{1-x}Ga_xSb$ is 0.2 molefraction in equilibrium to the solution.

3. RESULTS AND DISCUSSIONS

3.1. $In_{1-x}Ga_xSb_{1-y}Bi_y$ quaternary crystals
3.1.1. Growth morphology

The roughly polished $(1\bar{1}0)$ surface of an InGaSbBi crystal is shown in Figure 3. The initial compositional ratio of a source solution was In:Ga:Sb:Bi = 48:2:25:25 (at.%). Temperature was lowered from 480°C at a speed of 0.6°C/h for 101 h. A seed crystal of silk-hat shape located at the left hand side was InSb, and an InGaSbBi crystal grew to the right hand side direction. From the observations by x-ray Laue patterns and growth morphology, the orientation of crystal was different from $(1\bar{1}0)$ at the end of the grown crystal, but, about 13 mm thickness of crystal was single.

As seen faintly in Figure 3, white and small areas were present in the InSb seed crystal. To confirm the growth morphology more clearly, an x-ray topograph of the seed region was taken using the $K_{\alpha 1}$ line from a Ag target as given in Figure 4. After the crystal was polished to 400μm thick, an x-ray of (220) diffraction was detected. A broken line indicates the boundary between InSb and InGaSbBi. Many black areas were clearly seen and corresponded to the white and small areas in Figure 3. Hereafter, we call these specular areas domains. They were about 2 mm long and 100~200μm wide near the interface, but they reduced gradually in size towards the left side from the growth interface. At a distance of about 6.5 mm far from the interface, the domains disappeared suddenly. Although many vertical lines were

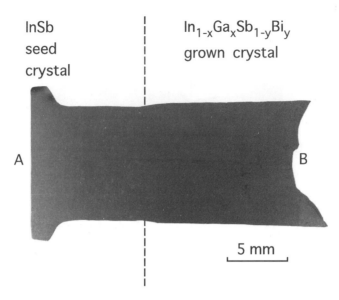

Figure 3. Roughly polished ($1\bar{1}0$) surface of an $In_{1-x}Ga_xSb_{1-y}Bi_y$ crystal. The initial compositional ratio in the solution was In:Ga:Sb:Bi = 48:2:25:25 (at.%). Crystal growth started at 480°C and the growth period was 101 h.

Figure 4. X-ray topograph of an InSb seed crystal. The $K_{\alpha 1}$ line of a Ag target was used.

580

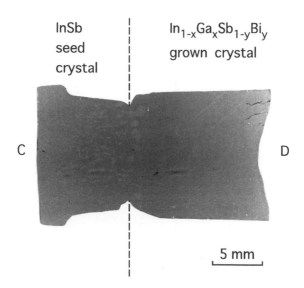

InSb
seed
crystal

$In_{1-x}Ga_xSb_{1-y}Bi_y$
grown crystal

C

D

5 mm

Figure 5. Roughly polished $(1\bar{1}0)$ surface of another $In_{1-x}Ga_xSb_{1-y}Bi_y$ crystal grown from the solution of In:Ga:Sb:Bi = 46:4:25:25 (at.%). Crystal growth started at 495°C and growth period was 133 h.

seen on the x-ray photograph, they were wrinkles of emulsion on the film.

Figure 5 shows an InGaSbBi quaternary crystal grown from the solution of different ratio of In:Ga:Sb:Bi = 46:4:25:25 (at.%). Compared with the grown crystal shown in Figure 3, the Ga compositional ratio in the solution was larger by 2 at.%, crystal growth started at 495°C which was higher by 15°C, and the period of growth (133 h) was longer by 32 h. The length of single crystal reached about 12.5 mm. One can see that the rotary Bridgman method served as a good way of growing InGaSbBi quaternary crystals with higher Ga ratio. In this crystal, many domains were seen not only in the InSb seed but also in a part of the grown crystal near the interface.

3.1.2. Compositional profiles

Figure 6 indicates compositional ratios of Ga (x) and Bi (y) in the InGaSbBi sample shown in Figure 3. Solid and opened circles denote x and y values, respectively. The measurement was carried out outside the domains along the line AB. In the grown InGaSbBi crystal, the Ga compositional ratio had the maximum value of 0.13 at the interface and gradually decreased to 0.037 at a distance of 11 mm. This was because the segregation coeffcient of Ga was larger than unity. On the contrary, the ratio of Bi was very low, and there was a slight increase owing to the segregation coefficient less than unity. Therefore, it was found that the Bi composition in the solution played a role as a solvent and further made the melting point of solution reduce below 525°C of that of InSb. In the InSb seed, both Ga and Bi compositions were incorporated. The Ga compositional ratio decreased gradually from the interface toward the end of the seed, and finally became zero

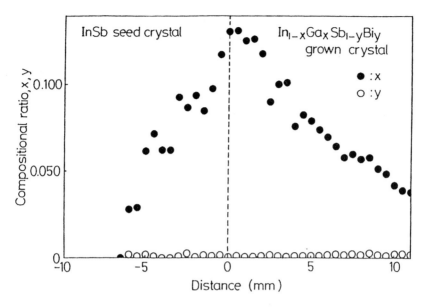

Figure 6. Compositional ratios of Ga (x) and Bi (y) along the growth direction of $In_{1-x}Ga_xSb_{1-y}Bi_y$ crystal shown in Figure 3.

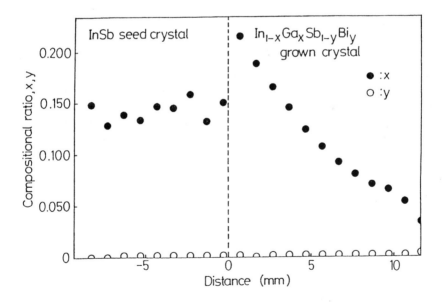

Figure 7. Compositional ratios of Ga (x) and Bi (y) along the growth direction of $In_{1-x}Ga_xSb_{1-y}Bi_y$ crystal shown in Figure 5.

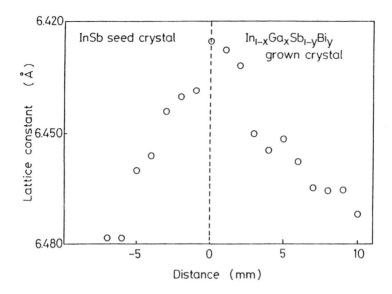

Figure 8. Lattice constant profile of the crystal shown in Figure 3.

at the distance of -6.5 mm; here the growth direction used was possitive. This distance corresponded to the position where the domains disappeared suddenly.

Figure 7 shows compositional profiles of the sample grown from the source solution contained 4 at.% Ga, as seen in Figure 5. The measurement was carried out outside the domains along the line CD. The value of Ga ratio at the interface was 0.21, which was higher than that of the crystal grown from the 2 at.% Ga solution as indicated in Figure 6. The difference in Ga ratio at the interface resulted from the Ga compositional ratio in the solution. In the inside of grown InGaSbBi, the Ga ratio decreased steadily to 0.033 with growing. On the contrary, the Ga compositional profile in the InSb seed crystal was very different from the corresponding one in Figure 6. The value of Ga reduced abruptly to about 0.15 in compositional ratio near the interface, and it was kept almost constant till the seed end; therefore, the incorporation of Ga was promoted by both higher growth temperature and longer growth period.

3.1.3. Lattice constant

Figure 8 represents lattice constant of the sample of 2 at.% Ga solutions. The lattice constant increased from the interface towards both the InGaSbBi grown crystal and the seed. In order to compare the compositional profile shown in Figure 6, the value of lattice constant was plotted so as to decrease along the Y axis. In the InSb seed crystal, the lattice constant decreased with the Ga compositional ratio. In other words, the incorporation of Ga component altered the lattice constant. This suggested that the Ga atoms were substituted for the In ones in InSb. In the InGaSbBi grown crystal, the profile of lattice constant was modulated by the incorpotation of Bi. This indicated that the lattice constant increased with the value of Bi.

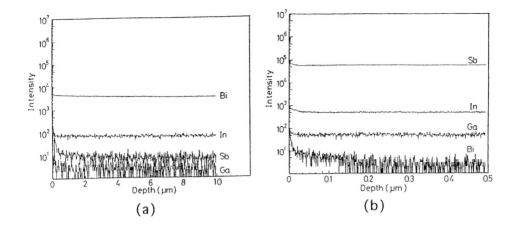

Figure 9. Compositional profiles in the crystal shown in Figure 4 analyzed by SIMS: (a) inside, (b) outside of the domain.

3.1.4. Formation of domains in the InSb seed crystal

On the formation of domains found in the InSb seed crystal, their compositions were investigated by SIMS. Figures 9(a) and 9(b) indicate respective depth profiles in the inside and the outside of a domain. Although quantitative values can not be decided due to the lack of standard samples, SIMS measurement has higher sensitivity than EDS. Cs^+ ions were implanted for 6000 s, but the dipped depth was quite different. The domain was about 20 times easier to be dipped. The atomic bond energy of Bi-Bi is 25.0 kcal/mole and that of Sb-Sb is 30.2 kcal/mole. [27] Although the values of another atomic bonds such as In-In, Ga-Ga, In-Ga, Ga-Sb were not known, the above results suggest that the Bi-rich domain had a weeker atomic bond than that of InGaSbBi crystal. The counting rate of Bi in the domain reached to nealy 5×10^3, though it was extremely low in the outside of the domain. On the other hand, the value of In was not zero in the domain, and it decreased to about 1/10 compared with that of the outside. The compositions of Sb and Ga were below the detection limit of the apparatus. This indicated that the domain was mainly composed of Bi and In.

EPMA measurement was performed in order to obtain the atomic ratio of the compositions in the domains. Figure 10 shows compositional ratios of Ga (x) and Bi (y) along the distance in the domains in the cystal shown in Figure 3. The Ga and Sb atomic ratios were almost zero, and the compositional ratios of In and Bi were nearly unity. This indicated that the InBi compound was formed in every domain.

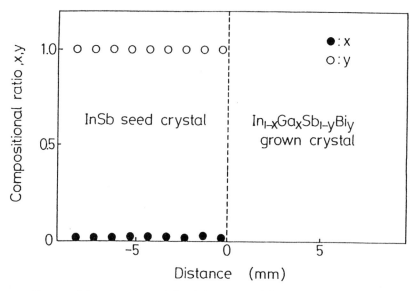

Figure 10. Compositional ratios of Ga (x) and Bi (y) in the domains along the distance in the crystal shown in Figure 3 analyzed by EPMA.

3.2. $InSb_{1-y}Bi_y$ and $In_{1-x}Ga_xSb$ ternary crystals

In order to investigate the incorporation of Ga and Bi into the InSb seed more clearly, $InSb_{1-y}Bi_y$ and $In_{1-x}Ga_xSb$ ternary crystals were grown and compared with the compositional profiles of InGaSbBi.

The value of Bi compositional ratio (y) in the $InSb_{1-y}Bi_y$ ternary crystal is measured along the growth direction by EPMA and is demonstrated in Figure 11. The initial compositional ratio in the solution was In:Ga:Sb:Bi = 50:0:25:25 (at.%). Crystal growth started at 474°C and the growth period was 182 h. Since the equilibrium segregation coefficient of Bi in InSb was less than unity [23], the value of y increased gradually from 0.002 to about 0.005 in molefraction with crystal growth. On the contrary, the Bi composition was not present in the InSb seed. These results indicated that the Bi atoms could not permeate into the InSb seed if there was no Ga.

Figure 12 indicates Ga compositional profiles on the $(1\bar{1}0)$ etched surface of the sample grown by the rotary Bridgman method. The zero position shown with a dotted line is the interface between the InSb seed and the InGaSb grown crystal. Open circles show the Ga compositional profile before annealing. The Ga compositional ratio at the interface was 0.16; it gradually decreased toward the end of the InGaSb grown crystal and finally became 0.08 at 10 mm from the interface because the segregation coefficient of Ga was larger than unity. The presence of Ga was also found in the InSb seed side till 8.3 mm from the interface. This means that Ga was incorporated in the InSb seed during InGaSb growth. From the Boltzmann-Matano analysis [28], the apparent diffusion coefficient of Ga was estimated to be 10^{-8} to 10^{-7} cm^2/s. This value was extremely high in contrast to self-diffusion coefficients of In and Sb in InSb [29], which ranged from 10^{-16} to 10^{-14} cm^2/s at

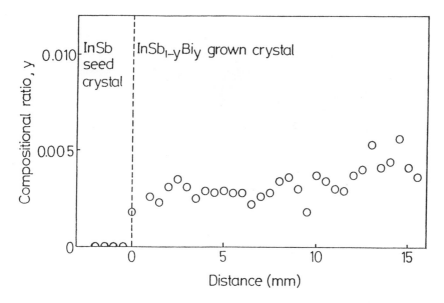

Figure 11. Compositional ratio of Bi (y) along the growth direction of the InSb$_{1-y}$Bi$_y$ crystal. The initial compositional ratio in the solution was In:Ga:Sb:Bi = 50:0:25:25(at.%). Crystal growth started at 474°C and the growth period was 182 h.

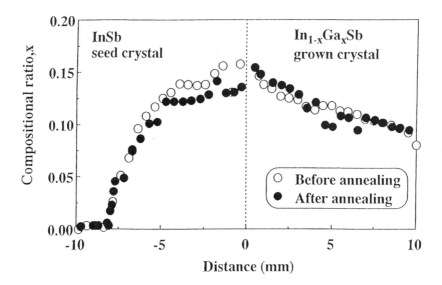

Figure 12. Ga compositional profiles in the ($1\bar{1}0$) etched surface of the In$_{1-x}$Ga$_x$Sb crystal grown by the rotary Bridgman method. Dotted line indicates the interface between the InSb seed and the InGaSb grown crystal. Open circles and closed circles show Ga compositional profiles before and after annealing.

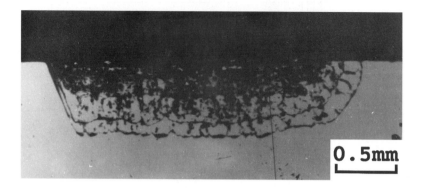

Figure 13. Photograph of the $(1\bar{1}0)$ cross section of the sample taken under an infrared microscope. The compositional ratio in the solution was In:Ga:Sb = 83.2:4.4:12.4 (at.%), and the holding temperature and the holding period were 380°C and 30 min, respectively.

475 to 517°C. This result suggested that Ga permeated rapidly in the InSb seed crystal.

To investigate whether the permeation of Ga in the InSb substrate took place without the solution, the grown sample was annealed for 100 h at 480°C in the argon atmosphere. The Ga compositional profile after annealing is shown with the closed circles. The change of the profile shape was negligibly small. Results show that the presence of In-Ga-Sb solution on the seed was necessary for Ga to permeate quickly into InSb.

3.3. Ga permeation into InSb

Using the conventional LPE apparatus, the permeation experiment of Ga into an InSb substrate was carried out at the constant temperature to make this phenomenon more clear [30]. Figure 13, an infrared photograph, shows the $(1\bar{1}0)$ cross section of the sample. The holding temperature and the holding period were 380°C and 30 min, respectively. A degenerated region with a thickness of about 560μm is evident. The boundary at the left hand side of the photograph is a straight line, and its angle measured from the $(\bar{1}\bar{1}1)$ growth plane is 70°. This corresponds to the $(\bar{1}11)$ plane. On the other hand, a smoothly curved line appears at the right side of the crystal, demonstrating that the permeation of Ga depends on the orientation.

It is necessary to investigate whether the substrate was given some influence or not during contact with the solution. Figure 14 shows a $(1\bar{1}0)$ cross section of the sample taken under the infrared microscope. The holding temperature and the holding period were 480°C and 30 min, respectively. A Te-doped InSb substrate

Figure 14. Photograph of the (1$\bar{1}$0) cross section of the sample taken under an infrared microscope. A Te-doped InSb substrate grown by the Czochralski method was used. The compositional ratio of the solution was In:Ga:Sb = 67.6:3.8:28.6 (at.%), and the holding temperature and the holding period were 480°C and 30 min, respectively.

grown by the Czochralski method was used. The middle region was a degenerated region. The thickness of this region was about 1.2 mm, which was larger than that in Figure 13. Outside of the region, horizontal impurity striations due to crystal rotation were clearly seen. Before the experiment, these striations existed in the whole region of the substrate. By contact with the solution, however, they vanished completely in the degenerated region. This clearly indicates that the InSb substrate was affected by the In-Ga-Sb solution; thus the degenerated region was formed.

The Ga compositional profiles measured along the depth direction in the samples are summarized in Figure 15. Open circles, closed triangles, and open triangles represent the results under the conditions that the holding period was 30 min and the holding temperatures were 380, 430, and 480°C, respectively. Closed circles show the result when the holding period decreased to 15 min at a holding temperature of 380°C. The initial Ga compositional ratio was nearly 0.2, which was the solid composition in equilibrium with the solution. By increasing the holding temperature and the holding period, Ga permeated a more longer distance. This is because the diffusion coefficient increases as the holding temperature increases.

The reason why the Ga permeates so rapidly in the InSb substrate is considered as follows. If droplets of the In-Ga-Sb solution move through the InSb substrate, and if $In_{1-x}Ga_xSb$ is grown followed by the dissolution of a small amount of InSb to resaturate the solution, the Ga compositional profile should have the same tendency for the initial 15 min when the holding period is changed from 30 to 15 min under

588

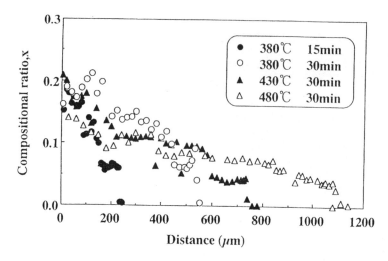

Figure 15. Ga compositional profiles. Open circles, closed triangles, and open triangles represent the results obtained under the conditions that the holding period was 30 min and the holding temperature was 380, 430, and 480°C, respectively. Closed circles show the result when the holding period decreased to 15 min at 380°C.

the same temperature. But, as shown in Figure 15, the Ga composition gradually decreased and approached zero at 560 μm from the interface for the holding period of 30 min and at 250 μm for 15 min. The initial Ga profiles for 15 min did not follow the profile for 30 min. This suggests that the permeation process takes place at first. After that, regrowth begins during the cooling process from the holding temperature to the room temperature. The permeation profile remains in the regrowth process, because the effective segregation coefficient approaches unity when the growth rate is large. As respective enthalpies of the formation energy of InSb and GaSb are 8.0 and 10.5 Kcal/mole [31], bonding energy of In-Sb is weaker than that of Ga-Sb. Therefore, the In-Sb bond is easily broken by Ga, as a result, a disorder of the crystal structure such as dislocation is introduced. The permeation under this situation will become more easier. The shape of the permeation profile, not a simple form, shows a plateau. The reason is not clear at present.

Several inclusions existed in the InSb substrate, as seen in Figures 13 and 14. Figures 16(a) and 16(b) show the inclusions taken under the SEM and the profiles of In, Ga, and Sb compositions measured along the A-B line across an inclusion by the EPMA. Outside of the inclusion, three compositions of In, Ga, and Sb existed. As the Sb concentration was 0.5 mole fraction, an $In_{1-x}Ga_xSb$ was formed. This suggested that Ga atoms were substituted for the In ones in InSb. On the other hand, In composition was observed only in the inclusion. The incorporation of Ga into InSb produced the excess In atoms. As all In atoms cannot move back to the solution, the excess In remains in the permeated region. As a result, precipitation

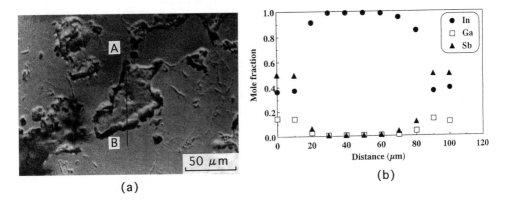

Figure 16. (a) Photograph near several inclusions taken by a scannning electron microscope. (b) Profiles of In, Ga, and Sb measured across an inclusion by an electron probe micro analysis.

of In occurred.

4. SUMMARY

The rotary Bridgman method was applied to grow InGaSbBi quaternary bulk single crystals on the InSb seed for the first time. In order to suppress the occurrence of constitutional supercooling at the growth interface, the Bridgman method was modified so as to make the ampoule rotate at high speed. By inclining the transparent resistance furnace, about 60%~90% of the seed surface was covered with the solution. This brought about the relative motion between the crystal and the solution and consequently the stirring effect was enhanced. Crystal growth was performed by lowering the solution temperature at the speed of 0.6°C/h with the ampoule rotation rate of 100 rpm. By this method, $In_{1-x}Ga_xSb_{1-y}Bi_y$ ($0 < x \leq 0.21$, $0 < y \leq 0.005$) single crystals more than 10 mm in length were grown. This proved that the rotary Bridgman method was a good technique to grow quaternary single crystals.

An optical microscope, an infrared microscope, x-ray topograph, a four-crystal x-ray diffractometry, EPMA, EDS, and SIMS measurements were performed to investigate the quality of crystals. Owing to the segregation phenomena, the Bi compositional ratio increased and the Ga ratio decreased as the crystal grew. The lattice constant changed according to the Ga and Bi compositions. In the InSb seed, permeation of both Ga and Bi compositions occurred and InBi domains were formed. For comparison, $InSb_{1-y}Bi_y$ ($0 < y \leq 0.05$) and $In_{1-x}Ga_xSb$ ($0 < x \leq 0.16$) ternary crystals were grown. Although Ga permeated into the InSb seed during the growth of InGaSb, Bi was not incorporated during the InSbBi growth. The contribution of Ga was necessary for the permeation of Bi into the InSb seed crystal. The incorporation of Ga produced the excess In, and as a result InBi domains were formed.

The In-Ga-Sb solution was put on the InSb substrate covered by the carbon plate with a stripe extending in the $[1\bar{1}0]$ direction. By increasing both the holding temperature and the holding period, the permeation distance increased. The apparent diffusion coefficient of Ga ranged between 10^{-8} and 10^{-7} cm^2/s, which was extremely high in contrast to self-diffusion coefficients of In and Sb in InSb. Incorporation of Ga atoms into InSb produced the excess In atoms. It brought about precipitation of In.

REFERENCES

1. M.Akiyama, T.Kawarada and K.Kaminishi, J. Cryst.Growth 68 (1984)21.
2. T.Soga, S.Hattori, S.Sakai and M.Umeda, J.Cryst.Growth 77 (1986) 498.
3. T.Soga, S.Sakai, M.Umeda and S.Hattori, Jpn.J.Appl.Phys. 26 (1987) 252.
4. T.Kusunoki, C.Takenaka and N.Nakajima, J.Cryst.Growth 112 (1991) 33.
5. K.Nakajima, T.Kusunoki and C.Takenaka, J.Cryst.Growth 113 (1991) 485.
6. T.Hibiya, H.Watanabe, H.Matsumoto and N.Iwata, J.Electrochem.Soc. 134 (1987) 981.
7. B.Joukoff and A.M.Jean-Louis, J.Crystal Growth 12 (1972) 169.
8. K.J.Bachmann, F.A.Thiel, H.Schreiber and J.J.Rubin, J.Electron. Mater. 9 (1980) 445.
9. A.Watanabe, A.Tanaka and T.Sukegawa, Jpn.J.Appl.Phys.32 (1993) L793.
10. R.Singh, A.F.Witt and H.C.Gatos, J.Electrochem.Soc. 121 (1974) 380.
11. K.M.Kim, J.Cryst.Growth 44 (1978) 403.
12. M.Kumagawa, H.Oka, N.V.Quang and Y.Hayakawa, Jpn.J.Appl.Phys. 19 (1980) 753.
13. Y.Hayakawa, Y.Sone, K.Tatsumi and M.Kumagawa, Jpn.J.Appl.Phys. 21 (1982) 1273.
14. Y.Hayakawa and M.Kumagawa, Jpn.J.Appl.Phys. 22 (1983) 206.
15. Y.Hayakawa, Y.Sone, F.Ishino and M.Kumagawa, Jpn.J.Appl.Phys. 22 (1983) 1069.
16. Y.Hayakawa and M.Kumagawa, Cryst.Res. Tech. 20 (1985) 3.
17. Y.Hayakawa and M.Kumagawa, Jpn.J.Appl.Phys.24, Suppl.24-1 (1985) 166.
18. T.Tsuruta, Y.Hayakawa and M.Kumagawa, Jpn.J.Appl.Phys. 27, Suppl.27-1 (1988) 47.
19. T.Tsuruta, Y.Hayakawa and M.Kumagawa, Jpn.J.Appl.Phys. 28 Suppl. 28-1 (1989) 36.
20. T.Tsuruta, N.Nishida, Y.Hayakawa and M.Kumagawa, in Proceedings of the 4th Asia Pacific Physics Conference 1990, Vol.1, p.499.
21. T.Tsuruta, K.Yamashita, S.Adachi, Y.Hayakawa and M.Kumagawa,Jpn.J.Appl.Phys. 31, Suppl. 31-1 (1992) 23.
22. M.Kumagawa, T.Ozawa and Y.Hayakawa, Appl.Surface Sci. 33/34 (1988) 611.

23. T.Ozawa, Y.Hayakawa and M.Kumagawa, J.Cryst.Growth 109 (1991) 212.
24. T.Ozawa, Y.Hayakawa and M.Kumagawa, J.Cryst.Growth 115 (1991) 728.
25. Y.Hayakawa, M.Ando, T.Matsuyama, E.Hamakawa, T.Koyama, S.Adachi, K.Takahashi, V.G.Lifshits and M.Kumagawa, J.Appl.Phys. 76 (1994) 858.
26. S.Szapiro, J.Phys.Chem.Solids 41 (1980) 279.
27. L.Pauling, The nature of the chemical bond (Cornel university press,U.S.A, 1960) p.72.
28. P.D.Shewmon, Diffusion in Solids (McGraw-Hill, New York, 1963), p.29.
29. D.L.Kendall and R.A.Huggins, J.Appl.Phys. 40 (1969) 2750.
30. Y.Hayakawa, E.Hamakawa, T.Koyama and M.Kumagawa, J.Cryst.Growth 163 (1996) 85.
31. G.S.Stringfellow, J.Phys.Chem.Solids 33 (1973) 665.

Subject Index